ALLE · ZEIT · WACH · 1842

Angelo Azzi Zdenek Drahota Sergio Papa (Eds.)

Molecular Basis of Membrane-Associated Diseases

With 121 Figures

Springer-Verlag Berlin Heidelberg New York
London Paris Tokyo Hong Kong

Prof. Dr. ANGELO AZZI
Institut für Biochemie und Molekularbiologie
Universität Bern
Bühlstraße 28
CH-3012 Bern

Prof. ZDENEK DRAHOTA
Institute of Physiology
Czechoslovak Academy of Sciences
Videnska 1083
CS-142 20 Praha 4

Prof. SERGIO PAPA
Institute of Medical Biochemistry
and Chemistry
University of Bari
Piazza G. Cesare
I-70124 Bari

ISBN-13: 978-3-642-74417-4 e-ISBN-13: 978-3-642-74415-0
DOI: 10.1007/978-3-642-74415-0

Library of Congress Cataloging-in-Publication Data. Molecular basis of membrane-associated diseases. Based on a symposium held in Prague on 14–16 July 1988. Includes index. 1. Cell membranes–Abnormalities–Congresses. 2. Mitochondrial membranes–Abnormalities–Congresses. 3. Lysosomal storage diseases–Congresses. I. Azzi, A. (Angelo) II. Drahota, Z. III. Papa, S. [DNLM: 1. Cell Membrane–pathology–congresses. 2. Diseases–etiology–congresses. 3. Membranes–pathology–congresses. QS 532.5.M3 M718 1988] RB152.M64 1989 616.07 89-11270

Softcover reprint of the hardcover 1st edition 1989

Typesetting: International Typesetters Inc. Makati, Philippines

2131/3145-543210 – Printed on acid-free paper

Preface

This book is a result of a symposium, held in Prague on 14–16 July 1988 under the auspices and with support from UNESCO and the International Biomedical Institute, Bari, Italy.

The topic, the role of membranes in disease, was chosen because in recent years the field has been expanding tremendously and an even larger development is foreseen in the future.

Membranes are the common denominator of a variety of disorders, ranging from some myopathies to cancer, from biliary cirrhosis to defects in hormone response. Membranes are in fact such an essential, ubiquitous and at the same time delicate structure that its alterations are not only very frequent, much more than it has been considered until now, but also a serious origin of disease.

The interaction-recognition phenomena, either between cells or with ligands, such as hormones or drugs, have as a basic element the plasma membrane. Energetic transformations are based essentially at the mitochondrial, catabolic processes at the lysosome or peroxisome level. To conceive these processes as separated is an oversimplification, since an interplay exists among the different cell membranous systems. Indeed this has been one other reason to try to bring together scientists of different but related disciplines and the result of this effort is summarized in this book.

Some of the chapters, conceived as advanced review articles rather than experimental papers try to give an overview of the problems of growth differentiation and controls which have to deal with the plasma membrane: Growth factors and hormone receptors represent a fascinating world, projected into the future.

A series of chapters are dedicated to a new category of diseases, based on inherited mitochondrial alterations, and their attack by sophisticated diagnostic tools, ranging from magnetic resonance in vivo to molecular genetic approaches.

The third focus of this book is given to the classical lysosomal and the neoclassical peroxisomal dysfunctions: The understanding of these diseases becomes continuously deeper and more exciting.

Although the basic knowledge, the diagnosis and the classification of membrane-based diseases become every day more complete, the perspective of a molecular therapy begins only now. However, it is only through basic understanding of the elementary phenomena that a rationalization of the present therapies and the conception of new drugs can take place.

This is in fact the last reason for having assembled so many outstanding scientists and for having collected their manuscripts in this book, to offer an interdisciplinary scientific basis to those who would like a more complete description of the elementary biological processes of nature and to propose solutions to correct for their possible alterations.

Bern/Praha/Bari, Spring 1989

Angelo Azzi
Zdenek Drahota
Sergio Papa

Contents

1 Plasma Membrane and Control of Cell Growth and Differentiation

2 Regulation of Membrane Channels

5 Cancer and Energy Metabolism

6 Lysosomes and Peroxisomes in Health and Disease

List of Contributors

You will find the addresses at the beginning of the respective contribution

1 Plasma Membrane and Control of Cell Growth and Differentiation

Insulin Receptor Kinase and Insulin Action

E. Van Obberghen, R. Ballotti, Y. Le Marchand-Brustel and J.C. Scimeca

1 Introduction

Insulin has been shown to elicit its numerous biological effects by interacting with a specific cell-surface receptor. The insulin receptor structure has been elucidated through a variety of techniques (Van Obberghen 1984), and more recently, the amino acid sequence of the human insulin receptor precursor has been unravelled by recombinant DNA technology (Ullrich et al. 1985; Ebina et al. 1985). Despite this progress the molecular mechanism of insulin action is still not entirely comprehended regarding the events following receptor binding and leading to the ultimate cellular responses. However, over the years considerable evidence has been gathered indicating that reversible phosphorylation contributes to the mechanism of insulin action (Denton 1986). Further, a promising discovery was made by the demonstration that the insulin receptor is an insulin-sensitive protein kinase (Kasuga et al. 1982a, b, c; Van Obberghen and Kowalski 1982; Petruzzelli et al. 1982). This observation is of particular interest for our understanding of insulin-regulated processes, since it is now recognized that phosphorylation-dephosphorylation of proteins is a mechanism whereby many cellular functions are regulated by hormones and neurotransmitters. Furthermore, protein kinases are also constituents of receptors for several growth factors, implying that receptor kinase activity may represent a general mechanism in transmembrane signalling of hormones and growth factors.

2 Insulin Receptor Phosphorylation

In intact cells, insulin stimulates the phosphorylation of its receptor β-subunit (Van Obberghen and Kowalski 1982; Kasuga et al. 1982a). In these experiments, cells were preincubated with $[^{32}P]P_i$ to label cellular ATP solubilized, and the glycoproteins purified on WGA-agarose. Immunoprecipitation by antibodies to insulin receptors followed by SDS/PAGE under reducing conditions and autoradiography revealed a labeled band (M_r 95 k), the phosphorylation of which was stimulated by insulin. Its identity with insulin receptor β-subunit was established based on its appropriate electrophoretic mobility and on the fact that it was not precipitated with nonimmune serum. In intact cells phosphoamino acid analysis of the insulin receptor β-subunit showed phosphorylation of serine, threonine, and

INSERM U 145, Faculté de Médecine, Nice, France

A. Azzi, Z. Drahota and S. Papa
Molecular Basis of Membrane-Associated Diseases

tyrosine under basal conditions. Insulin induced a rapid, severalfold increase in ^{32}P incorporation on tyrosine, followed by a slower rise in labeling of phosphoserine (Kasuga et al. 1982c; Gazzano et al. 1983; White et al. 1985a; Ballotti et al. 1987).

Subsequently, insulin-stimulated phosphorylation of the insulin receptor β-subunit was demonstrated in cell-free systems using [γ-^{32}P]ATP and purified receptors (Van Obberghen and Kowalski 1982; Kasuga et al. 1982b; Van Obberghen et al. 1983; Shia and Pilch 1983; Petruzzelli et al. 1982, 1984). These purified receptor preparations exhibited insulin-stimulated protein kinase activity which catalyzed phosphorylation of both the β-subunit and exogenous substrates. With highly purified receptor, the phosphorylation occurred exclusively on tyrosine residues under basal conditions, and insulin stimulatory action was accounted for by a severalfold increase in phosphotyrosine. Thus, the tyrosine kinase appeared as a constituent of the insulin receptor. Further, the β-subunit contains an ATP-binding site as demonstrated by covalent affinity labeling (Van Obberghen et al. 1983; Shia and Pilch 1983; Roth and Cassell 1983). The simultaneous presence of phosphorylation sites and an ATP-binding site on the receptor β-subunit indicates that the insulin receptor acts as its own tyrosine kinase. Further proof is the demonstration that the insulin-binding activity and insulin-dependent tyrosine-kinase activity copurified to homogeneity at a constant stoichiometric ratio (Petruzzelli et al. 1984). In addition, insulin binds to and promotes phosphorylation of the insulin receptor precursor (M_r 210 k) (Rees-Jones et al. 1983).

Following the identification of the insulin receptor protein kinase activity, its biochemical properties have been investigated in detail. For the original references see in Gammeltoft and Van Obberghen (1986).

3 Insulin Receptor Tyrosine Kinase and Hormone Action

3.1 Characteristics of the Receptor Tyrosine Kinase

Since the discovery that the insulin receptor is an insulin-dependent protein tyrosine kinase, it was anticipated that this receptor enzymatic function was involved in insulin action (Gazzano et al. 1983). For the validity of this contention at least the following four criteria were expected to be fulfilled.

First, the insulin dose-response relationship of the kinase should be within the physiological range and correlate with that of binding to receptor. Several authors found that the kinase activation was half-maximal at an insulin concentration of 2–5 nM (ED_{50}), which corresponded to the apparent K_d of the receptor-insulin complex of the same solubilized receptor preparations (Kasuga et al. 1982a; Shia and Pilch 1983; Petruzzelli et al. 1984; Sadoul et al. 1985). In contrast, a dissociation between dose-response curves of insulin binding and kinase activation was observed with soluble receptors from rat liver and human erythrocytes (Grigorescu et al. 1983). Here, the apparent K_d exceeded the ED_{50} by a factor of 3–10, suggesting that the phenomenon of "spare receptors" observed for other insulin actions is also applicable for kinase activation. It is not clear whether these findings are explained by differences in tissues, purification procedures, or assay methods. In conclusion, in most instances the receptor kinase is activated by insulin concentrations within a physiological range corresponding to receptor binding.

Secondly, the receptor kinase should be capable of phosphorylating cellular substrates other than the receptor itself, in order to propagate the insulin response. The insulin receptor kinase can phosphorylate a number of substrates on tyrosine in vitro, although none of the proteins tested is proven to be physiologically relevant substrates. The two first "putative" substrates described were a 110–120 kD protein, and a 185 kD protein. In 1985 two laboratories, independently, identified in purified glycoproteins from rat liver and rabbit brown adipose tissue, a cellular protein "substrate" of M_r 110 k for the insulin receptor kinase (Sadoul et al. 1985; Rees-Jones and Taylor 1985). This glycoprotein appears as a monomeric structure, and is not part of the insulin receptor itself. Phosphorylation of the M_r 110 k protein and of the receptor β-subunit was stimulated by insulin in a remarkably similar dose-dependent fashion ($ED_{50} \simeq 1$ nM). Further, kinetic studies suggested that phosphorylation of the M_r 110 k protein occurred after activation and phosphorylation of the insulin receptor kinase. The nature and function of this endogenous substrate was as yet unknown. In the same period, a different putative substrate for the insulin receptor kinase was identified in a hepatoma cell line, Fao (White et al. 1985b). This M_r 185 k substrate does not contain carbohydrate moieties, and appears to be monomeric. Since the reports on these two "putative" substrates, a number of other phosphoproteins have been described including a M_r 15 k protein that may play a role in insulin-mediated glucose transport (Bernier et al. 1987; Table 1). However, at present it remains to be shown whether any of these proteins has a physiological significance. One is forced to admit that despite intensive efforts extremely little is known concerning putative substrates. This is likely due to the fact that they are rare and labile. Further, as expected from the large array of biological responses induced by insulin, a whole series of nonabundant regulatory proteins likely exists to account for the metabolic and growth-promoting effects of insulin.

Table 1. Endogenous substrates for insulin receptor kinase

kD	Origin	Nature	Authors
185	Fao hepatoma (rat)	Cytosolic protein	White et al. (1985a)
160	Adipocyte (rat)	Plàsma membrane glycoprotein	Yu et al. (1987)
120	Brown adipose tissue (rabbit) Hepatocyte (rat) Hepatoma H35 (rat)	Plasma membrane glycoprotein	Sadoul et al. (1985) Rees-Jones and Taylor (1985)
46	Adipocyte (rat)	Plasma membrane glycoprotein	Haring et al. (1987)
15	Adipocyte (3T3 Ll, mouse)	Cytosolic protein	Bernier et al. (1987)

The third criterion is reversibility of insulin receptor phosphorylation. To exert a regulatory function, the phosphorylated and activated receptor kinase should return to basal activity through a dephosphorylation reaction. Lectin-purified receptor preparations were found to contain phosphatase activity, which slowly reduced the ^{32}P content of phosphorylated receptor, and which was insulin-independent (Kowalski et al. 1983). Exposure of phosphorylated insulin receptor to alkaline phosphatase resulted in the removal of about 50% of the β-subunit phosphotyrosine, and about 65% reduction in kinase activity (Yu and Czech 1984). Thus, the insulin receptor kinase can be deactivated through dephosphorylation of tyrosine residues.

The fourth criterion pertains to the specificity of an insulin effect on its receptor kinase. Several insulin analogues stimulated receptor phosphorylation with potencies relative to porcine insulin, which were identical to their relative binding affinities and potencies in other assay systems (Kasuga et al. 1982a; Grigorescu et al. 1983). Finally, polyclonal antisera to insulin receptor, which exert insulinlike effects in several cell types, were also able to stimulate the receptor tyrosine kinase (Gammeltoft and Van Obberghen 1986; Gherzi et al. 1987). In conclusion, the insulin effect on receptor phosphorylation has the affinity and specificity of a typical insulin receptor-mediated event.

Taken together, the kinase activity of the insulin receptor seems to be a fundamental receptor property, since whenever insulin receptors are present, insulin-stimulated autophosphorylation occurs (Kasuga et al. 1982a; Van Obberghen and Kowalski 1982; Shia and Pilch 1983; Grigorescu et al. 1983; Petruzzelli et al. 1984; Gammeltoft et al. 1983; Gazzano et al. 1985; Tanti et al. 1986). An important feature of the insulin receptor tyrosine kinase is that receptor autophosphorylation on one or more tyrosyl residues activates the receptor kinase towards exogenous substrates without affecting the insulin-binding characteristics (Stadtmauer and Rosen 1983; Yu and Czech 1984).

3.2 Role of Insulin Receptor Tyrosine Kinase in Hormone Action

Since its discovery in 1982 the protein tyrosine kinase activity of the insulin receptor has been presumed to function as transducer of insulin action. A large body of evidence has led to a general consensus for a role of insulin receptor kinase in hormone action (Table 2). A first series of suggestive observations was provided by studies showing that alterations in insulin action are associated with parallel alterations in insulin receptor tyrosine kinase activity. Thus, the receptor kinase is impaired in various insulin-resistant states including the syndrome of extreme insulin resistance type A (Grunberger et al. 1984), melanoma cell cultures (Häring et al. 1984), gold-thioglucose obese mice (Le Marchand-Brustel et al. 1985), and streptozotocin diabetic rats (Kadowaki et al. 1984). Conversely, insulin receptor kinase is hyperactive in insulin hyperresponsive adipocytes of young obese Zucker rats (Debant et al. 1987). Further, insulinomimetic agents (vanadate, lectins, trypsin) increase receptor autophosphorylation (Tamura et al. 1983; Roth et al. 1983). Introduction into mammalian cells of a monoclonal antibody, which inhibits insulin receptor kinase, blocks the rapid effects of insulin (Morgan and Roth 1987). Conversely, microinjection of antiphosphotyrosine antibodies, which

Table 2. Insulin receptor kinase and insulin action

		Insulin receptor kinase activity	Insulin response
Insulino-mimetic agents	– Antireceptor antibodies	↗	↗
	– Vanadate	↗	↗
	– Trypsine	↗	↗
	– Lectine	↗	↗
Aggregation of receptor		↗	↗
Physiopathologic states a) Insulin resistance	Diabetes	↘	↘
	Obesity	↘	↘
	Syndrome A	↘	↘
b) Increased insulin response	Young Zucker rat	↗	↗
Transfection of eukaryotic cells with mutated receptor	ATP binding site –	↘	↘
	TYR 1150–1151 deletion	↘	↘
Injection of antibody	Monoclonal to receptor β-subunit	↘	↘
	Antiphosphotyrosine	↗	↗

stimulate the insulin receptor kinase, induces enhanced insulin-evoked glucose transport and amino acid uptake (Fig. 1; Ballotti et al. 1989). The mechanism by which antiphosphotyrosine antibodies increase the insulin receptor kinase activity, and augment glucose and amino acid uptake is not established. The observed effects could be due to interaction of the antibodies either with the phosphoreceptor itself or with cellular substrates phosphorylated on tyrosine residues. However, in contrast to some of our other phosphotyrosine antibodies, the antibodies used in this particular study recognize only the insulin receptor and the EGF receptor, and fail to identify cellular substrates. Therefore, we favor the idea that the observed effects are due to interaction with the insulin receptor rather than with cellular substrates. It is tempting to suggest that these antibodies enhance selectively the phosphorylation of a particular autophosphorylation site implicated in positive regulation of the insulin receptor kinase activity. However, since our antibodies did not change the general pattern of autophosphorylation sites, this hypothesis seems remote. Therefore, we favor the concept that the enhanced kinase activity evoked by our antiphosphotyrosine antibodies is due either to the induction of a favorable conformational change in the kinase domain of the insulin receptor, or to an increased aggregation of insulin receptor molecules with resulting increased kinase activity.

The most convincing and elegant evidence for the idea that insulin action depends on receptor protein tyrosine kinase activity comes from mutagenesis experiments involving the receptor. Thus, insulin receptors mutated on the ATP-binding site lack protein tyrosine kinase activity, and fail to mediate insulin post-receptor effects including glucose transport, glycogen synthesis, S6 kinase activity and thymidine uptake (Chou et al. 1987).

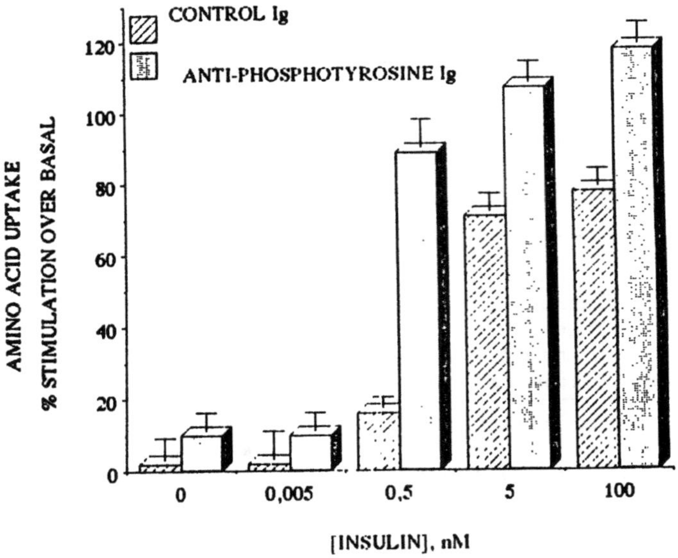

Fig. 1. Effect of antiphosphotyrosine antibody on insulin-stimulated amino acid transport. Normal Ig injected cells or antiphosphotyrosine Ig injected cells were incubated with increasing concentrations of insulin for 8 h, then ^{14}C AIB was added for 15 min. Radioactivity was measured by liquid scintillation counting. Results shown are the mean ± SEM of four determinations

4 Seryl (and Threonyl) Phosphorylation of Insulin Receptors

In intact cells, the rapid insulin-stimulated phosphorylation of its receptor on tyrosine is followed by a slower serine phosphorylation (Kasuga et al. 1982c; Gazzano et al. 1983; White et al. 1985a; Pang et al. 1985; Ballotti et al. 1987). Further, with partially purified insulin receptor, insulin-stimulated phosphorylation of both tyrosine and serine on its receptor (Kasuga et al. 1982c; Zick et al. 1983) as well as on exogenous substrates (Gazzano et al. 1983; Ballotti et al. 1986) was shown. The serine kinase activity appears to be noncovalently associated with the receptor, and is removed during further purification, because highly purified receptor displayed solely tyrosine kinase activity. The relationship between the two protein kinase activities associated with the receptor and their cellular role remains to be established. At least two possibilities exist, one in which both kinase activities serve separate cellular functions, and another one with sequential kinase activation (Gazzano et al. 1983). According to the first model, the tyrosine kinase would be involved in insulin's growth-promoting action similar to tyrosine phosphorylations mediating cellular responses to growth factors and to cellular and retroviral oncogene proteins (Hunter and Cooper 1985). In contrast, the serine kinase activities would play a role in insulin's metabolic actions. All kinases involved in the control of intermediary metabolism are indeed serine- or threonine-specific. In the second model, the two types of kinases are activated sequentially. Insulin binding to receptor leads to activation of the constituent tyrosine kinase, which induces activation of the receptor-associated serine kinase(s) and this accounts for the generation of cellular responses to insulin.

In contrast to insulin receptor autophosphorylation on tyrosine residues, phosphorylation of insulin receptor β-subunit on serine and threonine residues results in a decrease in receptor tyrosine kinase activity. This has been observed in the following situations: (1) intact cells treated either with phorbol esters, which are thought to act through protein kinase C (Jacobs et al. 1983; Häring et al. 1986), or with agents leading to an increase in cellular cAMP (Stadtmauer and Rosen 1986); (2) purified insulin receptors exposed to cAMP-dependent protein kinase (Roth and Beaudoin 1987) or protein kinase C (Bollag et al. 1986).

The role of serine and threonine phosphorylation of the insulin receptor remains a matter of speculation. It is tempting to suggest that at least part of the antagonistic action of some hormones, which oppose insulin's effects and act through cAMP-dependent protein kinase or protein kinase C, is mediated by seryl and threonyl β-subunit phosphorylation with concomitant decrease in receptor tyrosine kinase activity. Further, we like to think that the seryl and threonyl insulin receptor phosphorylation might also be the end point of a negative feedback loop in the regulation of insulin action. According to this idea, the activated insulin receptor tyrosine kinase would phosphorylate and activate a series of functional substrates, one of which would be a serine/threonine kinase. This kinase would phosphorylate the insulin receptor on serine/threonine residues leading to reduced tyrosine kinase activity and, consequently, to a decreased insulin signal. A working hypothesis taking into account the receptor tyrosine kinase activity and receptor seryl phosphorylation is schematized in Fig. 2.

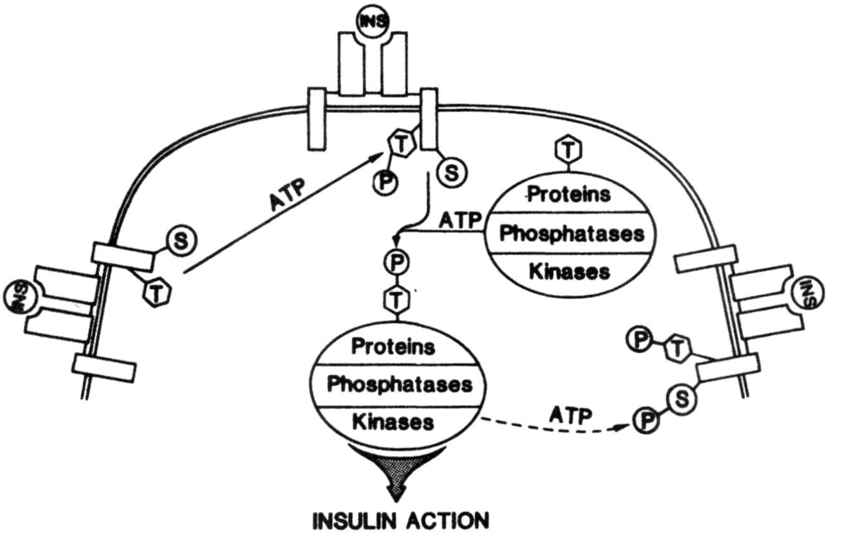

Fig. 2. Insulin receptor tyrosine kinase and receptor seryl phosphorylation

5 Signal Transduction

A large body of evidence establishes the insulin receptor tyrosine kinase as a fundamental property of the receptor, and indicates that this receptor enzymatic function is essential for generation of the metabolic and growth-promoting effects of insulin. The insulin receptor displays two functional domains, an extracellular insulin binding α-subunit and an insulin-responsive protein kinase contained within the intracellular domain of the β-subunit. At present it is not established how hormone recognition at the cell surface transmits a signal to the cytoplasmic receptor domain through a unique transmembrane stretch. the simplest mechanism would be that interaction of insulin with the receptor α-subunit triggers a conformational change which is propagated at the level of the contact region between α-β subunits resulting in activation of the receptor kinase (Fig. 3). How the activated receptor kinase transduces the hormone signal is also not known presently. We favor the idea that the insulin-stimulated receptor kinase leads to phosphorylation of cellular protein substrates, which are likely to belong to two broad categories, those generating metabolic effects of insulin, and those resulting in growth-promoting effects.

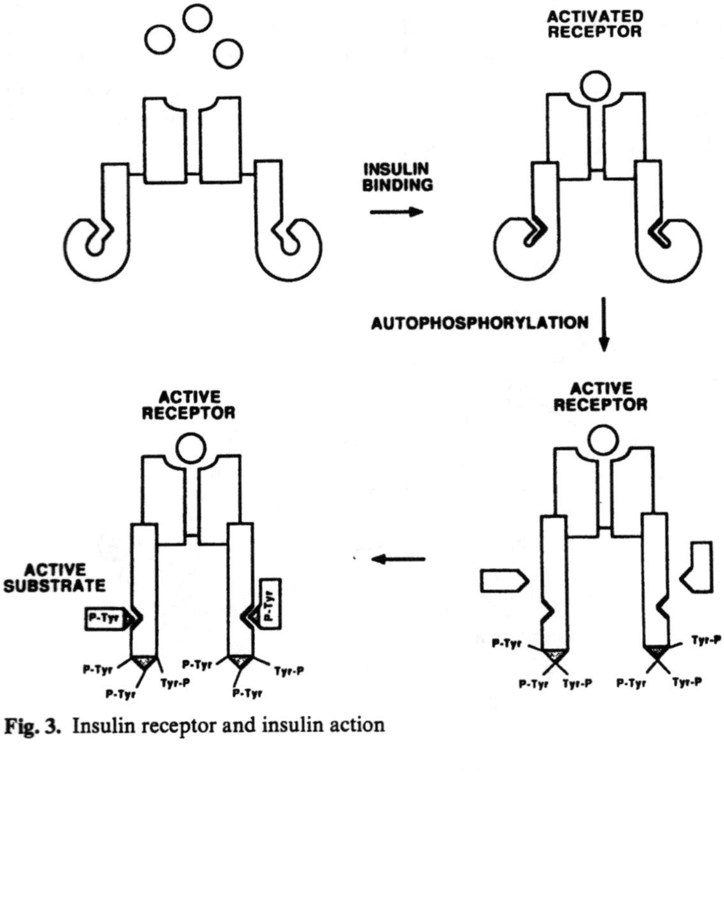

Fig. 3. Insulin receptor and insulin action

References

Ballotti R, Kowalski A, Le Marchand-Brustel Y, Van Obberghen E (1986) Presence of an insulin-stimulated serine kinase in cell extracts from IM-9 cells. Biochem Biophys Res Commun 139:179–185

Ballotti R, Kowalski A, White MF, Le Marchand-Brustel Y, Van Obberghen E (1987) Insulin stimulates tyrosine phosphorylation of its receptor β subunit in intact rat hepatocytes. Biochem J 241:99–104

Ballotti R, Scimeca JC, Kowalski A, Van Obberghen E (1988) Antiphosphotyrosine antibodies modulate insulin receptor kinase activity and insulin action. Cell Signall 1:195–204

Bernier M, Laird DM, Lane MD (1987) Insulin-activated tyrosine phosphorylation of a 15-kilodalton protein in intact 3T3-L1 adipocytes. Proc Natl Acad Sci USA 84:1844–1848

Bollag GE, Roth RA, Beaudoin J, Mochly-Rosen D, Koshland DE, Jr. (1986) Protein kinase C directly phosphorylates the insulin receptor in vitro and reduces its protein-tyrosine kinase activity. Proc Natl Acad Sci USA 83:5822–5824

Chou CK, Dull TJ, Russell DS, Gherzi R, Lebwohl D, Ullrich A, Rosen OM (1987) Human insulin receptors mutated at the ATP-binding site lack protein tyrosine kinase activity and fail to mediate postreceptor effects of insulin. J Biol Chem 262:1842–1847

Debant A, Guerre-Millo M, Le Marchand-Brustel Y, Freychet P, Lavau M, Van Obberghen E (1987) The adipocyte insulin receptor kinase from young obese Zucker rats is hyperresponsive to insulin. Am J Physiol 252:E273–E278

Denton RM (1986) Early events in insulin actions. In: Greengard P, Robison A (eds) Advances in cyclic nucleotide and protein phosphorylation research, vol 20. Raven, New York, 293–341

Ebina Y, Ellis L, Jarnagin K, Edery M, Graf L, Clauser E, Ou JH, Masiarz F, Kan YW, Goldfine ID, Roth RA, Rutter WJ (1985) The human insulin receptor cDNA: the structural basis for hormone-activated transmembrane signalling. Cell 40:747–758

Gammeltoft S, Van Obberghen E (1986) Protein kinase activity of the insulin receptor. Biochem J 235:1–11

Gammeltoft S, Kowalski A, Fehlmann M, Van Obberghen E (1983) Insulin receptors in rat brain: insulin stimulates phosphorylation of its receptors β-subunit. FEBS Lett 172:87–90

Gammeltoft S, Haselbacher GK, Humbel RE, Fehlmann M, Van Obberghen E (1985) Two types of receptors for insulin-like growth factors in mammalian brain. EMBO J 4:3407–3412

Gazzano H, Kowalski A, Fehlmann M, Van Obberghen E (1983) Two different protein kinase activities are associated with the insulin receptor. Biochem J 216:575–582

Gazzano H, Halban P, Prentki M, Ballotti R, Brandenburg D, Fehlmann M, Van Obberghen E (1985) Identification of functional insulin receptors on membranes from an insulin producing cell line (RINm5F). Biochem J 226:867–872

Gherzi R, Russell DS, Taylor SI, Rosen OM (1987) Reevaluation of the evidence that an antibody to the insulin receptor is insulinomimetic without activating the protein tyrosine kinase activity of the receptor. J Biol Chem 262:16900–16905

Grigorescu F, White MF, Kahn CR (1983) Insulin binding and insulin-dependent phosphorylation of the insulin receptor solubilized from human erythrocytes. J Biol Chem 258:13708–13716

Grunberger G, Zick Y, Gorden P (1984) Defect in phosphorylation of insulin receptors in cells from an insulin resistant patient with normal insulin binding. Science 223:932–934

Häring HU, White MF, Kahn CR, Kasuga M, Lauris V, Fleischmann R, Murray M, Pawelek J (1984) Abnormality of insulin binding and receptor phosphorylation in an insulin-resistant melanoma cell line. J Cell Biol 99:900–908

Häring HU, Kirsch D, Obermaier B, Ermel B, Machicao F (1986) Tumor-promoting phorbol esters increase the Km of the ATP binding site of the insulin receptor kinase from rat adipocytes. J Biol Chem 261:3869–3875

Häring HU, White MF, Machicao F, Ermel B, Schleicher E, Obermaier B (1987) Insulin rapidly stimulates phosphorylation of a 46-kDa membrane protein on tyrosine residues as well as phosphorylation of several soluble proteins in intact fat cells. Proc Natl Acad Sci USA 84:113–117

Hunter T, Cooper JA (1985) Protein-tyrosine kinases. Annu Rev Biochem 54:897–930

Jacobs S, Sahyoun NE, Saltiel AR, Cuatrecasas P (1983) Phorbol esters stimulate the phosphorylation of receptors for insulin and somatomedin C. Proc Natl Acad Sci USA 80:6211–6213

Kadowaki T, Kasuga M, Akanuma Y, Ezaki O, Takaku F (1984) Decreased autophosphorylation of the insulin receptor-kinase in streptozotocin-diabetic rats. J Biol Chem 259:14208–14216

Kasuga M, Karlsson FA, Kahn CR (1982a) Insulin stimulates the phosphorylation of the 95,000-dalton subunit of its own receptor. Science 215:185–187

Kasuga M, Zick Y, Blithe DL, Crettaz M, Kahn CR (1982b) Insulin stimulates tyrosine phosphorylation of the insulin receptor in a cell-free system. Nature (London) 298:667–669

Kasuga M, Zick Y, Blithe DL, Karlsson FA, Häring HU, Kahn CR (1982c) Insulin stimulation of phosphorylation of the beta subunit of the insulin receptor. Formation of both phosphoserine and phosphotyrosine. J Biol Chem 257:9891–9894

Kowalski A, Gazzano H, Fehlmann M, Van Obberghen E (1983) Dephosphorylation of the hepatic insulin receptor: absence of phosphatase activity in purified receptors. Biochem Biophys Res Commun 117:885–893

Le Marchand-Brustel Y, Grémeaux T, Ballotti R, Van Obberghen E (1985) Insulin receptor kinase is defective in skeletal muscle of obese mice. Nature (London) 315:676–679

Morgan DO, Roth RA (1987) Acute insulin action requires insulin receptor kinase activity: introduction of an inhibitory monoclonal antibody into mammalian cells blocks the rapid effects of insulin. Proc Natl Acad Sci USA 84:41–45

Pang DT, Sharma BR, Schafer JA, White PF, Kahn CR (1985) Predominance of tyrosine phosphorylation of insulin receptors during the initial response of intact cells to insulin. J Biol Chem 260:7131–7136

Petruzzelli LM, Ganguly S, Smith CJ, Cobb MH, Rubin CS, Rosen OM (1982) Insulin activates a tyrosine-specific protein kinase in extracts of 3T3-L1 adipocytes and human placenta. Proc Natl Acad Sci USA 79:6792–6796

Petruzzelli L, Herrera R, Rosen OM (1984) Insulin receptor is an insulin-dependent tyrosine protein kinase: copurification of insulin-binding activity and protein kinase activity to homogeneity from human placenta. Proc Natl Acad Sci USA 81:3327–3331

Rees-Jones RW, Taylor SI (1985) An endogenous substrate for the insulin receptor-associated tyrosine kinase. J Biol Chem 260:4461–4467

Rees-Jones RW, Hedo JA, Zick Y, Roth J (1983) Insulin-stimulated phosphorylation of the insulin receptor precursor. Biochem Biophys Res Commun 116:417–422

Roth RA, Beaudoin J (1987) Phosphorylation of purified insulin receptor by cAMP kinase. Diabetes 36:123–126

Roth RA, Cassell DJ (1983) Insulin receptor: Evidence that it is a protein kinase. Science 219:299–301

Roth RA, Cassell DJ, Maddux DA, Goldfine ID (1983) Regulation of insulin receptor kinase activity by insulin mimickers and an insulin antagonist. Biochem Biophys Res Commun 115:245–252

Sadoul JL, Peyron JF, Ballotti R, Debant A, Fehlmann M, Van Obberghen E (1985) Identification of a cellular 110,000 Da protein substrate for the insulin receptor kinase. Biochem J 227:887–892

Shia MA, Pilch PF (1983) The β-subunit of the insulin receptor is an insulin-activated protein kinase. Biochemistry 22:717–721

Stadtmauer LA, Rosen OM (1983) Phosphorylation of exogenous substrates by the insulin receptor-associated protein kinase. J Biol Chem 258:6682–6685

Stadtmauer L, Rosen OM (1986) Increasing the cAMP content of IM-9 cells alters the phosphorylation state and protein kinase activity of the insulin receptor. J Biol Chem 261:3402–3407

Tamura S, Fujita-Yamaguchi Y, Larner J (1983) Insulin-like effect of trypsin on the phosphorylation of rat adipocyte insulin receptor. J Biol Chem 258:14749–14752

Tanti JF, Grémeaux T, Brandenburg D, Van Obberghen E, Le Marchand-Brustel Y (1986) Brown adipose tissue in lean and obese mice: insulin-receptor binding and tyrosine kinase activity. Diabetes 35:1243–1248

Ullrich A, Bell JR, Chen EY, Herrera R, Petruzzelli LM, Dull TJ, Gray A, Coussens L, Liao YC, Tsubokawa M, Mason A, Seeburg PH, Grunfeld C, Rosen OM, Ramachandran J (1985) Human insulin receptor and its relationship to the tyrosine kinase family of oncogenes. Nature (London) 313:756–761

Van Obberghen E (1984) The insulin receptor: its structure and function. Biochem Pharmacol 33:59–65

Van Obberghen E, Kowalski A (1982) Phosphorylation of the hepatic insulin receptor: stimulating effect of insulin on intact cells and in a cell-free system. FEBS Lett 143:179–182

Van Obberghen E, Rossi B, Kowalski A, Gazzano H, Ponzio G (1983) Receptor-mediated phosphorylation of the hepatic insulin receptor: evidence that the Mr 95 000 receptor subunit is its own kinase. Proc Natl Acad Sci USA 80:945–949

White MF, Maron R, Kahn CR (1985a) Insulin rapidly stimulates tyrosine phosphorylation of a Mr-185,000 protein in intact cells. Nature (London) 318:183–186
White MF, Takayama S, Kahn CR (1985b) Differences in the sites of phosphorylation of the insulin receptor in vivo and in vitro. J Biol Chem 260:9470–9478
Yu KT, Czech MP (1984) Tyrosine phosphorylation of the insulin receptor β subunit activates the receptor-associated tyrosine kinase activity. J Biol Chem 259:5277–5286
Yu KT, Khalaf N, Czech MP (1987) Insulin stimulates a membrane-bound serine kinase that may be phosphorylated on tyrosine. Proc Natl Acad Sci USA 84:3972–3976
Zick Y, Grunberger G, Podskalny JM, Moncada V, Taylor SI, Gorden P, Roth J (1983) Insulin stimulates phosphorylation of serine residues in soluble insulin receptors. Biochem Biophys Res Commun 116:1129–1135

Structural and Functional Aspects of Signal Transduction by Receptor Tyrosine Kinases

Y. YARDEN

1 The RTK Family of Receptors

Cells derived from all three embryonic germ layers appear to share a common mechanism of growth regulation that involves activation of membranal protein kinases specific for tyrosine residues. The prototypic structure of these enzymes includes an extracellular ligand-binding domain connected through a single membrane-spanning stretch to a large cytoplasmic tail that carries the kinase function. Detailed structural analyses revealed that most of the receptor tyrosine kinases (RTK's) are encoded for by a family of related genes that fall into three subclasses as shown in the following table (Yarden and Ullrich 1988a):

	Receptors	Ligands	Oncogenic version
RTK subclass:			
Subclass I	EGF-R	EGF, TGF-α	v-erbB
	neu	?	*neu*
Subclass II	I-R	Insulin	
	IGF-1-R	IGF-1	
Subclass III	PDGF-R	PDGF	
	CSF-1-R	CSF-1	v-fms
	c-kit	?	v-kit

Perhaps the best exemplification of the pivotal role of this family of hormone receptors in cellular regulation is their involvement in malignant transformation by either one of three mechanisms:

1. Autocrine/paracrine loop whereby the peptide growth factor (e.g., TGF-α, PDGF) is secreted by the transformed cell leading to a constitutive growth signal (Sporn and Roberts 1985).
2. Structurally altered receptor molecules whose modulatory domains are modified but the catalytic regions kept intact (e.g., the protein products of the viral oncogenes *erb-B* and *fms*), presumably yielding deregulated mitogenic stimuli (Downward et al. 1984; Sherr et al. 1985).
3. Overexpression of some RTK molecules (e.g., EGF-receptor, *neu*) in certain human malignancies is thought to contribute to overgrowth through a yet unidentified mechanism (Libermann et al. 1985; Slamon et al. 1987).

Department of Chemical Immunology, Weizmann Institute of Science, Rehorot, 76100, Israel

A. Azzi, Z. Drahota and S. Papa
Molecular Basis of Membrane-Associated Diseases

Importantly, the *neu* (Coussens et al. 1985; Bargmann et al. 1986; Yamamoto et al. 1986) and *kit* (Yarden et al. 1987) proto-oncogenes are related to EGF receptor and CSF-1 receptor, respectively. Yet, they are not capable of binding these ligands, suggesting that their own ligands are yet to be identified. Thus, new hormones may be tracked through reverse genetics of their putative receptors. Once worked out, such an approach may be employed for the elucidation of the function of other, more distantly related transmembrane tyrosine kinases, such as the proteins encoded by *ros, met*, and *eph* oncogenes.

Molecular analysis, by means of site-directed mutagenesis, of structure-function relationships within the RTK molecule has recently started to sharpen our view of signal transduction by growth factors and their receptors (Yarden and Ullrich 1988b). Perhaps the most important lesson learned is that the tyrosine-kinase function of the RTK molecule is ubiquitous for cellular activation (Rosen 1987; Schlessinger 1988). Thus, the RTK molecule can be viewed as a transmembrane allosteric enzyme whose activity is regulated by noncatalytic domains including the ligand-binding domain (positive regulation) and cytoplasmic noncatalytic molecular segments (mostly negative regulation).

2 Signal Transduction by RTK's: Definition of the Problem

A central question arising from the above mentioned genetic and biochemical analyses is how the formation of the ligand-receptor complex outside of the cell translates into activation of the cytoplasm-facing enzymatic function? A simple-minded model would attribute signal transfer to the single transmembranal α-helix of the prototypic RTK molecule. However, close inspection of the corresponding domains in all RTK molecules reveals that they are flanked by proline residues at the extracellular side and charged amino acids at the cytoplasmic side, so that vertical translocations would be retarded by a high free-energy barrier. In addition, the hydrophobic stretches show no conservation of primary structure as would be required for an intramolecular mechanism of signal transfer. Thus, the single α-helix may not be rigid enough to provide the needed molecular scaffolding. To demonstrate this latter point, an analogy can be made to another family of receptors that mediate hormonal signals, the G-protein coupled receptors (e.g., rhodopsin, adrenergic and muscarinic cholinergic receptors).

Despite a remarkable diversification of cellular functions elicited through activation of yet different guanine nucleotide regulatory proteins (G-proteins), all the G-protein coupled receptors share a common basic structure: each contains seven stretches of 20–28 hydrophobic amino acids that likely represent membrane spanning regions (Dixon et al. 1986; Yarden et al. 1986). Moreover, amino acid identities in functionally related receptors are concentrated within these transmembrane regions. Importantly, the seven α-helices basic structure is remarkably evolutionarily conserved and appears in the receptors for the yeast mating factors a and α and in bacteriorhodopsin, a pigment protein of archaebacterium. Only in the latter case is the spatial organization of the seven-helices structure known (Stoeckenius and Bogomolni 1982), an annular structure composed of all the seven helices nests the retinal group. A similar pocket may accomodate binding of adrenergic or cholinergic ligands, so that an overall conformational change will be then transmitted into the cytoplasm. That an occupied receptor is conformably changed at the cytoplasmic face is exemplified by its preferential accessibility to specific protein kinases (Benovic et al. 1986), and by the ligand-promoted interaction of the third cytroplasmic loop with a G-protein (Kobilka et al. 1988). Moreover, site-directed mutagenesis and limited proteolysis suggest that many or even all the helices contribute to forming the ligand-binding pocket (Lefkowitz and Caron 1988). In conclusion, multiple membrane spans are indispensable for a transmembranal rigid structure that channels the molecular signal down to the G-protein.

3 Signal Transduction by EGF Receptor

In addressing the question of signal transduction through a single membrane span, EGF receptor emerged to be an appropriate model system. On living cells EGF induces receptor redistribution into small clusters that eventually internalize through coated pits. However, the requirement for cytoskeletal elements and metabolic energy for the macroaggregation process as well as its relation to cellular activation is not clear. The first indication for reversible receptor microaggregation in a cell-free system was obtained by measuring the rotational diffusion of occupied receptors in plasma membrane vesicles (Zidovetzki et al. 1986). However, biochemical evidence of the requirement of receptor interaction for kinase activation was first obtained with detergent-solubilized receptor fractions (Yarden and Schlessinger 1987a). Three lines of experimental evidence supported an intermolecular activation process:

1. Kinase activation exhibited bimolecular reaction order;
2. Crosslinking of nonoccupied receptors partially activated the kinase function;
3. Prevention of molecular interactions abolished receptor activation.

Furthermore, receptor aggregates exhibited higher affinity to EGF and together with the kinetic measurements suggested that minimal oligomerization, namely dimerization of receptors, which is driven by ligand binding, is sufficient for kinase activation. Structural evidence in support of this mechanism was then obtained by employing a nondenaturing gel system that preserved the oligomeric receptor (Yarden and Schlessinger 1987b). This approach confirmed the existence of a noncovalently held receptor dimer, as predicted by the biochemical approach, and further supported the following model: EGF receptor exists, in detergent solution, in equilibrium between monomeric and dimeric forms. EGF binding, by favoring the latter, shifts the equilibrium toward the dimer form which is the active state of the intrinsic kinase. In contrast to an exactly opposing model, based on sucrose gradient fractionation of the receptor forms (Biswas et al. 1985), a more recent report, also utilizing sucrose gradients, confirmed the dimerization-activation model (Boni-Schnetzler and Pilch 1987). Further support of this model was obtained by chemical crosslinking of the ligand-induced receptor dimers in vitro

(Glenney et al. 1988; Cochet et al. 1988; Northwood and Davis 1988) and also on living cells (Fanger et al. 1986; Cochet et al. 1988). The extent and relevance of receptor dimerization in vivo is difficult to assess with the chemical crosslinking method. Alternatively, microinjection of site-specific antibodies and construction of dimerization-negative receptor mutants may prove useful.

4 Signal Transduction by Insulin Receptor

The receptor for insulin, due to its heterotetrameric structure, presents additional complexity as to the question of signal transduction. Although its predominant structure is β-S-S-α-S-S-α-S-S-β, other receptor forms may exist (Kasuga et al. 1982). Nevertheless, a chimeric insulin receptor (extracellular domain) – EGF receptor (cytoplasmic domain) retains insulin responsiveness of the kinase function (Riedel et al. 1986), indicating that similar transduction mechanisms are utilized by both receptors, despite differences in molecular organization. A major step toward the elucidation of the mechanism of signal transduction was the elaboration of a technique that enables isolation of functional $\alpha\beta$ heterodimers by a combination of mild reducing and alkaline conditions (Sweet et al. 1986; Boni-Schnetzler et al. 1986). In analogy to a monomer of EGF receptor, the isolated dilute $\alpha\beta$ heterodimer exhibits no insulin-dependent phosphorylation and displays lower binding affinity to insulin (Boni-Schnetzler et al. 1987). Insulin responsiveness is restored in concentrated solution of $\alpha\beta$ heterodimers and is associated with the induction of both covalent and noncovalent interactions between $\alpha\beta$ receptor halves (Boni-Schnetzler et al. 1986; Morrison et al. 1988). More recent evidence appears to support the possibility that the covalent associations are not essential for the ligand-induced activation of the kinase. In conclusion, it is clear that an analogy exists between the mechanisms of signal transduction by the receptors for insulin and EGF.

5 Open Questions and Prospective

It should be emphasized that although the evidence for an ollasteric-dimerization mechanism of RTK activation appears convincing, most of it was derived from in vitro systems. The extrapolation to the living cell, although likely, should await appropriate experimental support, and even at the in vitro level many questions have yet to be resolved. Some of the main open issues are discussed below:

1. Mechanistic Aspects. How is the kinase function of the dimerized receptor being activated? Is it due merely to the close apposition of two identical domains or are other molecular mechanisms involved? Perhaps the most compelling possibility, which is indirectly supported by the occurrence of C-terminally proteolized and activated receptors, is that the kinase function of the monomeric receptor is inhibited by the flexible carboxy-terminal tail by folding over the catalytic site. Once dimerized the tails would be retracted and the inhibition released. Another question relates to the spontaneous dissociation of the dimer: Will it now give rise to activated monomeric receptors? Although fragmentary, the experimental evidence (Yarden and Schlessinger 1987b; Boni-Schnetzler et al. 1986) seems to support such a "collision-coupling" type of receptor activation. Potentially, this mechanism may result in an amplification of a signal obtained in response to partial receptor occupation. The last mechanistic question deals with this latter point: Do both partners of the dimer have to be occupied? Experiments with insulin receptor (Pang and Shafer 1984; Sweet et al. 1986) clearly indicate that the dimeric receptor is monovalent. The mechanism by which the second binding site is closed is presently unknown, but in combination with a collision-coupling mechanism it could lead to a significant enhancement of sensitivity in terms of hormone concentration; one occupied receptor, through multiple encounters with unoccupied partners, may potentially activate the whole cell.

2. *Is there a dimerization site on the receptor molecule?* Apparently the dimerization function is intrinsic to the receptor molecule, although heterologous ligands, by means of covalent modifications and cytoskeletal elements, probably affect it. It is then important to ask, which receptor domain mediates the intermolecular interactions? Experiments with various truncated forms of EGF receptor, including most of the extracellular domain or parts of the cytoplasmic moiety (Basu et al. 1986; Yarden and Schlessinger 1987b), failed to detect dimerization. In addition, a truncated β chain of insulin receptor appears to be dimerization incompetent. One possibility is that the dimer is stabilized by multiple interactions along the surfaces of the participating species. Most likely, the initial contact is established at or close to the ligand-binding pocket and then propagates down in a "zipperlike" mechanism. A critical point, according to this model, is the membrane-spanning α-helices. In addition, the cysteine-rich regions, present at the extracellular domains of EGF and insulin receptors, may also contribute to the noncovalent interactions.

3. *Relevance to Cellular Activation.* A large body of evidence appears to support the possibility that receptor aggregation is involved in signal transduction in intact cells. Thus monoclonal and polyclonal antireceptor antibodies mimic at least part of the pleiotropic cellular response (Kahn et al. 1981; Schreiber et al. 1981; Chandler et al. 1985; Fernandez-Pol 1985). Although receptor aggregation appears to be sufficient for activation of the kinase function in vivo (Defize et al. 1986), it is probably insufficient for the stimulation of DNA synthesis (Yarden et al. 1982; Defize et al. 1986). Other studies showed that only a small fraction of EGF receptors is involved in the stimulation of DNA synthesis, and that this class of receptors has high affinity to the ligand (Kawamoto et al. 1983; Gregoriou and Rees 1984) and restricted lateral mobility (Rees et al. 1984). It is then appealing, but yet unproved, that this small fraction of receptors is composed of high-affinity oligomerized receptors that transmit the mitogenic signal through activation of the tyrosine kinase.

Finally, receptor dimerization that was initially found in vitro should be further characterized and extended to the living cell. Conceivably, modulation of this receptor function through homologous and heterologous effectors as well as by genetic factors such as deletions, mutations, or overexpression of the receptor on the cell surface may be involved in normal and aberrant cell function.

Acknowledgments. I would like to thank Drs. Joseph Schlessinger and Axel Ullrich for many discussions and continuous help and Marilyn Daly for her expertise and patience in typing this paper.

References

Basu M, Sen-Majumdar A, Basu A, Murthy U, Das M (1986) Regulation of kinase and intermolecular bonding in intact and truncated epidermal growth factor receptor. J Biol Chem 261:12879–82

Bargmann CI, Hung MC, Weinberg RA (1986) The neu oncogene encodes an epidermal growth factor receptor-related protein. Nature 319:226–230

Benovic JL, Kuhn H, Weyand I, Codina J, Caron MG, Lefkowitz RJ (1987) Functional desensitization of the isolated beta-adrenergic receptor by the beta-adrenergic receptor kinase: potential role of an analog of the retinal protein arrestin (48-kDa protein). Proc Natl Acad Sci USA 84:8879–8882

Biswas R, Basu M, Sen-Majundar A, Das M (1985) Intrapeptide autophosphorylation of the epidermal growth factor receptor: regulation of kinase catalytic function by receptor dimerization. Biochemistry 24:3795–3802

Boni-Schnetzler M, Pilch PF (1987) Mechanism of epidermal growth factor receptor autophosphorylation and high-affinity binding. Proc Natl Acad Sci USA 84:7832–7836

Boni-Schnetzler M, Rubin JB, Pilch PF (1986) Structural requirement for the transmembrane activation of the insulin receptor kinase. J Biol Chem 261:15281–15287

Boni-Schnetzler M, Scott W, Waugh SM, DiBella E, Pilch PF (1987) The insulin receptor: structural basis for high affinity ligand binding. J Biol Chem 262:8395–8401

Chandler LP, Chandler CE, Hosang M, Shooter EM (1985) A monoclonal antibody which inhibits epidermal growth factor binding has opposite effects on the biological action of epidermal growth factor on different cells. J Biol Chem 260:3360–3367

Cochet C, Kashles O, Chambaz EM, Barrello I, King CR, Schlessinger J (1988) Demonstration of epidermal growth factor-induced receptor dimerization in living cell using a chemical covalent cross-linking agent. J Biol Chem 263:3290–3295

Coussens L, Yang-Fang TL, Liao YC, Chen E, Gray A, McGrath J, Seeburg PH, Libermann TA, Schlessinger J, Francke U, Levinson A, Ullrich A (1985) Tyrosine kinase receptor with extensive homology to EGF receptor shares chromosomal location with neu oncogene. Science 230:1132–1139

Defize LHK, Moolenaar WH, van der Saag PT, de Laat SW (1986) Dissociation of cellular responses to epidermal growth factor using anti-receptor monoclonal antibodies. EMBO J 5:1187–1192

Dixon RAF, Kobilka BK, Strader DJ, Benovic JL, Dohlman HG, Frielle T, Bolanowski MA, Bennett CD, Rands E, Diehl RE, Mumford RA, Slater EE, Sigal IS, Caron MG, Lefkowitz RJ, Strader D (1986) Cloning of the gene and cDNA for mammalian β-adrenergic receptor and homology with rhodopsin. Nature (London) 321:75–79

Downward J, Yarden Y, Mayes E, Scarce G, Toty N, Stockwell P, Ullrich A, Schlessinger J, Waterfield MD (1984) Close similarity of epidermal growth factor receptor and v-erb-B oncogene protein sequences. Nature (London) 307:521–527

Fanger BO, Austin KS, Earp HS, Cidlowski JA (1986) Cross-linking of epidermal growth factor receptors in intact cells: detection of initial stages of receptor clustering and determination of molecular weight of high affinity receptors. Biochemistry 25:6414–6420

Fernandez-Pol JA (1985) Epidermal growth factor receptor of A431 cells: characterization of a monoclonal anti-receptor antibody noncompetitive agonist of epidermal growth factor action. J Biol Chem 260:5003–5011

Glenny JR Jr, Chen WS, Lazar CS, Walton GM, Zokas LM, Rosenfeld MG, Gill GN (1988) Ligand-induced endocytosis of the EGF receptor is blocked by mutational inactivation and by microinjection of anti-phosphotyrosine antibodies. Cell 52:675–684

Gregoriou M, Rees AR (1984) Properties of a monoclonal antibody to epidermal growth factor receptor with implication for the mechanism of action of EGF. EMBO J 3:929–937

Kahn CR, Baird KL, Flier JS, Grunfield C, Harmon JT, Harrison LC, Karlsson FA, Kasuga M, King GL, Lang U, Podskalny JM, Van-Obberghen E (1981) Insulin receptors, receptor antibodies, and the mechanism of insulin action. Recent Prog Horm Res 34:477–568

Kasuga M, Hedo JA, Yamada KM, Kahn CR (1982) The structure of insulin receptor and its subunits: evidence for multiple nonreduced forms and a 210,000 possible proreceptor. J Biol Chem 257:10392–10399

Kawamoto T, Sato JD, Le A, Polikoff J, Sato GH, Mandelsohn J (1983) Growth stimulation of A431 cells by epidermal growth factor: identification of high affinity receptors for epidermal growth factor by anti-receptor monoclonal antibodies. Proc Natl Acad Sci 80:1337–1341

Kobilka BK, Kobilka TS, Daniel K, Regan JW, Caron MG, Lefkowitz RJ (1988) Chimeric α_2-, β_2-adrenergic receptors: delineation of domains involved in effector coupling and ligand binding specificity. Science 240:1310–1316

Lefkowitz RJ, Caron MG (1988) Adrenergic receptors: models for the study of receptors coupled to guanine nucleotide regulatory proteins. J Biol Chem 263:4993–4996

Libermann TA, Nussbaum HR, Razon N, Kris R, Lax I, Soreq H, Whittle N, Waterfield MD, Ullrich A, Schlessinger J (1985) Amplification, enhanced expression and possible rearrangement of EGF receptor gene in primary human brain tumors of glial origin. Nature (London) 313:144–147

Morrison BD, Swanson ML, Sweet LJ, Pessin JE (1988) Insulin-dependent covalent reassociation of isolated $\alpha\beta$ heterodimeric insulin receptors into an $\alpha_2\beta_2$ heterotetrameric disulfide linked complex. J Biol Chem 263:7806–7813

Northwood IC, Davis RJ (1988) Activation of the epidermal growth factor receptor tyrosine protein kinase in the absence of receptor oligomerization. J Biol Chem 263:7450–7453

Pang DT, Shafer JA (1984) Evidence that insulin receptor from human placenta has a high affinity for only one molecule of insulin. J Biol Chem 259:8589–8596

Rees AR, Gregoriou M, Johnson P, Garland PB (1984) High affinity epidermal growth factor receptors on the surface of A431 cells have restricted lateral diffusion. EMBO J 3:1843–1847

Riedel H, Schlessinger J, Ullrich A (1986) A chimeric receptor allows insulin to stimulate tyrosine kinase activity of epidermal growth factor receptor. Nature (London) 324:68–70

Rosen OM (1987) After insulin binds. Science 237:1452–1458

Schlessinger J (1988) The epidermal growth factor receptor as a multifunctional allosteric protein. Biochemistry 27:3119–3123

Schreiber AB, Lax I, Yarden Y, Eshhar Z, Schlessinger J (1981) Monoclonal antibodies against receptor for epidermal growth factor induce early and delayed effects of epidermal growth factor. Proc Natl Acad Sci USA 78:7535–7539

Sherr CJ, Rettenmier CW, Sacca R, Roussel MF, Look AT, Stanley R (1985) The c-fms proto-oncogene product is related to the receptor for the mononuclear phagocyte growth factor, CSF-1. Cell 41:665–676

Slamon DJ, Clark GM, Steven GW, Levin WJ, Ullrich A, McGuire WL (1987) Human breast cancer: correlation of relapse and survival with amplification of HER-2-neu oncogene. Science 235:177–181

Sporn MB, Roberts AB (1985) Autocrine growth factors and cancer. Nature (London) 313:745–747

Stoeckenius W, Bogomolniu RA (1982) Bacteriorhodopsin and related pigments of holobacteria. Annu Rev Biochem 51:587–615

Sweet LJ, Morrison BD, Pessin JE (1986) Isolation of functional $\alpha\beta$ heterodimers from the purified human placental $\alpha_2\beta_2$ heterotetrameric insulin receptor complex. J Biol Chem 262:6939–6942

Yamamoto T, Ikawa S, Akiyama T, Samba K, Nomura N, Miyajima N, Saito T, Toyoshima K (1986) Similarity of protein encoded by the human c-erb-B-2 gene to epidermal growth factor receptor. Nature (London) 319:230–234

Yarden Y, Schlessinger J (1987a) Epidermal growth factor induces rapid, reversible aggregation of the purified epidermal growth factor receptor. Biochemistry 26:1443–1451

Yarden Y, Schlessinger J (1987b) Self-phosphorylation of epidermal growth factor receptor: evidence for a model of intermolecular allosteric activation. Biochemistry 26:1434:1442

Yarden Y, Ullrich A (1988a) Growth factor receptor tyrosine kinases. Annu Rev Biochem 57:443–478

Yarden Y, Ullrich A (1988b) Molecular analysis of signal transduction by growth factors. Biochemistry 27:3113–3119

Yarden Y, Schreiber AB, Schlessinger J (1982) A nonmitogenic analogue of epidermal growth factor induces early responses mediated by epidermal growth factor. J Cell Biol 92:687–693

Yarden Y, Kuang WJ, Yang-Feng T, Coussens L, Munemitsu S, Dull TJ, Chen E, Schlessinger J, Francke U, Ullrich A (1987) Human proto-oncogene c-kit: a new cell surface receptor tyrosine kinase for an unidentified ligand. EMBO J 6:3341–3351

Yarden Y, Rodriguez H, Wong SKF, Brandt DR, May DC, Burnier J, Harkins RN, Chen EY, Ramachandran J, Ullrich A, Ross REM (1986) The avian β-adrenergic receptor: primary structure and membrane topology. Proc Natl Acad Sci USA 83:6795–6799

Zidovetzki R, Yarden Y, Schlessinger J, Jovin TM (1986) Microaggregation of hormone-occupied epidermal growth factor receptors on plasma membrane preparations. EMBO J 5:247–250

Nerve Growth Factor (NGF): Physiological Functions and Regulation of Its Synthesis

D. LINDHOLM[1], R. HEUMANN, G. D. BORASIO and H. THOENEN

1 Introduction

The physiological role of NGF as a neurotrophic factor in the peripheral nervous system is well established. NGF is essential for the development and maintenance of specific functions of peripheral sympathetic and neural-crest derived sensory neurons. NGF acts as a retrograde messenger between the target organs and the innervating neurons. The regionally different quantities of NGF in target tissues reflect the density of innervation by NGF-responsive neurons, which transport NGF in membrane-confined compartments to the perikaryon after NGF is bound via specific membrane receptors. Although the physiological function of NGF is well documented and a large number of NGF-specific effects on target neurons have been identified, there are two main questions which remain to be elucidated, namely the nature of the second messenger(s) evolving from the binding of NGF to its receptor and the mode of regulation of NGF synthesis in the target tissues. A better understanding of this regulation would provide the prerequisite for a rational approach to the pharmacological modification of NGF synthesis. This is of particular interest because it has recently become apparent that NGF is essential for the preservation of the function of the cholinergic neurons of the basal forebrain nuclei. These neurons are consistently affected in Alzheimers's disease and their impaired function seems to be largely responsible for the cognitive deficits of this disease.

2 Physiological Functions of NGF

Earlier work on NGF was mainly focussed on the role of NGF in embryonic development and differentiation of the peripheral nervous system (Levi-Montalcini 1987). NGF has been shown to regulate the survival of peripheral sympathetic and neural-crest derived sensory neurons during restricted periods of development (cf. Levi-Montalcini and Angeletti 1968; Greene and Shooter 1980; Thoenen and Barde 1980). This has been conclusively demonstrated in experiments on neonatal rodents where administration of anti-NGF antibodies leads to

Department of Neurochemistry, Max-Planck-Institute for Psychiatry, Am Klopferspitz 18 A,D-8033 Martinsried/Munich, FRG
[1]To whom correspondence should be addressed

A. Azzi, Z. Drahota and S. Papa
Molecular Basis of Membrane-Associated Diseases

enhanced cell death in the NGF-responsive neuronal population, whereas exogenous NGF can rescue neurons which normally would die during the period of natural cell death. In addition, NGF is also required for differentiation and maintenance of function of the peripheral sympathetic and sensory neurons in adult life. Thus, NGF has been shown to regulate synthesis of specific enzymes which are involved in neurotransmitter production, such as tyrosine hydroxylase and dopamine β-hydroxylase as well as a series of neuron-specific peptides (cf. Otten 1984; Johnson E.M. et al. 1986; Thoenen et al. 1987). In the peripheral target fields NGF is synthesized in very low amounts and is retrogradely transported to the cell bodies of NGF-responsive neurons after being taken up via specific cell surface receptors (cf. Schwab et al. 1982). It has also been shown that the density of innervation of target organs by NGF-responsive neurons correlates with the levels of NGF and its mRNA (Korsching and Thoenen 1983; Heumann et al. 1984; Shelton and Reichardt 1984). Moreover, interruption of the retrograde axonal transport of NGF following axotomy or by pharmacological interference (e.g. colchicine treatment which results in disassembly of microtubuli) leads to degenerative changes in the NGF-responsive neurons, demonstrating that NGF acts as a retrograde messenger between the periphery and the innervating neurons (cf. Thoenen et al. 1987).

In contrast to sympathetic neurons in the peripheral nervous system, central adrenergic neurons neither respond to NGF nor carry NGF receptors (Konkol et al. 1978; Schwab et al. 1979). However, evidence for a trophic role of NGF in the central nervous system became apparent when it was shown that the cholinergic neurons of the basal forebrain nuclei express NGF receptors and exhibit specific retrograde transport of NGF (Schwab et al. 1979; Seiler and Schwab 1984). The cholinergic neurons have also been shown to respond to NGF with an increase in the levels of choline acetyltransferase (ChAT), the enzyme responsible for synthesis of the cholinergic neurotransmitter acetylcholine. This has been shown both in neonatal rats (Gnahn et al. 1983), in aggregate cultures of fetal forebrain (Honegger and Lenoir 1982), and in adult rats after partial fimbria lesions, i.e. of the septo-hippocampal connection (Hefti et al. 1984). The cell bodies of the cholinergic neurons are situated in the medial septal nuclei, in the diagonal band of Broca and in the basal nuclei of Meynert. Their fiber tracts project to the hippocampus, neocortex and olfactory bulb (cf. Cuello and Sofroniew 1984) and these innervated areas exhibit relatively high levels of NGF and NGF-mRNA as compared to other brain regions (Korsching et al. 1985; Whittemore et al. 1986; Shelton and Reichardt 1986). In analogy to the situation in the peripheral nervous system, there is a close correlation between the levels of NGF and its mRNA and the density of cholinergic innervation. Even within the rat hippocampus there are differences between NGF levels of various regions, e.g. the CA3, CA4 and dentate gyrus region have higher levels of NGF protein compared with the CA1-CA2 region and this is reflected in the density of cholinergic innervation (cf. Korsching 1986; Thoenen et al. 1987). Recent autoradiographic receptor binding studies on sections from adult rat brain have clearly demonstrated the presence of NGF receptors on these cholinergic neurons (Riopelle et al. 1987; Raivich and Kreutzberg 1987). Moreover, NGF levels and ChAT activity in rat hippocampus

4 Regulation of NGF Synthesis and Implications for Treatment of Alzheimer's Disease

In Alzheimer's disease the cholinergic neurons of the basal forebrain nuclei are consistently affected and there is an atrophy and/or loss of these neurons during the course of this disease (Coyle et al. 1983; Pearson et al. 1983). Moreover, the impaired synthesis of acetylcholine is one of the earliest detectable signs of Alzheimer's disease and correlates well with the cognitive deficits observed (Francis et al. 1985). In view of the fact that these cholinergic neurons are also the target of NGF in the central nervous system, the hypothesis was put forward that NGF could play a role in this disease (Hefti and Weiner 1986). The idea that Alzheimer's disease results from an insufficient supply of NGF for these neurons has recently been disputed by Goedert and co-workers (1986). These authors demonstrated that the levels of NGF-mRNA in the cerebral cortex of patients with Alzheimer's disease are not different from the corresponding levels found in age-matched controls. Nevertheless, the following observations support the notion that pharmacological amounts of NGF are of potential value in treatment of Alzheimer's disease by supporting the cholinergic neurons undergoing atrophy or degeneration. Thus, it has been demonstrated that chronic administration of NGF over a period of 4 weeks can improve memory function, as shown by behavioural tests, and can counteract cholinergic cell body atrophy observed in a subpopulation of aged rats (Fischer et al. 1987). Moreover, studies on septo-hippocampal lesions clearly show that exogenous NGF has a beneficial effect in preventing cholinergic cell degeneration in the septal area after a transection of the fimbria (Hefti 1986; Kromer 1987; Williams et al. 1986). Likewise, NGF injections (Will and Hefti 1985) or transplantation of fetal cholinergic neurons into the hippocampus (Dunnett et al. 1982; Gage and Björklund 1986) can improve both learning and memory deficits observed in rats after such a lesion.

In view of these findings it seems justified to consider NGF supplementation as a possible therapeutic approach for the treatment of patients suffering from Alzheimer's disease. This could be achieved by either local administration of recombinant human NGF, of peptides with NGF-like functions or by enhancing endogenous NGF synthesis by pharmacological agents. However, a prerequisite for such an intervention is a better understanding of the physiological mechanisms by which NGF is regulated in vivo. Recent studies in the peripheral nervous system have shown that specific polypeptide factors are involved in the control of NGF synthesis in the nonneuronal cells. Using the rat sciatic nerve as a model system under conditions of degeneration and regeneration after a nerve lesion, we were able to show that macrophages play an important role in NGF regulation (Heumann et al. 1987; Thoenen et al. 1988). Interleukin-1, a lymphokine produced by these cells, was found to be the responsible agent increasing NGF-mRNA levels in the nonneuronal cells (Lindholm et al. 1987). Recent studies indicate that the action of IL-1 involves activation of phospholipase A2 and that IL-1 has a dual effect on NGF-mRNA. Thus, IL-1 enhances NGF gene transcription and also increases the stability of NGF-mRNA in cultured sciatic nerve fibroblasts (Lindholm et al. 1988).

The studies on the regulation of NGF synthesis in the central nervous system are less advanced compared with the peripheral nervous system. We have recently obtained evidence that IL-1 and other mesenchymal growth factors can influence NGF-mRNA levels in cultured rat astrocytes. We are currently studying the complex interplay of various growth factors in controlling NGF-mRNA synthesis in the central nervous system both in vivo and in vitro. Moreover, since it has very recently been reported that hippocampal neurons can also express NGF-mRNA (Rennert and Heinrich 1986; Ayer-LeLievre et al. 1988) this adds a new dimension to studies concerned with the regulation of NGF synthesis in the central nervous system.

References

Auburger G, Heumann R, Hellweg R, Korsching S, Thoenen H (1987) Developmental changes of nerve growth factor (NGF) and its mRNA in the rat hippocampus: comparison with choline acetyltransferase. Dev Biol 120:322–328

Ayer-LeLievre C, Olson L, Ebendal T, Seiger A, Persson H (1988) Expression of the β-nerve growth factor gene in hippocampal neurons. Science 240:1339–1341

Bar-Sagi D, Feramisco JR (1985) Microinjection of the ras oncogene protein into PC12 cells induces morphological differentiation. Cell 42:841–848

Chandler CE, Parson LM, Hosang M, Shooter EM (1984) A monoclonal antibody modulates the interaction of nerve growth factor with PC12 cells. J Biol Chem 259:6882–6889

Coyle JT, Price DL, de Long MR (1983) Alzheimer's disease: a disease of cortical cholinergic innervation. Science 219:1184–1189

Cuello AC, Sofroniew MV (1984) The anatomy of the CNS cholinergic neurons. Trends Neurosci 7:74–78

Dunnett SB, Low WC, Iversen SD, Stenevi U, Björklund A (1982) Septal transplants restore maze learning in rats with fornix-fimbria lesions. Brain Res 251:335–348

Fischer W, Wictorin K, Björklund A, Williams LR, Varon S, Gage FH (1987) Amelioration of cholinergic neuron atrophy and spatial memory impairment in aged rats by nerve growth factor. Nature (London) 329:65–68

Francis PT, Palmer AM, Sioms NR, Bowen DM, Davison AN, Esiri MM, Neary D, Smowden JS, Wilcock GK (1985) Neurochemical studies of early-onset Alzheimer's disease. Possible influence of treatment. Lancet 4:7–11

Gage FH, Björklund A (1986) Cholinergic septal grafts into hippocampal formation improve spatial learning and memory in aged rats by an atropine-sensitive mechanism. J Neurosci 6:2837–2847

Gnahn H, Hefti F, Heumann R, Schwab M, Thoenen H (1983) NGF-mediated increase of choline acetyltransferase (ChAT) in the neonatal forebrain: evidence for a physiological role of NGF in the brain? Dev Brain Res 9:45–52

Godfrey EW, Shooter EM (1986) Nerve growth factor receptors on chick embryo sympathetic cells: binding characteristics and development. J Neurosci 6:2543–2550

Goedert M, Fine A, Hunt SP, Ullrich A (1986) Nerve growth factor mRNA in peripheral and central rat tissues and in the human central nervous system: lesion effects in the rat brain and levels in Alzheimer's disease. Mol Brain Res 1:85–92

Green SH, Rydel RE, Connolly JL, Greene LA (1986) PC12 mutants that possess low- but not high-affinity nerve growth factor receptors neither respond to nor internalize nerve growth factor. J Cell Biol 102:830–843

Greenberg ME, Greene LA, Ziff EB (1985) Nerve growth factor and epidermal growth factor induce rapid transient changes in proto-oncogene transcription in PC12 cells. J Biol Chem 260: 14101–14110

Greene LA, Shooter EM (1980) The nerve growth factor: biochemistry, synthesis and mechanism of action. Annu Rev Neurosci 3:353–402

Hagag N, Halegoua S, Viola M (1986) Inhibition of growth factor-induced differentiation of PC12 cells by microinjection of antibody to ras p21. Nature (London) 319:680–682
Hefti F (1986) Nerve growth factor (NGF) promotes survival of septal cholinergic neurons after fimbrial transections. J Neurosci 6:2155–2162
Hefti F, Weiner WJ (1986) Nerve growth factor and Alzheimer's disease. Ann Neurol 20:275–281
Hefti F, Dravid A, Hartikka J (1984) Chronic intraventricular injections of nerve growth factor elevate hippocampal choline acetyltransferase activity in adult rats with septo-hippocampal lesions. Brain Res 293:305–311
Heumann R, Schwab M, Thoenen H (1981) A second messenger required for nerve growth factor biological activity? Nature (London) 292:838–840
Heumann R, Korsching S, Scott J, Thoenen H (1984) Relationship between levels of nerve growth factor (NGF) and its messenger RNA in sympathetic ganglia and peripheral target tissues. EMBO J 3:3183–3189
Heumann R, Lindholm D, Bandtlow C, Meyer M, Radeke MJ, Misko TP, Shooter E, Thoenen H (1987) Differential regulation of mRNA encoding nerve growth factor and its receptor in rat sciatic nerve during development, degeneration and regeneration: role of macrophages. Proc Natl Acad Sci USA 84:8735–8739
Honegger P, Lenoir D (1982) Nerve growth factor (NGF) stimulation of cholinergic telencephalic neurons in aggregating cell cultures. Dev Brain Res 3:229–239
Hosang M, Shooter EM (1985) Molecular characteristics of nerve growth factor receptors on PC12 cells. J Biol Chem 260:655–662
Johnson D, Larahan A, Buck CR, Sehgal A, Morgan C, Mercer E, Bothwell M, Chao M (1986) Expression and structure of the human NGF receptor. Cell 47:545–554
Johnson EM Jr, Rich KM, Yip HK (1986) The role of NGF in sensory neurons in vivo. Trends Neurosci 1:33–37
Konkol RJ, Mailman RB, Bendeich EG, Garrison AM, Mueller RA, Breese GR (1978) Evaluation of the effects of nerve growth factor and anti-nerve growth factor on the development of central catecholaminergic neurons. Brain Res 144:277–285
Korsching S (1986) Nerve growth factor in the central nervous system. Trends Neurosci 11/12:570–573
Korsching S, Thoenen H (1983) Nerve growth factor in sympathetic ganglia and corresponding target organs of the rat: correlation with density of sympathetic innervation. Proc Natl Acad Sci USA 80:3513–3516
Korsching S, Auburger G, Heumann R, Scott J, Thoenen H (1985) Levels of nerve growth factor and its mRNA in the central nervous system of the rat correlate with cholinergic innervation. EMBO J 4:1389–1393
Kromer LF (1987) Nerve growth factor treatment after brain injury prevents neuronal death. Science 235:214–216
Levi-Montalcini R (1987) The nerve growth factor: thirty-five years later. EMBO J 6:1145–1154
Levi-Montalcini R, Angeletti PU (1968) Nerve growth factor. Physiol Rev 48:534–569
Lindholm D, Heumann R, Meyer M, Thoenen H (1987) Interleukin-1 regulates synthesis of nerve growth factor in non-neuronal cells of rat sciatic nerve. Nature (London) 330:658–659
Lindholm D, Heumann R, Hengerer B, Thoenen H (1988) J Biol Chem 263:16348–16351
Milbrandt J (1987) A nerve growth factor-induced gene encodes a possible transcriptional regulatory factor. Nature (London) 238:797–799
Otten U (1984) Nerve growth factor and the peptidergic sensory neurons. Trends Pharmacol 7:307–310
Pearson RCA, Sofroniew MV, Cuello AC, Powell TPS, Eckenstein F, Esiri MM, Wilcock GK (1983) Persistence of cholinergic neurons in the basal, nuclei in a brain with senile dementia of the Alzheimer's type demonstrated by immunohistochemical staining for choline acetyltransferase. Brain Res 289:375–379
Radeke MJ, Misko TP, Hsu C, Herzenberg LA, Shooter EM (1987) Gene transfer and molecular cloning of the rat nerve growth factor receptor: a new class of receptors. Nature (London) 325:593–597
Raivich G, Kreutzberg GW (1987) The localization and distribution of high affinity ßNGF binding sites in the central nervous system of the adult rat. A light microscopic autoradiographic study using (125I) ßNGF. Neuroscience 20:23–36
Rennert PD, Heinrich G (1986) Nerve growth factor mRNA inbrain: localization by in situ hybridization. Biochem Biophys Res Commun 138:813–818

Riopelle RJ, Verge VMK, Richardson PM (1987) Properties of receptors for nerve growth factor in the mature rat nervous system. Mol Brain Res 3:45–53

Ross AH, Grob P, Bothwell M, Elder DE, Ernst CS, Marano N, Ghrist BFD, Slemp CC, Herlyn M, Atkinson B, Koprowski H (1984) Characterization of nerve growth factor receptor in neural crest tumours using monoclonal antibodies. Proc Natl Acad Sci USA 81:6681–6685

Schwab M, Otten U, Agid Y, Thoenen H (1979) Nerve growth factor (NGF) in the rat CNS: absence of specific retrograde axonal transport and tyrosine hydroxylase induction in locus coeruleus and substantia nigra. Brain Res 168:473–483

Schwab ME, Heumann R, Thoenen H (1982) Communication between target organs and nerve cells: retrograde axonal transport and site of action of nerve growth factor. Cold Spring Harbor Symp Quant Biol 46:125–134

Seiler M, Schwab ME (1984) Specific retrograde transport of nerve growth factor (NGF) from neocortex to nucleus basalis in the rat. Brain Res 300:33–39

Shelton DL, Reichardt LF (1984) Expression of the nerve growth factor gene correlates with the density of sympathetic innervation in effector organs. Proc Natl Acad Sci USA 81:7951–7955

Shelton DL, Reichardt LF (1986) Studies on the expression of the β nerve growth factor (NGF) gene in the central nervous system: level and regional distribution of NGF mRNA suggest that NGF functions as a trophic factor for several distinct populations of neurons. Proc Natl Acad Sci USA 83:2714–2718

Stacey DW, Watson T, Kung H-F, Curran T (1987) Microinjection of transforming ras protein induces c-fos expression. Mol Cell Biol 7:523–527

Sutter A, Riopelle RJ, Harris-Warrick RM, Shooter EM (1979) Nerve growth factor receptors: characterization of two distinct classes of binding sites on chick embryo sensory cells. J Biol Chem 254:5972–5982

Thoenen H, Barde Y-A (1980) Physiology of nerve growth factor. Physiol Rev 60:1284–1335

Thoenen H, Bandtlow C, Heumann R (1987) The physiological function of nerve growth factor in the central nervous system: comparison with the periphery. Rev Physiol Biochem Pharmacol 109:145–178

Thoenen H, Bandtlow C, Heumann R, Lindholm D, Meyer M, Rohrer H (1988) Nerve growth factor: cellular localization and regulation of synthesis. Cell Mol Neurobiol 8:35–40

Whittemore SR, Ebendal T, Lärkfors L, Olson L, Seiger A, Strömberg I, Persson H (1986) Developmental and regional expression of β nerve growth factor messenger RNA and protein in the rat central nervous system. Proc Natl Acad Sci USA 83:817–821

Will B, Hefti F (1985) Behavioural and neurochemical effects of chronic intraventricular injections of nerve growth factor in adult rats with fimbria lesions. Behav Brain Res 17:17–24

Williams LR, Varon S, Peterson GM, Wictorin K, Fischer W, Björklund A, Gage FH (1986) Continuous infusion of nerve growth factor prevents basal forebrain neuronal death after fimbria fornix transection. Proc Natl Acad Sci USA 83:9231–9235

Signal Transduction by Growth Factor Receptors

W.H. MOOLENAAR

1 Introduction

The proliferation of cells in vivo and in culture is tightly regulated by polypeptide growth factors. Like all polypeptide hormones, growth factors initiate their action by binding to specific, high affinity receptor molecules on the cell surface. Interest in growth factors and their receptors has been dramatically intensified by the discovery that at least some of these molecules show a striking structural homology with certain viral oncogene products. For example, the *sis* oncogene encodes a platelet-derived growth factor (PDGF) molecule, while the *erb-B* oncogene product is a truncated form of the receptor for epidermal growth factor (EGF) (Waterfield et al. 1983; Schlessinger 1986). Inappropriate expression of the cellular counterparts of viral oncogenes is thought to be responsible for the initiation and maintenance of malignant growth. Therefore, the study of the mode of action of growth factors will undoubtedly yield important insight into the mechanisms underlying carcinogenesis.

2 Signal Pathways in Growth Factor Action

Understanding the mode of action of growth factors requires the identification of intracellular signals that are essential for the stimulation of DNA synthesis and cell division. Following growth factor binding, the activated receptor mediates a cascade of rapid biochemical and physiological changes in the cell, which ultimately (after 10–20 h) lead to enhanced DNA synthesis. As a rule, the growth factor has to be present throughout the entire "pre-replicative" G0/G1 phase (usually 8–10 h) for commitment to DNA synthesis to occur. It appears that there are at least three potential signal pathways for mediating growth factor action: (1) activation of a protein tyrosine kinase activity that is intrinsic to the receptors for EGF and PDGF, (2) the phospholipase C-mediated breakdown of inositol phospholipids generating various signal molecules, and (3) increases in cytoplasmic free Ca^{2+} and pH (pH_i).

There is increasing evidence that in addition to the "classical" mitogens, such as EGF and PDGF, certain Ca^{2+}-mobilizing neurotransmitters can also function as growth factors. For example, the peptides bombesin, substance P and vasopressin are mitogenic for quiescent fibroblasts (Zachary et al. 1987), while our own

The Netherlands Cancer Institute, Plesmanlaan 121, 1066 CX Amsterdam, The Netherlands

A. Azzi, Z. Drahota and S. Papa
Molecular Basis of Membrane-Associated Diseases

results indicate that bradykinin and histamine, acting through H-1 receptors, stimulate the growth of certain human carcinoma cells (Tilly et al. 1987, and unpublished results).

3 Inositol Lipid Hydrolysis

There is now a general agreement that some, but not all, growth factor receptors evoke the phospholipase C-mediated breakdown of inositol phospholipids, resulting in the formation of diacylglycerol which binds to and directly activates protein kinase C (Nishizuka 1984) and inositol-1,4,5-trisphosphate [IP_3(1,4,5)] which triggers the release of Ca^{2+} from internal stores (Berridge and Irvine 1984), as well as various other inositol phosphates (Fig. 1).

We have analyzed and compared the actions of EGF and bradykinin or histamine on both inositol phosphate formation and calcium signalling in responsive human carcinoma cells (Tilly et al. 1987). It appears that EGF is a rather weak inducer of phospholipase C activity. In A431 cells, which overexpress the EGF receptor, EGF evokes a rather small and variable release of intracellularly stored Ca^{2+}, while accumulation of IP_3(1,4,5) is barely detectable. In contrast, EGF-induced increases in other inositol phosphates such as IP_1 and IP_2 are readily detectable.

Bradykinin and histamine, on the other hand, are among the most potent activators of phospholipase C, resulting in large Ca^{2+} signals and prominent increases in inositol phosphate levels.

Bradykinin is mitogenic for skin fibroblasts (see review by Zachary et al. 1987) while our own results (unpublished) indicate that bradykinin can also function as a mitogen for human A431 carcinoma cells. Bradykinin stimulation of A431 cells rapidly activates phospholipase C, resulting in the formation of inositol phosphates and a transient rise in $[Ca^{2+}]i$ (Tilly et al. 1987; Hepler et al. 1987). Almost immediately after hormone addition the levels of IP_3(1,4,5) start to increase, reaching a peak value of -10 times the basal level within 10-15 s which coincides with the $[Ca^{2+}]i$ peak. The bradykinin-induced formation of IP_3(1,4,5) is a very transient phenomenon, and the results support the view that the (1,4,5) isomer is

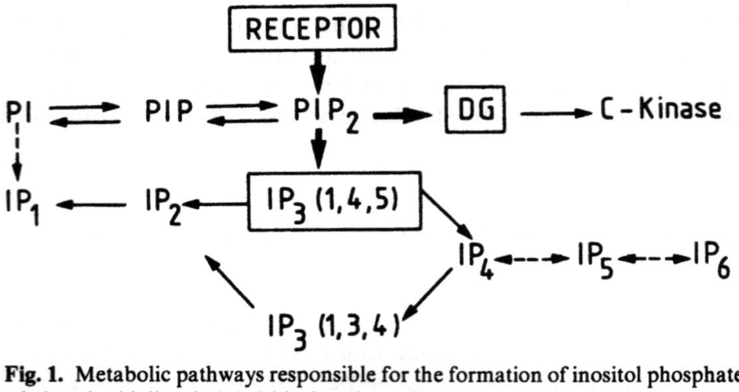

Fig. 1. Metabolic pathways responsible for the formation of inositol phosphates from the breakdown of phosphatidylinositol-4, 5-bisphosphate (PIP_2). See text for details

rapidly metabolized either to IP_2 through phosphatase action to $IP_3(1,3,4)$ via phosphorylation/dephosphorylation with IP_4 as intermediate. The accumulation of inositol phosphates in response to bradykinin levels off in less than 2 min. This apparent desensitization of the response is not observed with EGF as a stimulus (Tilly et al. manuscripts submitted). In general, the bradykinin-dependent pattern of stimulation is significantly different from that observed in EGF-stimulated A431 cells, suggesting that there are separate mechanisms of inositol lipid hydrolysis involved.

A hitherto unrecognized growth factor is histamine, which stimulates the growth of human HeLa cells by activating the Ca^{2+}-mobilizing H1 receptor (Tilly et al. manuscript in preparation). Histamine is formed by decarboxylation of the amino acid histidine and is an active mitogenic amine involved in a number of biological processes including neurotransmission, inflammation and allergic reactions. Recently, a number of studies have suggested a possible role of histamine in tumor development and cell proliferation (for review, see Bartholeyns and Fozard 1985). In a recent series of experiments, we have examined the action of the H1 receptor on inositol phosphate metabolism and growth stimulation in human HeLa cells. Histamine induces a dose-dependent accumulation of inositol phosphates that parallels its mitogenic activity. Both the formation of inositol phosphates and the stimulation of cell proliferation by histamine are blocked by the H1 receptor antagonist pyrilamine (Fig. 2). Histamine-induced inositol phosphate formation in Hela cells lasts for at least several hours. This would imply that the H1 receptor also mediates a prolonged production of diacylglycerol and concomitant activation of protein kinase C. Since a number of other mitogens, including thrombin and bombesin, similarly evoke phosphoinositide turnover, it seems that any agonist capable of stimulating phosphoinositide hydrolysis may ultimately function as a growth factor.

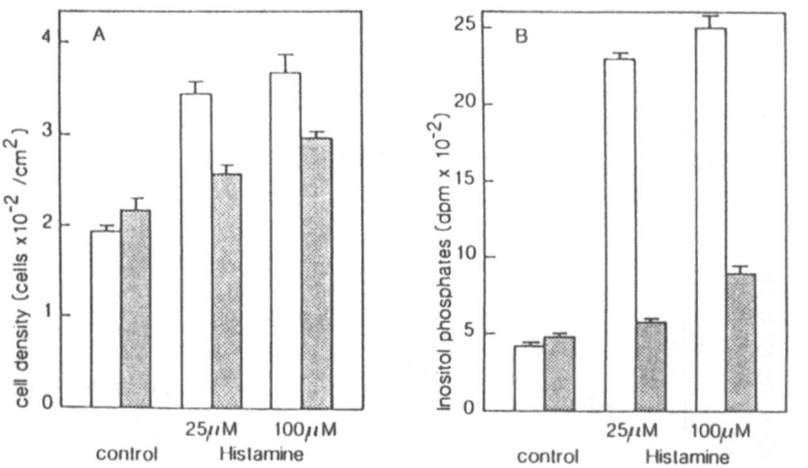

Fig. 2. Histamine-induced cell proliferation (**A**) and inositol phosphate formation (**B**) in HeLa cells in the presence (*closed bars*) or absence (*open bars*) of 250 nM pyrilamine. Inositol phosphates were determined as in Tilly et al. (1987). In **A**, histamine and pyrilamine were present for 48 h prior to cell counting

increase in parallel during postnatal development, further supporting a role of NGF in this septo-hippocampal system (Auburger et al. 1987). Likewise, fimbria lesion experiments have shown that exogenous NGF administrated through an infusion pump can rescue cholinergic neurons of the basal forebrain nuclei which would otherwise degenerate following the transection (Hefti 1986; Kromer 1987; Williams et al. 1986).

3 Mechanism(s) of Action of NGF

Although many actions of NGF on target neurons have been described, the overall mechanism by which NGF exerts its effects is still unknown. This also holds true for the most important long-term effect of NGF, namely its survival-promoting activity on NGF-responsive neurons. NGF receptors on sensory (Sutter et al. 1979) and sympathetic neurons (Godfrey and Shooter 1986) and on PC12 pheochromocytoma cells (Hosang and Shooter 1985) have been identified and antibodies to the NGF receptor have been produced (Chandler et al. 1984; Ross et al. 1984). Moreover, binding studies have shown the existence of both high- and low-affinity NGF receptors on target cells which are encoded by the same mRNA. However, the mechanism for conversion between the two forms is not known but the high-affinity receptor seems to be responsible for most of the biological action of NGF, at least in PC12 cells (Green et al. 1986). Recently, both human and rat NGF receptors have been cloned (Johnson D. et al. 1986; Radeke et al. 1987) and the molecular data demonstrate that the NGF receptor belongs to a novel class of membrane receptors which lack an intrinsic tyrosine kinase activity. NGF is also known not to affect neuronal proliferation which is in contrast to most growth factors which have a mitogenic activity on their target cells (cf. Thoenen et al. 1987).

Previous studies have shown that intracellular microinjection of NGF into the cytoplasm of PC12 cells, bypassing the receptor, fails to elicit differentiation which is readily seen with NGF in the culture medium (Heumann et al. 1981). Conversely, microinjected, function-blocking, anti-NGF antibodies do not inhibit NGF action on PC12 cells. These studies led to the conclusion that NGF acts by a distinct second messenger system, the nature of which remains to be elucidated (Heumann et al. 1981). The results of recent studies have suggested a role for the ras oncogene product in NGF signal transduction. Introduction of purified, activated, ras protein into PC12 cells can elicit differentiation (Bar-Sagi and Feramisco 1985), and microinjected, anti-ras antibodies are able to block the effect of NGF on these cells (Hagag et al. 1986). The exact site of action of ras or of ras-like proteins in the signal transduction cascade induced by NGF is not known. Further downstream, transcriptional activators are most probably involved in regulating NGF-specific gene expression. Interestingly, NGF (Greenberg et al. 1985) as well as ras protein (Stacey et al. 1987) are known to transiently increase expression of the c-fos proto-oncogene product which is thought to be involved in transcriptional regulation. Milbrandt (1987) has recently described another putative transcriptional activator gene whose transcription is strongly induced in PC12 cells by short-term NGF treatment.

4 Tyrosine Kinase Activation

To explore the importance of the receptor tyrosine kinase in the action of EGF, we have used transfected NIH-3T3 cells expressing either the normal human EGF receptor or a receptor mutated at Lys 721, a key residue in the presumed ATP-binding region (collaboration with J. Schlessinger). The wild-type receptor responds to EGF by causing inositol phosphate formation (Fig. 3), rises in Ca^{2+} and pH_i and DNA synthesis. In marked contrast, the kinase-deficient mutant fails to evoke any of these responses (Moolenaar et al. 1988; Fig. 3).

It thus appears that activation of the receptor tyrosine kinase is a crucial signal that initiates the multiple, postreceptor effects of EGF that lead to DNA synthesis. In other words, EGF-induced phosphorylation of specific cellular substrates on tyrosine residues is a critical step in the pleiotropic response to EGF. The inability of the kinase-deficient receptor to stimulate inositol lipid hydrolysis (Fig. 3) is particularly intriguing, since it suggests that one or more of the key regulatory proteins in the phosphoinositide signalling cascade (e.g. G-proteins, phospholipase-C) may serve as a substrate for the EGF-R kinase. Such a mechanism contrasts to the accepted mode of action of Ca^{2+}-mobilizing hormones, such as histamine and bradykinin, whose receptors are linked to G-protein intermediates, without any evidence for tyrosine phosphorylation being required for the activation process.

5 Activation of Na^+/H^+ Exchange

Most, if not all, mitogens activate the otherwise quiescent Na^+/H^+ exchanger in the plasma membrane of their target cells (Moolenaar 1986). This Na^+/H^+ exchanger, which is amiloride-sensitive, is normally involved in the close regulation of pH_i, while under appropriate conditions growth factors activate the Na^+/H^+ exchanger

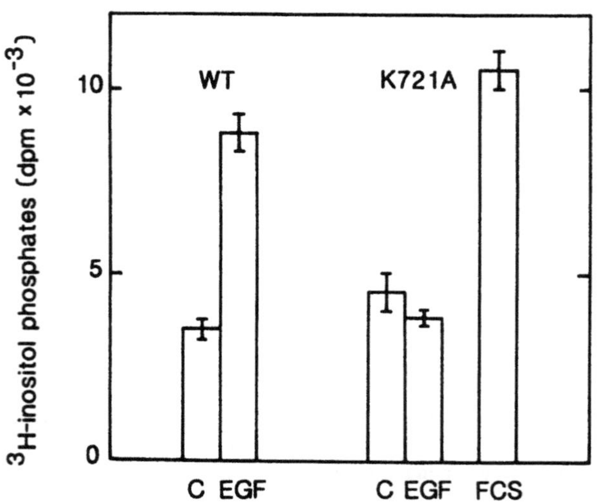

Fig. 3. Inositol phosphate formation induced by EGF interacting with the wild-type (**WT**) or kinase-defective (*K721A*) human EGF receptor expressed in 3T3 cells. FCS, fetal calf serum (7.5% v/v). For further details, see Moolenaar et al. (1988)

to raise steady-state pH_i by 0.2–0.3 units. This rise in pH_i is thought to be permissive for DNA synthesis (Chambard and Pouyssegur 1986; Moolenaar 1986).

How does receptor occupancy lead to activation of Na^+/H^+ exchange? The underlying kinetic mechanism is now fairly well understood. By comparing the pH_i dependence of Na^+/H^+ exchange in quiescent and stimulated cells, it turns out that growth factors and activators of kinase C, such as phorbol esters, act by increasing the sensitivity of the Na^+/H^+ exchanger for cytoplasmic H^+. This alkaline shift in pH_i sensitivity of 0.2–0.3 units may reflect an increase in the apparent affinity of the regulatory or "modifier" site for H^+. One could imagine that some ionizable group at the regulatory site acquires a greater pKa because its immediate environment becomes more negatively charged, for example by phosphorylation either via kinase C or through the intrinsic receptor tyrosine kinase.

The involvement of the Na^+/H^+ exchanger in such a carefully regulated process as cell growth raises the intriguing possibility that the exchanger in autonomously growing tumour cells may be subject to uncontrolled activation. Indeed, recent work by Bierman et al. (1987) on pH_i regulation in embryonal carcinoma (EC) cells shows that the Na^+/H^+ exchanger in mouse P19 EC cells is constitutively activated and fails to respond to any extracellular stimulus. These pluripotent cells have no requirement for exogenous growth factors and exhibit a highly transformed phenotype. Upon differentiation, however, the cells lose their transformed phenotype and they become dependent on the presence of external growth factors. This differentiation process, induced by either retinoic acid or DMSO, is accompanied by "deactivation" of the Na^+/H^+ exchanger and a resultant drop in steady-state pH_i of up to 0.5 units (Bierman et al. in preparation). Kinetic analysis indicates that the relatively high resting pHi of undifferentiated EC cells is attributable to an alkaline shift in the pHi sensitivity of the Na^+/H^+ exchange rate as compared to that in the differentiated cells. It thus seems as if signal pathways, normally utilized by growth factor receptors to activate the Na^+/H^+ exchanger, are constitutively operative in autonomously growing, undifferentiated EC cells. The biochemical nature of this pathway remains to be elucidated.

6 Concluding Remarks

In recent years much has been learned about the various molecular events that are induced by growth factor-receptor interaction. Although uncertainty still exists about the biological significance of many of the receptor-linked "early events", it is obvious that site-directed mutagenesis of cloned receptor and G-protein cDNA's followed by transfection into appropriate target cells will be of great help in analyzing the role of the various signal pathways in growth control.

Another challenge for future research is to examine how the signals generated by growth factor receptors are modified or overridden by the action of transforming oncogene products, some of which are known to be tyrosine protein kinases or G-like proteins.

Acknowledgements. Research related to this paper was carried out by A.J. Bierman, L.H.K. Defize and B.C. Tilly at the Hubrecht Laboratory in Utrecht, the Netherlands. Mrs. Gea de Jong-Meijerink is acknowledged for preparing the manuscript.

References

Bartholeyns J, Fozard JR (1985) Role of histamine in tumor development. Trends Pharmacol Sc 6:123–125

Berridge MJ, Irvine RF (1984) Inositol trisphosphate, a novel second messenger in cellular signal transduction. Nature (London) 312:315–321

Bierman AJ, Tertoolen LGJ, de Laat SW, Moolenaar WH (1987) The Na^+/H^+ exchanger is constitatively activated in P19 embryonal carcinoma cells. J Biol Chem 262:9621–9628

Chambard JC, Pouyssegur J (1986) Intracellular pH controls growth factor-induced ribosomal protein S6 phosphorylation and protein synthesis in the G0----G1 transition of fibroblasts. Exp Cell Res 164:282–294

Hepler JR, Nakahata N, DiGuiseppi J, Herman B, Earp JS, Harden TK (1987) Inositol phosphate production by EGF in A431 cells. J Biol Chem 262:2951–2956

Moolenaar WH (1986) Effects of growth factors on intracellular pH regulation. Annu Rev Physiol 48:363–376

Moolenaar WH, Bierman AJ, Tilly BC, Verlaan I, Honegger AM, Ullrich A, Schlessinger J (1988) A point mutation at the ATP-binding site of the EGF-receptor abolishes signal transduction. EMBO J 7:707–710

Nishizuka Y (1984) The role of protein kinase C in cell surface signal transduction and tumour promotion. Nature (London) 304:645–648

Schlessinger J (1986) Allosteric regulation of the EGF receptor kinase. J Cell Biol 103:2067–2072

Tilly BC, Van Paridon PA, Verlaan I, Wirtz KWA, de Laat SW, Moolenaar WH (1987) Inositol phosphate formation in bradykinin-stimulated A431 cells. Biochem J 244:129–135

Waterfield MD, Scrace GT, Whittle N, Johnson A, Wasteson A, Westermark B, Heldin CH, Huang JS, Deuel TF (1983) PDGF is structurally related to the putative transforming protein of simian sarcoma virus. Nature (London) 304:35–39

Zachary I, Woll PJ, Rozengurt E (1987) A role of neuropeptides in the control of cell proliferation. Dev Biol 124:295–308

Altered Regulation of Receptor-Associated Tyrosine Kinases in Human Tumors

P.M. Comoglio[1], M.F. Di Renzo, G. Gaudino, S. Giordano, L. Naldini and C. Ponzetto

1 Growth Factors, Receptors and Oncogenes

The discovery of a link between known oncogenes and growth factors, their receptors and their signalling mechanisms has renewed hopes that the basic features of cancerous growth can be understood. Such an understanding could provide novel means of diagnosing and treating cancer. It is now established that several of the so-called activated oncogenes are altered versions of genes encoding proteins involved in the intercellular and intracellular signalling pathways (for reviews, see Bishop 1983; Hunter 1984; James and Bradshow 1984; Ben-Neriah and Baltimore 1986).

An activated oncogene is, for example, a deregulated gene that encodes a growth-stimulating factor: the cell constitutively produces an excess of the growth factor, and if provided with the appropriate receptor, autostimulates its own growth, proliferating in an uncontrolled manner (Heldin and Westermark 1984; Cuttitta et al. 1985a). The phenomenon has been termed "autocriny". Another example of activation is the structural alteration of a gene that encodes a growth factor receptor. Under normal conditions the receptor transduces to the cell interior the signal that triggers proliferation, when exclusively stimulated by the specific growth factor. Conversely, the activated oncogene encodes an altered receptor, which is continuously excited, independently of the presence or absence of the growth signal. In this case as well, the cells proliferate in an uncontrolled manner. To date, this type of activation has been proved for the epidermal growth factor receptor (Downward et al. 1984) and for the receptor of the hemopoietic growth factor CSF-1 (Sherr et al. 1985). Oncogenes also encode proteins involved in the cascade of intracellular events that follow the growth factor-receptor binding and ensuing excitement. This cascade of events is responsible for the transduction of the extracellular signal to the cytoplasm and nucleus and triggers the metabolic reactions that lead to cell division.

On the basis of the functions attributed to the proteins they encode, oncogenes have been divided into classes. The oncogenes of class 2 are those that encode growth factors; those of class 1 encode plasma membrane proteins that function as receptors for growth factors (Hunter 1984). In addition, three other classes of oncogenes have been defined, including genes coding for, respectively, proteins

University of Torino, Medical School, Torino 10126, Italy
[1]To whom correspondence should be addressed: Dept. Biomedical Sciences, C.so M. D'Azeglio 52, 10126 Torino, Italy

A. Azzi, Z. Drahota and S. Papa
Molecular Basis of Membrane-Associated Diseases

located in the plasma membrane, in the cytoplasm, or in the nucleus; the biochemical function of these molecules, which are likely to be involved in the chain of intracellular events mediating signal transduction and amplification, is still largely unknown (Bishop 1983; Hunter 1984).

2 The Oncogene-Coded Tyrosine Kinases

About one-third of the known oncogenes encode proteins that phosphorylate tyrosine residues on other proteins. These enzymes are called tyrosine-specific protein kinases. The reaction they catalyze is the transfer of the terminal phosphoryl group of ATP to the hydroxyl group of a tyrosine residue in a protein. Until the discovery that retrovirus oncogene products could phosphorylate tyrosine residues, the known sites of protein phosphorylation were limited to serine and threonine.

Tyrosine phosphorylation is an important property of the molecules involved in the control of cell proliferation. In fact, after the discovery of the enzymatic activity of the proteins coded by viral oncogenes, it was demonstrated that some growth factor receptors are themselves tyrosine kinases. So far, these receptors are among the very few proteins in animal cells whose tyrosine kinase activity has been conclusively ascertained. These proteins, like those coded by the class 1 oncogenes (Hunter and Cooper 1985), are located on the plasma membrane. It is now known that class 1 oncogenes include the activated version of genes that encode growth factor receptors (Downward et al. 1984; Sherr et al. 1985).

Under normal conditions, the receptors perform their enzyme functions exclusively when they are excited by binding of their appropriate growth factors. Tyrosine phosphorylation of intracellular targets is in fact part of the mechanism of signal transduction that induces the cell to proliferate (Bertics and Gill 1985; Schlessinger 1986). This is an efficient system for amplifying the extracellular signal and for initiating a cascade of intracellular metabolic events. The receptors have internal mechanisms that insure that the signal is transient (Schlessinger 1986). In this manner cell proliferation occurs only in the presence of the appropriate growth factor and is an event limited in time.

The kinases coded by the activated oncogenes, however, phosphorylate their targets in a constitutive manner, namely continuously and independently of the presence or absence of the growth factor (Kris et al. 1985; Di Renzo et al. 1986). As in the case of autocrine stimulation, the mechanisms that control proliferation are short-circuited. This is nevertheless not due to the continuous signal evoked by the growth factor, but to the presence in the membrane of a "broken" receptor, blocked in its active form. Thus, the metabolic reactions that are brought about by phosphorylation are continuously sustained and the cell proliferates indefinitely.

Some activated class 1 oncogenes have been isolated, cloned and compared to the homologous genes encoding normal receptors. It has been established through sequence analysis that the former are generally "truncated" versions of the latter. In general, the deletions pertain to segments coding parts of the receptor that have inhibitory functions on tyrosine kinase activity (for review, see Hunter 1987).

3 Proteins Phosphorylated on Tyrosine as Neoplastic Cell Markers

Neither free phosphotyrosine nor tyrosine phosphorylated proteins are easily detectable in normal cells under physiological conditions. In the absence of stimulation, the tyrosine kinase activity associated with the cell growth factor receptors is repressed. The activity of the receptor is expressed only after binding to the specific factor. Under this condition there is the simultaneous phosphorylation of the receptor itself (autokinase activity) and of its intracellular targets (Ek and Heldin 1984; Giordano et al. 1988a). Since the receptor is autoregulated, this activity is only transiently maintained. The average duration of the phosphorylated status of the tyrosine residue is kept short by the presence of specific tyrosine phosphatases (Nelson and Branton 1984); the receptor and its targets are rapidly dephosphorylated. The result is that, in a normal cell, only a minimal amount of tyrosine phosphorylated proteins are detectable, during the brief period of time that follows stimulation by a growth factor. Since the growth factor does not react with all the cells simultaneously, the individual phosphorylation reactions are staggered and it is practically impossible to measure the proteins containing phosphorylated tyrosine residues in normal cells, except under controlled experimental conditions of "in vitro" culture (i.e. after cell synchronization; Sturani et al. 1986a; Zippel et al. 1986).

In cells transformed by class 1 or class 2 oncogenes, however, due either to the continuous presence of growth factors or to the loss of regulation of the receptors, there is a net, detectable, accumulation of molecules phosphorylated at tyrosine. In our laboratory we have been able to demonstrate that, when adequate technology is available, it is possible to take advantage of this accumulation to identify neoplastic cells (Giordano et al. 1987).

4 Antiphosphotyrosine Antibodies as Probe for Growth Factor Receptors and Class 1 Oncoproteins

Polyclonal and monoclonal antibodies were obtained, able to react with tyrosine phosphorylated proteins, but which do not recognize those phosphorylated on serine or threonine. Considering that the latter constitute more that 99.9% of cellular phosphoproteins, the antiPtyr antibodies represent an extremely sensitive and selective probe for identifying tyrosine phosphorylated growth factor receptors, the products of class 1 oncogenes, and their intracellular targets (Ross et al. 1981; Frackelton et al. 1983; Ek and Heldin 1984; Comoglio et al. 1984; Ohtsuka et al. 1984; Wang 1985; Di Renzo et al. 1986).

When conjugated to a fluorescent tracer or to a chromogen, phosphotyrosine antibodies are able to discriminate, with diagnostic accuracy, between neoplastic cells and their normal counterparts, when the malignant transformation is associated with the expression of a class 1 or 2 oncogene. In immunoblot assays such probes can be used to identify receptors for growth factors or products of oncogenes endowed with tyrosine kinase activity. Phosphotyrosine antibodies allowed us to study the receptor autophosphorylation/dephosphorylation process triggered by PDGF (Sturani et al. 1986a; Zippel et al. 1986), and the modulation exerted by protein kinase-C on the receptor for PDGF and EGF (Sturani et al. 1986b). With this technique it was also possible not only to measure the expression of the class 1 oncogene carried by oncogenic RNA viruses, but to detect intracellular targets of their tyrosine kinase activity (Comoglio et al. 1984; Di Renzo et al. 1986; Naldini et al. 1986). Among the latter are plasma membrane and cytoskeletal proteins crucially involved in the processes regulating cell-substratum adhesion (Cirillo et al. 1984; Marchisio et al. 1984; Tarone et al. 1985).

5 Activation of the Protein Tyrosine Kinase Coded by c-abl in Hemopoietic Malignancies

The deregulated expression of cellular oncogenes has been implicated in the pathogenesis of some human cancers associated with chromosomal abnormalities. Perhaps the best example of such a malignancy is chronic myelogenous leukemia (CML), which is characterized by the presence of the Philadelphia chromosome (Ph′) in the leukemic cells of 95% of all patients (Sandberg and Hossfeld 1970).

The Philadelphia chromosome is a shortened chromosome 22, resulting from a reciprocal translocation between chromosome 9 (q34) and chromosome 22 (q11). At the molecular level the formation of the Philadelphia chromosome determines the translocation and the transcriptional activation of the oncogene c-abl (de Klein et al. 1982; Heisterkamp et al. 1983a). This is the cellular homologue of the transforming gene of the Abelson murine leukemia virus (AMuLV) (Goff et al. 1980). It encodes a tyrosine kinase which is believed to be part of an as yet unidentified growth factor receptor complex (Ben-Neriah and Baltimore 1986).

Molecular cloning of the sequences spanning the 9;22 junction, together with Southern analysis of the DNA of numerous CML patients, have shown that the breakpoint on chromosome 9 falls within the far 5′ end of the c-abl gene (Collins and Groudine 1983; Mes-Masson et al. 1986; Bernards et al. 1987). The 5′-truncated c-abl gene is fused to chromosome 22 within a 5.8 kb region of a transcriptionally active locus (breakpoint cluster region: bcr, Heisterkamp et al. 1983b, 1985; Hariharan and Adams 1987), coding for a protein of unknown function (Bernards et al. 1987).

The newly created bcr-abl gene, which contains bcr sequences at the 5′ end and abl sequences at the 3′ end, is expressed in Ph′ positive (Ph +) cells into a novel chimeric bcr-abl mRNA (Morris et al. 1986; Collins et al. 1984, 1987; Leibowitz et al. 1985a; Shtivelman et al. 1985; Grosveld et al. 1986), which directs the synthesis of a bcr-abl fusion protein of 210 kD (P210) (Ben-Neriah et al. 1986a).

P210 is characterized by constitutively elevated tyrosine kinase activity, comparable to that of the transforming protein of the Abelson MuLV and of other acutely transforming retroviruses (Konopka et al. 1984). This strongly suggests an involvement of this hybrid bcr-abl gene product in the malignant transformation of myeloid stem cells in CML.

An understanding of the precise role played by the activated c-abl gene in the pathogenesis of CML is complicated by the biphasic clinical course of the disease, which indicates that the fully malignant phenotype is the result of a multistep process.

CML begins with an early chronic phase of approximately 1 to 4 years, characterized by the abnormal accumulation of Ph+ granulocytes, which is invariably followed by an acute phase, or blast crisis, in which the disease resembles acute leukemia. At this stage the Ph+ cells are no longer capable of differentiation. Instead, there is a clonal malignant expansion of Ph+ immature myeloid or B lymphoid cells (Champlain and Golde 1985).

Molecular studies of CML patients in chronic and acute phase have shown that the bcr-abl restriction enzyme patterns of the DNA from their leukemic cells vary between patients, but are identical within one individual in both phases of the disease, whether the blast crisis is of myeloid or lymphoid type (Zaleberg et al. 1986).

These results argue in favor of the clonal origin of the disease, and suggest that the Ph′ translocation occurs in an early stem cell, still capable of myeloid or lymphoid differentiation. They also indicate that the second event required for progression into blast crisis does not appear to involve any additional molecular alteration of the bcr-abl locus.

Accordingly, the same size bcr-abl mRNA is detected in blast crisis and chronic phase cells. Quantitatively, the amount of the bcr-abl mRNA present in the two phases of the disease varies considerably between CML patients, and the level of expression does not correlate with the stage of the disease (Gale and Canaani 1984).

In the majority of patients additional chromosomal abnormalities are detected during blast crisis (Champlain and Golde 1985). This probably leads to the activation of other genes which, superimposing their effects upon the Ph′ genetic defect, may promote the transition into blast crisis and determine whether the disease will evolve into a myeloid or lymphoid acute phase.

The concept that the recombination of bcr and abl sequences must play a critical role in an earlier stage of the disease is further strengthened by the observation that the bcr-abl rearrangement has also been detected in CML cases which were cytogenetically Ph negative (Ph–) (Bartram et al. 1986; Morris et al. 1986). Here, the c-abl-bcr fusion on the apparently normal chromosome 22 resulted from an interstitial insertion of a small sequence from the tip of chromosome 9 into bcr (Bartram et al. 1986; Morris et al. 1986).

Finally, activation of the c-abl gene by recombination with sequences of chromosome 22 located upstream the bcr region seems to be involved in the pathogenesis of at least some cases of Ph+ acute lymphoblastic leukemia (ALL) (de Klein et al. 1986; Erikson et al. 1986).

The association of the Ph′ chromosome with ALL is not as frequent as with CML, since only 15% of the cases of ALL are ph+. Within these cases some carry the typical bcr-abl rearrangement and express P210, but a subset does not show any bcr rearrangement. In these patients the chromosome 22 breakpoint appears to be located further upstream, to the left of the breakpoint cluster region (Erikson et al. 1986). It has been recently shown that leukemic cells from these patients express a new altered c-abl protein (P190), of lower molecular weight than P210, which still displays enhanced in vitro tyrosine kinase activity, but with different specificity (Clark et al. 1987).

In conclusion, two similar but distinct molecular defects involving activation of the c-abl gene seem to play a role in determining the establishment of an acute lymphoid versus a chronic myeloid leukemia.

6 P210, the Hybrid bcr-abl Tyrosine Kinase

A structurally altered c-abl protein (P210) of higher molecular weight (210 kD versus 145 kD of the normal c-abl protein) has been identified in CML-derived cell lines and in bone marrow cells from Ph+ CML patients in the blast crisis phase of the disease (Konopka et al. 1984, 1985; Naldini et al. 1986; Shtivelman et al. 1986; Ben-Neriah 1986a,b; Maxwell et al. 1987). P210 is precipitated by antibodies prepared against c-abl as well as against bcr determinants (Ben-Neriah et al. 1986a). This confirms that the protein represents the translation product of the bcr-abl hybrid transcript. P210 expression is not restricted to myeloid cells: it has been shown in Ph + B lymphoid cell lines (Konopka et al. 1985) and in somatic cell hybrids of mouse fibroblasts and CML cells (Kozbor et al. 1986).

P210 is overexpressed, relative to c-abl, in leukemic cells in the acute phase, while it is difficult to detect in Ph'-positive CML cells during the benign phase of the disease (although these are known to contain the hybrid bcr-abl mRNA). It has been suggested that this is due to an inhibitory factor present in mature myeloid cells (Maxwell et al. 1987). P210 has in fact been detected in cultures of $Ph^{+}B$ lymphocytes from a patient in benign phase, indicating that the protein is present in an enzymatically active form in the early phase of CML (Maxwell et al. 1987).

P210 shows cis-acting and trans-acting in vitro tyrosine phosphorylation activity similar to that of P160, the v-abl transforming protein (Davis et al. 1985). Furthermore, unlike the normal c-abl protein and similar to P160, P210 is phosphorylated on tyrosine in vivo (Konopka et al. 1984). This property has allowed demonstration of the presence of P210 in CML cell lines and bone marrow samples through the use of phosphotyrosine-specific antibodies in Western blotting (Naldini et al. 1986; Huhn et al. 1987). Using the same antibodies it was also possible to visualize a number of additional phosphotyrosine-containing proteins in the same cells (Naldini et al. 1986; Huhn et al. 1987). These represent potential substrates for P210, and at least one of them appears to correspond to P36, a phosphoprotein found in fibroblasts transformed by viral tyrosine kinases (Naldini et al. 1986; Saggioro et al. 1986; Di Renzo et al. 1986).

Although the molecular mechanisms through which transforming tyrosine kinases alter the control of cell growth are not known, it is generally thought that they interact with growth factor-mediated pathways through their constitutively high enzymatic activity. Accordingly, transformation by AMuLV renders myeloid and lymphoid cells independent of growth factors (Cook et al. 1985, 1987; Pierce et al. 1985; Mathey-Prevot et al. 1986; Rovera et al. 1987).

The association of P210, a constitutively active tyrosine kinase, with CML cells, strongly suggests that such an enzymatic activity may be involved in their uncontrolled proliferation, perhaps releasing them from their dependence on specific limiting growth factors.

Recent gene transfer experiments involving expression of the bcr-abl protein in NIH/3T3 fibroblasts, show that P210 differs from P160 v-abl, since it is not capable of causing fibroblast transformation (Daley et al. 1987). However, preliminary results seem to indicate that P210 is transforming, when cells of the hematopoietic lineage are used as recipient of the bcr-abl constructs (Daley et al. 1987). Thus, the capability of the bcr-abl gene product to mediate malignant transformation seems to be restricted to a very specific cell type, and probably to cells belonging to a very narrow developmental window within the hematopoietic lineage.

7 P190, the c-abl Variant Present in Acute Lymphoblastic Leukemia (ALL)

The Ph′ chromosome, in addition to being associated with 95% of the cases of CML, has been identified in other malignant hematopoietic disorders, including acute lymphoblastic leukemia (ALL), where it is present in 15–25% of the adult and 2–6% of the childhood cases (de Klein et al. 1986).

Investigation of the chromosomal breakpoints in Ph+ ALL has revealed that these patients can be divided into two subgroups. In some of the cases a typical bcr-abl rearrangement, similar to that observed in CML, is present. These cases are defined ALL Ph+ bcr+. In other cases the breakpoint cluster region does not appear to be involved in the 9;22 translocation (ALL Ph+ bcr–) (de Klein et al. 1986). Analysis of the DNA of several Ph+ bcr– ALL patients suggests that the translocation point may be a site of breakpoint clustering for ALL, similar to the bcr region for CML (Leibowitz et al. 1985b).

A new variant c-abl protein, of MW approximately 190 kD (P190), has also been identified in Ph′+ bcr– ALL clinical samples (Clark et al. 1987; Kurzrock et al. 1987; Chan et al. 1987). This protein shows in vitro tyrosine kinase activity comparable to P210. Nevertheless, P190 and P210 appear to have distinct substrate specificity, as indicated by the fact that their autophosphorylation sites clearly differ (Clark et al. 1987). Structural differences between P190 and P210 probably affect their interactions with specific substrates in different target cells, determining genetic changes which may be responsible for the development of an acute lymphoid versus a chronic myeloid malignancy.

Accordingly, it has been proposed that bcr+ P210+ ALLs are probably lymphoid blast crises following a clinically silent chronic phase of CML arising in multipotential stem cells. Conversely, bcr– P190+ cases are thought to be de novo acute lymphoblastic leukemias arising in more restricted precursors (Clark et al. 1987; Kurzrock et al. 1987; Chan et al. 1987).

8 Bombesin is a Potent Growth Factor Which Mediates Autocriny

Cell proliferation is regulated by signals transmitted by other cells. As mentioned above, it is possible that genes that control some of these signals are deregulated: cells autonomously produce the growth factor to which they are responsive and are autostimulated, bypassing the normal feedback controls (Heldin and Westermark 1984)

An example of this pathology is, in humans, small-cell lung carcinoma (microcytoma). These tumor cells synthesize a 14-amino acid peptide called bombesin. Bombesin is an extremely potent growth factor to which the microcytoma cells are themselves sensitive. The result is a short circuit that keeps tumor cells in continuous proliferation (Cuttitta et al. 1985b).

The tetradecapeptide bombesin and bombesin-like peptides have been detected in various normal mammalian tissues (Warthon et al. 1978; Walsh et al. 1979; Moody and Pert 1979), together with specific cell surface receptors (Jensen et al. 1978; Moody et al. 1978; Rozengurt and Sinnet-Smith 1983; Zachary and Rozengurt 1985). Bombesin was initially known as a neurohormone stimulating secretion, smooth muscle contraction and affecting temperature control and behavior in mammals (Gibbs et al. 1979; Tache et al. 1980; Swope and Schonbrunn 1984). More recently, it has been shown that bombesin has also a potent mitogenic activity on Swiss 3T3 fibroblasts (Rozengurt and Sinnet-Smith 1983) and normal human bronchial epithelial cells (Willey et al. 1984). In human small-cell lung carcinoma bombesin-related peptides appear to act in an autocrine circuit stimulating growth (Cuttitta et al. 1985b). Furthermore, high levels of bombesin-like immunoreactivity were reported in rat mammary tumors (Gaudino et al. 1986) and human lung or thyroid carcinomas (Moody et al. 1981; Erisman et al. 1982; Matsubayashi et al. 1984). The broad tissue specificity of this peptide and its mitogenic activity on different cell types are suggestive of a role for bombesin in growth control, both of normal and transformed cells.

Known growth factors trigger one or more of the following events in target cells: (1) activation of receptor-associated tyrosine protein kinase activity (James and Bradshow 1984), (2) stimulation of polyphosphoinositide breakdown, leading to protein kinase C activation and Ca^{2+} mobilization (Brown et al. 1984; Majerus et al. 1984), (3) alkalinization of cytoplasm through the activation of a Na^+/H^+ antiport (Moolenaar et al. 1983; L'Allemain et al. 1984; Glaser et al. 1985), and (4) expression of c-fos and c-myc proto-oncogenes (Greenberg and Ziff 1984; Heldin and Westermark 1984). Bombesin and bombesin-related peptides, such as GRP, elicit a complex array of early biological responses (Rozengurt 1986) including enhanced phosphoinositide turnover and release of calcium from intracellular stores (Brown et al. 1984; Heslop et al. 1986; Mendoza et al. 1986), activation of protein kinase-C (Zachary et al. 1986), stimulation of sodium-proton exchange (Mendoza et al. 1986), and rapid and transient expression of c-fos and c-myc proto-oncogenes (Letterio et al. 1986; Palumbo et al. 1986). We have shown that bombesin and GRP also induce activation of a tyrosine protein kinase (Cirillo et al. 1986), strengthening the concept that tyrosine phosphorylation may play an essential role in mitogenic response pathways (Hunter 1987).

9 Bombesin Triggers a Receptor-Associated Protein Tyrosine Kinase

To assess whether the mitogenic effect of bombesin involves the activation of a protein tyrosine kinase we used purified antibodies capable of recognizing phosphotyrosine residues in proteins (Comoglio et al. 1984; Naldini et al. 1986; Zippel et al. 1986). As mentioned, these antibodies (P-Tyr antibodies) have previously been used to detect tyrosine protein kinases associated with growth factor receptors through the identification of the autophosphorylated receptors and their cellular substrates in mitogen-stimulated cells (Frackelton et al. 1983, 1984; Ek and Heldin 1984; White et al. 1985; Cirillo et al. 1986).

As described in detail elsewhere (Cirillo et al. 1986), the interaction of bombesin with its specific receptor in the plasma membrane of Swiss 3T3 cells is rapidly followed by the phosphorylation on tyrosine of an M_r 115,000 membrane protein (p115), detectable in bombesin-stimulated intact cells by Western blotting analysis employing P-Tyr antibodies. The extent of phosphorylation was dependent on both time and dose of exposure to bombesin. With a fully mitogenic concentration (3 nM), maximal tyrosine phosphorylation of p115 was achieved as soon as after 3 min of stimulation at 37° C. The dose dependency for p115 tyrosine phosphorylation was closely related to that observed for the mitogenic activity of bombesin on Swiss 3T3 cells. P115 tyrosine phosphorylation was mediated by binding of bombesin to its specific receptor since no such effect was detectable in cell types not bearing the bombesin receptor (e.g. Balb/C mouse fibroblasts).

When p115 was immunoprecipitated with P-Tyr antibodies from bombesin-stimulated cells and exposed to (γ-^{32}P)ATP and Mn^{2+}, it was phosphorylated on tyrosine by an associated protein kinase activity.

The bombesin dependence of the tyrosine protein kinase activity associated with p115 was shown to be mediated by the presence of specific and saturable binding sites for bombesin in the immunoprecipitate. Hence, the bombesin-dependent p115 tyrosine phosphorylation in immunoprecipitates probably mimics an early event taking place after the interaction of bombesin with its target cells.

P115 could be a substrate of an unidentified tyrosine protein kinase activated by bombesin or may be a kinase itself that, in analogy to other tyrosine protein kinases, is capable of autophosphorylation. The relationships of p115 with the bombesin receptor and/or with the tyrosine protein kinase activity will be elucidated upon the purification of the protein.

In conclusion, it seems clear that the neuropeptide bombesin shares with other, better known growth factors the property of inducing a receptor-associated tyrosine protein kinase activity in target cells. In the case of PDGF, EGF and insulin, the protein kinase activity has been shown to be intrinsic to the receptor molecule, which then autophosphorylates on tyrosine upon ligand binding (Heldin and Westermark 1984; Hunter and Cooper 1985). Whether the bombesin receptor carries both the ligand binding site and the catalytic domain on the same molecule or whether it is a multisubunit assembly, remains to be ascertained.

10 Protein Tyrosine Kinase Associated with the Bombesin Receptor Complex in Small-Cell Lung Carcinomas

The human tumor known as small-cell lung carcinoma (SCLC) is characterized by cells closely related to the neuroendocrine component of the respiratory epithelium (Yesner 1978). Therefore, unlike the other major histological types of lung cancer, it exhibits ectopic synthesis of numerous peptide hormones and growth factors (Moody et al. 1981; Wood et al. 1981; Erisman et al. 1982; Little et al. 1983; Iannuzzi and Scoggin 1986). Among these, bombesin-like peptides, such as the Gastrin Releasing Peptide (GRP), known to be powerful growth stimulators of many cell types, have been proposed as possible autocrine growth factors for SCLC (Anastasi et al. 1971; McDonald et al. 1979). In fact, a specific receptor-binding activity was demonstrated in many cell lines derived from these tumors (Moody et al. 1983) and a monoclonal antibody to the binding moiety of bombesin/GRP was found to prevent growth of tumor cells in vitro and of xenografts in vivo (Cuttitta et al. 1985b).

We used P-Tyr antibodies and Sepharose-immobilized bombesin to investigate the occurrence of p115 tyrosine phosphorylation in small-cell lung carcinoma lines. Four cell lines derived from "classic" SCLC (Cuttitta et al. 1985a), which produce detectable levels of bombesin (Moody et al. 1983), were compared with the "variant" SCLC cell line GLC-1, which is devoid of bombesin-like immunoreactivity (de Leij et al. 1985).

A protein of 115 kD, phosphorylated on tyrosine, was detectable in the four classic SCLC lines studied, namely NCI-H128, NCI-N592, GLC-8 and SHP77, but not in the "variant" line that does not produce bombesin. This molecule (SCLC p115) co-migrated with the mouse p115 protein phosphorylated on tyrosine in response to bombesin, it bound, after detergent treatment, to bombesin-Sepharose beads and was phosphorylated on tyrosine after both immunoprecipitation with P-Tyr antibodies or binding to bombesin-Sepharose (Gaudino et al. 1988a, b).

The p115 detected in normal fibroblasts becomes phosphorylated in response to the addition of bombesin and the phosphorylation is transient (Cirillo et al. 1986). The p115 found in SCLC cells is phosphorylated without the addition of exogenous ligand. This may be explained by autocrine activation, via permanent occupation of receptor binding sites, as suggested by the observation that no available binding sites for the radiolabeled ligand were measurable, and that the kinase activity was not affected by addition of exogenous bombesin.

Whether in classic SCLC cells the binding of bombesin-related peptides to their receptors takes place at the cell surface, at the end of a complete autocrine circuit, or occurs inside the cell, in the exocytotic pathway, still remains to be established. Moreover, a partial lesion of the bombesin-sensitive protein tyrosine kinase leading to receptor uncoupling and constitutive activation cannot be excluded.

11 Protein Tyrosine Kinase Activated in Human Tumors

Recently, P-Tyr antibodies have been used to screen a number of cell lines derived from human tumors and a few containing high levels of proteins phosphorylated on tyrosine were found (Giordano et al. 1987). A tyrosine-phosphorylated protein with the M_r of 145,000 (p145) was unique to a gastric carcinoma line.

P145 was detected by antibodies to phosphotyrosine (P-Tyr) in Western blot analysis (Giordano et al. 1988a). It was not observed in normal human fibroblasts grown in vitro, nor in samples prepared with a comparable number of epithelial cells scraped from the gastric mucosa of patients affected by different pathologies (Giordano et al. 1987). To assess the possible identity between p145 and known growth factor receptors, tumor cells were stimulated by EGF, PDGF, insulin and bombesin under conditions known to induce or to enhance the autophosphorylation on tyrosine of their corresponding receptors. When EGF was tested, the 180,000 M_r EGF receptor became phosphorylated and, thus, detectable in Western blots with P-Tyr antibodies. The electrophoretic migration of the EGF receptor was clearly different from that of p145. The addition of the other growth factors had no effect on the phosphorylation of p145. In cells metabolically labelled with ^{32}P-ortophosphate the protein is phosphorylated on tyrosine and on serine.

After immunoprecipitation with P-Tyr antibodies, p145 displays a strong associated protein kinase activity in vitro (Giordano et al. 1988a). In fact, when immunoprecipitates were exposed to (γ-^{32}P)ATP to assay the kinase activity, p145 became phosphorylated on tyrosine. In analogy to the known tyrosine kinases and in view of the fact that p145 is virtually the only protein which becomes phosphorylated on tyrosine during the assay, it is possible that p145 could autophosphorylate itself. There is an apparent paradox in this observation, since p145 immunoprecipitated with P-Tyr antibodies must, by definition, already be phosphorylated on tyrosine prior to the in vitro kinase assay. This, however, could be explained by the occurrence of multiple sites of autophosphorylation, as observed also in the case of the EGF receptor and other tyrosine kinases. Moreover, it has already been shown that tyrosine kinases precipitated by P-Tyr antibodies phosphorylate themselves in vitro.

The addition to the immunocomplexes of EGF, PDGF, insulin or bombesin did not induce the appearance of any extra band phosphorylated on tyrosine, or affect the phosphorylation of p145 itself. The possible identity between p145 and some growth factor receptors was further investigated (Giordano et al. 1988a). Antibodies against the growth factor receptors and some of the oncogene products did not cross-react with p145.

12 P145 Displays Features of a Growth Factor Receptor

The biological functions of p145 are not yet known. However, the molecule has some of the features peculiar to growth factor receptors: it spans the plasma membrane and it is associated with a tyrosine kinase activity. Both properties suggest a role in transducing signals from an external stimulus.

P145 is a cysteine-rich transmembrane glycoprotein. The presence of oligosaccharide moieties linked to p145 was shown by its binding with lentil lectin bound to Sepharose beads, followed by specific elution with alpha-methyl-D-mannoside. P145 was also metabolically labelled by 3H-glucosamine as assessed by immunoprecipitation with P-Tyr antibodies. In addition, the extracellular domain can be labelled by ^{125}I, under non-permeating conditions. Further evidence confirming that a domain of p145 is exposed at the cell surface came from experiments of partial proteolysis. The extracellular domain can be cleaved by mild trypsin treatment of intact cells.

Tyrosine phosphorylation of p145 in live cells is sustained by an extracellular factor. A short acid treatment has been shown to be effective in removing EGF or PDGF molecules already bound to receptors and to inhibit the growth factor-induced tyrosine kinase activity of some receptors. The effect is rapid and reversible. The extent of tyrosine phosphorylation of p145 of gastric carcinoma cells after acid treatment was evaluated on Western blots probed with P-Tyr antibodies. The dephosphorylation of p145 after acid treatment was almost complete between 5 and 15 min. Furthermore, neutralization of pH was followed by rephosphorylation on tyrosine of p145 within 30 min. In order to rule out the possible interference of the treatment on the tyrosine kinase activity, mouse fibroblasts transformed by v-src were treated with acid: the activity of $pp60^{src}$ was unaffected (Giordano et al. 1988a). These data suggest that the dephosphorylation of p145 may be due to the fact that acid interferes with events taking place at the cell surface, most likely the binding of an extracellular factor.

13 Tyrosine Phosphorylation of p145 in Live Cells is Sustained by an Autocrine Loop

The gastric carcinoma cell line studied has the ability to grow in serum-free medium. Since in several instances such an ability is mediated by growth factor(s) produced by the cells themselves, experiments were performed to interfere with such a putative autocrine stimulation (Giordano et al. 1988b). Gastric carcinoma cells were treated with mild acid solution (see above) which has been shown to rapidly remove bound EGF or PDGF molecules from their receptors and to switch off the growth factor-induced tyrosine kinase activity. The effect is rapid and reversible (Zippel et al. 1986).

Following acid treatment, dephosphorylation of p145 was almost complete within 5–15 min. The acid treatment appears to interfere with events taking place at the cell surface, most likely the binding of an extracellular factor.

Evidence suggesting the presence of a factor secreted by the gastric carcinoma cells themselves was obtained by studying the effect of conditioned medium. P145 is rephosphorylated if the cells are allowed to recover in serum-free medium at 4°C for a minimum of 15 min. The addition of conditioned medium restores the tyrosine phosphorylation of p145 within 5 min at the same temperature.

14 Peculiarity of the p145 Tyrosine Kinase of Gastric Carcinoma

The possible identity between p145 and the known tyrosine kinases is unlikely on the basis of the following data. (1) Tyrosine phosphorylation of p145, measured both in vivo and in vitro, was not affected by the addition of known growth factors such as EGF, PDGF, insulin or bombesin. (2) No cross-reactions were found between p145 and antibodies against the EGF and PDGF receptors. Moreover, p145 did not cross-react with antibodies against the tyrosine kinases coded by the viral oncogenes v-src, v-abl, v-fes, v-fps, or by the human proto-oncogenes c-trk and c-erb-B2. (3) The gastric carcinoma cell line does not express c-fms, c-kit and insulin receptor m-RNA as shown by Northern blot analysis (G. Manenti, T. Dragani, P. Rossino, S. Giordano, M. Pierotti and P.M. Comoglio, in preparation). (4) The molecular weight of p145 and, above all, the electrophoretic shift to an M_r of 190,000 (when analyzed in SDS-PAGE under non-reducing conditions) are unique. The latter property is particularly relevant to differentiate p145 from the already known c-onc proteins displaying structural homologies with growth factor receptors. These are the proteins coded by c-met (Park et al. 1987), c-trk (Martin et al. 1986a,b), c-kit (Yarden et al. 1987), c-fms (Sherr et al. 1985) and c-erb-B2, none of which, like PDGF and EGF receptors, shows a similar M_r shift when run under non-reducing conditions. (5) P145 is not part of the receptors for insulin or IGF-1. These molecules have indeed a multichain structure (two "heavy" and two "light" chains) but clearly differ from p145, as their mobility in SDS-PAGE under reducing or non-reducing conditions is different from that of p145. (6) P145 is covalently linked to a "light" chain of 45,000 M_r. Experiments were performed by metabolic labelling of cells with either ^{35}S-methionine or ^{35}S-cysteine. A band of the approx. M_r of 190,000 was precipitated and resolved in non-reducing gels. The gel fragment corresponding to this band was excised, the proteins were eluted, treated with 2-mercaptoethanol, and analyzed again in a second gel under reducing conditions. This procedure resolved the 190,000 M_r band in two bands of M_r 145,000 and 50,000. These data are consistent with the presence of a disulphide-linked complex formed by p145 and a light chain of approximately M_r 50,000. The existence of an extra polypeptide was shown by cross-linking the surface protein before the immunoprecipitation with P-Tyr antibodies. The cross-linker covalently bound the two chains in a stable complex of the apparent M_r of 190,000 (Giordano et al. 1988b).

Acknowledgements. The original work reviewed in this paper has been supported by the Italian National Research Council (C.N.R.) Special Project "Oncologia", grants No. 86.00369.44 and 86.00394.44. The support given by the Associazione Italiana per la Ricerca sul Cancro is also gratefully acknowledged.

References

Anastasi A, Erspamer V, Bucci M (1971) Isolation and structure of bombesin and alytesin, two analogous active peptides from the skin of the European amphibian bombina and alytes. Experientia 27:166–167

Bartram CR, Kleihauer E, de Klein A, Grosveld G, Teyssier JR, Heisterkamp N, Groffen J (1986) C-abl and bcr are rearranged in a Ph[1]-negative CML patient. EMBO J 4:683–686

Ben-Neriah Y, Baltimore D (1986) Normal and transforming N-terminal variants of c-abl. In: Kahn P, Graf T (eds) Oncogene and growth control. Springer, Berlin Heidelberg New York, pp 106–114

Ben-Neriah Y, Daley GQ, Mes-Masson AM, Witte ON, Baltimore D (1986a) The chronic myelogenous leukemia-specific P210 proteins is the product of the bcr/abl hybrid gene. Science 233:212–214

Ben-Neriah Y, Bernards A, Paskind M, Daley GQ, Baltimore D (1986b) Alternative 5′ exons in c-abl mRNA. Cell 44:577–586

Bernards A, Rubin CM, Westbrook CA, Paskind M, Baltimore D (1987) The first intron in the human c-abl gene is at least 200 kilobases long and is a target for translocations in chronic myelogenous leukemia. Mol Cell Biol 7:3231–3236

Bertics PJ, Gill GN (1985) Self phosphorylation enhances the protein-tyrosine kinase activity of the epidermal growth factor receptor. J Biol Chem 260:14642–14647

Bishop JM (1983) Cellular oncogenes and retroviruses Annu Rev Biochem 52:301–354

Brown KD, Blay J, Irvin RF, Heslop JP, Berridge MJ (1984) Reduction of epidermal growth factor receptor affinity by heterologous ligands: evidence for a mechanism involving the breakdown of phosphoinositides and the activation of protein kinase C. Biochem Biophys Res Commun 123:377–384

Champlain RE, Golde DW (1985) Chronic myelogenous leukemia: recent advances. Blood 65:1039–1097

Chan LC, Karhi KK, Rayter SI, Heisterkamp N, Eridani S, Powles R, Lawler SD, Groffen J, Foulkes JG, Greaves MF, Wiedemann LM (1987) A novel abl protein expressed in Philadelphia chromosome positive acute lymphoblastic leukemia. Nature (London) 325:635–637

Cirillo D, Comoglio PM, Naldini L, Marchisio PC, Pasquali Ronchetti Y (1984) Visualization of phosphotyrosine containing molecules within the detergent insoluble cell matrix of v-src transformed fibroblasts. Cell Biol Int Rep 8:959–969

Cirillo DM, Gaudino G, Naldini L, Comoglio PM (1986) Receptor for bombesin with associated tyrosine kinase activity. Mol Cell Biol 6:4641–4649

Clark SS, McLaughlin J, Crist WM, Champlin R, Witte ON (1987) Unique forms of the abl tyrosine kinase distinguish Ph^1-positive CML from Ph^1-positive ALL. Science 235:85–88

Collins SJ, Groudine MT (1983) Rearrangement and amplification of c-abl sequences in the human chronic myelogenous leukemia cell line K-562. Proc Natl Acad Sci USA 80:4813–4817

Collins S, Kubonishi I, Myoshi I, Groudine MT (1984) Altered transcription of the c-abl oncogene in K-562 and other chronic myelogenous leukemia cells. Science 225:72–74

Collins S, Coleman H, Groudine M (1987) Expression of bcr and bcr-abl fusion transcripts in normal and leukemic cells. Mol Cell Biol 7:2870–2876

Comoglio PM, Di Renzo MF, Tarone G, Giancotti FG, Naldini L, Marchisio PC (1984) Detection of phosphotyrosine containing proteins in the detergent-insoluble fraction of RSV-transformed fibroblasts by azobenzene phosphonate antibodies. EMBO J 3:483–489

Cook WD, Metcalf D, Nicola NA, Burgess AW, Walker F (1985) Malignant transformation of a growth factor-dependent myeloid cell line by Abelson virus without evidence of an autocrine mechanism. Cell 41:677–683

Cook WD, Fazekas de St. Groth B, Miller JFAP, McDonald HR, Gabathuler R (1987) Abelson virus transformation of an interleukin 2-dependent antigen-specific T-cell line. Mol Cell Biol 7:2631–2635

Cuttitta F, Carney DN, Mulshine J, Moody TW, Fedorko J, Fischler A, Minna JD (1985a) Autocrine growth factors in human small cell lung cancer. Cancer Surv 4:707–727

Cuttitta F, Carney DN, Mulshine J, Moody TW, Fedorko J, Fischler A, Minna JD (1985b) Bombesin-like peptides can function as autocrine growth factors in human small-cell lung cancer. Nature (London) 316:823–826

Daley GQ, McLaughlin J, Witte ON, Baltimore D (1987) The CML-specific P210 bcr/abl protein, unlike v-abl, does not transform NIH/3T3 fibroblasts. Science 237:532–535

Davis RL, Konopka JB, Witte ON (1985) Activation of the c-abl oncogene by the viral transduction or chromosomal translocation generates altered c-abl proteins with similar in vitro kinase properties. Mol Cell Biol 5:204–213

de Klein A, van Kessel AG, Grosveld G, Bartram CR, Hagemeijer A, Bootsma D, Spurr NK, Heisterkamp N, Groffen J, Stephenson JR (1982) A cellular oncogene is translocated to the Philadelphia chromosome in chronic myelocytic leukemia. Nature (London) 300:765–767

de Klein A, Hagemeijer A, Bartram CR, Houwen R, Hoefsloot L, Carbonell F, Chan L, Barnett M, Greaves M, Kleihauer E, Heisterkamp N, Groffen J, Grosveld G (1986) bcr rearrangement and translocation of the c-abl and oncogene in Philadelphia positive acute lymphoblastic leukemia. Blood 68:1369–1375

de Leij L, Postmus PE, Buys CM, Elema JD, Ramaekers F, Poppema S, Brouwer M, van der Veeh AY, Mesander G, The H (1985) Characterization of three new variant type cell lines derived from small cell carcinoma of the lung. Cancer Res 45:6024–6029

Di Renzo MF, Ferracini R, Naldini L, Giordano S, Comoglio PM (1986) Immunological detection of proteins phosphorylated at tyrosine in cells stimulated by growth factors or transformed by retroviral-oncogene-coded tyrosine kinases. Eur J Biochem 158:383–391

Downward J, Yarden Y, Mayes E, Scrace J, Totty N, Stockwell P, Ullrich A, Schlessinger J, Waterfield MD (1984) Close similarity of epidermal growth factor receptor and v-erb-B oncogene protein sequences. Nature (London) 307:521–527

Ek B, Heldin CH (1984) Use of an antiserum against phosphotyrosine for the identification of phosphorylated components in human fibroblasts stimulated by platelet-derived growth factor. J Biol Chem 259:5287–5294

Erikson J, Griffin CA, Ar-Rushdi A, Valtieri M, Hoxie J, Finan J, Emanuel BS, Rovera G, Nowell PC Croce CM (1986) Heterogeneity of chromosome 22 breakpoint in Philadelphia-positive (Ph^+) acute lymphocytic leukemia. Proc Natl Acad USA 83:1807–1811

Erisman MD, Linnoila RI, Hernandez O, Di Augustine RP, Lazarus LH (1982) Human lung small-cell carcinoma contains bombesin. Proc Natl Acad Sci USA 79:2379–2383

Frackelton AR, Ross A, Eisen HN (1983) Characterization and use of monoclonal antibodies for isolation of phosphotyrosil proteins from retrovirus transformed cells and growth factor-stimulated cells. Mol Cell Biol 3:1343–1352

Frackelton AR Jr., Tremble PM, Williams LT (1984) Evidence for the platelet-derived growth factor-stimulated tyrosine phosphorylation of the platelet-derived growth factor receptor in vivo. J Biol Chem 259:7909–7915

Gale RP, Canaani E (1984) An 8-kilobase abl RNA transcript in chronic myelogenous leukemia. Proc Natl Acad Sci USA 81:5648–5652

Gaudino G, De Bortoli ME, Lazarus LH (1986) A bombesin related peptide in experimental mammary tumors in rats. Ann NY Acad Sci 464:450–453

Gaudino G, Cilli M, Gandino L, Rossino P, Mondino A, Comoglio PM (1988a) A tyrosine protein kinase activated by bombesin in normal fibroblasts and small cell carcinomas. Ann NY Acad Sci (in press)

Gaudino G, Naldini L, Cirillo D, Rossino P, Comoglio PM (1988b) Activation of the tyrosine kinase associated with the bombesin receptor complex in small cell lung carcinomas. Proc Natl Acad Sci USA 85:2166–2170

Gibbs JD, Fauser J, Rowe AJ, Rolls BJ, Rolls ET, Madison SJ (1979) Bombesin suppress feeding in rats. Nature (London) 282:208–210

Giordano S, Di Renzo MF, Cirillo D, Naldini L, Chiado'-Piat L, Comoglio PM (1987) Proteins phosphorylated on tyrosine as markers of human tumor cell lines. Int J Cancer 39:482–487

Giordano S, Di Renzo MF, Ferracini R, Chiado'-Piat L, Comoglio PM (1988a) P145 a protein with associated tyrosine kinase activity in a human gastric carcinoma cell line. Mol Cell Biol 8:3510–3517

Giordano S, Di Renzo MF, Narsimhan RP, Tamagnone L, Gerbaudo E, Chiado'-Piat L, Comoglio PM (1988b) Evidence for autocrine activation of a tyrosine kinase in a human gastric carcinoma cell line. J Cell Biochem (in press)

Glaser L, Whiteley B, Rothenberg P, Cassel D (1985) Mitogenic polypeptides control ion flux in responsive cells. Bioassays 1:16–20

Goff SP, Gilboa E, Witte ON, Baltimore D (1980) Structure of the Abelson murine leukemia virus genome and the homologous cellular gene: studies with cloned viral DNA. Cell 22:777–785

Greenberg ME, Ziff EB (1984) Stimulation of 3T3 cells induces transcription of the c-fos proto-oncogene. Nature (London) 311:433–438

Grosveld G, Verwoerd T, van Agthoven T, de Klein A, Ramachandran KL, Heisterkamp N, Stam K, Groffen J (1986) The chronic myelocytic cell line K562 contains a breakpoint in bcr and produces a chimeric brc/c-abl transcript. Mol Cell Biol 6:607–616

Hariharan JI, Adams JM (1987) cDNA sequence for human bcr, the gene that translocates to the abl oncogene in chronic myeloid leukaemia. EMBO J 6:115–119

Heisterkamp N, Stam K, Groffen J, de Klein A, Grosveld G (1985) Structural organization of the bcr gene and its role in the Ph' translocation. Nature (London) 315:758–761

Heisterkamp N, Stephenson JR, Groffen J, Hansen PF, de Klein A, Bartram CR, Grosveld G (1983a) Localization of the c-abl oncogene adjacent to a translocation break point in chronic myelocytic leukaemia. Nature (London) 306:239–242

Heisterkamp N, Groffen J, Stephenson JR (1983b) The human v-abl cellular homologue. J Mol Appl Genet 2:57–68

Heldin CH, Westermark B (1984) Growth factors: mechanism of action and relation to oncogenes. Cell 37:9–20

Heslop P, Blakeley D, Brown K, Irvine B, Berridge M (1986) Effects of bombesin and insulin on inositol (1,4,5) trisphosphate and inositol (1,3,4) trisphosphate formation in Swiss 3T3 cells. Cell 47:703–709

Huhn RD, Posner MR, Rayter SI, Foulkes JG, Frackelton AR (1987) Cell lines and peripheral blood leukocytes derived from individuals with chronic myelogenous leukemia display virtually identical proteins phosphorylated on tyrosine residues. Proc Natl Acad Sci USA 84:4408–4412

Hunter T (1984) The proteins of oncogenes. Sci Am 251:60–69

Hunter T (1987) A thousand and one protein kinases. Cell 50:823–829

Hunter T, Cooper J (1985) Protein tyrosine kinases. Annu Rev Biochem 54:897–930

Iannuzzi MC, Scoggin CH (1986) Small cell lung cancer. Am Rev Respirat Dis 134:593–608

James R, Bradshow RA (1984) Polypeptide growth factors. Annu Rev Biochem 53:259–292

Jensen RT, Moody T, Pert C, Rivier JE, Gardner JD (1978) Interaction of bombesin and litorin with specific membrane receptors on pancreatic acinar cells. Proc Natl Acad Sci USA 75:6139–6143

Konopka JB, Watanabe SM, Witte ON (1984) An alteration of the human c-abl protein in K562 leukemia cells unmasks associated tyrosine kinase activity. Cell 37:1035–1042

Konopka JB, Watanabe SM, Singer JW, Collins SJ, Witte ON (1985) Cell lines and clinical isolates derived from Ph^1-positive chronic myelogenous leukemia patients express c-abl proteins with a common structural alteration. Proc Natl Acad Sci USA 82:1810–1814

Kozbor D, Giallongo A, Sierzega ME, Konopka JB, Witte ON, Showe LC, Croce CM (1986) Expression of a translocated c-abl gene in hybrids of mouse fibroblasts and chronic myelogenous leukaemia cells. Nature (London) 319:331–333

Kris R, Lax I, Gullick M, Waterfield M, Ullrich A, Fridkin M, Schlessinger J (1985) Antibodies against a synthetic peptide as a probe for the kinase activity of the avian EGF receptor and v-erbB protein. Cell 40:619–625

Kurzrock R, Shtalrid M, Romero P, Kloetzer WS, Talpas M, Trujillo JM, Blik M, Beran M, Gutterman JU (1987) A novel c-abl protein product in Philadelphia-positive acute lymphoblastic leukaemia. Nature (London) 325:631–635

L'Allemain G, Paris S, Pouyssegur JJ (1984) Growth factor action and intracellular Ph regulation in fibroblasts. Evidence for a major role of the Na^+/H^+ antiport. J Biol Chem 259:5809–5815

Leibowitz D, Cubbon R, Bank A (1985a) Increased expression of a novel c-abl related RNA in K562 cells. Blood 65:526–529

Leibowitz D, Schaefer-Rego K, Popenoe DW, Mears JG, Bank A (1985b) Variable breakpoints on the Philadelphia chromosome in chronic myelogenous leukemia. Blood 66:243–245

Letterio J, Coughlin SR, Williams LT (1986) Pertussis toxin-sensitive pathway in the stimulation of c-myc expression and DNA synthesis by bombesin. Science 234:1117–1119

Little CD, Nau MM, Carney DN, Gazdar AF, Minna JD (1983) Amplification and expression of the c-myc oncogene in human lung cancer cell lines. Nature (London) 306:194–196

Majerus PW, Neufeld E, Wilson DB (1984) Production of phosphoinositide-derived messengers. Cell 37:701–703

Marchisio PC, Di Renzo MF, Comoglio PM (1984) Immunofluorescence localization of phosphotyrosine containing proteins in RSV-transformed mouse fibroblasts. Exp Cell Res 154:112–124

Martin Zanca D, Hughes SH, Barbacid M (1986a) A human oncogene formed by the fusion of truncated tropomyosin and protein tyrosine kinase sequences. Nature (London) 319:743–748

Martin Zanca D, Mitra G, Long LK, Barbacid M (1986b) Molecular characterization of the human trk oncogene. Cold Spring Harbor Symp Quant Biol 51:983–992

Mathey-Prevot B, Nabel G, Palacios R, Baltimore D (1986) Abelson virus abrogation of interleukin-3 dependence in a lymphoid cell line. Mol Cell Biol 6:4133–4135

Matsubayashi S, Yanaihara C, Ohrubo M, Fukata S, Hayashi Y, Tamai H, Nakagawa T, Miyauchi A, Kuma K, Abe K, Suzuki T, Yanaihara N (1984) Gastrin-releasing peptide immunoreactivity in medullary thyroid carcinoma. Cancer 53:2472–2477

Maxwell SA, Kurzrock R, Parsons SJ, Talpaz M, Gallick GE, Kloetzer WS, Arlinghaus RB, Kouttab NM, Keating MJ, Gutterman JU (1987) Analysis of $P210^{bcr\text{-}abl}$ tyrosine protein kinase activity in various subtypes of Philadelphia chromosome-positive cells from chronic myelogenous leukemia patients. Cancer Res 47:1731–1739

McDonald TJ, Jornvall H, Nilsson G, Vagne M, Ghatei M, Bloom SR, Mutt V (1979) Characterization of a gastrin releasing peptide from porcine non-antral gastric tissue. Biochem Biophys Res Commun 90:227–233

Mendoza S, Schneider J, Lopez-Rivas A, Sinnett-Smith J, Rozengurt E (1986) Early events elicited by bombesin and structurally related peptides in quiescent Swiss 3T3 cells J Cell Biol 102:2223–2233

Mes-Masson AM, McLaughlin J, Daley GQ, Paskind M, Witte ON (1986) Overlapping cDNA clones define the complete coding region for the $P210^{c\text{-}abl}$ gene product associated with chronic myelogenous leukemia cells containing the Philadelphia chromosome. Proc Natl Acad Sci USA 83:9768–9772

Moody TW, Pert CB (1979) Bombesin-like peptides in rat brain: quantitation and biochemical characterization. Biochem Biophys Res Commun 90:7–14

Moody TW, Pert CB, Rivier J, Brown MR (1978) Bombesin: specific binding to rat brain membranes. Proc Natl Acad Sci USA 75:5372–5376

Moody TW, Pert CB, Gazdar AF, Carney DN, Minna JD (1981) High levels of intracellular bombesin characterize human small-cell lung carcinoma. Science 214:1246–1248

Moody TW, Bertness V, Carney DN (1983) Bombesin-like peptides and receptors in human tumor cell lines. Peptides 4:683–686

Moolenaar WH, Tsien RY, van der Saag PT, de Laat SW (1983) Na^+/H^+ exchange and cytoplasmatic pH in the action of growth factors in human fibroblasts. Nature (London) 304:645–648

Morris CM, Reeve AE, Fitzgerald PH, Hollings PE, Beard MEJ, Heaton DC (1986) Genomic diversity correlates with clinical variation in Ph^1-negative chronic myeloid leukaemia. Nature (London) 320:281–283

Naldini L, Stacchini A, Cirillo D, Aglietta M, Gavosto F, Comoglio PM (1986) Phosphotyrosine antibodies identify the $p210^{c\text{-}abl}$ tyrosine kinase and proteins phosphorylated on tyrosine in human chronic myelogenous leukemia cells. Mol Cell Biol 6:1803–1811

Nelson RL, Branton PE (1984) Identification, purification and characterization of phosphotyrosine-specific protein phosphatases from cultured chicken embryo fibroblasts. Mol Cell Biol 4:1003–1012

Ohtsuka M, Ihara S, Ogawa R, Watanabe T, Watanabe Y (1984) Preparation and characterization of antibodies to O-phosphotyrosine and their use for identification of phosphotyrosine-containing proteins. Int J Cancer 34:855–861

Palumbo A, Rossino P, Comoglio PM (1986) Bombesin stimulation of c-fos and c-myc gene expression in cultures of Swiss 3T3 cells. Exp Cell Res 167:276

Park M, Dean M, Kaul K, Braun MJ, Gonda MA, Vande Woude G (1987) Sequence of MET protooncogene cDNA has features characteristic of the tyrosine kinase family of growth factor receptors. Proc Natl Acad Sci USA 84:6379–6383

Pierce JH, Di Fiore PP, Aaronson SA, Potter M, Pumphrey J, Scott A, Ihle J (1985) Neoplastic transformation of mast cells by Abelson-MuLV: abrogation of IL-3 dependence by a nonautocrine mechanism. Cell 41:685–693

Ross AH, Baltimore D, Eisen HN (1981) Phosphotyrosine-containing proteins isolated by affinity chromatography with antibodies to a synthetic hapten. Nature (London) 294:654–656

Rovera G, Valtieri M, Mavilio F, Reddy EP (1987) Effect of Abelson murine leukemia virus on granulocytic differentiation and interleukin-3 dependence of a murine progenitor cell line. Oncogene 1:29–35

Rozengurt E (1986) Early signals in the mitogenic response. Science 234:161–166

Rozengurt E, Sinnet-Smith J (1983) Bombesin stimulation of DNA synthesis and cell division in cultures of Swiss 3T3 cells. Proc Natl Acad Sci USA 80:2936–2940

Saggioro D, Ferracini R, Di Renzo MF, Chieco-Bianchi L, Comoglio PM (1986) Protein phosphorylation at tyrosine residues in v-abl transformed mouse lymphocytes and fibroblasts. Int J Cancer 37:623–628

Sandberg AA, Hossfeld DK (1970) Chromosomal abnormalities in human neoplasia. Annu Rev Med 21:379–408

Schlessinger J (1986) Regulation of cell growth by the EGF receptor. In: Kahn P, Graf T (eds) Oncogenes and growth control. Springer, Berlin Heidelberg New York, pp 77–84

Sherr CJ, Rettenmier CW, Sacca R, Russel M, Look T, Stanley R (1985) The c-fms proto-oncogene product is related to the receptor for the mononuclear phagocyte growth factor CSF-1. Cell 41:665–676

Shtivelman E, Lifshitz D, Gale RP, Canaani E (1985) Fused transcript of abl and bcr genes in chronic myelogenous leukaemia. Nature (London) 315:550–554

Shtivelman E, Lifshitz B, Gale RP, Roe BA, Canaani E (1986) Alternative splicing of RNAs transcribed from the human abl gene and from the bcr-abl fused gene. Cell 47:277–284

Sturani E, Toschi L, Zippel R, Comoglio PM, Alberghina L (1986a) Phosphorylation of the PDGF receptor in mouse Swiss 3T3 fibroblasts under different growth conditions. In: Baserga R (ed) Biological regulation of cell proliferation. Raven, New York, pp 303–306

Sturani E, Vicentini LM, Zippel R, Toschi L, Pandiella-Alonso A, Comoglio PM, Meldolesi J (1986b) PDGF-induced receptor phosphorylation and phosphoinositide hydrolysis are unaffected by protein kinase C activation in mouse Swiss 3T3 and human skin fibroblasts. Biochem Biophys Res Commun 137:345–350

Swope SL, Schonbrunn A (1984) Bombesin stimulates insulin secretion by a pancreatic islet cell line. Proc Natl Acad Sci USA 81:1822–1826

Tache Y, Vale W, Rivier WJ, Brown M (1980) Brain regulation of gastric secretion: influence of neuropeptides. Proc Natl Acad Sci USA 77:5515–5519

Tarone G, Cirillo D, Giancotti F, Comoglio PM, Marchisio PC (1985) Rous sarcoma virus transformed fibroblast adhere primarily at discrete protrusions of the ventral membrane called podosomes. Exp Cell Res 159:141–157

Walsh JH, Wong HC, Dockray GJ (1979) Bombesin-like peptides in mammals. Fed Proc 38:2315–2319

Wang JYJ (1985) Isolation of antibodies for phosphotyrosine by immunization with a v-abl oncogene-encoded protein. Mol Cell Biol 5:3640–3643

Warthon JJ, Polak M, Bloom SR, Ghatei MA, Solcia E, Brown MR, Pearse AGE (1978) Bombesin-like immunoreactivity in the lung. Nature (London) 273:769–770

White MF, Maron R, Kahn CR (1985) Insulin rapidly stimulates tyrosine phosphorylation of a M_r-185,000 protein in intact cells. Nature (London) 318:183–186

Willey JC, Lechner JF, Harris CC (1984) Bombesin and C-terminal tetradecapeptide of gastrin-releasing peptide are growth factors for normal human bronchial epithelial cells. Exp Cell Res 153:245–248

Wood SM, Wood JR, Ghatei MA, O'Shaughnessy D, Bloom SR (1981) Bombesin, somatostatin and neurotensin-like immunoreactivity in bronchial carcinoma. J Clin Endocrinol Metab 53:1310–1312

Yarden Y, Kuang WJ, Yang-Feng T, Coussens L, Munemitsu S, Dull TJ, Chen E, Schlessinger J, Francke U, Ullrich A (1987) Human proto-oncogene c-kit: a new cell surface receptor tyrosine kinase for an unidentified ligand. EMBO J 6:3341–3352

Yesner R (1978) Spectrum of lung cancer and ectopic hormones. In: Sommers SC, Rosen PP (eds) Pathology annual, pt 1. Appleton, New York, pp 217–240

Zachary J, Rozengurt E (1985) High-affinity receptors for peptides of the bombesin family in Swiss 3T3 cells. Proc Natl Acad Sci USA 82:7616–7620

Zachary I, Sinnett-Smith J, Rozengurt EJ (1986) Early events elicited by bombesin and structurally related peptides in quiescent Swiss 3T3 cells. J Cell Biol 102:2211–2222

Zaleberg JR, Friedlander ML, Minden MD (1986) Molecular evidence for the clonal origin of blast crisis in chronic myeloid leukemia. Br J Cancer 53:459–464

Zippel R, Toschi G, Naldini L, Sturani E, Alberghina L, Comoglio PM (1986) In vivo phosphorylation and dephosphorylation of the platelet-derived growth factor receptor studied by immunoblot analysis with phosphotyrosine antibodies. Biochim Biophys Acta 881:54–61

The Role of Plasma Membrane Phospholipids in Cellular Pathology

A. STRUNECKÁ, J. PALEČEK, L. KRPEJŠOVÁ, N. I. EL DESOUKI, P. FOLK and J. MÁCHA

It has been documented by many authors that phospholipids are in a highly dynamic state altering membrane functions and participating actively in the regulation of cell metabolism. Phospholipids serve also as important mediators in the transduction of extracellular signals. We present two examples from our study of human red blood cell (RBC). These examples document the possible involvement of plasma membrane phospholipids in cell pathology.

One of the best-established features of RBC plasma membrane is the asymmetric distribution of phospholipids between the two leaflets of the bilayer. The outer leaflet contains mostly phosphatidylcholine (PC) and sphingomyelin (SM), the greater part of phosphatidylethanolamine (PE) is located in the inner leaflet, phosphatidylserine (PS) is localized exclusively in the inner leaflet (Verkleij et al. 1973). Very little is known regarding the functional significance of such asymmetry of phospholipids.

We present our investigation of RBC of a patient with an autoerythrocyte sensitization syndrome. The syndrome expresses itself in episodes of spontaneous bruising with swelling and pain. Gardner and Diamond (1955) ascribed the recurrence of a painful erythematous bruising to autosensitization against patient RBC. In accordance with this conclusion a positive skin response after intradermal injection of an autologous erythrocyte suspension was observed by many authors (for a review, see Hersle and Mobacken 1969; Pevny et al. 1982). Groch et al. (1966) analyzed the possible biochemical factor responsible for the development of the cutaneous lesions and identified PS as a provocative factor.

On the other hand, Ratnoff and Agle (1968), who examined the most extensive set of patients, found a negative skin response to autologous whole blood in one-third of their patients. These authors concluded that psychological factors play a critical role in the pathogenesis of this syndrome and suggested the denomination, psychogenic purpura. This view has been lately supported by other authors (Stocker et al. 1977; Hanna et al. 1981; Steyrer et al. 1982; Sörensen et al. 1985).

We have investigated the possibility suggested by Groch et al. (1966) that PS might be the component of RBC evoking the symptoms. Skin tests to PS have not been routinely used for diagnosis and therefore the positive evidence supporting Groch's assumption is sporadic (Waldorf and Lipkin 1968; Hersle and Mobacken 1969; Boxley 1971). Since in RBC PS is localized exclusively in the inner leaflet of the phospholipid bilayer (Gordesky and Marinetti 1973; Verkleij et al. 1973; Zwaal

Department of Physiology and Developmental Biology, Charles University, Viničná 7, 128 00 Prague 2, Czechoslovakia

A. Azzi, Z. Drahota and S. Papa
Molecular Basis of Membrane-Associated Diseases

et al. 1975), we checked the integrity of RBC plasma membrane using toluidine blue staining. Even after 1-h incubation in 1% toluidine blue in ACD no damage of cells was ascertained manifesting itself through leakage of stain into the cytoplasm. Also, the measurement of osmotic resistance did not show any difference between control and patient values. We supposed therefore that there is no significant destruction of RBC plasma membrane and that PS might be localized in the outer leaflet of the plasma membrane. This possibility was tested by three techniques: by merocyanine binding, by using TNBS and by the assay of phospholipid degradation by phospholipase A_2.

We employed the fluorescent probe merocyanine 540 since there has been a suggestion that this probe shows enhanced fluorescence in the presence of anionic phospholipids (Williamson et al. 1982, 1985). Since it does not penetrate into the inner leaflet of the plasma membrane, it can be used for the detection of anionic phospholipids in the outer leaflet. In our experiments we observed that merocyanine alone does not fluoresce and when applied on the spots of separated phospholipids on TLC plates, the fluorescence appeared in the spots of PS. Control RBC did not bind any probe and no fluorescence was observed.

After incubation of RBC of patient with merocyanine, approximately 80% of cells bound merocyanine. We tried to repeat this observation with patient RBC stored for 1–5 days in a refrigerator. RBC were incubated for 2 h at 37°C either in medium or in patient plasma (hematocrit 50%). When RBC of the patient were washed and stored and then incubated in medium they did not bind merocyanine, but if the incubation was performed in the patient plasma, fluorescence appeared. It was possible to induce merocyanine binding in homologous RBC after their incubation with the patient plasma (Table 1).

Our observations of merocyanine binding as the supposed indicator of changes in PS redistribution into the outer leaflet of the plasma membrane were supported by our results with trinitrobenzene sulfonic acid (TNBS). This probe reacts with amino groups of phospholipids in the outer leaflet. It has been used for the demonstration of phospholipid asymmetry in RBC (Gordesky and Marinetti 1973). The amount of phospholipids reacting with TNBS was estimated from the difference in the phospholipid content of a sample incubated without TNBS and one incubated with TNBS. Data in Table 2 show that while in control RBC no PS reacts with TNBS, in patient RBC 52.6% of PS reacted with the probe.

Degradation with bee venom phospholipase A_2 was measured after 20 min of its action. Some lysis (about 7% measured as hemoglobin release) of RBC was observed. Moreover, it cannot be excluded that lysoPC might exert some effect on RBC structure. Therefore, we did not follow up the longer duration of phospholipase A_2 action and the data obtained are used to confirm that some part of PS is degraded by this enzyme (Table 3).

From all these results it is evident that in RBC of the patient part of PS is distributed in the outer leaflet of the phospholipid bilayer.

The maintenance of phospholipid asymmetry in RBC plasma membrane might be the result of two mechanisms: The first and probably more important is the interaction of phospholipids with the cytoskeleton. The second one is the action of a head group selective ATP-dependent phospholipid pump which moves PS to the inner leaflet (Williamson et al. 1987). The changes in the redistribution of PS

Table 1. Merocyanine 540 binding to RBC observed as fluorescence of cells in a fluorescence microscope

Conditions	Patient RBC	Homologous RBC
Fresh smear or RBC	+++	---
Fresh RBC, 2-h incubation in serum of patient	+++	++
Stored RBC, then incubated 2 h in medium	---	---
Stored RBC, then incubated 2 h in patient plasma	++	++

Table 2. Reaction of RBC phospholipids with TNBS[a]

	Control RBC		Patient RBC	
	Fresh	Fresh + plasma of patient	Fresh	Stored + plasma
PE	18.9 ± 0.5	18.0 ± 0.7	12.9 ± 0.3	19.7 ± 0.5
PS	0	28.5 ± 0.9	52.6 ± 0.2	31.5 ± 0.9

[a] 2 ml of washed RBC were incubated in 18 ml buffered Tris saline, pH 8.2, containing 5.5 mM glucose and 0.5% bovine serum albumin with or without 5 mM TNBS for 2 h. The data were calculated as the difference between both incubations and expressed as percentage of phospholipids reacting with TNBS.

Table 3. Degradation of phospholipids by bee venom phospholipase A_A[a]

	Control RBC	Patient RBC
PE	11.2 ± 0.7	17.4 ± 0.7
PC	52.1 ± 0.6	50.8 ± 0.4
PS	0	18.9 ± 0.5
SM	0	0

[a] 2 ml of RBC were incubated in 2 ml medium with 2 mM Ca^{2+}, pH 7.4, 37°C, 15 min, then 40 μg bee venom phospholipase A_2 was added. After 20 min the incubation was stopped by the addition of EDTA (final concentration 5 mM). Data are expressed as percentage of phospholipid content degraded by phospholipase A_2.

in RBC, which we observed in our patient, induced us to try to detect possible changes in the RBC cytoskeleton by indirect immunofluorescence. Our goal was to elaborate suitable and accessible techniques of cytoskeleton visualization in human blood cells, especially in connection with the possibility of routine examination of blood smears for clinical purposes. A detailed description of the methods, the control of specificity of utilized antisera and especially the quite surprising detection of tubulin in human RBC are discussed in Paleček et al. (1988).

Control RBC showed tubulin and actin in the form of a ring over the whole plasma membrane (Fig. 1A, B). Up to 50% of the patient RBC showed irregularily organized aggregates of tubulin and actin in their cytoplasm (Fig. 2A, B).

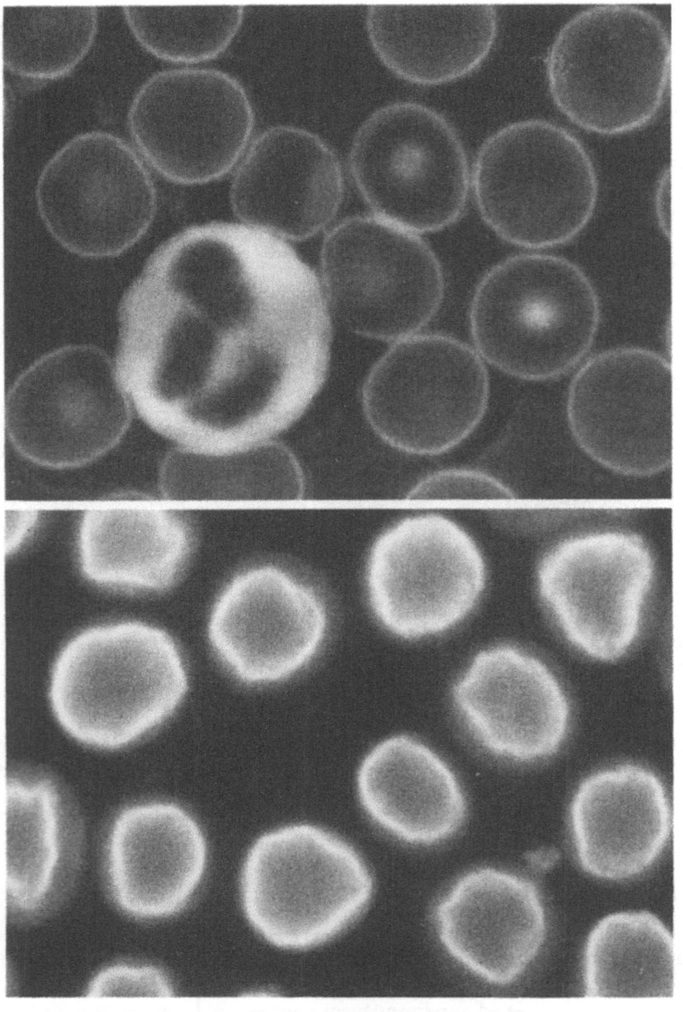

Fig. 1. Tubulin (A) and actin (B) in the RBC of a control

Separation of membrane proteins by SDC-PAGE according to Laemmli (1970) is demonstrated on Fig. 3. The pattern of proteins of patient RBC shows a defect in protein 4.1.

We explain the observed cytoskeleton disorganization as the possible consequence of the missing bonds between cytoskeletal proteins and PS in the inner leaflet of the plasma membrane. The defect in protein 4.1.a and thus the weaker binding of PS to cytoskeleton enables PS to move to the outer leaflet. The external factor (thermolabile factor present in patient plasma) probably binds PS and prevents its flip-flop back into the inner leaflet.

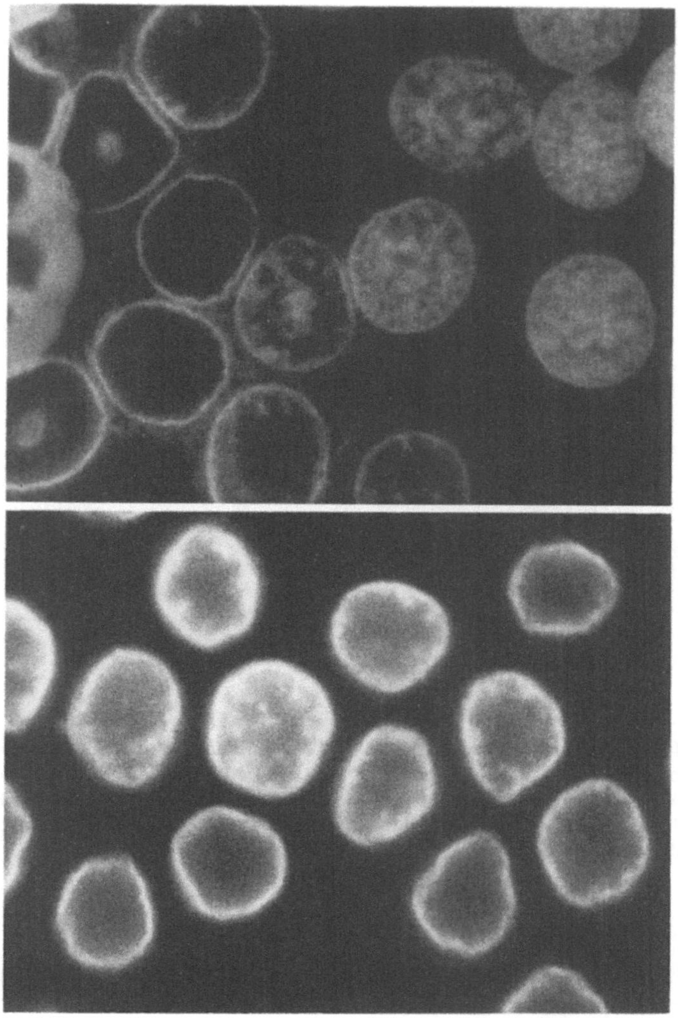

Fig. 2. Distribution of tubulin (**A**) and actin (**B**) in RBC of the patient with AES syndrome. In both cases redistribution of fluorescence in the vicinity of the plasma membrane takes place

The exposure of PS in the outer leaflet might have severe pathological consequences. PS redistribution and some changes in the pattern of cytoskeletal proteins were described in connection with other pathological changes, e.g. in sickle cells and in hereditary spherocytosis (for a review, see Schwartz et al. 1984). Sickle cells or asymmetric ghosts bind to a monolayer of vascular endothelial cells in a five to ten times greater number than ghosts or RBC with a normal phospholipid distribution (Schlegel et al. 1985). The enrichment of the outer leaflet of plasma membrane by synthetic PS derivates increases binding to macrophages

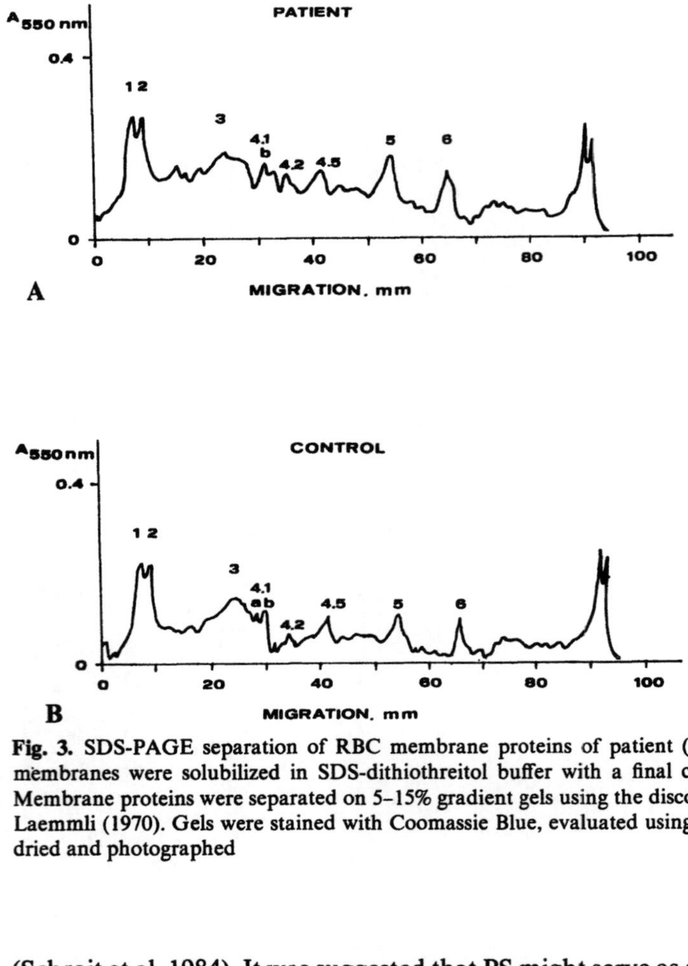

Fig. 3. SDS-PAGE separation of RBC membrane proteins of patient (**A**) and control (**B**). RBC membranes were solubilized in SDS-dithiothreitol buffer with a final concentration of 3% SDS. Membrane proteins were separated on 5–15% gradient gels using the discontinuous buffer system of Laemmli (1970). Gels were stained with Coomassie Blue, evaluated using a scanning densitometer, dried and photographed

(Schroit et al. 1984). It was suggested that PS might serve as a signal for triggering macrophage recognition.

The second example of possible involvement of phospholipids in cell pathology is connected with the phosphoinositide signalling system (Fig. 4). Molecules of phospholipid in plasma membrane, phosphatidylinositol 4,5-biphosphate (PIP_2), are split by the action of phosphodiesterase. Both products are supposed to act as the second messengers: 1,4,5-inositol triphosphate mobilizes calcium ions from endoplasmic reticulum, 1,2-diacylglycerol (DAG) activates protein kinase C (Berridge and Irvine 1984). There is no doubt that this system is very important in all processes requiring calcium ions. In RBC there is little evidence for the involvement of receptors in their physiology. However, it is interesting that they possess all enzymes of the phosphoinositide signalling system. Phosphoinositides in RBC have very high turnover rates (Fig. 5) (Strunecká and Folk 1988). In RBC only one stimulus evoking the diester breakdown of phosphoinositides has been found. It is the ionophor A 23 187.

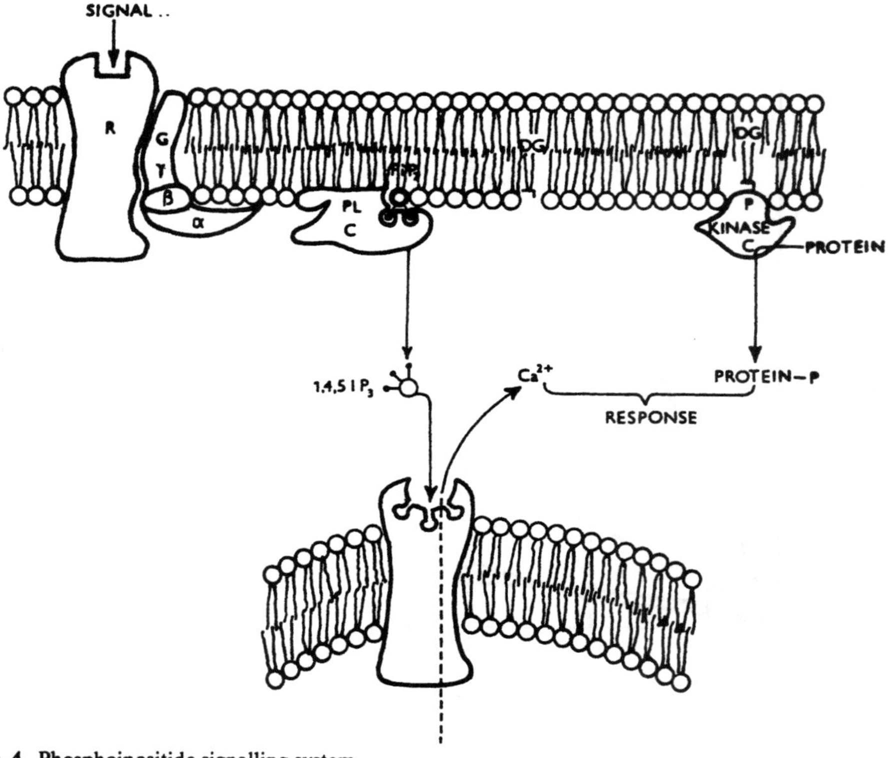

Fig. 4. Phosphoinositide signalling system

The effect of A 23 187 on intact human erythrocytes is shown in Table 4. Isolated human erythrocytes were incubated at 37° C in the presence of $^{32}P_i$ in media with 10 mM EDTA (to buffer divalent cations), 0.15 mM Mg^{2+}, 75 mM K^+, and varying Ca^{2+} concentrations. The composition of the media has been chosen to prevent undesirable side effects of the ion movements. The concentrations of Mg^{2+} and K^+ are set at levels, at which no net changes in their intracellular content will occur. In all experiments A 23 187 free acid was added for 10 min to 10 μM final concentration. The presence of 0.01% DMSO (used as solvent for A 23 187) had no effect on the parameters followed in our experiments (data not shown).

The A 23 187 treatment evoked significant changes of phospholipid metabolism. It induced a decrease in concentrations of PIP_2 and phosphatidylinositol 4-phosphate (PIP) to 17.7 and 70.9% of controls, respectively. With $^{32}P_i$ as precursor, the S.A. of PIP_2 remained unchanged, whereas the S.A. (specific activity) of PIP decreased to 30.3% of control. Thus, the breakdown of PIP must be accompanied by replacement of PIP with lower specific activity, probably originating from monoesteric breakdown of PIP_2. Alternatively, when two metabolic pools of this phospholipid are postulated, the breakdown of the pool with high S.A. would explain the decrease in total S.A. Simultaneously, activation of

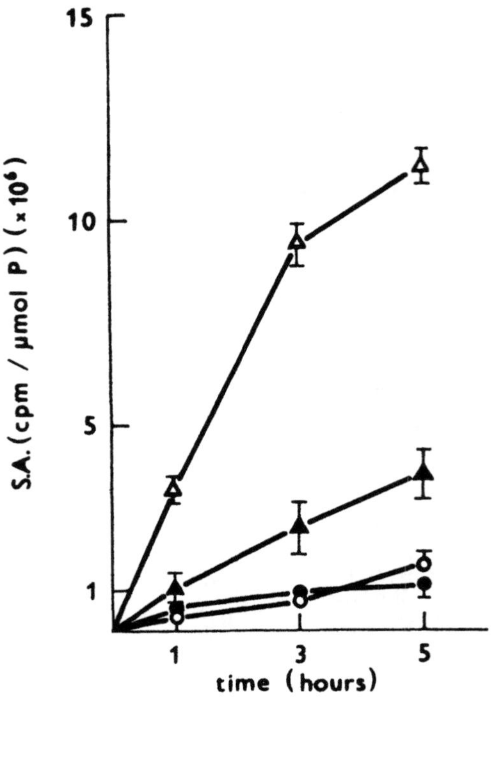

Fig. 5. Time course of $^{32}P_i$ incorporation into phospholipids of human erythrocytes. △ PIP_2; ○ lyso PC; ▲ PIP

Table 4. Effect of A 23 187 on the turnover of phosphoinositides and PA of intact human erythrocytes at varying Ca^{2+} concentrations[a]

PL	Activity (% of controls)			Specific activity (% of controls)		
	Without Ca^{2+}	0.04 µM Ca^{2+}	1.5 mM Ca^{2+}	Without Ca^{2+}	0.04 µM Ca^{2+}	1.5 mM Ca^{2+}
PIP_2	94.6 ± 7.1	99.2 ± 3.6	17.8 ± 3.7*	108.2 ± 9.6	96.7 ± 11.2	100.9 ± 4.1
PIP	104.3 ± 10.0	94.5 ± 10.1	21.5 ± 10.7*	100.1 ± 6.4	90.9 ± 8.1	30.3 ± 12.4**
PA	121.9 ± 23.9	105.3 ± 9.1	425.6 ± 23.5**	94.5 ± 6.1	89.7 ± 11.1	238.5 ± 35.0**

[a] Intact erythrocytes were incubated for 240 min in the presence of $^{32}P_i$ in Ca^{2+} buffered media. 10 µM A 23 187 was added 10 min before the end of the incubation. The results are expressed as % of values found in control incubations ± SEM. The data are from 4 separate experiments. Where indicated, the effect of the ionophore is significant: *($P < 0.01$), **($P < 0.001$).

synthesis of PA was found, resulting in increase in concentration of this phospholipid.

There is still an open question as to whether phosphodiesterase is activated under physiological conditions. This enzyme in RBC is membrane-bound (for a review, see Irvine 1982). At low ionic strengths it is very sensitive to Ca^{2+}, but Downes and Michell (1982) concluded that at physiological ionic strengths, the Ca^{2+} concentration required for its activation is too high. We can hardly accept their opinion that mature RBC keep the phosphoinositide signalling system and the high turnover rate of phosphoinositides as a relict from precursor cells. A number of diseases are thought to involve the increased permeability for Ca^{2+} across the plasma membrane (Al-Jobore et al. 1984).

The physiological function of RBC is to transport oxygen and carbon dioxide. In the course of their journey through the circulation RBC are subjected to continuous shape deformations. Also, the loss of phosphoinositides during aging or metabolic depletion is connected with the shape changes.

We investigated whether it is possible to mimic these shape changes by the products of PIP_2 breakdown, i.e. IP_3 and/or diacylglycerol. The cells were incubated in medium with a very low Ca^{2+} concentration (0.04 μM) corresponding to the cytosolic free Ca^{2+} level. The cells were permeabilized with saponin. Five μM IP_3 evoked the disorganization of the spectrin network. The antibody against spectrin was used for the visualization of the spectrin network (Fig. 6). The changes in the shape of RBC were observed, i.e. the speculation and the formation of vesicles. The vesicles do not contain spectrin. Figure 7 shows the vesicle visualization using antibody against tubulin. Since IP_3 in RBC is rapidly metabolized, 10 min after the IP_3 addition the cells returned to a normal appearance. DAG evoked the RBC fusion.

The physiological stimulus evoking the breakdown of PIP_2 in RBC remains to be found. Our results demonstrate that both PIP_2 and PS are closely connected with the regulation of membrane skeleton network. Our full understanding of these interactions might enable new approaches to be made in the therapy of some hematological disorders.

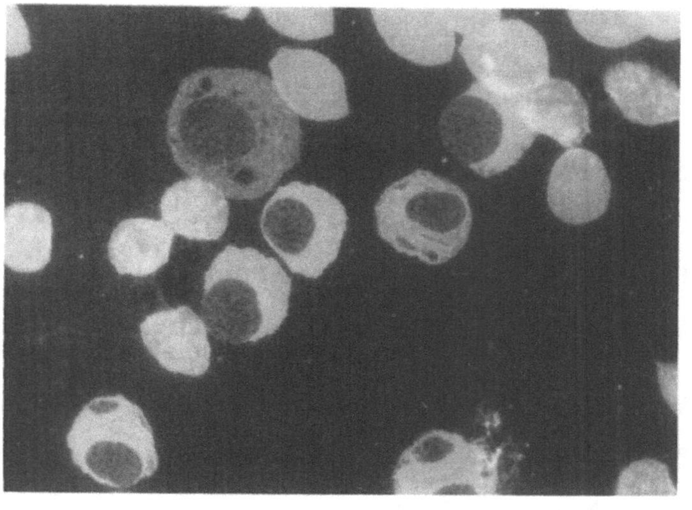

Fig. 6. The effect of 5 μMIP_3 on spectrin organization. RBC were incubated in medium with 0.04 μM Ca^{2+}, permeabilized with saponin (20 μg/ml) and IP_3 was added. Antibody against spectrin was used for immunofluorescence staining of spectrin

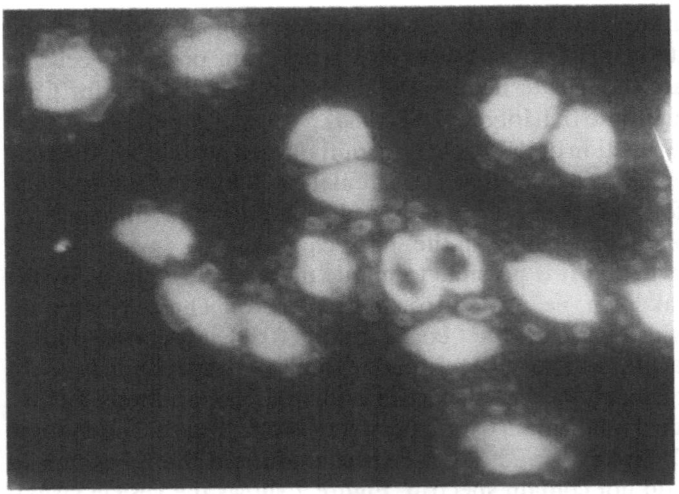

Fig. 7. Formation of vesicles after the action of 5 $\mu M IP_3$. For immunofluorescence staining monoclonal antibody against α - tubulin was used

References

Al-Jobore A, Minocherhomjee AM, Villalobo A, Roufogalis BD (1984) Active calcium transport in normal and abnormal human erythrocytes. In: Kruckeberg WC (ed) Erythrocyte membranes, vol 3: Recent clinical and experimental advances. Liss, New York, pp 243–292

Berridge MJ, Irvine RF (1984) Inositol trisphosphate, a novel second messenger in cellular signal transduction. Nature (London) 312:5992:315–321

Boxley LD (1971) Autoerythrocyte sensitization (painful bruising) syndrome. Proc R Soc London 64:24–25

Downes P, Michell RH (1982) Phosphatidylinositol 4-phosphate and phosphatidylinositol 4,5-biphosphate:lipids in search of a function. Cell Calcium 3:467–502

Gardner FH, Diamond LK (1955) Autoerythrocyte sensitization: a form of purpura producing painful bruising following autosensitization to red blood cells in certain women. Blood 10:675–690

Gordesky SE, Marinetti GV (1973) The asymmetric arrangement of phospholipids in the human erythrocyte membrane. Biochem biophys Res Commun 50:1027–1031

Groch GS, Finch SC, Rogoway W, Fisher DS (1966) Studies in the pathogenesis of autoerythrocyte sensitization syndrome. Blood 28:19–33

Hanna WT, Fitzpatrick MR, Kraus S, Dunn CHDR (1981) Psychogenic purpura (autoerythrocyte sensitization). S Med J 74:538–442

Hersle K, Mobacken H (1969) Autoerythrocyte sensitization syndrome (painful bruising syndrome). Report of two cases and review of the literature. Br J Dermatol 81:574–587

Irvine RF (1982) The enzymology of stimulated inositol lipid turnover. Cell Calcium 3:295–309

Laemmli UK (1970) Clevage of structural proteins during the bacteriophage T4. Nature (London) 227:680–685

Paleček J, Mácha J, Strunecká A (1988) Methods for the visualization of tubulin vimentin, actin and spectrin in human and frog blood smear cells. Histochemistry (in press)

Pevny R, Metz J, Wernze H, Deuchert K, Kustner H (1982) Das erythrozytäre Sensibilisierungssyndrom (Gardner-Diamond-Syndrom, Painful Bruising Syndome Sharp). Hautartzt 33:251–256

Ratnoff OD, Agle DP (1968) Psychogenic purpura. A re-evaluation of the syndrome of autoerythrocyte sensitization. Medicine 47:475–500

Schlegel RA, Prendergast TW, Williamson P (1985) Membrane phospholipid asymmetry as a factor in erythrocyte-endothelial cell interactions. J Cell Physiol 123:215–218
Schroit AJ, Tanaka Y, Madsen J, Fidler IJ (1984) The recognition of RBC by macrophages: role of phosphatidylserine and possible implications of membrane phospholipid asymmetry. Biol Cell 51:227–238
Schwartz RS, Chiu DTY, Lubin B (1984) Studies on the organization of plasma membrane phospholipids in human erythrocytes. In: Kruckeberg WC (ed) Erythrocyte membranes, vol 3: Recent clinical and experimental advances. Liss, New York, pp 89–122
Sörensen RU, Newman AJ, Gordon EM (1985) Psychogenic purpura in adolescent patients. Clin Pediat 24:700–704
Steyrer K, Gschnait F, Lindlbauer R (1982) Psychogene purpura. Z Hautkrankh 58:9–14
Stocker WW, McIntyre OR, Clendenning WE (1977) Psychogenic purpura. Arch Dermatol Res 113:606–609
Strunecká A, Folk P (1988) Phospholipid biosynthesis in mature human erythrocytes. Gen Physiol Biophys 7:205–216
Verkleij AJ, Zwaal RFA, Roelofsen B, Comfurius P, Kastelijn D, Van Deenen LLM (1973) The asymmetric distribution of phospholipids in the human red cell membrane. A combined study using phospholipases and freeze-etch electron microscopy. Biochim Biophys Acta 323:178–193
Waldorf DS, Lipkin G (1968) Sensitization to electrocytes: a distinctive syndrome. J Am Med Assoc 203:597–599
Williamson P, Bateman J, Kozarsky K, Mattocks K, Hermanowicz N, Choe HR, Schlegel RA (1982) Involvement of spectrin in the maintenance of phase-state asymmetry in the erythrocyte membrane. Cell 30:725–733
Williamson P, Algarin L, Bateman J, Choe HR, Schlegel RA (1985) Phospholipid asymmetry in human erythrocyte ghosts. J Cell Physiol 123:209–214
Williamson P, Antia R, Schlegel RA (1987) Maintenance of membrane phospholipid asymmetry. FEBS Lett 219:316–320
Zwaal RFA, Roelofsen B, Comfurius P, Van Deenen LLM (1975) Organization of phospholipids in human red cell membranes as detected by the action of various purified phospholipases. Biochim Biophys Acta 406:83–96

Changes in the Structure and Function of the Red Cell Membrane Skeleton and Hemolytic Anemias

P. JAROLIM, S.-C. LIU, Th. L. COETZER, M. LAHAV and J. PALEK

Summary

This chapter reviews three molecular defects of the red cell membrane skeleton. In all three cases, the immediate cause of hemolysis are the abnormalities of skeletal protein interactions. However, only in one case, the primary cause of hemolysis is a hereditary defect of a molecule of the membrane skeleton, while in the other two cases the skeletal defect is either potentiated by an acquired disorder or caused by a mutation in the hemoglobin molecule. The first case reviewed is a strictly hereditary defect of the skeleton: A mutation of the protein spectrin which is expressed clinically as a severe elliptocytic hemolytic anemia. The second one involves a combination of an inherited spectrin defect with an acquired circulatory disorder which manifests as a severe life-threatening hemolytic anemia. Finally, an example is given of a defect in skeletal protein interactions which is not caused by mutations of skeletal proteins. In this case, irreversible hemichromes, the final products of intracellular hemoglobin denaturation, were found to release hemin which decreases the skeletal stability by weakening the spectrin-protein 4.1-actin interaction.

1 Structure and Function of the Membrane Skeleton

Since several excellent reviews on the structure of the red cell membrane skeleton are available [1–4], only a brief overview of its structure is given here. Two subunits of the fibre-like protein spectrin, α- and β-spectrin, are present in the skeleton as heterodimers that associate at their head regions into twice as long tetramers and, to a lesser extent, hexamers and higher oligomers. Spectrin tetramers are the basic functional units of spectrin in the red cell membrane skeleton. In the normal red cell membrane, spectrin tetramers are assembled into a relatively uniform hexagonal lattice [5,6]. The spectrin network is interconnected by actin and protein 4.1 at the tail regions of the spectrin heterodimers [7] and the basic structure of the membrane skeleton is completed by binding of ankyrin to β-spectrin close to the spectrin self-association site [1–3]. The skeleton is attached to the membrane by several linkages, including interaction of ankyrin with the anion channel protein,

Institute of Hematology and Blood Transfusion, 128 20 Prague 2, Czechoslovakia, and the Department of Biomedical Research, St. Elizabeth's Hospital of Boston, Tufts University School of Medicine, Boston, MA 02135, USA

A. Azzi, Z. Drahota and S. Papa
Molecular Basis of Membrane-Associated Diseases

band 3 [8], interactions of protein 4.1 with the integral proteins glycophorin A and C, and with the anion channel protein [9]. The ankyrin-band 3 interaction is further strengthened by protein 4.2 [10,11]. Moreover, further connections are provided by a direct interaction of spectrin and protein 4.1 with the negatively charged lipids of the inner membrane lipid layer [12–15]. The basic structure of the red cell membrane skeleton is schematically depicted in Fig. 1. Moreover, several minor proteins have been described, the role of which in the red cell membrane skeleton is less evident [16–22].

The major role of the red cell membrane skeleton is to maintain the cell deformability and integrity during the passage of erythrocytes through the microcirculation [23]. It has also been shown to play a role in the maintenance of lipid asymmetry, and hence of cell shape, and also in the regulation of the lateral mobility of integral proteins [24]. Generally, failure of the skeletal proteins to associate would lead to a decrease in membrane stability. On the other hand, excessive protein associations, such as the crosslinking of proteins, might severely curtail membrane deformability.

2 Abnormalities of the Spectrin Molecule

Several abnormalities of the content and/or structure of the protein constituents of the red cell membrane skeleton have been described in patients with hereditary hemolytic anemias (reviewed in [25,26]). These defects are responsible for a shape change and/or a shortened survival of red cells in the circulation. Of particular interest is a group of mutations in the region of the α-spectrin molecule which contains the site where spectrin heterodimers associate into tetramers, the α^I domain [27].

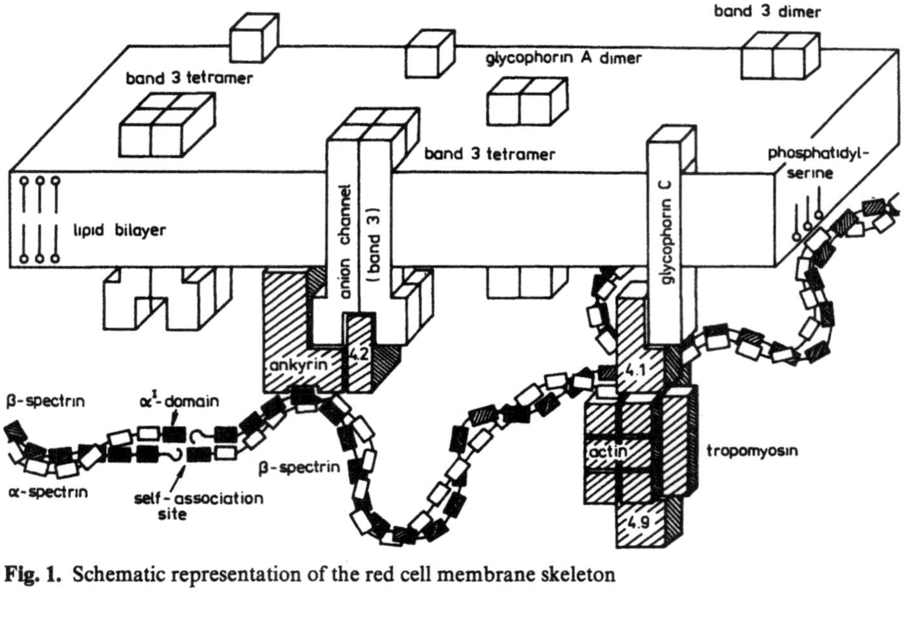

Fig. 1. Schematic representation of the red cell membrane skeleton

These mutations are designated according to the molecular weight of the most prominent abnormal tryptic cleavage product that is formed during limited tryptic digestion of a crude spectrin extract. Molecular defects of the domain I of α-spectrin generally cause hereditary elliptocytosis of various severities or, when combined with a spectrin deficiency, lead to hereditary pyropoikilocytosis [28,29]. A number of mutants of the domain I of α-spectrin have been described [25].

3 Severe Hemolytic Anemia Due to a Spectrin Mutation

A spectrin mutant causing elliptocytic transformation of erythrocytes is the mutant called spectrin $\alpha^{I/74}$ [30–31]. It may be detected by non-denaturing electrophoresis of the low ionic strength spectrin extracts and by one- or two-dimensional electrophoresis of the cleavage products of the limited tryptic digestion of the spectrin extracts. In the heterozygous case, the clinical expression of this mutant is a mild to severe hereditary elliptocytosis [28,31] or, when combined with spectrin deficiency, a hereditary pyropoikilocytosis [32]. The very rare homozygotes suffer from a life-threatening transfusion-dependent hemolytic anemia [24,25,33].

We recently had an opportunity to study a large kindred with this mutation. In all affected members of the family, the clinical expression was particularly severe for the heterozygous state, requiring transfusions and inevitably leading to splenectomy (Jarolim et al. unpublished). The fragmentation of the isolated skeletons, Triton shells, was remarkably fast showing the extremely low stability of the membrane skeletons.

When the low ionic strength spectrin extracts were subjected to non-denaturing gel electrophoresis, we detected over 50% of spectrin dimers. This proves the total inability of the mutant spectrin to self-associate into tetramers. This molecular defect is an example of a strictly hereditary mutation of the membrane skeletal protein.

4 Severe Hemolytic Anemia Due to the Combination of a Spectrin Mutation with an Acquired Thrombotic Microangiopathy

Among the defects of the spectrin molecule, the α-spectrin mutation designated Sp $\alpha^{I/65}$ [34,35] is typically associated with a mild hereditary elliptocytosis [28,29]. In the heterozygous form, this spectrin mutation leads to only mild alterations of spectrin self-association. Consequently, it presents as an asymptomatic condition characterized by elliptocytosis with occasional rod-shaped cells [24,25].

However, we recently had the opportunity to study two patients with this mutation who presented with a severe, life-threatening hemolysis and red cell morphology similar to that in hereditary poikilocytosis [24,25,36]. The first patient was referred for chronic renal transplant rejection with interstitial fibrosis and vascular intimal proliferation. In the other patient, thrombotic thrombocytopenic purpura [37,39] was diagnosed. Hence, both patients had a secondary thrombotic microangiopathy [40] superimposed on an inherited spectrin mutation.

The term thrombotic microangiopathic hemolytic anemia, or briefly thrombotic microangiopathy [40], comprises the whole spectrum of clinical syndromes that are characterized by red cell fragmentation with hemolysis, thrombocytopenia and diffuse abnormalities of the arteriolar and precapillary endothelium. While the pathogenesis in various types of thrombotic microangiopathy may differ, the impact of these syndromes on the red cell morphology is similar. The main feature is the presence of fragmented forms with occasional classic helmet cells, microspherocytes, and large polychromatophils or orthochromatic erythroblasts.

However, in both our patients, the erythrocyte abnormalities were more severe than in most patients with renal transplant rejection or with thrombotic thrombocytopenic purpura [41] and, at the same time, the clinical course by far exceeded the usual manifestations of anemias associated with the Sp $\alpha^{I/65}$ mutation. These cases serve as an example of how an underlying red cell membrane skeletal defect may aggravate the clinical course of thrombotic microangiopathic hemolytic anemias.

5 Hemolytic Anemia Due to the Effect of Hemoglobin Denaturation Products on Protein Associations in the Red Cell Membrane Skeleton

As an illustration of an acquired abnormality of protein interactions in an otherwise normal red cell skeleton, we present here the studies of the effects of hemoglobin oxidation products on red cell skeletal stability and membrane protein interactions.

Oxidant damage to hemoglobin leads to formation of methemoglobin and reversible and irreversible hemichromes which precipitate and form Heinz bodies [42,43]. Heinz bodies attach to the membrane and cause injury leading to hemolysis [44]. To localize the molecular site of such injury and elucidate its mechanism, we studied the effect of methemoglobin, reversible hemichromes and irreversible hemichromes on the red cell membrane skeleton.

The methemoglobin effect was identical to that of native hemoglobin: It stabilized the membrane skeleton and promoted formation of spectrin tetramers from spectrin dimers [45]. In contrast, irreversible hemichromes reduced the skeletal stability in a concentration-dependent fashion by decreasing the formation of the spectrin-protein 4.1-actin complex. On the other hand, irreversible hemichromes had no effect on spectrin tetramer association in solution, binding of spectrin to inside-out vesicles and spectrin-actin binding in the absence of protein 4.1 [46,47]. The effect of irreversible hemichromes was not prevented by dithiothreitol, suggesting that it is not mediated by release of free radicals or oxidation. At the same time, it was prevented by removal of free hemin by passing irreversible hemichromes over a Dowex AG 1-X8 column [48]. Incubation of irreversible hemichromes led to release of up to 5% of free hemin. Pure hemin in these concentrations had the same effect on red cell skeletal stability [49] and spectrin-protein 4.1-actin interaction as irreversible hemichromes. In striking contrast, virtually no hemin was released either from hemoglobin or methemoglobin, even after long periods of incubation.

We conclude that the deleterious effect of irreversible hemichromes on red cell membranes is mediated by the release of free hemin, which may play a role in the destruction of erythrocytes in the so-called Heinz body containing hemolytic anemias or hereditary hemolytic anemias with unstable hemoglobins. At the same time, the absence of hemin release from methemoglobin is in good correlation with the absence of hemolysis in congenital methemoglobinemias. Thus, these results may serve as an example of a defect in skeletal protein interactions which is acquired during the aging of the red cell.

These three examples illustrate the role of both intrinsic defects and external disturbances to the function of the red cell membrane skeleton for pathogenesis of a group of hemolytic anemias. Further characterization of the molecular defects may bring important insight to the understanding of the ethiology, pathogenesis and mode of inheritance of these disorders and, moreover, provide important diagnostic clues.

References

1. Cohen CM (1983) The molecular organization of the red cell membrane skeleton. Sem Hematol 20:141–158
2. Marchesi VT (1985) Stabilizing infrastructure of cell membranes. Annu Rev Cell Biol 1:531–561
3. Bennett V (1985) The membrane skeleton of human erythrocytes and its implications for more complex cells. Annu Rev Biochem 54:273–304
4. Lazarides E (1987) From genes to structural morphogenesis: the genesis and epigenesis of a red blood cell. Cell 51:345–356
5. Byers TJ, Branton D (1985) Visualization of the protein associations in the erythrocyte membrane skeleton. Proc Natl Acad Sci USA 82:6153–6157
6. Liu SC, Derick LH, Palek J (1987) Visualization of the hexagonal lattice in the erythrocyte membrane skeleton. J Cell Biol 104:361–376
7. Cohen CM, Foley SF (1984) Biochemical characterization of complex formation by human erythrocyte spectrin, protein 4.1 and actin. Biochemistry 23:6091–6098
8. Hargreaves WR, Giedd KN, Verkleij A, Branton D (1980) Reassociation of ankyrin with band 3 in erythrocyte membranes and in lipid vesicles. J Biol Chem 255:11965–11972
9. Pasternack GR, Anderson RA, Leto TL, Marchesi VT (1985) Interactions between protein 4.1 and band 3. J Biol Chem 269:3676–3683
10. Korsgren C, Cohen CM (1986) Purification and properties of human erythrocyte band 4.2. J Biol Chem 261:5536–5543
11. Korsgren C, Cohen CM (1988) Associations of human erythrocyte band 4.2: Binding to ankyrin and to the cytoplasmic domain of band 3. J Biol Chem (in press)
12. Cohen AM, Liu SC, Derick LH, Palek J (1986) Ultrastructural studies of the interaction of spectrin with phosphatidylserine liposomes. Blood 68:920–926
13. Cohen AM, Liu SC, Lawler J, Derick L, Palek J (1988) Identification of the protein 4.1 binding site to phosphatidylserine vesicles. Biochemistry 27:614–619
14. Rybicki AC, Heathe R, Lubin B, Schwartz RS (1988) Human erythrocyte protein 4.1 is a phosphatidylserine binding protein. J Clin Invest 81:255–260
15. Burn P (1988) Amphitropic proteins: a new class of membrane proteins. Trends Biochem Sci 13:79–83
16. Gardner K, Bennett V (1986) A new erythrocyte membrane-associated protein with calmodulin binding activity. J Biol Chem 261:1339–1348
17. Gardner K, Bennett V (1987) Modulation of spectrin-actin assembly by erythrocyte adducin. Nature (London) 328:359–362
18. Miche SM, Mooseker M, Morrow JS (1987) Erythrocyte adducin: a calmodulin-regulated actin-bundling protein that stimulates spectrin-actin binding. J Cell Biol 105:2837–2845

19. Fowler V (1986) An actomyosin contractile mechanism for erythrocyte shape transformations. J Cell Biochem 31:1-9
20. Fowler V, Bennett V (1984) Erythrocyte membrane tropomyosin. J Biol Chem 259:5978-5989
21. Fowler V, Davis JQ, Bennett V (1987) Human erythrocyte myosin: identification and purification. J Cell Biol 100:47-55
22. Hall TG, Bennett V (1987) Regulatory domains of erythrocyte ankyrin. J Biol Chem 262:10537-10545
23. Mohandas N, Chasis JA, Shohet SB (1983) The influence of membrane skeleton on red cell deformability, membrane material properties and shape. Sem Hematol 20:225-240
24. Tsuji A, Ohnishi S (1986) Restriction of the lateral motion of band 3 in the erythrocyte membrane by the cytoskeletal network: dependence on spectrin association state. Biochemistry 25:6133-6139
25. Palek J (1987) Hereditary elliptocytosis, spherocytosis and related disorders: consequences of a deficiency or a mutation of membrane skeletal proteins. Blood Rev 1:147-168
26. Palek J (1985) Hereditary elliptocytosis and related disorders. Clin Haematol 14:45-87
27. Palek J, Liu S-C, Lawler J, Coetzer TL (1986) Molecular defects of alpha spectrin in hereditary elliptocytosis and pyropoikilocytosis. In: Bennett V, Cohen CM, Lux SE, Palek J (eds) Membrane skeletons and cytoskeletal-membrane associations. UCLA Symp Mol Cell Biol, New Ser, vol 38. Liss, New York, pp 357-369
28. Coetzer T, Lawler J, Prchal JT, Palek J (1987) Molecular determinants of clinical expression of hereditary elliptocytosis and pyropoikilocytosis. Blood 70:766-772
29. Palek J, Coetzer T (1987) Clinical expression of alpha spectrin mutants in hereditary elliptocytosis. Blood Cells 13:237-50
30. Lawler J, Liu SC, Palek J, Prchal J (1984) A molecular defect of spectrin in a subset of patients with hereditary elliptocytosis. J Clin Invest 73:1688-1695
31. Lawler J, Liu SC, Palek J, Prchal J (1982) Molecular defect of spectrin in hereditary pyropoikilocytosis: alterations in the trypsin-resistant domain involved in spectrin self-association. J Clin Invest 70:1019-1030
32. Liu SC, Palek J, Prchal J, Castleberry RP (1981) Altered spectrin dimer-dimer association and instability of erythrocyte membrane skeletons in hereditary pyropoikilocytosis. J Clin Invest 68:597-605
33. Palek J, Coetzer T, Lahav M, Jarolim P, Lawler J, Wang W, Prchal JT (1987) Clinical and biochemical expression of double heterozygotes for α-spectrin mutants that involve the spectrin heterodimer self-association site. Blood 70 (Suppl 1):55a
34. Lawler J, Coetzer TL, Palek J, Jacob HS, Luban N (1985) Sp alphaI/65: a new variant of the alpha subunit of spectrin in hereditary elliptocytosis. Blood 66:706-709
35. Alloisio N, Guetarni D, Morle L, Pothier B, Ducluzeau MT, Soun A, Colonna P, Clerc M, Philippe N, Delaunay J (1986) Sp alphaI/65 hereditary elliptocytosis in North Africa. Am J Hematol 23:113-122
36. Dacie J (1985) The haemolytic anaemias, 3rd edn. Livingstone, Edinburgh, pp 134-215
37. Kwaan HC (1987) Introduction: thrombotic microangiopathy. Sem Hematol 24:69-70
38. Byrnes JJ, Moake JL (1986) Thrombotic thrombocytopenic purpura and the hemolytic-uraemic syndrome: evolving concepts and therapy. Clin Haematol 15:413-442
39. Kwaan HC (1987) Clinicopathologic features of thrombotic thrombocytopenic purpura. Sem Hematol 24:71-81
40. Kwaan HC (1987) Miscellaneous secondary thrombotic microangiopathy. Sem Hematol 24:141-147
41. Brain MC, Dacie JV, Hourihane DOB (1962) Microangiopathic haemolytic anaemia: the possible role of vascular lesions in pathogenesis. Br J Haematol 8:358-374
42. Peisach J, Blumberg WE, Rachmilewitz EA (1975) The demonstration of ferrihemochrome intermediates in Heinz body formation following the reduction of oxyhemoglobin A by acetylphenylhydrazine. Biochim Biophys Acta 393:404-418
43. Winterbourn CC, Carrell RW (1974) Studies on hemoglobin denaturation and Heinz body formation in the unstable hemoglobins. J Clin Invest 54:678-689
44. Jacob HS (1970) Mechanism of Heinz body formation and attachment to red cell membrane. Sem Hematol 7:341-354

45. Liu SC, Palek J (1984) Hemoglobin enhances the self-association of spectrin heterodimers in human erythrocytes. J Biol Chem 259:11556–11562
46. Lahav M, Liu SC, Jarolim P, Palek J (1987) Red cell membrane skeletal damage during hemoglobin denaturation. Blood 70 (Suppl 1): 40a
47. Jarolim P, Lahav M, Liu SC, Palek J (1988) Effect of hemoglobin oxidation products on red cell membrane skeletal stability and membrane protein interactions. Proc 14th Int Congr Biochemistry, Prague 1988
48. Liu SC, Zhai S, Palek J (1988) Detection of hemin release during hemoglobin S denaturation. Blood (in press)
49. Liu SC, Zhai S, Lawler J, Palek J (1985) Hemin-mediated dissociation of erythrocyte membrane skeletal proteins. J Biol Chem 260:12234–12239

Regulation of Intracellular Ca^{2+} in Pancreatic Acinar Cells by Membrane Pumps and Channels

R. L. DORMER

Summary

Measurement and manipulation of cytoplasmic free Ca^{2+} in intact pancreatic acinar cells have suggested its primary role in regulating secretion in response to acetylcholine and cholecystokinin. Evidence suggests that Ca^{2+} is initially released from an intracellular store but increased Ca^{2+} entry is required for maintained stimulation. The mechanism of increased Ca^{2+} entry is unknown: no evidence for voltage- or receptor-operated channels has been found. Stimulation of Ins(1,4,5)P_3 and Ins(1,3,4,5)P_3 formation within 5 s of receptor occupancy supports recent data suggesting their involvement in activating Ca^{2+} entry. Intracellular Ca^{2+} release is accompanied by net loss of cellular Ca^{2+} released into the ductal secretions and actively extruded across the basolateral membrane. A Ca^{2+}, Mg-ATPase and Na^+/Ca^{2+} exchange has been demonstrated in purified plasma membranes but their properties do not allow a definition of which processes are most important physiologically.

Subcellular fractionation studies have suggested that the site of the intracellular Ca^{2+} store is a component of the rough endoplasmic reticulum, which, in the resting cell, is maintained by a Ca^{2+}, Mg-ATPase and released following stimulation, by the action of Ins(1,4,5)P_3. The mechanisms for Ca^{2+} release might include a Ca^{2+}-activated component. Heterogeneity of Ca^{2+} stores and their mechanism of release from subcompartments of the endoplasmic reticulum may provide a link between Ca^{2+} release at a site closely apposed to the basolateral membrane and its release at sites nearer the apical end of the cell, causing activation of secretion.

1 Introduction

The pancreatic acinar cell is a polarized epithelial cell which secretes both electrolytes and proteins, primarily in response to acetylcholine and cholecystokinin (CCK). In some species (cat, rat and guinea pig) secretion is also stimulated by secretin. Receptors for these secretagogues have been demonstrated, by ligand-binding and autoradiography, to be present on the basolateral membrane of the acinar cell [1,2], whereas secretion occurs at the apical membrane as a result of activation by intracellular regulators. It should be borne in mind, therefore, that the

Department of Medical Biochemistry, University of Wales College of Medicine, Heath Park, Cardiff, CF4 4XN, UK

A. Azzi, Z. Drahota and S. Papa
Molecular Basis of Membrane-Associated Diseases

site of generation of these regulators may be several microns from the site of their stimulatory action.

Electrolyte secretion is thought to conform to the general model [3,4] in which inward cotransport of Na^+, K^+ and Cl^- (in the ratio 1:1:2) is driven by the Na^+ gradient formed by Na^+, K^+-ATPase activity in the basolateral membrane [5]. As a result, intracellular Cl^- activity is above its electrochemical equilibrium potential such that the opening of putative Cl^- channels in the apical membrane causes Cl^- secretion followed by paracellular movement of Na^+ and water. In some species such as pig and man [6,7] basolateral K^+ channels have been demonstrated by patch-clamp techniques. The opening of these channels is controlled by membrane potential and Ca^{2+} on the internal face of the membrane and is thought to allow recycling of K^+ following stimulation of the cell. In rodent species, such as rat and mouse, non-selective cation channels regulated by intracellular Ca^{2+}, but not the selective K^+ channels, are present [8]; the role of these channels in the secretion of electrolytes and water is not clear.

Secretion of digestive enzymes, which are stored in secretory granules, also occurs at the apical membrane by exocytosis [9].

2 Intracellular Ca^{2+} as a Regulator of Pancreatic Secretion

With the development of sensitive indicators of ionized Ca^{2+} (see [10]), in particular the Ca^{2+}-activated photoproteins and the fluorescent indicators quin 2 and fura 2, which can respond to changes in Ca^{2+} concentration in the time scale of 0.1 s or less, over the range of 10^{-8}–10^{-6} M (10^{-5} M for photoproteins), it has become possible to determine directly the effects of pancreatic secretagogues on the cytoplasmic free Ca^{2+} concentration in acinar cells [11].

Using a method of hypotonic loading to introduce the Ca^{2+}-activated photoprotein aequorin into isolated rat acini [20], it was shown that the cells responded to carbachol with a transient increase in free Ca^{2+} which preceded the onset of amylase secretion. The use of quin 2, which is introduced into the cell as a permeant ester and then entrapped as a result of its hydrolysis to the free acid by intracellular esterase activity [12], greatly extended the number and detail of measurements made (see [11]). It was shown that with a maximally effective concentration of carbachol, the cytoplasmic free Ca^{2+} concentration increased after a lag of approximately 1 s to a peak at 5 s [13]. The concentration at which the free Ca^{2+} stabilized during sustained stimulation (at least 3 min) has been variously reported as indistinguishable from [14,15] and two to three times [13,16,17] the unstimulated level.

A second criterion which implicates Ca^{2+} as a primary regulator of secretion is to introduce it directly into the cell and demonstrate stimulation of secretion. Bivalent cation ionophores such as A23187 were first shown to stimulate enzyme secretion from the pancreas [18,19] and more recently used to demonstrate a positive correlation between increasing concentrations of cytoplasmic Ca^{2+} and amylase release [17,20]. In addition, Iwatsuki and Petersen [21] showed that iontophoresis of Ca^{2+} into the acinar cell caused changes in membrane potential and input resistance similar to those evoked by acetylcholine.

A third criterion demonstrating the importance of intracellular Ca^{2+} in regulating pancreatic secretion is that buffering the rise in free Ca^{2+} concentration should inhibit stimulation of secretion. EGTA or BAPTA, incorporated into isolated acini by hypotonic swelling [22] to concentrations estimated to be 1–2 mM, caused 80% inhibition of the maximum stimulation of amylase release by carbachol. Similar concentrations of quin 2 + BAPTA (incorporated as acetomethoxyesters) also inhibited carbachol-stimulated amylase release [13]. Injection of EGTA into acinar cells [23] reduced acetylcholine-induced membrane depolarization and resistance reduction.

Although the evidence described above firmly implicates intracellular Ca^{2+} as a primary regulator of pancreatic secretion stimulated by CCK and acetylcholine, inconsistencies remain. In some studies measuring cytoplasmic free Ca^{2+} concentration using quin 2, the dose dependence on secretagogues for stimulation of enzyme release and increased cytoplasmic Ca^{2+} were coincident [13,16,17]; in others, the lowest secretagogue concentrations causing increased secretion caused no detectable increase in cytoplasmic Ca^{2+} [14,16]. This does not appear to be due to lack of sensitivity or buffering of Ca^{2+} by quin 2, since a recent study by Bruzzone et al. [24] has demonstrated the same phenomenon using fura 2, which was present at approx. ten-fold lower concentration in the cytoplasm than previously used for quin 2.

A second inconsistency has been revealed in studies measuring stimulation of secretion by Ca^{2+}-ionophores [17]: for a similar rise in free Ca^{2+}, the ionophores stimulated less amylase release than did physiological agonists. This might be explained because ionophores enter cells [25] and may release Ca^{2+} from an inappropriate store or may desensitize or inhibit subsequent activation steps since they cause a prolonged rise in cytoplasmic Ca^{2+} rather than the transient rise evoked physiologically.

However, it is also possible that a second intracellular messenger system, independent of an increased cytoplasmic free Ca^{2+} concentration, is involved in the stimulation of secretion. A likely candidate has been proposed to be protein kinase C which is activated by diacylglycerol [26], a product of phospholipase C action. Exogenously added diacylglycerol, like phorbol esters, which also activate protein kinase C [26], stimulated enzyme release [15,17,27,28] without increasing cytoplasmic free Ca^{2+} [15,17]. These agents also potentiated Ca^{2+}-ionophore-induced stimulation of secretion [17,28] and increased the sensitivity of exocytosis to Ca^{2+} in acinar cells permeabilized by intense electric fields [29]. In addition, Bruzzone et al. [24] have presented evidence that at low secretagogue concentrations, which do not evoke detectable increases in cytoplasmic free Ca^{2+}, protein kinase C is translocated from a soluble to a particulate cellular compartment, indicating its activation.

3 Movements of Ca^{2+} in Pancreatic Acinar Cells

Measurements of cytoplasmic free Ca^{2+} and manipulation of intracellular Ca^{2+}, as discussed above, have given the clearest evidence to date that Ca^{2+} is a primary regulator of secretion but do not indicate the source of this Ca^{2+} nor the mechanisms by which Ca^{2+} moves into or out of the cell.

Earlier studies on the dependence of the stimulation of enzyme release on extracellular Ca^{2+}, in which only acute removal of Ca^{2+} was considered, demonstrated that stimulation would occur normally for 10–15 min in the absence of Ca^{2+} but thereafter declined to the unstimulated rate [30–33]. Studies of $^{45}Ca^{2+}$ fluxes on a variety of in vitro preparations consistently demonstrated an increased efflux of $^{45}Ca^{2+}$ from prelabelled cells within 1–2 min of stimulation [34–37]. This $^{45}Ca^{2+}$ efflux was independent of extracellular Ca^{2+} [34–36] and shown to represent net Ca^{2+} loss from the cell [37–39]. This phase of net efflux was followed after 10 min of sustained stimulation by net uptake. These studies supported the hypothesis [34] that Ca^{2+} was initially released from an intracellular site, which explained the phase of stimulation of secretion independent of extracellular Ca^{2+}, and were consistent with the requirement for Ca^{2+} to sustain a stimulated rate of secretion.

3.1 *Mechanism of Ca^{2+} Entry*

Ca^{2+} entry into pancreatic acinar cells has been studied either by the measurement of $^{45}Ca^{2+}$ influx or by electrophysiological techniques. In isolated acinar cells or acini, increased $^{45}Ca^{2+}$ influx in response to secretagogues has been demonstrated [37,38,40] and shown to possess different kinetic properties from $^{45}Ca^{2+}$ efflux [37]. This suggested that increased influx is not a simple reflection of Ca^{2+} exchange and may represent an increased permeability of the cell membrane to Ca^{2+}. Ca^{2+} entry into cells has generally been considered to be a passive process, due to the large inward electrochemical gradient (10^{-3} M outside: 10^{-7} M in the cytoplasm) and therefore not inhibited by metabolic poisons [41,42]. However, $^{45}Ca^{2+}$ influx into isolated pancreatic acini was shown to be markedly inhibited by antimycin A and the stimulation by cholinergic agonists abolished [37]. This suggested that a component of unstimulated uptake and Ca^{2+} entry stimulated by secretagogues is driven by an intracellular, ATP-dependent Ca^{2+} pump.

The mechanism by which Ca^{2+} crosses the basolateral membrane is unresolved. It was suggested by Petersen and Maruyama [43] that Ca^{2+} may enter through the non-selective cation channel. They recently reported that opening of this channel is controlled not only by intracellular Ca^{2+} but inhibited by ATP (4). The effect of metabolic poisons on $^{45}Ca^{2+}$ influx would, however, be inconsistent with this possibility since lowering cellular ATP levels might be expected to increase Ca^{2+} entry. It was previously shown that the Ca^{2+}-entry blocker, D600, had no effect on $^{45}Ca^{2+}$ influx by isolated acini [37]. Figure 1 shows the effects of two other blockers, nifedipine and cinnarizine. Nifedipine (10 μM) had no effect on $^{45}Ca^{2+}$ influx into rat pancreatic acini over a 10-min period of stimulation by a maximally effective concentration of carbachol (3 μM). Cinnarizine (10 μM), however, had a marked inhibitory effect, due in part to an increase in unstimulated $^{45}Ca^{2+}$ influx (170% of control). Cinnarizine may enter cells (see [44]) and therefore affect an intracellular Ca^{2+} pump important in controlling Ca^{2+} entry.

The lack of effect of D600 and nifedipine agrees with electrophysiological studies [45] in which no evidence has been obtained for secretagogue-evoked membrane potential and conductance changes having a Ca^{2+}-dependent com-

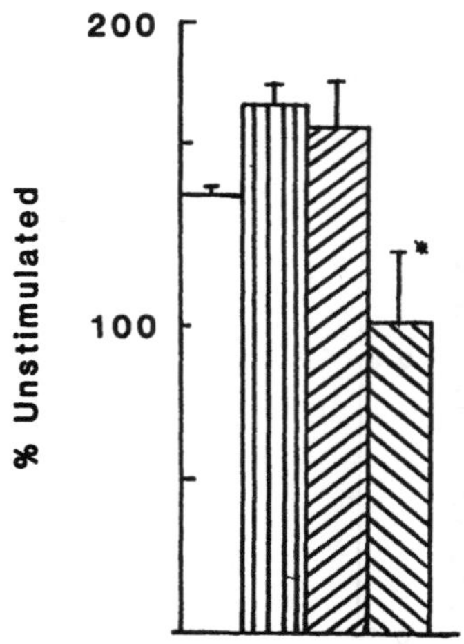

Fig. 1. Effect of Ca^{2+}-entry blockers on $^{45}Ca^{2+}$ uptake by rat pancreatic acini. Isolated rat pancreatic acini [20] were preincubated for 30 min at 37°C with Ca^{2+}-entry blockers; $^{45}Ca^{2+}$ (15 μCi/ml) was added in the presence or absence of 3 μM carbachol and $^{45}Ca^{2+}$ uptake measured after 10 min at 37°C as previously described [37]. Results are mean ± SEM for 4 pairs of determinations of stimulation of uptake. □, no addition; ▥, + 0.2% v/v ethanol (carrier for blockers); ▨, + 10 μM nifedipine; ▧ + 10 μM cinnarizine. *, $p < 0.05$ for difference from ethanol control

ponent. Indirect evidence from patch-clamp studies also indicated a lack of Ca^{2+} channels directly linked to receptors [46] in that, using cell-attached patches, Ca^{2+}-activated K^+ channels opened in response to a CCK analogue added outside the patch. This assumes that the agonist cannot diffuse into the patch and that the only mechanism for opening the K^+ channels is an increase in intracellular Ca^{2+}. Interestingly, the K^+ channel opening was dependent upon Ca^{2+} in the patch pipette (extracellular) suggesting that a component of Ca^{2+} entry, stimulated by an intracellular messenger, was required.

Recent evidence from studies of the activation of sea urchin eggs [47] and of K^+ channels in the lacrimal gland [48] has suggested that the intracellular messengers responsible for stimulating Ca^{2+} entry in these cells could be the inositol polyphosphates, Ins(1,4,5)P_3 and Ins(1,3,4,5)P_4. Activation of these cells in a Ca^{2+}-dependent manner required the combined action of both compounds. Both have been shown to increase in pancreatic acinar cells within 5 s of stimulation by secretagogues [49,50] but no data has been reported concerning their action on Ca^{2+} entry into these cells.

3.2 Mechanism of Ca^{2+} Extrusion

Ca^{2+} can only be extruded from cells by an active transport process. In the erythrocyte, this has been shown to be accomplished by a Ca^{2+}-activated, Mg^{2+}-dependent ATPase [51] but in other tissues ATPases activated by Ca^{2+}, present in the plasma membrane, do not appear to be obligatorily linked to Ca^{2+} transport by membrane vesicles [52,53]. In plasma membranes from pancreatic acinar cells an ATP phosphohydrolase, which is an ectoenzyme, has been demonstrated [54] and an ATPase [55] which has an approximately equal affinity for both Ca^{2+} (1.73 μM) and Mg^{2+} (2.98 μM). The difficulty of demonstrating a high-affinity Ca^{2+}-activated ATPase in the presence of millimolar Mg^{2+} concentrations is illustrated in Fig. 2. The concentration dependence of ATPase activity on Mg^{2+}, in a highly purified

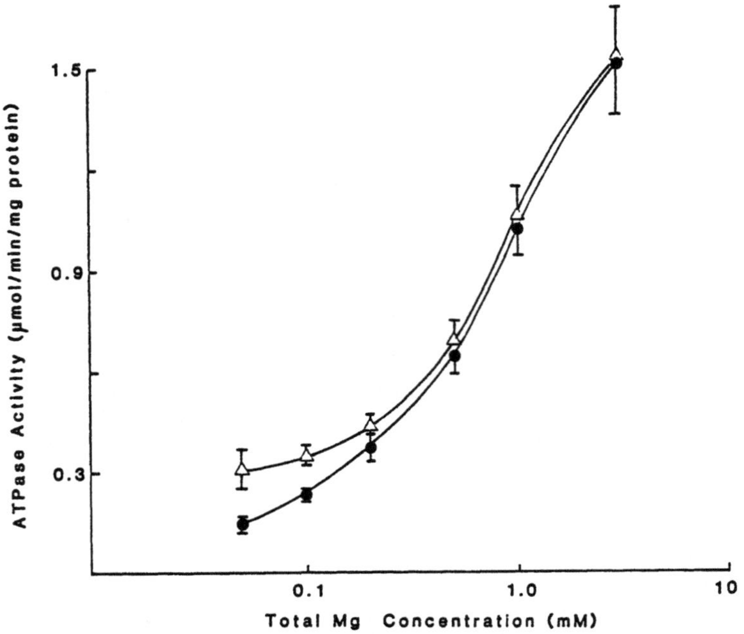

Fig. 2. ATPase activity in purified plasma membranes from rat pancreatic acini. Plasma membranes were purified from isolated rat pancreatic acini and ATPase activity measured as previously described [56,86]. ●, Mg-ATPase (3 mM ATP + concentration of total $MgCl_2$ shown); Δ, Mg-ATP + 50 μM free Ca^{2+} (1 mM EGTA/1 mM $CaCl_2$, pH 7.1). Each point is mean ± SEM for 3 different membrane preparations

plasma membrane preparation from rat pancreatic acini [56], shows that maximal activity was attained at not less than 3 mM, such that half-maximal activity was probably at 0.6–0.8 mM Mg^2. Addition of Ca-EGTA to give a free Ca^{2+} concentration of 51 μM caused an approximate doubling of the ATPase activity at 50 μM Mg^{2+}, indicating the presence of the non-specific divalent cation activated ATPase. At higher Mg^{2+} concentrations in the physiological range (0.5–3 mM), the activation by Ca^{2+} was less than 10% of the total ATPase and indistinguishable above 0.2 mM Mg^{2+}. Ca^{2+} transport by isolated plasma membrane vesicles has been studied [57] and shown to possess half-maximal activity at 0.88 μM free Ca^{2+} in the presence of 0.3 mM Mg^{2+}. The Ca^{2+}-ATPase (measured at 1–2 μM Mg^{2+}) was shown to be activated by calmodulin [55] but the effects of calmodulin on Ca^{2+} transport have not been reported. Thus the presence and physiological role of a Ca^{2+}-transporting ATPase in the basolateral membrane remains equivocal; the current data support the hypothesis that such a Ca^{2+} pump would play little part in regulating the cytoplasmic free Ca^{2+} in the resting cell [11].

NA^+/Ca^{2+} countertransport has also been described as a mechanism for extruding the Ca^{2+} from cells [58]. An activity has been described in plasma membranes purified from pancreatic acinar cells [59] which has half-maximal activity at 0.62 μM free Ca^{2+}. However, studies in whole cells have not demonstrated unequivocally that Na^+/Ca^{2+} exchange contributes to Ca^{2+} movements into or out of the acinar cell. Thus, although incubation of cells with ouabain for

30 min inhibited $^{45}Ca^{2+}$ efflux [39], replacement of extracellular Na^+ had no effect on either unstimulated $^{45}Ca^{2+}$ efflux or efflux stimulated by cholinergic agonists or CCK analogues [60,61]. In addition, after incubation for 30 min with ouabain in K^+-free medium, or in medium containing no Na^+, pancreatic acinar cells loaded with fura 2 showed a normal rise in cytoplasmic free Ca^{2+} and reduced this concentration to resting levels at the same rate as controls [61].

The studies discussed above were carried out using either purified plasma membranes, isolated acinar cells or acini, none of which discriminate definitively between Ca^{2+} movements across the basolateral or apical membranes.

Studies in vivo [62,63] and in vitro using bathed or perfused whole pancreas [64,65], in which it is possible to sample from the serosal and ductal sides separately, suggested that Ca^{2+} in the pancreatic juice is derived from at least two sources, a component which is directly correlated with stimulated protein secretion and a component independent of secretory activity, which correlated with serum levels [66] and was therefore presumed to enter the ducts by movement between the cells. The component stimulated by enzyme secretagogues also appeared to be derived from more than one source [63–65], one being the secretory granules and the other, movement via a paracellular route.

The existence of an increase in Ca^{2+} efflux across the basolateral membrane was inferred, but not directly measured in a study comparing $^{45}Ca^{2+}$ fluxes in intact rabbit pancreas and isolated fragments [65]. More recently, using the isolated rat pancreas perfused with $^{45}Ca^{2+}$ [67], a significant release of $^{45}Ca^{2+}$ into the pancreatic circulation was detected in response to carbachol, added during $^{45}Ca^{2+}$ washout following 60-min loading. This suggested that a component of the increased $^{45}Ca^{2+}$ efflux from isolated acinar cells represents efflux across the basolateral membrane.

4 Nature of the Ca^{2+} Store

Determination of changes in $^{45}Ca^{2+}$ and total Ca^{2+} content of subcellular fractions led to early proposals that mitochondria [68] or a component of a microsomal fraction [69] were sites of Ca^{2+} release following stimulation of acinar cells. More recently, loss of total Ca^{2+} from rough endoplasmic reticulum membranes, purified from carbachol-stimulated acini, was demonstrated [70]. This Ca^{2+} was not regained until receptor occupancy was blocked by atropine, which is consistent with the analysis of $^{45}Ca^{2+}$ fluxes [39] in which it had been suggested that the 'trigger pool' was not refilled during sustained stimulation. The hypothesis that the rough endoplasmic reticulum (ER) is the site of intracellular Ca^{2+} release was strengthened by the discovery of Ins(1,4,5)P_3 as a putative messenger linking receptor occupancy to Ca^{2+} release [71]. Streb et al. [72] first showed that Ins(1,4,5)P_3 released Ca^{2+} from a non-mitochondrial source in permeabilized acinar cells and this compound can release Ca^{2+} directly from purified rough ER membranes [49,73].

However, evidence has also been presented [30,74,75] that the initial site of Ca^{2+} release is from the inner surface of the plasma membrane. These two proposals may be partially resolved by the hypothesis that there is heterogeneity of

Ca^{2+} stores in the ER and that there is a Ca^{2+}-release compartment closely apposed to the basolateral membrane. In guinea-pig parotid acinar cells, Ca^{2+} release stimulated by Ins(1,4,5)P_3 was separated by density gradient centrifugation from Ca^{2+} release stimulated by 5^1-GTP [76]. The GTP-induced release, although its physiological relevance is not clear, was shown to occur in fractions consisting predominantly of rough ER, whereas Ins (1,4,5)P_3-induced Ca^{2+} release was most active in a fraction consisting of smooth-surfaced vesicles [76]. Volpe et al. [77] have demonstrated that in liver and pancreatic acinar cells, a protein with immunological similarity to calsequestrin [78] was located in apparently discrete organelles which they termed calciosomes. Their subcellular fractionation studies also showed that Ins(1,4,5)P_3-induced Ca^{2+} release was enriched in less dense membrane fractions. These studies provide evidence that Ins(1,4,5)P_3-induced Ca^{2+} release may therefore occur in discrete organelles or a subcompartment of the ER which is close to the plasma membrane. This would support the model proposed by Putney [79] and obviate the problem of how Ins(1,4,5)P_3 can diffuse away from its site of generation at the plasma membrane without being inactivated by 5^1 phosphomonoesterases [71].

The question remains as to whether Ca^{2+} release induced by Ins(1,4,5)P_3 is sufficient to explain the initial rise in cytoplasmic free Ca^{2+} and stimulation of secretion following activation by secretagogues. Data on Ins(1,4,5)P_3-induced Ca^{2+} release from purified rough ER membranes [49] suggested that the amount released was similar to that released from membranes isolated from carbachol-stimulated acini [70]. However, this does not indicate whether release from a site which may be close to the plasma membrane can raise the Ca^{2+} concentration at the apical end of the cell where secretion occurs. In the case of skeletal and cardiac muscle, it has been suggested that Ca^{2+} release from specific regions of the sarcoplasmic reticulum (SR) can be induced by Ca^{2+} [80,81]. Figure 3 shows that $^{45}Ca^{2+}$ release by purified rough ER membranes can be measured by the technique described by Meissner [80] for skeletal muscle SR. Vesicles are loaded passively in the presence of 5 mM $^{45}CaCl_2$ and $^{45}Ca^{2+}$ release initiated by dilution into media at 37°C. As shown in Fig. 3, as in SR, ^{45}Ca release is inhibited by 5 mM Mg^{2+} in the presence of 1 mM EGTA (free Ca^{2+} approx. 10^{-8} M). However, in the presence of 5 mM Mg^{2+}, 1 μM Ins(1,4,5)P_3 caused significant increase in $^{45}Ca^{2+}$ release by 2 min, and 5 μM A23187 caused approx. 80% of the $^{45}Ca^{2+}$ to be released within 30 s, showing that most of the $^{45}Ca^{2+}$ was within the membrane vesicles. In the absence of 5 mM Mg^{2+}, increasing the free Ca^{2+} to approx. 10^{-6} M caused less $^{45}Ca^{2+}$ release than at 10^{-8} M. This pattern resembles the less dense SR fraction designated 'control fraction' by Meissner [80]. In preliminary experiments (data not shown) a denser ER fraction was isolated from pancreatic acini which showed consistently more $^{45}Ca^{2+}$ release in the presence of 10^{-6} M free Ca^{2+} (35.7% released) than in 10^{-8} M (23.2% released).

Taken with the data on GTP-induced Ca^{2+} release from parotid rough ER [76] it is possible that Ins(1,4,5)P_3-induced Ca^{2+} release may occur in a limited compartment of ER or in a discrete organelle [77] at a site close to the basolateral membrane which begins a cascade of Ca^{2+} release from other parts of the rough ER by mechanisms involving either direct Ca^{2+} activation or a GTP-binding protein, thus allowing Ca^{2+} to be released at a site nearer the apical part of the cell.

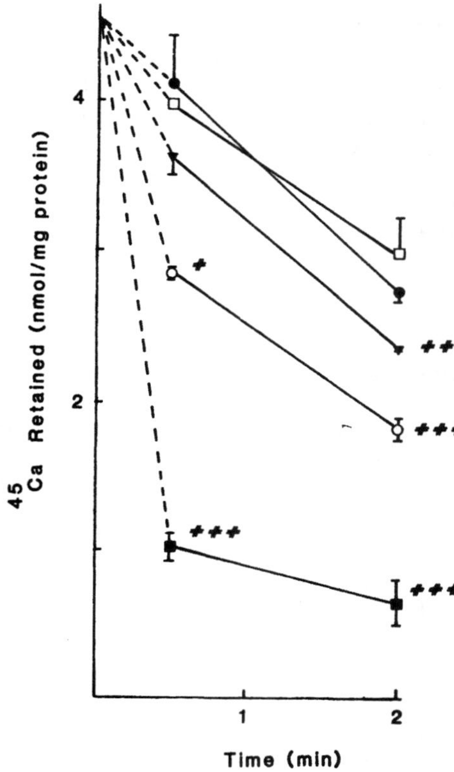

Fig. 3. $^{45}Ca^{2+}$ release from rough ER isolated from rat pancreatic acini. Rough ER membranes, isolated as previously described [87], were incubated at approx. 10 mg protein/ml in 100 mM imidazole-HCl, pH 6.8, containing 0.25 M sucrose, 100 mM KCl, 2 mM sodium azide + 5 mM $^{45}CaCl_2$ (33 μCi/μmol) for 2 h at room temperature. Aliquots (approx. 70 μg protein) were diluted 300-fold into the same buffer without $^{45}CaCl_2$ at 37°C and with additions shown below. $^{45}Ca^{2+}$ retained at times shown was determined by filtration and washing as previously described [87]. *Closed symbols*, + 5 mM $MgCl_2$, 1 mM EGTA: ●, no addition; ▼+ 1 μM Ins(1,4,5)P_3; ■, + 5 μM A23187. *Open symbols*, No $MgCl_2$: ○ + 1 mM EGTA; □, + 50 μM EGTA, 50 μM $CaCl_2$ (free Ca^{2+}, 10^{-6} M). Results are mean ± SEM for 3 determinations. #, $p < 0.05$; ##, $p < 0.01$; ###, $p < 0.001$ for difference from + 5 mM $MgCl_2$, 1 mM EGTA at equivalent time point

5 Conclusions

The importance of intracellular Ca^{2+} as a regulator of pancreatic secretion has been demonstrated by direct measurements and manipulation of its concentration in the intact cell. The initial phase of stimulation is probably caused by release of Ca^{2+} from internal stores by the action of Ins(1,4,5)P_3.

Recent evidence suggests heterogeneity of the Ca^{2+} store and it is proposed that Ins(1,4,5)P_3-induced Ca^{2+} release may initiate a cascade of Ca^{2+} release from rough ER allowing Ca^{2+} to be released at the apical part of the cell to activate Cl^- channel opening and exocytosis.

During sustained stimulation, increased Ca^{2+} entry into the cell occurs. The mechanism for this is not known, although it may involve the actions of inositol polyphosphates. Sustained Ca^{2+} entry must be buffered by the cell to avoid injurious levels of cytoplasmic free Ca^{2+} being attained. Evidence from studies of plasma membrane Ca^{2+} transport does not allow an unequivocal conclusion as to how Ca^{2+} may be pumped out of the cell. Alternatively, it has been suggested [11] that increased flux of Ca^{2+} through the secretory pathway, following uptake by the rough ER Ca^{2+}, Mg-ATPase could provide an additional buffering capacity.

An understanding of the regulation of these processes is fundamental to determining the aetiology of diseases such as cystic fibrosis, a common autosomal recessive genetic disease of exocrine epithelia [82]. Cystic fibrosis is characterized by disturbed autonomic regulation of fluid and protein secretion [83,84]. The biochemical defect is thought to lie in an altered intracellular regulator of Cl^- channel opening and exocytosis involving interaction between pathways activated by cyclic AMP and Ca^{2+} [85].

Acknowledgements. This work was supported by the Cystic Fibrosis Research Trust, UK, the Medical Research Council and the Welsh Scheme for the Development of Health and Social Science.

I am grateful to Drs. M.A. McPherson and F.M. Ashcroft for stimulating discussions.

References

1. Gardner JD, Jensen RT (1986) Receptors and cell activation with pancreatic enzyme secretion. Annu Rev Physiol 48:103–117
2. Williams JA, Sankaran H, Roach E, Goldfine ID (1982) Quantitative electron microscope autoradiographs of ^{125}cholecystokinin in pancreatic acini. Am J Physiol 243:G291–G296
3. Petersen OH (1986) Calcium-activated potassium channels and fluid secretion by exocrine glands. Am J Physiol 151:G1–G13
4. Petersen OH, Gallacher DV (1988) Electrophysiology of pancreatic and salivary acinar cells. Annu Rev Physiol 50:65–80
5. Bundgaard M, Moller M, Poulsen JH (1981) Localization of sodium pump sites in cat pancreas. J Physiol London 313:405–414
6. Maruyama Y, Petersen OH, Flanagan P, Pearson GT (1983) Quantification of Ca^{2+}-activated K^+ channels under hormonal control in pig pancreas acinar cells. Nature (London) 305:228–232
7. Petersen OH, Findlay I, Iwatsuki I, Singh J, Gallacher DV, Fuller CM, Pearson GT, Dunne MJ, Morris AP (1985) Human pancreatic acinar cells: studies of stimulus-secretion coupling. Gastroenterology 89:109–117
8. Maruyama Y, Petersen OH (1984) Single calcium-dependent cation channels in mouse pancreatic acinar cells. J Membrane Biol 81:83–87
9. Palade GE (1975) Intracellular aspects of the process of protein synthesis. Science 189:347–358
10. Dormer RL, Hallett MB, Campbell AK (1985) Measurement of intracellular free Ca^{2+}: importance during activation and injury of small cells In: Parratt JR (ed) Control and manipulation of calcium movement. Raven, New York, pp 1–27
11. Dormer RL, Brown GR, Doughney C, McPherson MA (1987) Intracellular Ca^{2+} in pancreatic acinar cells: regulation and role in stimulation of enzyme secretion. Biosci Rep 7:333–344
12. Tsien RY, Pozzan T, Rink TJ (1982) Calcium homeostasis in intact lymphocytes: cytoplasmic free calcium monitored with a new, intracellularly trapped fluorescent indicator. J Cell Biol 94:325–334
13. Ochs DL, Korenbrot JI, Williams JA (1985) Relation between free cytosolic calcium and amylase release by pancreatic acini. Am J Physiol 249:G389–G398
14. Bruzzone R, Pozzan T, Wollheim CB (1986) Caerulein and cabamylcholine stimulate pancreatic amylase release at resting cytosolic free Ca^{2+}. Biochem J 235:139–143
15. Pandol SJ, Schoeffield MS, Sachs G, Muallem S (1985) Role of cytosolic calcium in secretagogue-stimulated amylase release from dispersed acini from guinea-pig pancreas. J Biol Chem 260:10081–10086
16. Powers RE, Johnson PC, Houlihan MJ, Saluja AK, Steer ML (1985) Intracellular Ca^{2+} levels and amylase secretion in Quin-2-loaded mouse pancreatic acini. Am J Physiol 248:C535-C541
17. Merritt JE, Rubin RP (1985) Pancreatic amylase secretion and cytoplasmic free calcium. Effects of ionomycin, phorboldibutyrate and diacylglycerols alone and in combination. Biochem J 230:151–159
18. Eimerl S, Savion N, Heichal O, Selinger Z (1974) Induction of enzyme secretion in rat pancreatic slices using the ionophore A23187 and Ca^{2+}. J Biol Chem 249:3991–3993
19. Williams JA, Lee M (1974) Pancreatic acinar cells: use of a Ca^{2+} ionophore to separate enzyme release from earlier steps in stimulus-secretion coupling. Biochem Biophys Res Commun 60:542–548

20. Dormer RL (1983) Direct demonstration of increases in cytosolic free Ca^{2+} during stimulation of pancreatic enzyme secretion. Biosci Rep 3:233–240
21. Iwatsuki I, Petersen OH (1977) Acetylcholine-like effects of intracellular Ca^{2+} application in pancreatic acinar cells. Nature (London) 268:147–149
22. Dormer RL (1984) Introduction of calcium chelators into isolated rat pancreatic acini inhibits amylase release in response to carbamylcholine Biochem Biophys Res Commun 119:876–883
23. Laugier R, Petersen OH (1980) Effects of intracellular EGTA injection on stimulant-evoked membrane potential and resistance changes in pancreatic acinar cells. Pflüger's Arch 386:147–152
24. Bruzzone R, Regazzi R, Wollheim CB (1988) Caerulein causes translocation of protein kinase C in rat pancreatic acini without increasing cytosolic free Ca^{2+} Am J Physiol 255:G1–G7
25. Chandler DE, Williams JA (1977) Intracellular uptake and α-amylase and LDH releasing actions of divalent cation ionophore A23187 in dissociated pancreatic acinar cells. J Membrane Biol 32:201–230
26. Nishizuka Y (1984) The role of protein kinase C in cell surface signal transduction and tumour promotion. Nature (London) 308:693–698
27. Gunther GR (1981) Effects of 12-0-tetradecanoyl-phorbol-13-acetate on Ca^{2+} efflux and protein discharge in pancreatic acini. J Biol Chem 256:12040–12045
28. DePont JJHHM, Fleurens-Jakobs AMM (1984) Synergistic effect of A23187 and a phorbol ester on amylase secretion from rabbit pancreatic acini. FEBS Lett 170:64–68
29. Knight DE, Koh E (1984) Ca^{2+} and cyclic nucleotide dependence of amylase release from isolated rat pancreatic acinar cells rendered permeable by intense electric fields. Cell Calcium 5:401–418
30. Petersen OH, Ueda N (1976) Pancreatic acinar cells: the role of calcium in stimulus-secretion coupling. J Physiol London 254:583–606
31. Gardner JD, Costenbader CL, Uhlemann ER (1979) Effect of extracellular Ca^{2+} on amylase release from dispersed pancreatic acini. Am J Physiol 236:E754–E762
32. Scheele G, Haymovits A (1979) Cholinergic and peptide-stimulated discharge of secretory proteins in guinea-pig pancreatic lobules. J Biol Chem 254:10346–10353
33. Williams JA (1980) Regulation of pancreatic acinar cell function by intracellular calcium. Am J Physiol 238:G269–G279
34. Case RM, Clausen T (1973) Relationship between calcium exchange and enzyme secretion in isolated rat pancreas. J Physiol London 235:75–102
35. Williams JA, Chandler DE (1975) Ca^{2+} and pancreatic amylase release. Am J Physiol 228:1729–1732
36. Kondo S, Schulz I (1976) Calcium fluxes in isolated cells of rat pancreas. Effect of secretagogues and different calcium concentrations. J Membrane Biol 29:185–203
37. Dormer RL, Poulsen JH, Licko V, Williams JA (1981) Calcium fluxes in isolated pancreatic acini: effect of secretagogues. Am J Physiol 240:G38–G49
38. Renckens BAM, Schrijen JJ, Swarts HGP, DePont JJHHM, Bonting SL (1978) Role of calcium in exocrine pancreatic secretion. IV. Calcium movements in isolated acinar cells of rabbit pancreas. Biochim Biophys Acta 544:338–350
39. Stolze H, Schulz I (1980) Effect of atropine, ouabain, antimycin A and A32187 on "trigger Ca^{2+} pool" in exocrine pancreas. Am J Physiol 238:G338–G348
40. Kondo S, Schulz I (1976) Calcium ion uptake in isolated pancreatic cells induced by secretagogues. Biochim Biophys Acta 419:76–92
41. Borle AB (1969) Kinetic analyses of Ca^{2+} movement in Hela cell cultures: I: Ca^{2+} influx. J Gen Physiol 53:43–56
42. Lamb JF, Lindsay R (1971) Effect of sodium, metabolic inhibitors and ATP on calcium movements in L cells. J Physiol London 218:691–708
43. Petersen OH, Maruyama Y (1983) What is the mechanism of the calcium influx to pancreatic acinar cells evoked by secretagogues. Pflüger's Arch 396:82–84
44. Godfraind T, Miller R, Wibo M (1986) Calcium antagonism and calcium entry blockade. Pharmacol Rev 38:321–416
45. Petersen OH (1980) The electrophysiology of gland cells. Academic Press, New York London
46. Suzuki K, Petersen CCH, Petersen OH (1985) Hormonal activation of single K^+ channels via internal messenger in isolated pancreatic acinar cells. FEBS Lett 192:307–312
47. Irvine RF, Moor RM (1986) Microinjection of inositol 1,3,4,5-tetrakisphosphate activates sea urchin eggs by a mechanism dependent on external Ca^{2+}. Biochem J 240:917–920

48. Morris AP, Gallacher DV, Irvine RF, Petersen OH (1987) Synergism of inositol trisphosphate and tetrakisphosphate in activating Ca^{2+}-dependent K^+ channels. Nature (London) 330:653–655
49. Doughney C, Brown GR, McPherson MA, Dormer RL (1987) Rapid formation of inositol-1,4,5-trisphosphate in rat pancreatic acini stimulated by carbamylcholine. Biochim Biophys Acta 928:341–348
50. Trimble ER, Bruzzone R, Mechan CJ, Biden TJ (1987) Rapid increases in inositol 1,4,5-trisphosphate, inositol 1,3,4,5-tetrakisphosphate and cytosolic free Ca^{2+} in agonist-stimulated pancreatic acini of the rat. Effect of carbachol, caerulein and secretin. Biochem J 242:289–292
51. Schatzmann HJ (1982) The plasma membrane calcium pump of erythrocytes and other animal cells. In: Carafoli E (ed) Membrane transport of calcium. Academic Press, New York London, pp 41–108
52. Lin S-H (1985) Novel ATP-dependent calcium transport component from rat liver plasma membranes. The transporter and the previously reported ($Ca^{2+}+Mg^{2+}$)-ATPase are different proteins. J Biol Chem 260:7850–7856
53. Minami J, Penniston JT (1987) Ca^{2+} uptake by corpus luteum plasma membranes. Evidence for the presence of both a Ca^{2+}-pumping ATPase and Ca^{2+}-dependent nucleoside triphosphatase. Biochem J 242:889–894
54. Hamlyn JM, Senior AE (1983) Evidence that Mg^{2+}- or Ca^{2+}-activated adenosine triphosphatase in rat pancreas is a plasma membrane ecto-enzyme. Biochem J 214:59–68
55. Ansah TA, Molla A, Katz S (1984) Ca^{2+}-ATPase activity in pancreatic acinar plasma membranes. Regulation by calmodulin and acidic phospholipids. J Biol Chem 259:13442–13450
56. Dormer RL, Al-Mutairy AR (1987) Pancreatic acinar cell plasma-membrane enzymes involved in stimulus-secretion coupling. In: Reid E, Cook BMW, Luzio JP (eds) Cells, membranes and disease, including renal. Plenum, New York, pp 273–283
57. Bayerdorffer E, Eckhardt L, Haase W, Schulz I (1985) Electrogenic calcium transport in plasma membranes of rat pancreatic acinar cells. J Membrane Biol 84:45–60
58. Blaustein MP, Nelson MT (1982) Sodium-calcium exchange: its role in the regulation of cell calcium. In: Carafoli E (ed) Membrane transport of calcium. Academic Press, New York London, pp 217–236
59. Bayerdorffer E, Haase W, Schulz I (1985) Na^+/Ca^{2+} countertransport in plasma membrane of rat pancreatic acinar cells. J Membrane Biol 87:107–119
60. Williams JA (1980) Multiple effects of Na^+ removal on pancreatic secretion in vitro. Cell Tissue Res 210:295–303
61. Muallem S, Beeker T, Pandol SJ (1988) Role of Na^+/Ca^{2+} exchange and the plasma membrane Ca^{2+} pump in hormone-mediated Ca^{2+} efflux from pancreatic acini. J Membrane Biol 102:153–162
62. Goebell H, Steffen Ch, Bode Ch (1972) Stimulatory effect of pancreozymin-cholecystokinin on calcium secretion in pancreatic juice of dogs. Gut 13:477–482
63. Ceccarelli B, Clemente F, Meldolesi J (1975) Secretion of calcium in pancreatic juice. J Physiol London 245:617–638
64. Argent BE, Case RM, Scratcherd T (1973) Amylase secretion by perfused cat pancreas in relation to secretion of calcium and other electrolytes and as influenced by external ionic environment. J Physiol London 230:575–593
65. Schreurs VVAM, Swarts HGP, DePont JJHHM, Bonting SL (1975) Role of Ca^{2+} in exocrine pancreatic secretion. I. Calcium movements in the rabbit pancreas. Biochim Biophys Acta 404:257–267
66. Layer P, Holtz J, Goebell H (1983) Calcium secretion from the feline pancreas. Digestion 26:89–98
67. Dormer RL, Mann GE, Norman PSR (1987) Comparison of calcium efflux from isolated rat pancreatic acini and perfused pancreas: effects of carbachol. J Physiol London 390:100P
68. Clemente F, Meldolesi J (1975) Calcium and pancreatic secretion. Dynamics of subcellular calcium pools in resting and stimulated acinar cells. Br J Pharmacol 55:369–379
69. Dormer RL, Williams JA (1981) Secretagogue-induced changes in subcellular Ca^{2+} distribution in isolated pancreatic acini. Am J Physiol 240:G130–G140
70. Brown GR, Richardson AE, Dormer RL (1987) The role of a ($Ca^{2+}+Mg^{2+}$)-ATPase of the rough endoplasmic reticulum in regulating intracellular Ca^{2+} during cholinergic stimulation of rat pancreatic acini. Biochim Biophys Acta 902:87–92

71. Berridge MJ, Irvine RF (1984) Inositol trisphosphate, a novel second messenger in cellular signal transduction. Nature (London) 312:315–321
72. Streb H, Irvine RF, Berridge MJ, Schulz I (1983) Release of Ca^{2+} from a nonmitochondrial intracellular store in pancreatic acinar cells by inositol-1,4,5-trisphosphate. Nature (London) 306:67–69
73. Streb H, Bayerdorffer E, Haase W, Irvine RF, Schulz I (1984) Effect of inositol-1,4,5-trisphosphate on isolated subcellular fractions of rat pancreas. J Membrane Biol 81:241–253
74. Schulz I, Wakasugi H, Stolze H, Kribben A, Haase W (1981) Analysis of Ca^{2+} fluxes and their relation to enzyme secretion in dispersed pancreatic acinar cells. Fed Proc 40:2503–2510
75. Schulz I, Kimura T, Wakasugi H, Haase W, Kribben A (1981) Analysis of Ca^{2+} fluxes and Ca^{2+} pools in pancreatic acini. Philos Trans Soc Lond Ser B 296:105–113
76. Henne V, Piiper A, Soling H-D (1987) Inositol-1,4,5-trisphosphate and 5′-GTP induce calcium release from different intracellular pools. FEBS Lett 218:153–158
77. Volpe P, Krause K-H, Hashimoto S, Zorzato F, Pozzan T, Meldolesi J, Lew DP (1988) "Calciosome", a cytoplasmic organelle: the inositol-1,4,5-trisphosphate-sensitive Ca^{2+} store of nonmuscle cells? Proc Natl Acad Sci USA 85:1091–1095
78. MacLennan DH, Campbell KP, Reithmeier RAF (1983) Calsequestrin. In: Cheung WY (ed) Calcium and cell function, vol 4. Academic Press, New York London, pp 151–173
79. Putney JW (1986) A model for receptor-regulated calcium entry. Cell Calcium 7:1–12
80. Meissner G (1984) Adenine nucleotide stimulation of Ca^{2+}-induced Ca^{2+} release in sacroplasmic reticulum. J Biol Chem 259:2365–2374
81. Fabiato A (1983) Ca^{2+}-induced release of calcium from the cardiac sarcoplasmic reticulum. Am J Physiol 245:C1–C14
82. McPherson MA, Goodchild MC (1988) The biochemical defect in cystic fibrosis. Clin Sci 74:337–345
83. McPherson MA, Dormer RL (1987) The molecular and biochemical basis of cystic fibrosis. Biosci Rep 7:167–185
84. McPherson MA, Dormer RL (1988) Cystic fibrosis: a defect in stimulus response coupling. TIBS 13:10–13
85. McPherson MA, Shori DK, Dormer RL (1988) Defective regulation of apical membrane chloride transport and exocytosis in cystic fibrosis. Biosci Rep 8:27–33
86. Al-Mutairy AR, Dormer RL (1985) Isolation of plasma membranes from pancreatic acinar cells: demonstration of a Ca^{2+}-activated Mg^{2+}-ATPase activity. Biochem Soc Trans 13:900–901
87. Richardson AE, Dormer RL (1984) Calcium-ion-transporting activity in two microsomal subfractions from rat pancreatic acini. Modulation by carbamylcholine. Biochem J 219:679–685

Control of Protein Kinase C Function by Activators and Inhibitors

CH. W. MAHONEY and A. AZZI

1 General Characteristics of Protein Kinase C

Ca^{2+}- and phospholipid-dependent protein kinase (protein kinase C), discovered by Nishizuka and associates in 1977 (Takai et al. 1977; Inoue et al. 1977), has had a large impact on the biochemical research in the last 11 years because of its implication in numerous biological processes including tumor promotion (Hecker and Schmidt 1979), membrane transport, channel modulation (Liles et al. 1986; Witters et al. 1985), differentiation (Pahlman et al. 1983; Morin et al. 1987; Malkinson et al. 1987), muscle contraction (Ikebe et al. 1985; Nishikawa et al. 1984, 1985), synaptic communication (Tanaka et al. 1984, 1986), secretion (Ieyasu et al. 1982; Kajikawa et al. 1983), the activation of granulocytes (Serhan et al. 1983a, b; Fujita et al. 1984; Mahoney et al. 1986; Lüthy and Azzi 1987; Mahoney and Azzi 1988a and in press, b), the immune response (Patel et al. 1987), growth and differentiation (Cooper et al. 1982; Cochet et al. 1984; McCaffrey et al. 1984; Davis and Czech 1985) and platelet aggregation (Naka et al. 1983; Sano et al. 1983; Froscio et al. 1988; Watson et al. 1988; Mahoney and Azzi in press, b; for reviews, see Nishizuka 1984, 1986).

Protein kinase c (PKC) with an apparent molecular weight (M_r) of 84,000 kD has the unique feature to require Ca^{2+} and phospholipid for activation. Other kinases have been characterized as cAMP-activated (for review, see Beavo and Mumby 1982), cGMP-activated (for review, see Kuo and Shoji 1982), and Ca^{2+}-calmodulin-activated enzymes (cf. Schulman 1982). An important feature of protein kinase c is its activation by tumor promoters (e.g. the phorbol 12, 13 diesters) and by diacylglycerols (DAG). These two group of substances act by lowering the Ca^{2+} required for enzyme activation (Takai et al. 1979; Castagna et al. 1982). PKC, which is a cytosolic protein in the resting cell state ($\approx$ 80% cytosolic, $\approx$ 20% membrane associated) (Ashendel et al. 1983), upon cell stimulation (for instance by hormones, neurotransmitters, the phorbol 12, 13 diesters, DAG and the inhibitor staurosporine) most of the PKC becomes cell membrane-associated ($\approx$ 80% membrane associated, $\approx$ 20% cytosolic) (Kraft et al. 1982; Kraft and Anderson 1983; Wolf and Baggiolini 1988). The same event can be produced by a rise in internal Ca^{2+} (Melloni et al. 1985; Wolf et al. 1985; Gopalakrishna et al. 1986). The physiologically relevant stimulus is afforded, in many cases, by extracellular ligand binding to cell surface receptors; as a consequence, a phospholipase C becomes activated (via coupling with a G protein) and the resulting

Institute of Biochemistry and Molecular Biology, University of Bern, 3012 Bern, Switzerland

A. Azzi, Z. Drahota and S. Papa
Molecular Basis of Membrane-Associated Diseases

hydrolysis of phosphatidyl inositol diphosphate (PIP2) produces inositol triphosphate (IP3) and DAG. The water soluble IP3 can indirectly activate PKC by inducing a release of Ca^{2+} from intracellular non-mitochondrial stores (Michell 1983; Berridge 1987) and DAG can directly induce a redistribution and activation of the enzyme (Wolf et al. 1985; Gopalakrishna et al. 1986). Phorbol 12, 13 diester DAG competition studies suggest that both compounds interact with the same or adjacent binding sites on the enzyme-phospholipid complex (Sharkey et al. 1984; Sharkey and Blumberg 1985; Fig. 1).

Three PKC genes have been sequenced, corresponding to three isozymic forms of the protein (α,β,γ). Their deduced amino acid sequences (Parker et al. 1986; Coussens et al. 1986; Knopf et al. 1986) of the α,β and γ forms of PKC consist of 672, 673, and 693 amino acid residues and they contain some relevant information: a characteristic amino terminus cysteine-rich domain, a Ca^{2+} binding domain, and a conserved protein kinase active site domain in the carboxy terminus (Parker et al. 1986). The differential role of the three isozymes remains unclear, yet recent studies have reported differences in the enzymatic properties and distributions of these isozymes (Ohno et al. 1987). The α,β,γ forms of PKC can be readily separated by hydroxylapatite chromatography (Huang et al. 1986; Ido et al. 1987; F.L. Huang et al. 1987; Shearman et al. 1987). The differential distribution of α,β,γ PKC isoforms within the brain has been reported by several groups (Huang et al. 1987; Brandt et al. 1987; Shearman et al. 1987). Human neutrophils (Sekiguchi et al. 1987) and normal and ras-transformed 3T3 fibroblasts (McCaffrey et al. 1987) contain only α PKC. McCaffrey et al. (1987), in addition, were not able to find any differences between α PKC from normal and ras-transformed 3T3 fibroblasts.

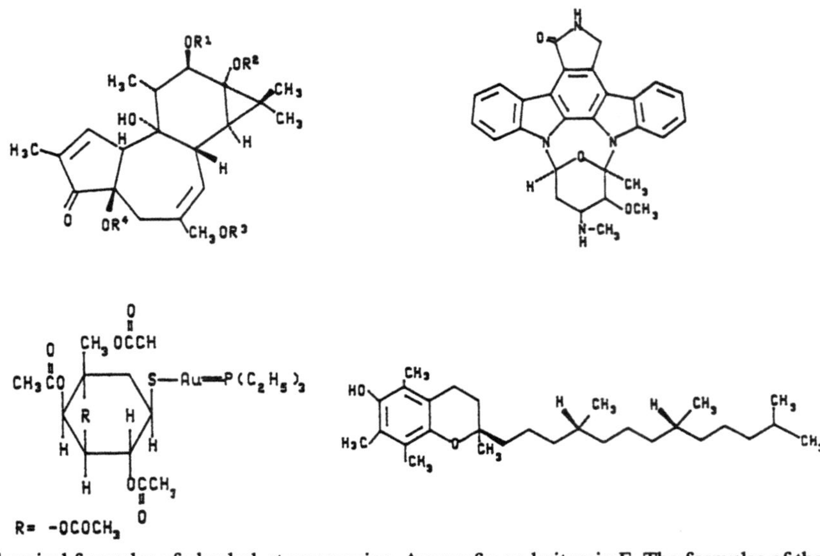

Fig. 1. Chemical formulas of phorbol, staurosporine, Auranofin and vitamin E. The formulas of the activator phorbol and of the inhibitors staurosporine, Auranofin and dl-α-tocopherol are shown

Ido et al. (1987) have provided preliminary evidence that the α PKC is able to phosphorylate the EGF receptor in membranes of epidermal carcinoma cells the most rapidly, with β PKC at an intermediate rate, and γ with the lowest rate. Auto-phosphorylation of α and γ PKC occurs only at serine residues, yet in the case of β PKC, threonine residue(s) can be phosphorylated as well (Huang et al. 1986). Since only 1–2 mol Pi/mol PKC are auto-incorporated, this suggests that in the case of α and γ PKC two serine residues are auto-phosphorylated, whereas for β PKC a single serine and threonine are modified.

The length and degree of saturation/unsaturation of the fatty ester side chains on DAG are critical in several respects for the binding to and activation of PKC (Mori et al. 1982). Generally, short saturated or unsaturated as well as long unsaturated fatty ester side chains are acceptable for these processes. The presence of a single, long-chain unsaturated fatty ester on DAG makes it possible for the second fatty ester to be a long-chain saturated fatty ester in order to maintain binding and activation properties with respect to PKC (Takai et al. 1979; Kishimoto et al. 1980; Lapetina et al. 1985; Davis et al. 1985; Go et al. 1987).

2 Activation of PKC

The physiological, basic activation of PKC has been reported above and can be ascribed to the interaction with three ligands, Ca^{2+}, phosphatidylserine (PS) and diacylglycerol. Other physiologically relevant situations or compounds can also be responsible for PKC activation.

2.1 Proteolysis

PKC is readily cleaved by a Ca^{2+}-activated intracellular protease into a 50 kD catalytically active (Ca^{2+} – phospholipid – independent) and a 30 kD M_r regulatory fragment (Inoue et al. 1977; Melloni et al. 1985; Nakadate et al. 1987). Proteolysis may represent the mechanism of down-regulation of PKC (Young et al. 1987; Woodgett and Hunter 1987). The 50 and 30 kD M_r catalytic and regulatory fragments can also be generated in vitro by limited tryptic digestion (Lee and Bell 1986; Huang and Huang 1986; Mochly-Rosen and Koshland 1987). Lee and Bell (1986) have shown that the Ca^{2+}, phosphatidylserine (PS), and PDB (Huang and Huang 1986) bind to the regulatory fragment. The 30 and the 50 kD regulatory and catalytic domains have been localized to the N and C terminal regions of the protein respectively (Parker et al. 1986). Current knowledge suggests that the active site of resting PKC located in the 50 kD domain is blocked by the 30 kD regulatory moiety and that on binding of Ca^{2+}, PS, and PDB or DAG to the latter, the active site becomes reactive to the substrates MgATP and protein acceptor.

2.2 Free Fatty Acids

Unsaturated fatty acids can directly activate PKC in the absence of phospholipid and calcium (Murakami and Routtenberg 1985). The three isozymes of PKC are activated by unsaturated free fatty acids (Sekiguchi et al. 1987) to a different extent.

The γ isoform shows 30–40% activation relative to Ca^{2+}, PS, DAG stimulated enzyme by arachidonic, oleic, or linoleic acid in the 25–100 μM concentration range, whereas the α isoform is activated by these three fatty acids (50–400 μM range) in the presence of Ca^{2+} to a similar degree as in the presence of Ca^{2+}, PS, and DAG. The β isoform shows intermediate activation properties by free fatty acid in the presence of Ca^{2+} (Sekiguchi et al. 1987). Although Sekiguchi et al. were able to find no significant difference in the Ca^{2+}, PS, DAG activation properties of the three isozymes, in contrast, Huang et al. (1986) found similar Ca^{2+}, PS, DAG activation properties (> 15-fold) for α and β PKC, but only a 4–8-fold activation for the γ form.

2.3 Tumor Promoters

The tumor promoters, phorbol diesters, teleocidin, and aplysiatoxin, as well as DAG are potent activators of PKC (Nishizuka 1986; Brockerhoff 1986; Jeffrey and Liskamp 1986). These compounds act by lowering the Ca^{2+} and phosphatidylserine requirement of the kinase (Castagna et al. 1982). Although these compounds have very different structures, they all have functional groups with similar three-dimensional coordinates that are critical for their interaction with PKC (Brockerhoff 1986; Jeffrey and Liskamp 1986). In the case of DAG, the carbonyl functional groups in positions 1 and 2 and the 3 hydroxy group are critical for binding to and activation of PKC. The carbonyl groups provide a hydrogen bond acceptor site with respect to a hydrogen bond donor on PKC and the 3 hydroxy group provides a hydrogen bond donor to an acceptor on the enzyme (Brockerhoff 1986). In the case of the phorbol diesters, the 9 α-OH serves as a hydrogen bond donor and the 12 and 13 carbonyl of the esters serve as hydrogen bond acceptors (Brockerhoff 1986). In addition, it has been demonstrated that DAG competes for the phorbol diester binding site on PKC (Sharkey et al. 1984). Other tumor promoters, which activate PKC, teleocidin and aplysiatoxin, also have analogous functional groups with similar three-dimensional coordinates (Brockerhoff 1986). Jeffrey and Liskamp (1986), through computer structural modeling, deduced that four oxygen groups in TPA (at C-3, C-4, C-9, and C-20) correspond to oxygen or nitrogen atoms in the three-dimensional structures of teleocidin and aplysiatoxin, suggesting that these atoms are critical in the interaction of these compounds with PKC. Hence, protein-lipid and protein-tumor promoter hydrogen bonding interactions appear to be critical in the activation of PKC by neutral lipid and the tumor promoters.

Chloroform has been found to activate PKC, but no competition with phorbol esters was shown (Roghani et al. 1987).

2.4 Heavy Metals

Lead (acetate, chloride, or citrate salts) was found to activate at picomolar concentrations (10^{-10} M) brain PKC (Markovac and Goldstein 1988). The maximal rate attained by the enzyme was ten-fold higher than that in the presence of 10^{-10} M $CaCl_2$ and maximal activation of kinase activity occurred at four orders of magnitude lower concentrations of lead as compared to Ca^{2+}. It is probable that the effect of lead is to bind to the Ca^{2+} site already with high affinity without the need of DAG or phorbol esters. Zinc appeared to be at least as efficient as calcium in promoting PKC activation.

3 Inhibitors of Protein Kinase C

3.1 Substances Affecting Ca^{2+} Binding

Some of the Ca^{2+} interacting drugs, such as trifluoperazine, and Ca^{2+} binding proteins (Kuo et al. 1984) can inhibit by competing for available free calcium.

Earlier reports indicate that a 17 kD Ca^{2+} binding protein from bovine brain (McDonald and Walsh 1985), calmodulin, S100, and troponin c can inhibit PKC activity (Albert et al. 1984).

3.2 Substances Affecting the Site of Interaction with the Protein Substrate

A number of compounds, due to their chemical nature, presumably react at the site of the enzyme responsible for binding the protein substrate that has to be phosphorylated, resulting in inhibition of the enzyme activity.

TLCK (N-tosyl-L-lysine chloromethyl ketone) and TPCK (tosyl-L-phenylalanine chloromethyl ketone), the anti-serine protease compounds, inhibit PKC presumably by going to a lysine or phenylalanine recognition site in the active site of the enzyme (Solomon et al. 1985). This suggestion is consistent with the known structural requirement of serine-X-lysine in the substrate for phosphorylation of serine by PKC (Kondo et al. 1987). Hucho et al. (1987) have reported the isolation of a 40 kD M_r protein from bovine brain cytosol which inhibits PKC activity.

House and Kemp (1987) have demonstrated that a synthetic peptide corresponding to residues 19–36 in PKC (part of the regulatory domain), which resembles a substrate phosphorylation site in its distribution of basic residues, can potently and specifically inhibit (Ki = 147 nM) PKC exogenous and auto-phosphorylation reactions. In addition, on substituting alanine 25 with a serine residue, the peptide becomes a phosphate acceptor peptide. Physiologically, residues 19–36 within PKC may play an inhibitory–regulatory role.

3.3 Substances Affecting the ATP Binding Site

Several inhibitors interact at the active site in the 50 kD catalytic domain: H-9 (Hidaka et al. 1984) and the K-252 compounds (Kase et al. 1987) inhibit by competing at the ATP binding site of the active site.

Staurosporine was first reported to potently inhibit PKC activity by Tamaoki et al. (1986). We have confirmed these results and have synthesized several secondary amine derivatives of staurosporine in order to generate potent yet more specific inhibitors of PKC relative to other protein kinases. N-propan-2-ol- and N-phenyl- derivatives of staurosporine are approximately two- to four-fold less inhibitory towards PKC than the parent comfound, staurosporine, yet both derivatives show approximately ten-fold higher specificity towards PKC relative to

Table 1. Specificity of inhibition of PKC by the secondary amine derivatives of staurosporine

	IC_{50} (nM) PKC	PKA	IC_{50}/catal. unit PKC	PKA	PKA/PKC
Staurosporine	3.5	14.7	0.11	71.9	153
N-propanol-	8.3	260	0.27	317	1174
N-phenyl-	16.8	732	0.55	893	1623

Assays: PKC: 0.304 μg enzyme; final volume, 0.25 ml; specific activity (presence of Ca^{+2} and PS), 121 nmol min^{-1} mg^{-1}; substrate, histone III-S. PKA: 10 μg PKA; final volume, 0.1 ml; specific activity (presence of 2 μM cAMP), 8.2 nmol min^{-1} mg^{-1}; substrate, histone III-S. The IC_{50}/catalytic unit ratio has dimensions of nM (inhibitor concentration) $pmol^{-1}$ min^{-1} mg^{-1} l^{-1}.

PKA, suggesting their usefulness in cell studies (Table 1; Mahoney et al. submitted).

The respiratory burst of neutrophils, which can be stimulated by DAG (Fujita et al. 1984), the PKC-specific phorbol diesters or f-met-leu-phe, is inhibited in a concentration-dependent manner by staurosporine with the IC_{50} of this process (9–10 nM) being similar to the IC_{50} of purified PKC (3.5 nM; Table 1). PKC activity appears to be essential throughout the respiratory burst, since addition of staurosporine during the up-, maximum-, or down-phase of the respiratory burst results in cessation of reactive oxygen species generation.

PKC has been implicated in agonist-receptor coupled platelet aggregation which is a critical reaction in normal blood clotting as well as in pathological processes. We have tested the role of staurosporine in platelet aggregation by using the inhibitor staurosporine with intact platelets. Staurosporine is able to inhibit platelet aggregation in a concentration-dependent manner when stimulated by phorbol 12, 13 dibutyrate (PDB; Mahoney and Azzi in press, b). Watson et al. (1988) have independently demonstrated that staurosporine is able to inhibit platelet aggregation as well as phorbol diester-specific phosphorylation of proteins (47 and 20 kD M_r) in intact platelets. These data provide further evidence for a regulatory role of PKC in platelet aggregation.

PKC has been implicated in tumor promotion by the tumor promoters, the phorbol diesters and related compounds, and in differentiation in neuroblastoma (Liu and Chen 1985; Ruusala et al. 1985). We have tested the role of PKC in differentiation using staurosporine as a potential differentiation inducing agent with a neuroblastoma cell line as a model system (Mahoney and Azzi in press, b). At very low concentrations of staurosporine (4–5 nM) a neuroblastoma cell line was induced to differentiate as determined by dendrite outgrowth processes developing over a period of several days. Further studies are in progress to delineate the role of PKC in differentiation in neuroblastoma cell lines.

3.4 Substances Affecting the Regulatory Domain

Many inhibitors of PKC have been reported and most of these compete for either the PDB/DAG or the PS sites in the hydrophobic regulatory domain (30 kD M_r) of PKC (Kuo et al. 1984; Mahoney et al. submitted). Almost all known inhibitors of PKC, with the exception of the staurosporines (Tamaoki et al. 1986; Mahoney et al. in press) and K-252 compounds (Nakanishi et al. 1986; Kase et al. 1986), which inhibit in nanomolar concentration ranges, inhibit PKC activity in the micromolar concentration ranges.

Sphingosine and lysosphingolipids (Hannun and Bell 1987), potential physiological-pathological regulators compete with PDB/DAG.

Aminoacridines have been shown (Hannun and Bell 1988) to be potent inhibitors of PKC, their inhibition occurring at concentrations comparable to those of DAG and PS. The inhibitory mechanism appears to be complex, involving both the catalytic and regulatory sites of the enzyme. The predominant effect is, however, inhibition of activation within the regulatory domain of the protein.

The gold-type anti-rheumatic drugs, Auranofin, Au-S-glucose, and Au-S-malate inhibit purified PKC (Mahoney and Azzi 1988a and in press, b) (IC_{50} = 3 -361 μM). Contrary to the phorbol diesters and staurosporine (Wolf and Baggiolini 1988) Auranofin does not induce cytosol to membrane translocation of PKC (Froscio et al. 1988). Auranofin in the 10–100 μM concentration range inhibits the production of superoxide in neutrophils that have been stimulated by PKC-specific PDB or f-met-leu-phe (Hafstrom et al. 1984; Parente et al. 1986) which suggests a role for PKC in superoxide production and a role or co-role for PKC in the mechanism of Auranofin therapy as an anti-rheumatic.

Auranofin has not been shown to inhibit NADPH oxidase of neutrophils, the key enzyme in the respiratory burst (Parente et al. 1986), which is consistent with an inhibitory role of Auranofin at the level of the PKC-mediated respiratory burst. The concentrations of Auranofin required to inhibit PKC kinase activity and the respiratory burst of neutrophils are in the same order of magnitude as those found in the serum (30–40 μM) (Danpure et al. 1979) and synovium (17–22 μM) (Gerber et al. 1972) of patients undergoing Auranofin treatment. Hence, the Auranofin inhibition of PKC obtained in vitro is of physiological significance.

Vitamin E (dl-α-tocopherol) has been known for many years as an antioxidant (for reviews, see McCay et al. 1982; Wasserman and Taylor 1972); it has been shown that vitamin E can induce differentiation in human neuroblastoma cell lines (Helson et al. 1982), that H_2O_2 production can be inhibited in human neutrophils isolated from individuals whose diet was supplemented with 1600 units of Vitamin E/day (Baehner et al. 1982), and that it can act as a platelet anti-aggregation agent (Steiner and Mower 1982). Since PKC has been implicated in tumor promotion, in differentiation, in granulocytes respiratory burst and platelet aggregation, and because of the hydrophobic nature of vitamin E and of the regulatory domain of PKC, we tested the possibility of a direct modulatory role for vitamin E on PKC. Ca^{2+}–PS–activated PKC is inhibited in a concentration-dependent manner by vitamin E with an approximate IC_{50} of 456 μM (Fig. 2; Mahoney and Azzi 1988b). To further elucidate the mechanism of this inhibition, increasing concentrations of PS were added to the vitamin E inhibition assays. Under these conditions there was no significant effect on the concentration-dependent inhibition curves, suggesting that vitamin E is not competing for PS sites in the enzyme-phospholipid complex. PKC activity, stimulated by PDB–PS– (in the presence and absence of added Ca^{2+}), yielded vitamin E concentration-dependent inhibition curves essentially identical to each other as well as to the Ca^{2+}–PS–activated vitamin inhibition

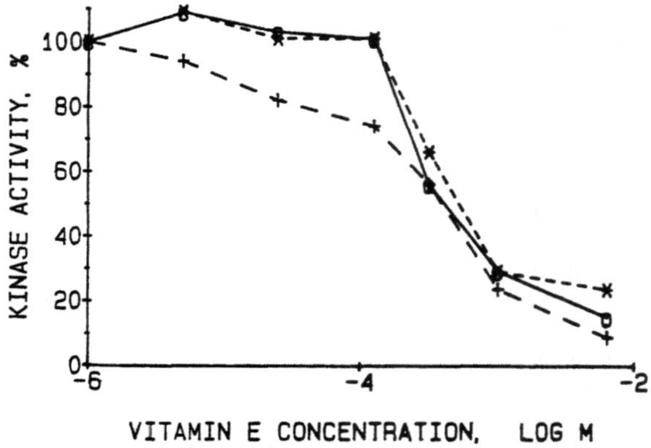

Fig. 2. Concentration-dependent inhibition of PKC kinase activity by vitamin E and lack of an effect with increasing concentrations of PS. Assays were carried out in the presence of 40 (o), 100 (*), or 250 (x) μg/ml PS and activation was also by 500 μM $CaCl_2$

curves (Mahoney and Azzi 1988b). These data indicate that Ca^{2+} and PDB also do not play a role in the interaction of vitamin E with PKC and its inhibition of the kinase activity. The concentration of vitamin E in plasma from normal humans is in the range of 12–35 μM (Stuart 1982), which is approximately 10–20-fold lower than the levels required here for 50% inhibition of purified PKC kinase activity. However, inhibition of kinase activity in our assays starts at greater than 10 μM in some cases and it may be that in intact cells a local concentration effect of vitamin E in the membrane may allow for modulation of PKC activity.

The water soluble vitamin E analogue, Trolox, is devoid of PKC inhibitory activity. However, after amidation of its carboxyl group Trolox becomes an inhibitor of PKC as potent as vitamin E. The oxidized product of vitamin E, α-tocopherylquinone, has a similar concentration-dependent inhibition profile as vitamin E, suggesting that the inhibition of kinase activity by vitamin E is not a redox mediated process (Mahoney and Azzi in press, a). It appears that vitamin E has both a role as antioxidant and as cellular regulator by modulating PKC activity.

4 Conclusions

The study of the regulation of PKC has a manifold significance: such a complicated protein, with binding sites for ATP, protein substrate, calcium, phosphatidylserine and DAG, with a regulatory and a catalytic domain, existing in a membrane-bound or a free form, with the possibility of autoregulation through phosphorylation, with several isozyme forms, offers a challenge to the enzymologist.

At a cellular level, the integration of this protein with the most important regulatory pathways of the cell, related to growth, differentiation, secretion, and cell activation, offers a unique target for studies on cell regulation. As a target of tumor promoters, inhibitors studies of PKC may offer as well a chance of pharmacological intervention in the process of neoplastic growth.

Acknowledgement. This work was supported by the Swiss National Science Foundation and the Schweizerische Krebsliga.

References

Albert KA, Wu WC, Nairn AC, Greengard P (1984) Inhibition by calmodulin of calcium/phospholipid-dependent protein phosphorylation. Proc Natl Acad Sci USA 81:3622–3625

Ashendel CL, Staller JM, Boutwell RK (1983) Identification of a calcium- and phospholipid-dependent phorbol ester binding activity in the soluble fraction of mouse tissues. Biochem Biophys Res Commun 111:340–345

Baehner RL, Boxer LA, Ingraham LM, Butterick C, Haak RA (1982) The influence of vitamin E on human polymorphonuclear cell metabolism and function. Ann NY Acad Sci 393:237–250

Beavo JA, Mumby MC (1982) Cyclic AMP-dependent protein phosphorylation. In: Nathanson JA, Kebabian JW (eds) Cyclic nucleotides, Pt 1, Biochemistry. Springer, Berlin Heidelberg New York pp 363–392

Berkow RL, Kraft AS (1985) Bryostatin, a non-phorbol macrocyclic lactone, activates intact human polymorphonuclear leukocytes and binds to the phorbol ester receptor. Biochem Biophys Res Commun 131:1109–1116

Berridge MJ (1987) Inositol triphosphate and diacylglycerol: two interacting second messengers. Annu Rev Biochem 56:159–193

Brandt SJ, Niedel JE, Bell RM, Young WS (1987) Distinct patterns of expression of different protein kinase C mRNAs in rat tissues. Cell 49:57–63

Brockerhoff H (1986) Membrane protein-lipid hydrogen bonding: evidence from protein kinase C, diglyceride, and tumor promoters. FEBS Lett 201:1–4

Castagna M, Takai Y, Kaibuchi K, Sano K, Kikkawa U, Nishizuka Y (1982) Direct activation of calcium-activated, phospholipid-dependent protein kinase by tumor-promoting phorbol esters. J Biol Chem 257:7847–7851

Cochet C, Gill GN, Meisenhelder J, Cooper JA, Hunter T (1984) C-kinase phosphorylates the epidermal growth factor receptor and reduces its epidermal growth factor-stimulated tyrosine protein kinase activity. J Biol Chem 259:2553–2558

Cooper RA, Braunwald AD, Kuo AL (1982) Phorbol ester induction of leukemic cell differentiation is a membrane-mediated process. Proc Natl Acad Sci USA 79:2865–2869

Costa MR, Catterall WA (1984) Phosphorylation of the alpha subunit of the sodium channel by protein kinase C. Cell Mol Neurobiol 4:291–297

Coussens L, Parker PJ, Rhee L, Yang Feng TL, Chen E, Waterfield MD, Francke U, Ullrich A (1986) Multiple, distinct forms of bovine and human protein kinase C suggest diversity in cellular signaling pathways. Science 233:859–866

Danpure CJ, Fyfe DA, Gumpel JM (1979) Distribution of gold among plasma fractions in rheumatoid patients undergoing chrysotherapy compared with its distribution in plasma incubated with aurothiomalate in vitro. Ann Rheum Dis 38:364–370

Davis RJ, Czech MP (1985) Platelet-derived growth factor mimics phorbol diester action on epidermal growth factor receptor phosphorylation at threonine-654. Proc Natl Acad Sci USA 82:4080–4084

Davis RJ, Ganong BR, Bell RM, Czech MP (1985) Structural requirements for diacylglycerols to mimic tumor-promoting phorboldiester action on the epidermal growth factor receptor. J Biol Chem 260:5315–5322

Delclos KB, Yeh E, Blumberg PM (1983) Specific labeling of mouse brain membrane phospholipids with [20-^{3}H]phorbol 12-p-azidobenzoate 13-benzoate, a photolabile phorbol ester. Proc Natl Acad Sci USA 80:3054–3058

Froscio M, Solanki V, Murray AW, Hurst NP (1988) Auranofin enhances phosphorylation of putative substrates of protein kinase C in human platelets. Biochem Pharmacol 37:366–368

Fujita I, Irita K, Takeshige K, Minakami S (1984) Diacylglycerol, 1-oleoyl-2-acetyl-glycerol, stimulates superoxide-generation from human neutrophils. Biochem Biophys Res Commun 120:318–324

Gerber RC, Paulus HE, Bluestone R, Lederer M (1972) Kinetics of aurothiomalate in serum and synovial fluid. Arthritis Rheum 15:625–629

Go M, Sekiguchi K, Nomura H, Kikkawa U, Nishizuka Y (1987) Further studies on the specificity of diacylglycerol for protein kinase C activation. Biochem Biophys Res Commun 144:598–605

Gopalakrishna R, Barsky SH, Thomas TP, Anderson WB (1986) Factors influencing chelator-stable, detergent-extractable, phorboldiester-induced membrane association of protein kinase C. Differences between Ca^{2+}-induced and phorbol ester-stabilized membrane bindings of protein kinase C. J Biol Chem 261:16438–16445

Hafstrom I, Seligmann BE, Friedman MM, Gallin JI (1984) Auranofin affects early events in human polymorphonuclear neutrophil activation by receptor-mediated stimuli. J Immunol 132:2007–2014
Hannun YA, Bell RM (1987) Lysosphingolipids inhibit protein kinase C: implications for the sphingolipidoses. Science 235:670–674
Hannun YA, Bell RM (1988) Aminoacridines, potent inhibitors of protein kinase C. J Biol Chem 263:5124–5131
Hansson A, Serhan CN, Haeggstrom J, Ingelman Sundberg M, Samuelsson B (1986) Activation of protein kinase C by lipoxin A and other eicosanoids. Intracellular action of oxygenation products of arachidonic acid. Biochem Biophys Res Commun 134:1215–1222
Hecker E, Schmidt R (1979) Phorbolesters – the irritants and cocarcinogens of Croton Tiglium L. Fortschr Chem Org Naturstoff 31:377–467
Helson L, Verma M, Helson C (1982) Vitamin E and neuroblastoma. Ann NY Acad Sci 393:226
Hidaka H, Inagaki M, Kawamoto S, Sasaki Y (1984) Isoquinoline sulfonamides, novel and potent inhibitors of cyclic nucleotide dependent protein kinase and protein kinase C. Biochemistry 23:5036–5041
House C, Kemp BE (1987) Protein kinase C contains a pseudosubstrate prototope in its regulatory domain. Science 238:1726–1728
Huang FL, Yoshida Y, Nakabayashi H, Huang KP (1987) Differential distribution of protein kinase C isozymes in the various regions of brain. J Biol Chem 262:15714–15720
Huang KP, Huang FL (1986) Conversion of protein kinase C from a Ca^{2+}-dependent to an independent form of phorbol ester-binding protein by digestion with trypsin. Biochem Biophys Res Commun 139:320–326
Huang KP, Nakabayashi H, Huang FL (1986) Isozymic forms of rat brain Ca^{2+}-activated and phospholipid-dependent protein kinase. Proc Natl Acad Sci USA 83:8535–8539
Hucho F, Kruger H, Pribilla I, Oberdieck U (1987) a 40 kD inhibitor of protein kinase C purified from bovine brain. FEBS Lett 211:207–210
Ido M, Sekiguchi K, Kikkawa U, Nishizuka Y (1987) Phosphorylation of the EGF receptor from A431 epidermoid carcinoma cells by three distinct types of protein kinase C. FEBS Lett 219:215–218
Ieyasu H, Takai Y, Kaibuchi K, Sawamura M, Nishizuka Y (1982) A role of calcium-activated, phospholipid-dependent protein kinase in platelet-activating factor-induced serotonin release from rabbit platelets. Biochem Biophys Res Commun 108:1701–1708
Ikebe M, Inagaki M, Kanamaru K, Hidaka H (1985) Phosphorylation of smooth muscle myosin light chain kinase by Ca^{2+}-activated, phospholipid-dependent protein kinase. J Biol Chem 260:4547–4550
Inoue M, Kishimoto A, Takai Y, Nishizuka Y (1977) Studies on a cyclic nucleotide-independent protein kinase and its proenzyme in mammalian tissues. II Proenzyme and its activation by calcium-dependent protease from rat brain. J Biol Chem 252:7610–7616
Jeffrey AM, Liskamp RM (1986) Computer-assisted molecular modeling of tumor promoters: rationale for the activity of phorbol esters, teleocidin B, and aplysiatoxin. Proc Natl Acad Sci USA 83:241–245
Kajikawa N, Kaibuchi K, Matsubara T, Kikkawa U, Takai Y, Nishizuka Y, Itoh K, Tomioka C (1983) A possible role of protein kinase C in signal-induced lysosomal enzyme release. Biochem Biophys Res Commun 116:743–750
Kase H, Iwahashi K, Matsuda Y (1986) K-252a, a potent inhibitor of protein kinase C from microbial origin. J Antibiot 39:1059–1065
Kase H, Iwahashi K, Nakanishi S, Matsuda Y, Yamada K, Takahashi M, Murakata C, Sato A, Kaneko M (1987) K-252 compounds, novel and potent inhibitors of protein kinase C and cyclic nucleotide-dependent protein kinases. Biochem Biophys Res Commun 142:436–440
Kawahara Y, Takai Y, Minakuchi R, Sano K, Nishizuka Y (1980) Phospholipid turnover as a possible transmembrane signal for protein phosphorylation during human platelet activation by thrombin. Biochem Biophys Res Commun 97:309–317
Kikkawa U, Ogita K, Ono Y, Asaoka Y, Shearman MS, Fujii T, Ase K, Sekiguchi K, Igarashi K, Nishizuka Y (1987) The common structure and activities of four subspecies of rat brain protein kinase C family. FEBS Lett 223:212–216
Kishimoto A, Takai Y, Mori T, Kikkawa U, Nishizuka Y (1980) Activation of calcium and phospholipid-dependent protein kinase by diacylglycerol, its possible relation to phosphatidylinositol turnover. J Biol Chem 255:2273–2276
Knopf JL, Lee MH, Sultzman LA, Kriz RW, Loomis CR, Hewick RM, Bell RM (1986) Cloning and expression of multiple protein kinase C cDNAs. Cell 46:491–502

Kondo H, Baba Y, Takaki K, Kondo K, Kagamiyama H (1987) Phosphorylation by protein kinase C of a synthetic heptapeptide bearing a lysine residue on the C terminal side of serine. Biochem Biophys Res Commun 142:155–161

Kraft AS, Anderson WB (1983) Phorbol esters increase the amount of Ca^{2+}, phospholipid-dependent protein kinase associated with plasma membrane. Nature (London) 301:621–623

Kraft AS, Anderson WB, Cooper HL, Sando JJ (1982) Decrease in cytosolic calcium/phospholipid-dependent protein kinase activity following phorbol ester treatment of EL4 thymoma cells. J Biol Chem 257:13193–13196

Kuo JF, Shoji M (1982) Cyclic GMP-dependent protein phosphorylation. In: Nathanson JA, Kababian JW (eds) Cyclic nucleotides, Pt 1, Biochemistry. Springer, Berlin Heidelberg New York, pp 393–424

Kuo JF, Schatzman RC, Turner RS, Mazzei GJ (1984) Phospholipid-sensitive Ca^{2+}-dependent protein kinase: a major protein phosphorylation system. Mol Cell Endocrinol 35:65–73

Lapetina EG, Reep B, Ganong BR, Bell RM (1985) Exogenous sn-1,2-diacylglycerols containing saturated fatty acids function as bioregulators of protein kinase C in human platelets. J Biol Chem 260:1358–1361

Lee MH, Bell RM (1986) The lipid binding, regulatory domain of protein kinase C. A 32-kDa fragment contains the calcium- and phosphatidylserine-dependent phorbol diester binding activity. J Biol Chem 261:14867–14870

Liles WC, Hunter DD, Meier KE, Nathanson NM (1986) Activation of protein kinase C induces rapid internalization and subsequent degradation of muscarinic acetylcholine receptors in neuroblastoma cells. J Biol Chem 261:5307–5313

Liu AY, Chen KY (1985) Differential effects of the tumor promoter phorbol-12-myristate-13-acetate on the morphological and biochemical differentiation of N-18 mouse neuroblastoma cells. J Cell Physiol 125:387–392

Lubin B, Chiu D (1982) Membrane phospholipid organization in pathologic human erythrocytes. Prog Clin Biol Res 97:137–150

Lüthy R, Azzi A (1987) Fluorescent and photoactive probes for the study of protein kinase C, Eur J Biochem 162:387–391

Mahoney CW, Azzi A (1988a) the Au(I) anti-rheumatic compounds inhibit protein kinase C. Experientia 44:A82

Mahoney CW, Azzi A (1988b) Vitamin E inhibits protein kinase C activity. Biochem Biophys Res Commun 154:694–697

Mahoney CW, Azzi A (in press,a) Structural requirements for vitamin E inhibition of protein kinase C activity. Experientia

Mahoney CW, Azzi A (in press,b) Ca^{2+}- and phospholipid dependent protein kinase (protein kinase C): a pleiotropic signal protein and its inhibition by staurosporine and Auranofin. In: Papa S (ed) Perspectives in molecular approaches to human diseases. Horwood, London

Mahoney CW, Lüthy R, Azzi A (1986) Membrane signal transduction via protein kinase C. In: Azzi A, Masotti L, Vecli A (eds) Membrane proteins. Springer, Berlin Heidelberg New York Tokyo, pp 56–66

Mahoney CW, Fredenhagen A, Peter H, Azzi A (in press) Staurosporine secondary amine derivatives: synthesis, fluorescent properties, and inhibitory action against protein kinase C. Hoppe-Seyler's Biol Chem

Malkinson AM, Girard PR, Kuo JF (1987) Strain-specific postnatal changes in the activity and tissue levels of protein kinase C. Biochem Biophys Res Commun 145:733–739

Markovac J, Goldstein GW (1988) Picomolar concentrations of lead stimulate brain protein kinase C. Nature (London) 334:71–73

McCaffrey PG, Friedman B, Rosner MR (1984) Diacylglycerol modulates binding and phosphorylation of the epidermal growth factor receptor. J Biol Chem 259:12502–12507

McCaffrey PG, Rosner MR, Kikkawa U, Sekiguchi K, Ogita K, Ase K, Nishizuka Y (1987) Characterization of protein kinase C from normal and transformed cultured murine fibroblasts. Biochem Biophys Res Commun 146:140–146

McCay PB, King MM, Poyer JL, Lai EK (1982) An update on antioxidant theory: spin trapping of trichloromethyl radicals in vivo. Ann NY Acad Sci 393:23–31

McDonald JR, Walsh MP (1985) Ca^{2+}-binding proteins from bovine brain including a potent inhibitor of protein kinase C. Biochem J 232:559–567

Melloni E, Pontremoli S, Michetti M, Sacco O, Sparatore B, Salamino F, Horecker BL (1985) Binding of protein kinase C to neutrophil membranes in the presence of Ca^{2+} and its activation by a Ca^{2+}-requiring proteinase. Proc Natl Acad Sci USA 82:6435–6439
Michell RH (1983) Polyphosphoinositide breakdown as the initiating reaction in receptor-stimulated inositol phospholipid metabolism. Life Sci 32:2083–2085
Mochly-Rosen D, Koshland DE Jr (1987) Domain structure and phosphorylation of protein kinase C. J Biol 262:2291–2297
Mori T, Takai Y, Yu B, Takahashi J, Nishizuka Y, Fujikura T (1982) Specificity of the fatty acyl moieties of diacylglycerol for the activation of calcium activated phospholipid-dependent protein kinase. J Biochem (Tokyo) 91:427–431
Morin MJ, Kreutter D, Rasmussen H, Sartorelli AC (1987) Disparate effects of activators of protein kinase C on HL 60 promyelocytic leukemia cell differentiation. J Biol Chem 262:11758–11763
Murakami K, Routtenberg A (1985) Direct activation of purified protein kinase C by unsaturated fatty acids (oleate and arachidonate) in the absence of phospholipids and Ca^{2+}. FEBS Lett 192:189–193
Naka M, Nishikawa M, Adelstein RS, Hidaka H (1983) Phorbol ester induced activation of human platelets is associated with protein kinase C phosphorylation of myosin light chains. Nature (London) 306:490–492
Nakadate T, Jeng AY, Blumberg PM (1987) Effect of phospholipid on substrate phosphorylation by a catalytic fragment of protein kinase C. J Biol Chem 262:11507–11513
Nakanishi S, Matsuda Y, Iwahashi K, Kase H (1986) K 252b c and d potent inhibitors of protein kinase C from microbial origin. J Antibiot 39:1066–1071
Nishikawa M, Sellers JR, Adelstein RS, Hidaka H (1984) Protein kinase C modulates in vitro phosphorylation of the smooth muscle heavy meromyosin light chain kinase. J Biol Chem 259:8808–8814
Nishikawa M, Shirakawa S, Adelstein RS (1985) Phosphorylation of smooth muscle myosin light chain kinase by protein kinase C. Comparative study of the phosphorylated sites. J Biol Chem 260:8978–8983
Nishizuka Y (1984) The role of protein kinase C in cell surface signal transduction and tumor promotion. Nature (London) 308:693–698
Nishizuka Y (1986) Studies and perspectives of protein kinase C. Science 233:305–312
Ohno S, Kawasaki H, Imajoh S, Suzuki K, Inagaki M, Yokokura H, Sakoh T, Hidaka H (1987) Tissue specific expression of three distinct types of rabbit protein kinase C. Nature (London) 325:161–166
Pahlman S, Ruusala AI, Abrahamsson L, Odelstad L, Nilsson K (1983) Kinetics and concentration effects of TPA-induced differentiation of cultured human neuroblastoma cells. Cell Diff 12:165–170
Parente JE, Wong K, Davis P (1986) Effect of gold compounds on NADPH oxidase system of human neutrophils. Inflammation 10:303–310
Parker PJ, Coussens L, Totty N, Rhee L, Young S, Chen E, Stabel S (1986) The complete primary structure of protein kinase C – the major phorbol ester receptor. Science 233:853–859
Patel MD, Samelson LE, Klausner RD (1987) Multiple kinases and signal transduction. Phosphorylation of the T cell antigen receptor complex. J Biol Chem 262:5831–5838
Roghani M, Da Silva C, Castagna M (1987) Tumor promoter chloroform is a potent protein kinase C activator. Biochem Biophys Res Commun 142:738–744
Ruusala AI, Mattsson M, Esscher T, Abrahamsson L, Jergil B, Pahlman S (1985) 12-O-tetradecanoyl-phorbol-13-acetate-induced differentiation of human neuroblastoma cells is not accompanied by an increase in the intracellular concentration of cyclic AMP. Brain Res 350:27–35
Sano K, Takai Y, Yamanishi J, Nishizuka Y (1983) A role of calcium-activated phospholipid-dependent protein kinase in human platelet activation. Comparison of thrombin and collagen actions. J Biol Chem 258:2010–2013
Schulman H (1982) Calcium-dependent protein phosphorylation. In: Nathanson JA, Kebabian JW (eds) Cyclic nucleotides, Ptl, Biochemistry. Springer, Berlin Heidelberg New York, pp 425–478
Sekiguchi K, Tsukuda M, Ogita K, Kikkawa U, Nishizuka Y (1987) Three distinct forms of rat brain protein kinase C: differential response to unsaturated fatty acids. Biochem Biophys Res Commun 145:797–802
Serhan CN, Smolen JE, Korchak HM, Weissmann G (1983a) Leukotriene B4 is a complete secretagogue in human neutrophils: Ca^{2+} translocation in liposomes and kinetics of neutrophil activation. Adv Prostaglandin Thromboxane Leukotriene Res 11:53–63

Serhan CN, Broekman MJ, Korchak HM, Smolen JE, Marcus AJ, Weissmann G (1983b) Changes in phosphatidylinositol and phosphatidic acid in stimulated human neutrophils. Relationship to calcium mobilization, aggregation and superoxide radical generation. Biochim Biophys Acta 762:420–428

Sharkey NA, Blumberg PM (1985) Kinetic evidence that 1,2-diolein inhibits phorbol ester binding to protein kinase C via a competitive mechanism. Biochem Biophys Res Commun 133:1051–1056

Sharkey NA, Leach KL, Blumberg PM (1984) Competitive inhibition by diacylglycerol of specific phorbol ester binding. Proc Natl Acad Sci USA 81:607–610

Shearman MS, Naor Z, Kikkawa U, Nishizuka Y (1987) Differential expression of multiple protein kinase C subspecies in rat central nervous tissue. Biochem Biophys Res Commun 147:911–919

Sigel E, Baur R (1988) Allosteric modulation by benzodiazepine receptor ligands of the GABA receptor channel expressed in *Xenopus* oocytes. J Neurosci 8:289–295

Solomon DH, O'Brian CA, Weinstein IB (1985) N-alpha-Tosyl-L-lysine chloromethyl ketone and N-alpha-tosyl-L-phenylalanine chloromethyl ketone inhibit protein kinase C. FEBS Lett 190:342–344

Steiner M, Mower R (1982) Mechanism of action of vitamin E on platelet function. Ann NY Acad Sci 393:289–299

Stuart MJ (1982) Vitamin E deficiency: its effect on platelet-vascular interaction in various pathologic states. Ann NY Acad Sci 393:277–288

Takai Y, Kishimoto A, Inoue M, Nishizuka Y (1977) Studies on a cyclic nucleotide-independent protein kinase and its proenzyme in mammalian tissues. I. Purification and characterization of an active enzyme from bovine cerebellum. J Biol Chem 252:7603–7609

Takai Y, Kishimoto A, Kikkawa U, Mori T, Nishizuka Y (1979) Unsaturated diacylglycerol as a possible messenger for the activation of calcium-activated, phospholipid-dependent protein kinase system. Biochem Biophys Res Commun 91:1218–1224

Tamaoki T, Nomoto H, Takahashi I, Kato Y, Morimoto M, Tomita F (1986) Staurosporine, a potent inhibitor of phospholipid/Ca^{++}-dependent protein kinase. Biochem Biophys Res Commun 135:397–402

Tanaka C, Taniyama K, Kusunoki M (1984) A phorbol ester and A23187 act synergistically to release acetylcholine from the guinea pig ileum. FEBS Lett 175:165–169

Tanaka C, Fujiwara H, Fujii Y (1986) Acetylcholine release from guinea pig caudate slices evoked by phorbol ester and calcium [published erratum appears in FEBS Lett 1986 May 26; 201(1):179]. FEBS Lett 195:129–134

Walter RD (1980) Effect of suramin on phosphorylation-dephosphorylation reactions in *Trypanosoma gambiense*. Mol Biochem Parasitol 1:139–142

Wasserman RH, Taylor AN (1972) Metabolic roles of fat-soluble vitamins D, E, and K. Annu Rev Biochem 41:179–202

Watson SP, McNally J, Shipman LJ, Godfrey PP (1988) The action of the protein kinase C inhibitor, staurosporine, on human platelets. Evidence against a regulatory role for protein kinase C in the formation of inositol trisphosphate by thrombin. Biochem J 249:345–350

Witters LA, Vater CA, Lienhard GE (1985) Phosphorylation of the glucose transporter in vitro and in vivo by protein kinase C. Nature (London) 315:777–778

Wolf M, Baggiolini M (1988) The protein kinase inhibitor staurosporine, like phorbol esters, induces the association of protein kinase C with membranes. Biochem Biophys Res Commun 154:1273–1279

Wolf M, Cuatrecasas P, Sahyoun N (1985) Interaction of protein kinase C with membranes is regulated by Ca^{2+}, phorbol esters, and ATP. J Biol Chem 260:15718–15722

Woodgett JR, Hunter T (1987) Immunological evidence for two physiological forms of protein kinase C. Mol Cell Biol 7:85–96

Young S, Parker PJ, Ullrich A, Stabel S (1987) Down-regulation of protein kinase C is due to an increased rate of degradation. Biochem J 244:775–779

2 Regulation of Membrane Channels

K^+ Channels: Structure, Function, Regulation, Molecular Pharmacology and Role in Diseased States

H. Bernardi, J.-N. Bidard, M. Fosset, M. Hugues, Ch. Mourre, H. Rehm, G. Romey, H. Schmid-Antomarchi, H. Schweitz, J.R. de Weille and M. Lazdunski

1 Introduction

This chapter will describe the molecular pharmacology and biochemistry of three types of K^+ channels: the calcium-activated potassium channels. ATP-regulated potassium channels and voltage-sensitive potassium channels.

2 Ca^{2+}-Activated K^+ Channels

The calcium-activated potassium channel (Ca^{2+}-activated K^+ channels) was the first of the K^+ channels for which a good molecular pharmacology was established.

Mammalian myotubes in culture contract spontaneously. In order to contract spontaneously they, of course, have to generate spontaneous action potentials. In order to generate spontaneous action potentials they have to have an after-hyperpolarization (ahp). It is during this hyperpolarization phase, which works as a pacemaker, that the sodium channel, which has been activated and then inactivated during the first action potential, is reactivated, thus allowing the generation of a new action potential. The Ca^{2+}-activated K^+ channels – like any other channel – is a protein. Perhaps it is a more intelligent protein than many others since it is activated by a combination of various membrane potentials and internal calcium concentrations. The Ca^{2+}-activated K^+ channel described here is sensitive to the toxin apamin (Fig. 1; Hugues et al. 1982b, d; Romey and Lazdunski 1984; Lazdunski 1983). This toxin from bee venom has 18 amino acids (Habermann 1972). Its active site is composed of two arginines (Vincent et al. 1975; Lazdunski 1982). The after-hyperpolarization, which is responsible for the pacemaking activity and which is associated with the activity of Ca^{2+}-activated K^+ channels, is blocked by apamin concentrations in the 1–10 nM range (Fig. 1). The toxin blocks Ca^{2+}-activated K^+ channels involved in the ahp but it does not block the other class of (big) Ca^{2+}-activated K^+ channels (Romey and Lazdunski 1984) (BK channels $\approx$ 250 pS). TEA (tetraethylammonium) blocks the big channels, but not the ahp (Romey and Lazdunski 1984). These channels are also blocked by a toxin recently discovered by C. Miller and his group, charybdotoxin (Miller et al. 1985), but the ahp is not affected. In the skeletal muscle cell the pacemaker channel, which is blocked by apamin, is not blocked by TEA or charybdotoxin. In contrast,

Centre de Biochimie du CNRS, Parc Valrose, 06034 Nice Cedex, France

A. Azzi, Z. Drahota and S. Papa
Molecular Basis of Membrane-Associated Diseases

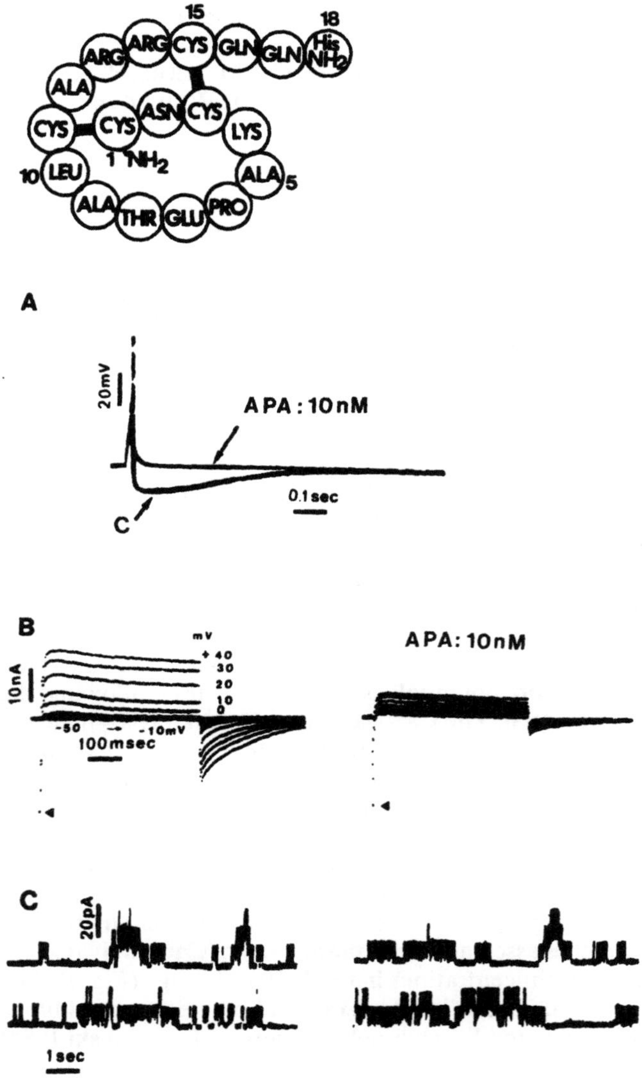

Fig. 1. *Above* The structure of apamin. **A** ahp in the control (**C**) is blocked by apamin (*APA*). **B** Voltage-clamp analysis of the effect of apamin on rat myosacs. Families of membrane currents are associated with different step depolarizations from 90 mV. Experiment done in 20 mM TEA. **C** Single channel recordings of the Ca^{2+}-dependent K^+ channel were made at +30 mV. *Left* Control; *right* absence of effect in 10 nM apamin

we have found recently (Van Renterghem et al. 1988) with an aortic cell line having a repetitive activity that the ahp, which is responsible for this repetitive activity, is blocked by charybdotoxin, but not by apamin.

One very interesting thing about the apamin-sensitive Ca^{2+}-activated K^+ channel in muscle concerns the transformation of its properties during the course of innervation. When the myotube is not innervated it contracts spontaneously and this contraction is essential for survival. Of course, later in development the skeletal muscle cell will be innervated and then spontaneous contraction is undesirable. At that stage contractions should only occur under the order of the nerve. Several things should then and do happen in this muscle cell during innervation in order to change its electrical phenotype. First, there is a change in the resting potential corresponding to a hyperpolarization. The reason for this hyperpolarization is to bring the resting potential farther from the threshold of the generation of the action potential, i.e. from the threshold of activation of the sodium channel. The second thing that one would expect is the disappearance of the apamin-sensitive channel which is an essential element of the repetitive electrical activity. This is exactly what happens (Schmid-Antomarchi et al. 1985).

If one keeps rat myotubes in vitro and one innervates them by co-culture with spinal cord cells, then after a while the action potential is no longer followed by an ahp (Schmid-Antomarchi et al. 1985; Fig. 2). In other words, as predicted, the innervation of the skeletal muscle cells has repressed the expression of the apamin-sensitive K^+ channel. Inversely, if one takes normal muscle it does not have spontaneous contractions, repetitive activity, nor any ahp following the action potential. Ten days after denervation of this muscle, the ahp appears and is completely blocked by apamin (Schmid-Antomarchi et al. 1985; Fig. 2). Thus, the clear message obtained from these data is that when the muscle is not innervated,

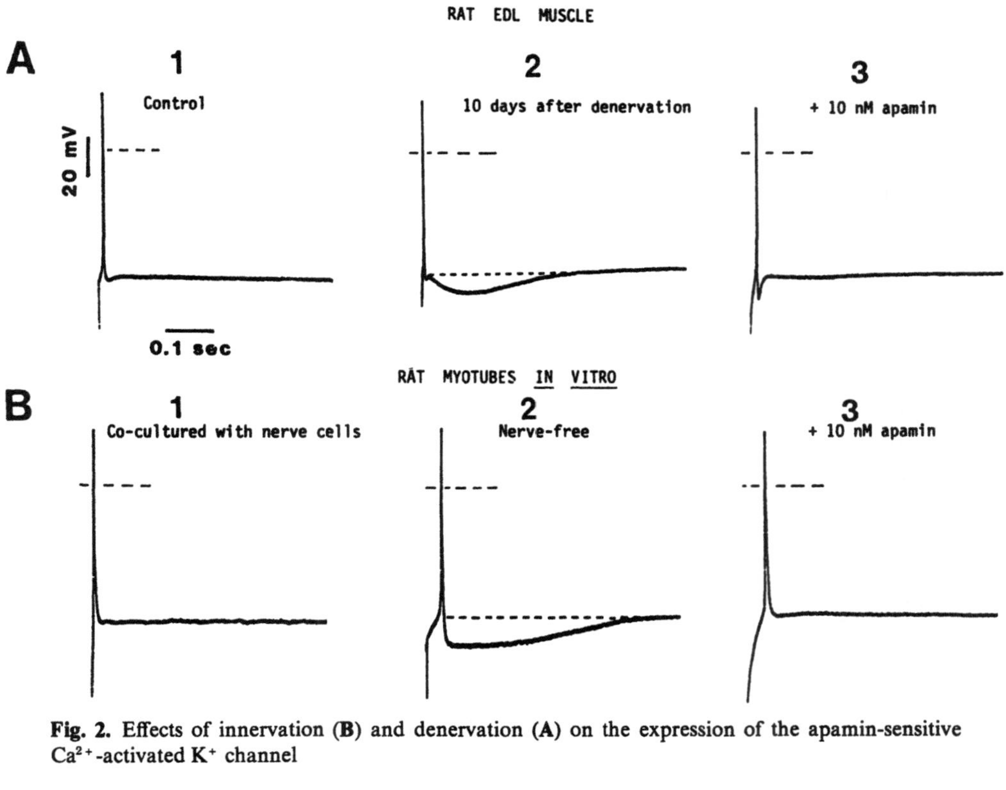

Fig. 2. Effects of innervation (**B**) and denervation (**A**) on the expression of the apamin-sensitive Ca^{2+}-activated K^+ channel

the apamin-sensitive Ca^{2+}-activated K^+ channel is expressed. When the muscle is innervated, the channel disappears. When the muscle is denervated, the Ca^{2+}-activated K^+ channel reappears.

All these observations focussed our attention on a frequent human muscle disease, myotonic muscle dystrophy or Steinert disease. Myotonic discharges occur and muscle relaxation is difficult. The normal innervated human muscle, like the rat muscle, when it is innervated, fails to express the apamin-sensitive Ca^{2+}-activated K^+ channel (Renaud et al. 1986). In Steinert disease, the muscle, although it is normally innervated, expresses, as expected, the Ca^{2+}-activated K^+ channel identified as an apamin receptor. Also, muscles of patients with Steinert disease are known to be more depolarized than those of normal muscles (Mourre et al. 1987). This is probably the molecular explanation of the myotonic symptoms of the disease. It seems that in Steinert disease there is a disorder not only in the muscle but also in the relationship between nerve and muscle (Renaud et al. 1986).

Apamin-sensitive channels are not only present in skeletal muscle. They are also present in smooth muscles where they participate in contraction. They are also present in the nervous system and ^{125}I-apamin has been used to locate the apamin receptor autoradiographically throughout the nervous system and at different steps of brain differentiation (Mourre et al. 1987). The apamin receptor is expressed very early during brain ontogenesis; earlier actually than calcium channels visualized as calcium channel blocker receptors or the sodium channels visualized as tetrodotoxin receptors. It is of course important to know whether these receptors correspond to the functional Ca^{2+}-activated K^+ channel. One method of determination is to measure deoxyglucose consumption in different parts of the brain after apamin injection (Mourre et al. 1988b). If apamin blocks Ca^{2+}-activated K^+ channels, this blockade will change the rhythm of firing in different parts of the brain and this in turn will change deoxyglucose consumption. This is an indirect way to determine whether apamin receptors are associated with channel activity. One can show, for example, when one injects apamin in very low concentrations (at which there are no visible effects on the animal) that glucose consumption increases tremendously in the auditory cortex, an area which is rich in apamin receptors. In general, the apamin receptors and the consumption of glucose have similar localizations (Mourre et al. 1988b).

What do we know about the molecular properties of this apamin-sensitive channel? The activity of the apamin-sensitive K^+ channel can be recorded in neuroblastoma cells. Figure 3 shows that it is a small channel with a conductance of about 9 pS. What do we know about the structure of the channel protein? Although apamin binds very tightly to its receptor, with a K_D of about 20 pM (Hugues et al. 1982a, b), the elucidation of the structure of the Ca^{2+}-activated K^+ channel that is blocked by apamin is difficult due to the low number of apamin binding sites in practically all cells with which we have worked. Therefore, purification is extremely difficult. Radiation-inactivation gave a molecular weight of about 250 000 and the main polypeptide chain which was affinity-labelled with the toxin has an M_r of about 30 000 (Hugues et al. 1982c; Schmid-Antomarchi et al. 1984). However, it appears now that there may be at least another polypeptide chain with a molecular weight of 90 000 (Seagar et al. 1985).

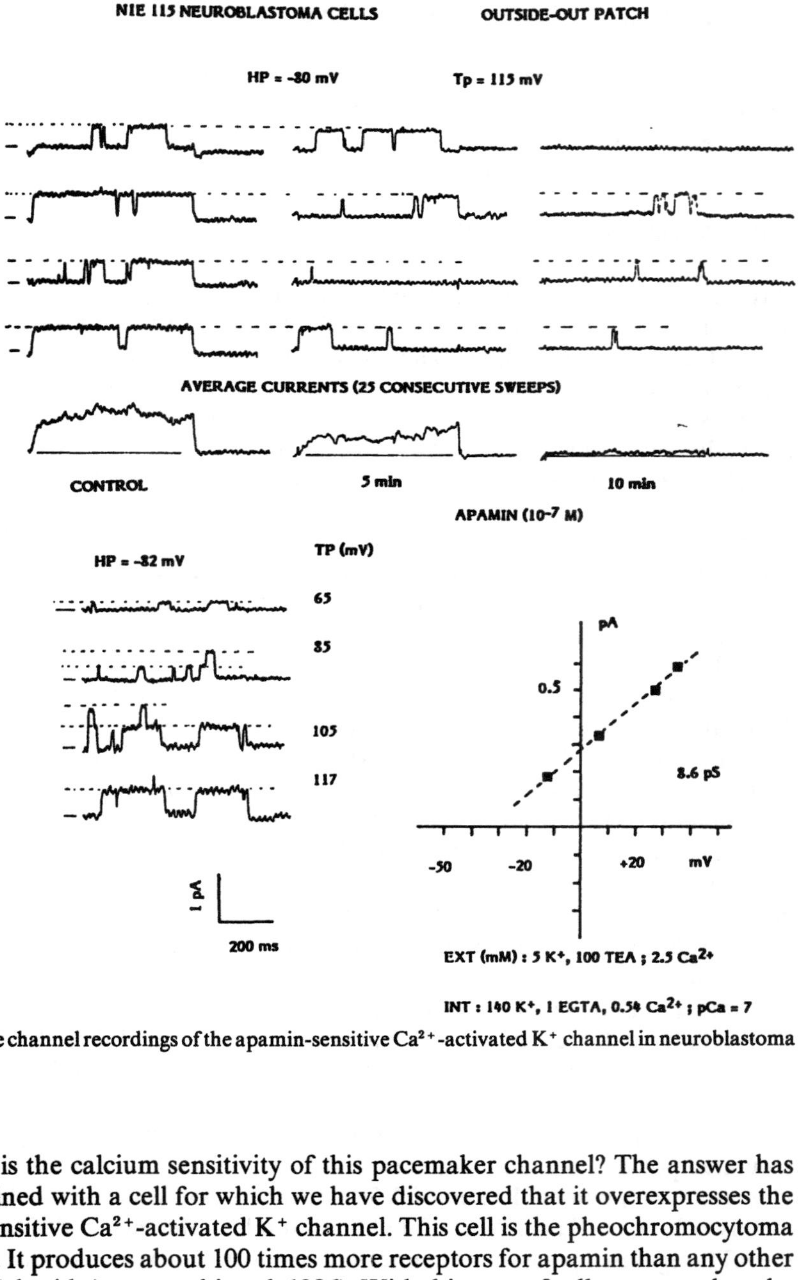

Fig. 3. Single channel recordings of the apamin-sensitive Ca^{2+}-activated K^+ channel in neuroblastoma cells

What is the calcium sensitivity of this pacemaker channel? The answer has been obtained with a cell for which we have discovered that it overexpresses the apamin-sensitive Ca^{2+}-activated K^+ channel. This cell is the pheochromocytoma cell, PC12. It produces about 100 times more receptors for apamin than any other cell type (Schmid-Antomarchi et al. 1986). With this type of cell one can relate the activity of the apamin-sensitive channel to the intracellular Ca^{2+} concentration. At 10^{-7} M $[Ca^{2+}]_i$, the channel is not activated but if the internal concentration increases only to 3×10^{-7} M, the channel is then fully activated. This Ca^{2+}-activated K^+ channel is activated at very low $[Ca^{2+}]_i$ concentrations and the dose-response curve for the internal Ca^{2+} concentration is very steep with a Hill coefficient of more than 3.

One of the important aspects of the apamin-sensitive Ca^{2+}-activated K^+ channel is that there exists an apamin-like molecule endogenous to the brain (Fosset et al. 1984).

3 ATP-Regulated K⁺ Channels

This class of K^+ channel is regulated by internal ATP concentrations. At a high ATP concentration the channel is closed. When the concentration of ATP is lowered, the channel is open. Therefore, a low ATP concentration will give a hyperpolarization, whereas a higher ATP concentration will lead to the closing of the potassium channel, thus producing a depolarization (Ashcroft et al. 1984; Cook and Hales 1984; Findlay et al. 1985; Rorsman and Trube 1985).

The interest for this channel has been recently shown to be very important, especially for β-cells. When a β-cell is perfused with glucose, it generates spontaneous action potentials. The β-cell has a normal potential near −70 mV. As soon as it is perfused with glucose, it depolarizes slowly, the depolarization reaches the threshold for activation of calcium channels, the cell than fires calcium action potentials, calcium permeates into the cell and insulin is released. Interestingly, not only glucose has this effect but also a class of pharmacologically important molecules, the sulfonylureas, which have now been used for more than 20 years for the treatment of diabetes type 2. Tolbutamide or any other sulfonylurea produce the same electrical effects as glucose perfusion, i.e. a slow depolarization leading to generation of calcium action potentials (Henquin and Meissner 1982). The easiest interpretation is the following for the glucose effect. When the cell is perfused with glucose, the ATP concentration inside of the cell increases, an increase of the ATP concentration will block the ATP-regulated potassium channels and then of course depolarization will occur. All these results taken together suggest that there is a link between glucose and tolbutamide or sulfonylureas, in general, and that glucose and sulfonylureas have at least indirectly the same type of target.

We have used the following strategy to show that sulfonylureas are blockers of ATP-regulated K^+ channels (Schmid-Antomarchi et al. 1987a, b). In normal insulinoma cells, the intracellular ATP concentration is 3 to 4 mM. These cells were loaded with $^{86}Rb^+$ then depleted of ATP with oligomycin. After treatment of the cells with oligomycin, the ATP concentration dropped drastically, the ATP-regulated potassium channel opened and $^{86}Rb^+$ efflux occurred (Fig. 4). Sulfonylureas are very potent blockers of ATP-dependent potassium channels measured by this technique ($K_d = 0.06$ nM for glibenclamide). Glibenclamide is a blocker of the ATP-regulated potassium channel that is more potent than tetrodotoxin for the sodium channel or more potent than 1,4-dihydropyridines for the calcium channel.

Results found by flux techniques have been verified by patch-clamp techniques (Schmid-Antomarchi et al. 1987b; Fig. 5). We have then used tritium-labelled glibenclamide and have identified the sulfonylurea receptors. There is an impressive correlation for a whole series of drugs of the sulfonylurea family between the binding and the blockade (Fig. 6). This observation makes us

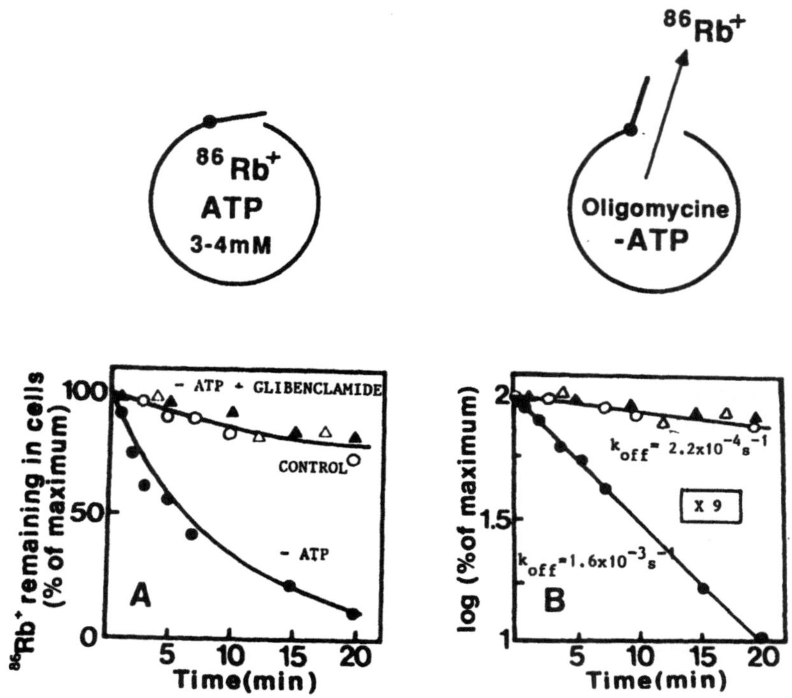

Fig. 4. Kinetics of $^{86}Rb^+$ efflux from RINm5F, effects of ATP depletion on $^{86}Rb^+$ efflux and blockade of the stimulated $^{86}Rb^+$ efflux by glibenclamide

confident that the receptor of sulfonylureas is really closely linked to the ATP-regulated potassium channel (Schmid-Antomarchi et al. 1987b).

We also wanted to know whether these ATP-regulated potassium channels, which are the targets of the very important anti-diabetic drugs, could be regulated by hormones which could themselves have a role in the secretion of insulin by the pancreas. Our interest was attracted by a hormone called galanin. This is a 29 amino-acid hormone which, when applied to the pancreas, decreases insulin secretion. We have now shown (De Weille et al. 1988) that galanin opens ATP-regulated potassium channels (and therefore decreases insulin secretion), while sulfonylureas block ATP-regulated potassium channels (and therefore increase insulin secretion) (De Weille et al. 1988).

Dose-response curves measured with the $^{86}Rb^+$ efflux assay are shown in Fig. 7.

Clearly, ATP-regulated potassium channels and their regulation might be involved in diabetes type 2.

The ATP-regulated potassium channel is not only present in β-cells but it is also present in cardiac cells (Noma 1983; Trube and Hescheler 1984). The properties of the sulfonylurea receptors associated with the ATP-regulated potassium channels and cardiac cells are the same in biochemical terms as they are in β-cells (Trube and Hescheler 1984). ATP-regulated potassium channels may be present in neurons. There are glucose-sensing neurons in the hypothalamus which

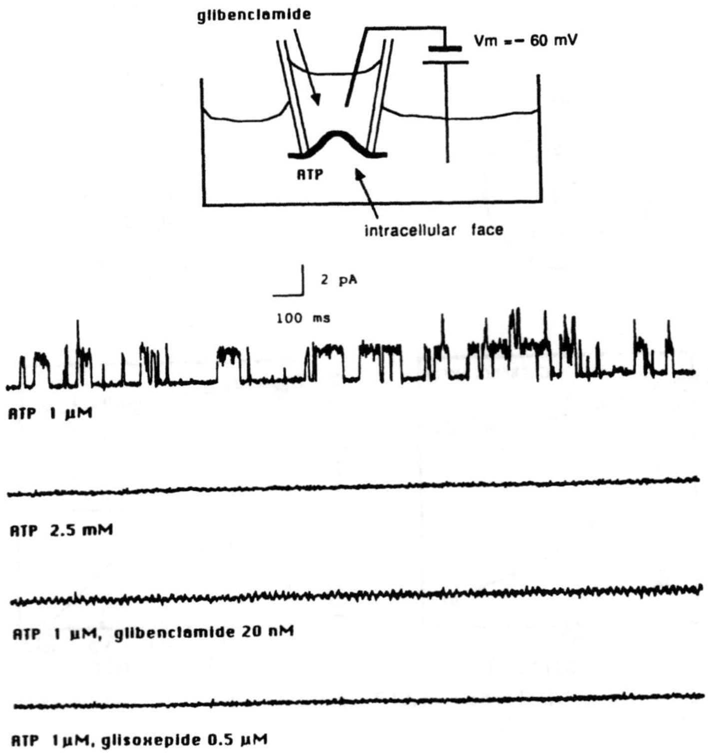

Fig. 5. Single channel currents recorded from inside-out patches of RINm5F cells. Membrane potential at −60 mV. ATP-regulated K^+ channels are opened at low $[ATP]_{in}$ concentrations and could be blocked at high (2.5 nM) $[ATP]_{in}$ concentrations, by glibenclamide or by glisoxepide

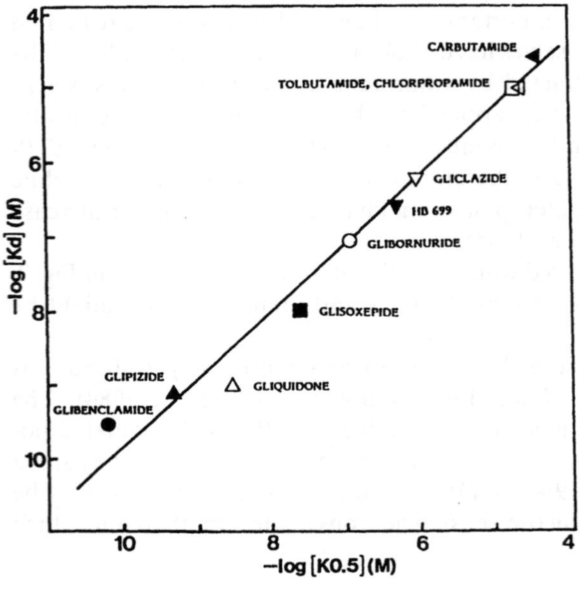

Fig. 6. Correlation between inhibition of various hypoglycemic drugs of specific [^{3}H]glibenclamide binding and $^{86}Rb^+$ efflux. K^d = half-inhibition of the ATP-sensitive rate of $^{86}Rb^+$ efflux

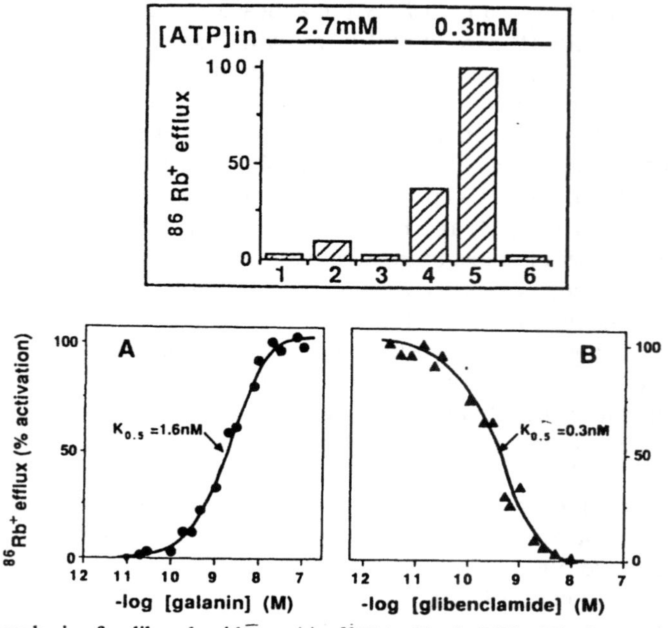

Fig. 7. *Above*: Activation by galanin of a glibenclamide-sensitive $^{86}Rb^+$ efflux in RINm5F cells. *Bars*: basal efflux in *1*, 2.7 mM $[ATP]_{in}$; *2*, 2.7 mM $[ATP]_{in}$ and 20 nM galanin; *3*, 2.7 mM $[ATP]_{in}$, 20 nM galanin and 20 nM glibenclamide; *4*, 0.3 mM $[ATP]_{in}$; *5*, 0.3 mM $[ATP]_{in}$ and 20 nM galanin; *6*, 0.3 mM $[ATP]_{in}$, 20 nM galanin and 20 nM glibenclamide. Galanin produced a 2.6-fold activation of $^{86}Rb^+$ efflux at $[ATP]_{in}$ = 0.3 mM and a 4-fold activation at $[ATP]_{in}$ = 2.7 mM. Time of $^{86}Rb^+$ efflux was 1 min. Experiments were carried out in duplicate. *Below*: Dose-response curves of activation by galanin of ATP-sensitive $^{86}Rb^+$ efflux and inhibition by glibenclamide of this $^{86}Rb^+$-activated efflux. Time of $^{86}Rb^+$ efflux was 1 min. $[ATP]_{in}$ depletion was for 20 min (then $[ATP]_{in}$ = 0.4 mM). Experiments were done in duplicate. **A** Activation of $^{86}Rb^+$ efflux measured in the presence of increasing concentrations of galanin. **B** Inhibition of $^{86}Rb^+$ efflux in the presence of 20 nM galanin and increasing concentrations of glibenclamide

control appetite and they may work by mechanisms similar to that in the β-cell. Moreover, one immediately thinks of the role of ATP-regulated potassium channels in ischemia of the brain [as well as of the heart (Fosset et al. 1988)].

4 Voltage-Dependent K+ Channels

There is a large variety of voltage-dependent potassium channels. However, in the past years there was really no adequate pharmacological tool available to study them. Usually TEA or 3,4-diaminopyridines were used, but these drugs act at high concentrations and are not completely selective. The situation has recently changed and new tools have been developed to study these channels. We have been actively working on a bee venom peptide called mast cell degranulating peptide (MCD) (Habermann 1972) because it degranulates mast cells. It is a 22 amino-acid peptide which, when injected intracerebroventricularly, is tremendously neurotoxic (Bidard et al. 1987a; Mourre et al. 1988a). On intracellular injection of MCD one sees hippocampal θ rhythm appearing in the rat which is associated with alert immobility (Bidard et al. 1987a); with higher doses of the peptide, epileptiform seizures and convulsions occur (Bidard et al. 1987a). Application of MCD to hippocampus slices produced long-term potentiation (Cherubini et al 1987). Long-term potentiation is associated with arousal, learning and memory. This MCD peptide blocks one type of voltage-dependent potassium channel (Dolly et al. 1987). The MCD peptide receptor has been identified in brain membranes (K_D of about 100 pM) and by quantitative autoradiography. MCD peptide receptors are present in most brain areas (Bidard et al. 1987b; Taylor et al 1984).

Dendrotoxins are another class of snake toxins that block voltage-sensitive potassium channels (Harvey and Anderson 1985). These toxins have sequence homologies with trypsin inhibitors but not with the MCD peptide and they facilitate neurotransmitter release (Harvey and Anderson 1985). A number of recent results obtained in different laboratories show that these toxins block one type of potassium channel (Harvey and Anderson 1985).

We have shown that the voltage-dependent potassium channel protein has two types of receptors, one for the MCD peptide, and one for the snake venom peptide dendrotoxin and that these receptors are in allosteric interaction (Fig. 8). MCD peptide binding is very strongly antagonized by dendrotoxin I in hippocampus (Bidard et al. 1987b).

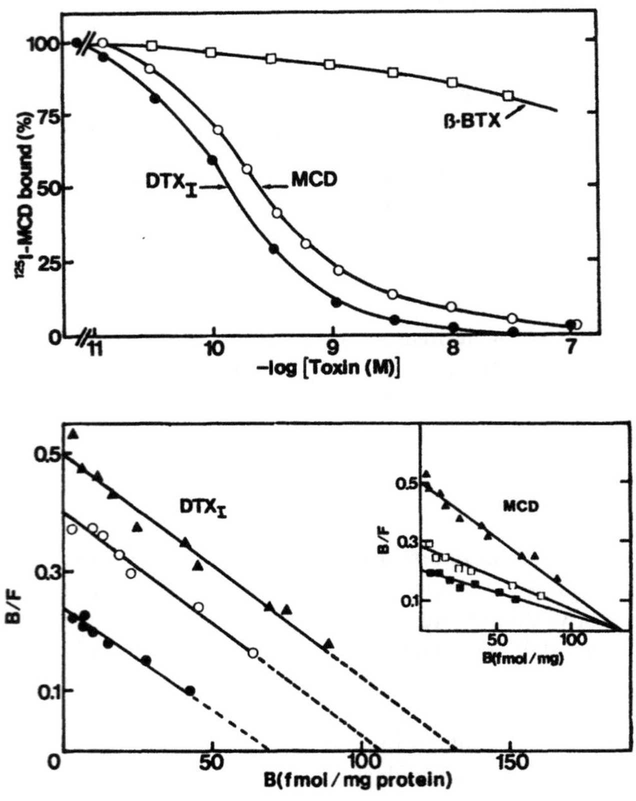

Fig. 8. Scatchard analysis of saturable binding to synaptosomal membranes. Inhibition of DTX_I. Synaptic membranes (150 µg protein/ml) were incubated with various concentrations of ^{125}I-MCD (1–200 pM). *Upper panel*: The effect of DTX_I (●), MCD (○) and β-BTX (□) on ^{125}I-MCD binding is shown. *Lower panel*: DTX_I was added at concentrations (nM) 0 (▲), 0.1 (○) and 0.32 (●). *Inset*: MCD was added at concentrations (nM) 0 (▲), 0.25 (□) and 0.5 (■). Specifically bound ^{125}I-MCD was quantified for duplicate samples and the Scatchard analysis of the free (*F*) and bound (*B*) ^{125}I-MCD concentrations were calculated

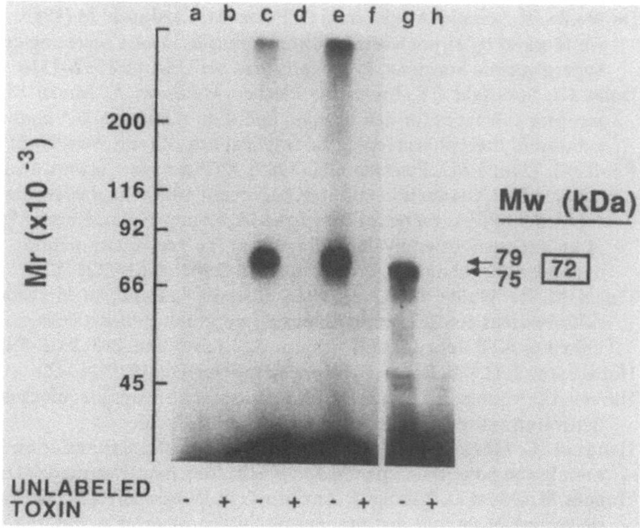

Fig. 9. Specific cross-linking (0.5 mg/ml dimethylsuberimidate) of ^{125}I-MCD (*g*) and ^{125}I-DTX (*c* and *e*) to synaptic membranes. Labelling was completely protected by unlabelled MCD and DTX. After correlation for the M_r of the grafted toxins, the M_r of the receptor protein appears to be 72 000

The same protein with the same molecular weight of 72 000 is labelled in affinity-labelling experiments by both the snake toxin and bee venom toxin MCD (Rehm et al. 1988; Fig. 9). Receptors for these toxins have now been completely purified (Rehm and Lazdunski 1988a). They co-purify and we have confirmed that the main protein component has a molecular weight of 72 000. This protein is actually heterogeneous and may contain different, very homologous peptides corresponding to K^+ channels with slightly different biophysical properties of conductance and activation and inactivation kinetics (Rehm and Lazdunski 1988b). This M_r is similar to the molecular weight of potassium channels cloned from *shaker* mutants in *Drosophila* (70 000) (Baumann et al. 1987; Tempel et al. 1987).

References

Ashcroft FM, Harrison DE, Ashcroft SJH (1984) Glucose induces closure of single potassium channels in isolated rat pancreatic β-cells. Nature (London) 312:446–448

Baumann A, Krah-Jentgens I, Müller R, Müller-Holtkamp F, Seidel R, Kecskemethy N, Casal J, Ferrus A, Pongs O (1987) Molecular organization of the maternal effect region of the *shaker* complex of *Drosophila*: characterization of an IA channel transcript with homology to vertebrate Na+ channel. EMBO J 6:3419–3429

Bidard JN, Gandolfo G, Mourre C, Gottesman C, Lazdunski M (1987a) The brain response to the bee venom peptide MCD. Activation and desensitization of an hippocampal target. Brain Res 418:235–244

Bidard JN, Mourre C, Lazdunski M (1987b) Two potent central convulsant peptides, a bee venom toxin, the MCD peptide, and a snake venom toxin, dendrotoxin I, known to block K+ channels, have interacting receptor sites. Biochem Biophys Res Commun 143:383–389

Cherubini E, Ben Ari Y, Gho M, Bidard JN, Lazdunski M (1987) Long-term potentiation of synaptic transmission in the hippocampus induced by a bee venom peptide. Nature (London) 328:70–73

Cook DL, Hales CN (1984) Intracellular ATP directly blocks K+ channels in pancreatic β-cells. Nature (London) 311:271–273

De Weille JR, Schmid-Antomarchi H, Fosset M, Lazdunski M (1988) ATP-sensitive K^+ channels that are blocked by hypoglycemic sulfonylureas in insulin secreting cells are activated by galanin, an hyperglycemic hormone. Proc Natl Acad Sci USA 85:1312–1316

Dolly JO, Stansfeld CE, Breeze A, Pelchen-Matthews A, Marsh SJ, Brown DA (1987). Neuronal acceptor sub-types for dendrotoxin and their relation to K^+ channels. In: Jenner PJ (ed) Neurotoxins and their pharmacological implications. Raven, New York, pp 81–96

Findlay I, Dunne MJ, Petersen OH (1985) ATP-sensitive inward rectifier and voltage- and calcium-activated K^+ channels in cultured pancreatic islet cells. J Membrane Biol 88:165–172

Fosset M, Schmid-Antomarchi H, Hugues M, Romey G, Lazdunski M (1984) The presence in pig brain of an endogenous equivalent of apamin, the bee venom peptide which specifically blocks Ca^{2+}-dependent K^+ channels. Proc Natl Acad Sci USA 81:7228–7232

Fosset M, de Weille JR, Green RD, Schmid-Antomarchi H, Lazdunski M (1988) Antidiabetic sulfonylureas control action potential properties in heart cells via high affinity receptors that are linked to ATP-dependent K^+ channels. J Biol Chem 263:7933–7936

Habermann E (1972) Bee and wasp venoms. Science 177:314–322

Harvey AL, Anderson A (1985) Dendrotoxins: snake toxins that block potassium channels and facilitate neurotransmitter release. Pharmacol Ther 31:33–55

Henquin JC, Meissner HP (1982) Opposite effects of tolbutamide and diazoxide on $^{86}Rb^+$ fluxes and membrane potential in pancreatic β-cells. Biochem Pharmacol 31:1407–1415

Hugues M, Duval D, Kitabgi P, Lazdunski M, Vincent JP (1982a) Preparation of a pure monoiododerivative of the bee venom neurotoxin apamin and its binding properties to rat brain synaptosomes. J Biol Chem 257:2762–2769

Hugues M, Romey G, Duval D, Vincent JP, Lazdunski M (1982b) Apamin as a selective blocker of the calcium-dependent potassium channel in neuroblastoma cells. Voltage-clamp and biochemical characterization of the toxin receptor. Proc Natl Acad Sci USA 79:1308–1312

Hugues M, Schmid H, Lazdunski M (1982c) Identification of a protein component of the Ca^{2+}-dependent K^+ channel by affinity labelling with apamin. Biochem Biophys Res Commun 107:1577–1582

Hugues M, Schmid H, Romey G, Duval D, Frelin C, Lazdunski M (1982d) The Ca^{2+}-dependent slow K^+ conductance in cultured rat muscle cells: characterization with apamin. EMBO J 9:1039–1042

Lazdunski M (1982) Etudes biochimiques du canal sodium voltage-sensible: structure, mécanisme et différenciation. C R Ste Biol 176:123

Lazdunski M (1983) Apamin, a neurotoxin specific for one class of Ca^{2+}-dependent K^+ channel. Cell Calcium 4:421–428

Miller C, Moczydlowski F, Latorre R, Phillips M (1985) Charybdotoxin, a protein inhibitor of single Ca^{2+}-activated K^+ channels from mammalian skeletal muscle. Nature (London) 313:316–318

Mourre C, Cervera P, Lazdunski M (1987) Autoradiographic analysis in rat brain of the postnatal ontogeny of voltage-dependent Na^+ channels, Ca^{2+}-dependent K^+ channels and slow Ca^{2+} channels identified as receptors for tetrodotoxin, apamin and (-)desmethoxyverapamil. Brain Res 417:21–32

Mourre C, Bidard JN, Lazdunski M (1988a) High affinity receptors for the bee venom MCD peptide. Quantitative autoradiographic localization at different stages of brain development and relationship with MCD neurotoxicity. Brain Res 446:106–112

Mourre C, Nehlig A, Lazdunski M (1988b) Cerebral glucose utilization after administration of apamin, a toxin active on Ca^{2+}-dependent K^+ channels. Brain Res 451:274–284

Noma A (1983) ATP-regulated K^+ channels in cardiac muscle. Nature (London) 305:147–148

Rehm H, Lazdunski M (1988a) Purification and subunit structure of a putative K^+ channel identified by its binding properties for dendrotoxin I. Proc Natl Acad Sci USA 85:4919–4923

Rehm H, Lazdunski M (1988b) Existence of different populations of the dendrotoxin I binding protein associated with neuronal K^+ channels. Biochem Biophys Res Commun 153:231–240

Rehm H, Bidard JN, Schweitz H, Lazdunski M (1988) The receptor site for the bee venom MCD peptide. Affinity labeling evidence for a common molecular target for MCD peptide and dendrotoxin I, a snake toxin active on K^+ channels. Biochemistry 27:1827–1832

Renaud JF, Desnuelle C, Schmid-Antomarchi H, Hugues M, Serratrice G, Lazdunski M (1986) Expression of apamin receptor in muscles of patients with myotonic muscular dystrophy. Nature (London) 319:678–680

Romey G, Lazdunski M (1984) The coexistence in rat muscle cells of two distinct classes of Ca^{2+}-dependent K^+ channels with different pharmacological properties and different physiological functions. Biochem Biophys Res Commun 118:669–674

Rorsman P, Trube G (1985) Glucose-dependent K^+ channels in pancreatic β-cells are regulated by intracellular ATP. Pflüger's Arch 405:305–309

Schmid-Antomarchi H, Hugues M, Norman RI, Ellory C, Borsotto M, Lazdunski M (1984) Molecular properties of the apamin-sensitive Ca^{2+}-dependent K^+ channel: radiation-inactivation, affinity labelling and solubilization. Eur J Biochem 142:1–6

Schmid-Antomarchi H, Renaud JF, Romey G, Hugues M, Schmid A, Lazdunski M (1985) The all-or-none role of innervation in expression of apamin receptor and of apamin-sensitive Ca^{2+}-activated K^+ channel in mammalian skeletal muscle. Proc Natl Acad Sci USA 82:2188–2191

Schmid-Antomarchi H, Hugues H, Lazdunski M (1986) Properties of the apamin-sensitive Ca^{2+}-activated K^+ channel in PC12 pheochromocytoma cells which hyper-produce the apamin receptor. J Biol Chem 261:8633–8637

Schmid-Antomarchi H, de Weille JR, Fosset M, Lazdunski M (1987a) The anti-diabetic sulfonylurea glibenclamide is a potent blocker of the ATP-modulated K^+ channel in insulin secreting cells. Biochem Biophys Res Commun 146:21–25

Schmid-Antomarchi H, de Weille JR, Fosset M, Lazdunski M (1987b) The receptor for antidiabetic sulfonylureas controls the activity of the ATP-modulated K^+ channel in insulin secreting cells. J Biol Chem 262:15840–15844

Seagar MJ, Labbé-Jullié C, Garnier C, Van Rietschoten J, Couraud F (1985) Photoaffinity labeling of components of the apamin-sensitive K^+ channel in neuronal membranes. J Biol Chem 260:3895–3898

Taylor JW, Bidard JN, Lazdunski M (1984) The characterization of high affinity binding sites in rat brain for the mast cell degranulating peptide from bee venom using the purified monoiodinated peptide. J Biol Chem 259:13957–13967

Tempel BL, Papazian D, Schwartz T, Jan Y, Jan L (1987) Sequence of probable potassium channel component encoded at *shaker* locus of *Drosophila*. Science 237:770–775

Trube G, Hescheler J (1984) Inward-rectifying channels in isolated patches of the heart cell membrane: ATP-dependent and comparison with cell-attached patches. Pflüger's Arch 401:178–184

Van Renterghem C, Romey G, Lazdunski M (1988) Vasopressin modulates the spontaneous electrical activity in aortic cells (A7r5) by acting on three different types of ionic channels. Proc Natl Acad Sci USA 85:9365–9369

Vincent JP, Schweitz H, Lazdunski M (1975) Structure-function relationships and site of action of apamin, a neurotoxic polypeptide of bee venom with an action on the central nervous system. Biochemistry 14:2521–2525

Voltage-Dependent Calcium Channels: Structure and Regulation in Normal and Abnormal States

M. M. HOSEY[1,3], F. C. CHANG[1], C. M. O'CALLAHAN[1], J. PTASIENSKI[1], J. BARHANIN[2] and M. LAZDUNSKI[2]

1 Introduction

Calcium (Ca) is an essential element for many cell processes. The major pathway by which Ca enters cells is through Ca channels located in cell membranes. Different types of Ca channels exist, and are generally classified as either voltage-dependent or receptor-dependent Ca channels. Voltage-dependent Ca channels are opened in response to a change in membrane potential, while receptor-dependent Ca channels are opened in response to the activation of an associated receptor. The subject of this review is a subclass of voltage-dependent Ca channels known as L-type or slow Ca channels [1–4]. These channels conduct a long-lasting current of large conductance [1] and are found in excitable (see refs. [2,3]) and nonexcitable [5] cells. L-type Ca channels are opened and closed in response to changes in membrane potential, but, in addition, they are *regulated* by receptor-dependent processes [3,6,7]. Considerable evidence from electrophysiological studies suggests that L-type Ca channels are regulated by a cAMP-dependent phosphorylation event. In cardiac cells, any agent that elevates cAMP, such as β-adrenergic agonists, increases the probability of opening of L channels in response to a given change in membrane potential [6,7]. L-type Ca channels in other cells, including skeletal muscle [8,9], also are regulated by cAMP. Application of the purified catalytic subunit of cAMP-dependent protein kinase mimics the effects of cAMP [10,11], and the inhibitor of this protein kinase prevents the effects [12]. Furthermore, the widely observed rundown of Ca channel activity over time in patch-clamp experiments has been slowed by cAMP-dependent protein kinase and ATP [13,14]. Taken together, the results support the hypothesis that certain L channels are regulated by the cAMP-dependent phosphorylation of the channel itself or a regulatory protein. However, until recently biochemical evidence to support this hypothesis has been lacking because the protein(s) that comprise Ca channels were not known. Recently, considerable progress has been made in our understanding of the structure and biochemical properties of L channels, and insights into the biochemical events involved in the regulation of the channels is now possible. This review will focus on our current understanding of the structure and regulation of Ca channels. Disease-related defects in L-type Ca channels will also be addressed.

[1]Department of Molecular and Cellular Biochemistry, University of Health Sciences/The Chicago Medical School, 3333 Green Bay Road, North Chicago, IL 60064, USA
[2]Centre de Biochimie du CNRS, Université de Nice, Parc Valrose, 06034 Nice, France
[3]To whom correspondence should be addressed at Department of Pharmacology, Northwestern University Medical School, 303 E Chicago Ave., Chicago, JL 60611

A. Azzi, Z. Drahota and S. Papa
Molecular Basis of Membrane-Associated Diseases

2 Purification and Characterization of L-Type Ca Channels

A large variety of potent and selective drugs have been invaluable as biochemical probes to elucidate the molecular properties of Ca channels. We will refer to these drugs as "Ca channel effectors". The most useful drugs for the purification of L channels have been: 1,4-dihydopyridine (DHP) derivatives such as nitrendipine, PN 200–110, and Bay k 8644, which have selective effects on L-type Ca channels, and the phenylalkylamines (PAA) such as verapamil, D-888 and LU47781 [3]. These drugs bind with high affinity (nM-pM) to distinct receptors on L channels, and it is now generally agreed that there is a close relationship between high affinity DHP binding sites and functional Ca channels [15–18].

The success in characterizing L-type channels was largely due to the availability of radiolabelled DHPs and the use of skeletal muscle transverse (T)-tubule membranes, which contain 50–100-fold more high affinity DHP receptors than any other source yet identified [19,20]. DHP-sensitive Ca currents [21] and DHP receptors [19,20] have both been localized to the T-tubule membrane system, and importantly, the K_d of 0.22 nM for DHP binding to the DHP receptors in isolated skeletal muscle membranes was found to be similar to the $K_{0.5}$ for DHPs to inhibit Ca channels in cultured skeletal muscle myoballs under depolarizing conditions [17]. These and other results suggest that the DHP receptors represent functional Ca channels, and have led to the use of skeletal muscle membranes as a model system for the isolation of L-type Ca channels.

The strategy utilized for the purification of L-type channels was suggested by the prior successful purification of the nicotinic acetylcholine receptor-associated ion channel and the voltage-dependent Na channels (reviewed in [22,23]). These proteins were first purified as high affinity neurotoxin receptors. Similarly, Ca channels were purified as high affinity DHP receptors, with the hope (now fulfilled) that the purified component(s) would constitute functional Ca channels. The preparations of DHP receptors purified from skeletal muscle contain two distinct large polypeptides with apparent M_r of 165 and 170 kD [24–29], which are designated α_1 and α_2 (Fig. 1). The α_2 peptide can be distinguished from the α_1 peptide by its characteristic electrophoretic behavior, in that under nonreducing conditions it migrates as a peptide of 170 kD, but under reducing conditions, it is cleaved to two components of 140 and ~30 kD (Fig. 1; the 30 kD peptide is not visible in this figure, but see ref. [37]). The 165 kD α_1 peptide, but not the 170 kD α_2, contains the receptors for Ca channel effectors and sites that may be involved in the regulation of Ca channels by phosphorylation (Fig. 2). Photoaffinity labelling studies demonstrated the specific labelling of the α_1 peptide by phenylalkylamines (Fig. 2), DHPs, diltiazem, and bepridil (24–31). Both the 170 and 165 kD peptides were found to be glycopeptides, although the 170 kD peptide was much more heavily glycosylated than the 165 kD peptide [28]. The α_1 peptide is much less stable than the α_2 peptide; this instability is the most likely reason that the α_1 peptide was not always observed in initial purification studies [32,33]. Purified DHP receptor preparations from skeletal muscle have been reconstituted into planar lipid bilayers by several groups and shown to contain Ca channel activity [34–36]. Purified preparations have also been incorporated into liposomes and the binding of Ca channel effectors and ^{45}Ca flux measured [37–39].

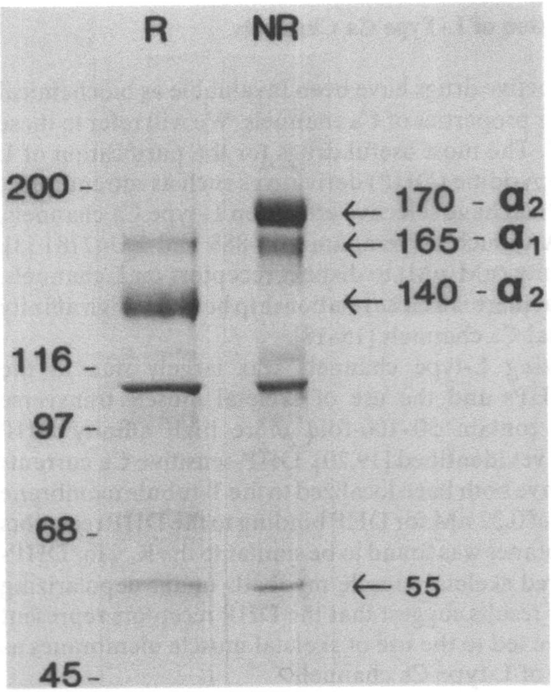

Fig. 1. Coomassie blue-stained SDS gel depicting the peptide content of partially purified L-type Ca channels from skeletal muscle. Channels were purified as high affinity DHP receptors as described [28]. Samples were electrophoresed under reducing (*R, left lane*) or nonreducing (*NR, right lane*) conditions. The mobility of the 165 kD α_1 peptide was essentially unchanged under either condition, while the mobility of the α_2 peptide migrates at 170 kD under NR conditions and 140 kD under R conditions

Recently, the cDNA for the α_1 peptide (DHP receptor) from skeletal muscle has been cloned [40]. An exciting finding was that the predicted amino acid sequence of the α_1 peptide has striking homology to that of the voltage-dependent Na channels [41,42]. The predicted amino acid sequences of the skeletal muscle α_1 peptide and the voltage-dependent Na channels possess four internal repeats, each of which is predicted to contain six membrane-spanning regions. In each case, the fourth membrane spanning segment in each internal repeat contains clusters of basic amino acids which have been suggested to comprise the voltage sensors [40–42]. The collected features of the α_1 peptide, particularly the demonstrations that it contains many of the receptors for Ca channel effectors *and* structurally resembles another protein known to constitute a functional ion channel, strongly suggest that it comprises the major functional unit of L-type Ca channels. Functional studies are needed to establish whether this is so.

The question as to whether or not α_1 and α_2 are subunits of the DHP/PAA sensitive Ca channel remains an issue. The co-purification, using different detergents and various purification protocols, of α_2 with DHP receptors from cardiac (see below) and skeletal muscle [24–29] might suggest that the peptides are functionally related. In addition to the two large peptides, the presence of a

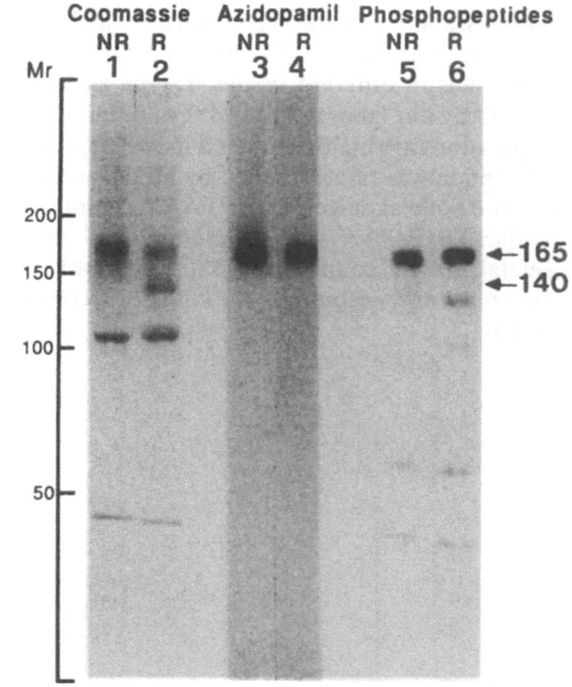

Fig. 2. Photoaffinity labelling and phosphorylation of the 165 kD component present in partially purified preparations of skeletal muscle L-type Ca channels. Membranes were photolabelled or not with the phenylalkylamine [^{3}H]azidopamil and Ca channels purified as described in Fig. 1. Channels from the nonlabelled membranes were phosphorylated with cAMP-dependent protein kinase after purification. Samples were analyzed after SDS-gel electrophoresis under reducing (*R*) and nonreducing (*NR*) conditions. *Lanes 1, 2* depict the peptide content of the partially purified samples visualized by Coomassie blue staining. *Lanes 3, 4* show the specifically photolabelled peptides visualized after fluorography. *Lanes 5, 6* show the phosphopeptides. Note that the 165 kD α_1 component whose mobility was unaffected by *R* or *NR* conditions was photolabelled by the phenylalkylamine and phosphorylated by cAMP-dependent protein kinase. No photolabelling or phosphorylation of the 170/140 α_2 component was observed under any conditions

nonglycosylated 55, and glycosylated ~32 and 27–29 kD peptides (designated as the β, γ, and δ subunits, respectively) have also been observed in some purified preparations [25–27]. However, the relative amounts of the different peptides in the purified preparations have been variable. This variability could suggest that the peptides are not subunits, or it could be caused by a differential susceptibility of the peptides to proteolysis or to other factors which could cause subunit dissociation during purification. Immunoprecipitation studies have demonstrated co-precipitation of the α_1, α_2, β, γ, and δ peptides using anti-α_1 and anti-β antibodies [25,43,44]. As all reported immunoprecipitations of these peptides have been performed with purified preparations, a more stringent test would be to demonstrate the immunoprecipitation of a complex from crude extracts.

Subtypes of L channels appear to exist since cardiac L channels differ in certain pharmacological and electrophysiological properties from skeletal muscle L channels, and in other aspects from neuronal L channels [2,45,46]. Taken together, these data suggest that there may be differences in the structures of components of L channels in different tissues. L-type Ca channels from the heart have been purified by our laboratory [47,48], and the results demonstrated that the purified preparations are highly enriched in two peptides of 185 and 140 kD (48). The 185 kD peptide was photolabelled by DHPs and PAAs, and was significantly larger than the skeletal muscle DHP/PAA receptor (Fig. 3). Other data showed that the cardiac DHP/PAA was structurally and immunologically different from its skeletal muscle counterpart [48]. The results provide evidence that structural differences may account for the known differences in channel properties in the two tissues.

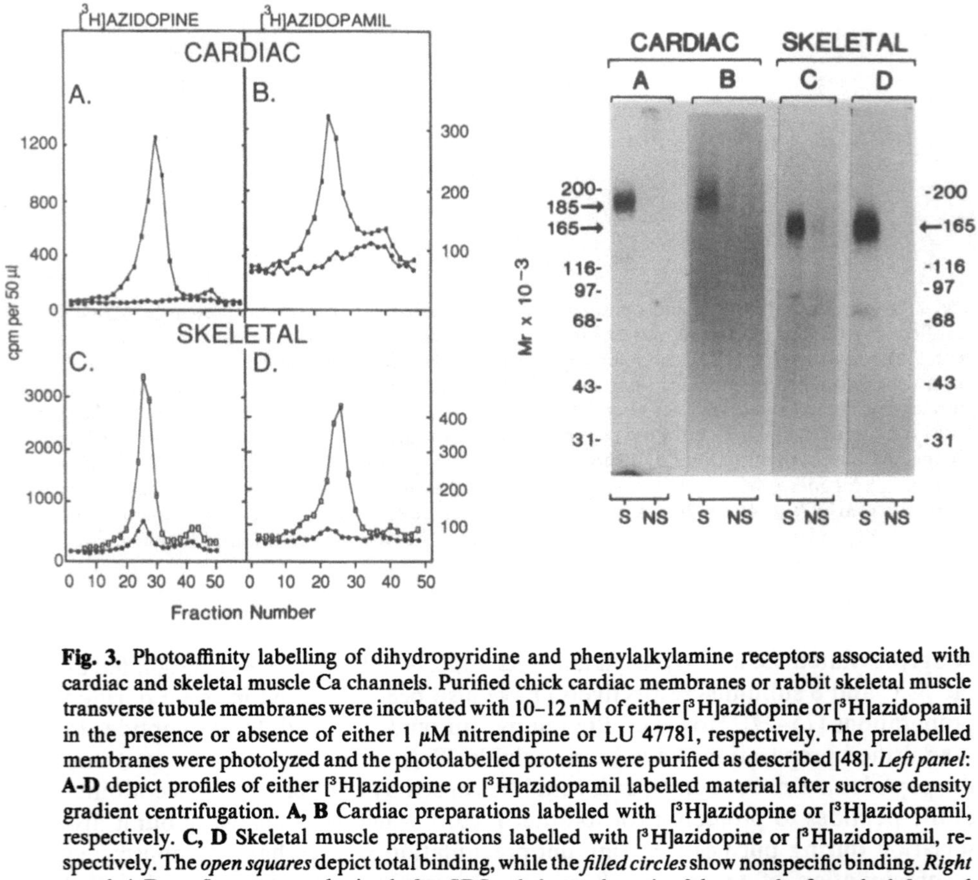

Fig. 3. Photoaffinity labelling of dihydropyridine and phenylalkylamine receptors associated with cardiac and skeletal muscle Ca channels. Purified chick cardiac membranes or rabbit skeletal muscle transverse tubule membranes were incubated with 10–12 nM of either [^{3}H]azidopine or [^{3}H]azidopamil in the presence or absence of either 1 μM nitrendipine or LU 47781, respectively. The prelabelled membranes were photolyzed and the photolabelled proteins were purified as described [48]. *Left panel*: **A-D** depict profiles of either [^{3}H]azidopine or [^{3}H]azidopamil labelled material after sucrose density gradient centrifugation. **A, B** Cardiac preparations labelled with [^{3}H]azidopine or [^{3}H]azidopamil, respectively. **C, D** Skeletal muscle preparations labelled with [^{3}H]azidopine or [^{3}H]azidopamil, respectively. The *open squares* depict total binding, while the *filled circles* show nonspecific binding. *Right panel*: **A-D** are fluorograms obtained after SDS-gel electrophoresis of the samples from the left panel containing the peak of binding activity. *S* and *NS* represent specific (from fractions depicted on the *left* in *open squares*) and nonspecific (from fractions depicted on the *left* with *closed circles*), respectively. Exposure times were 30 days for **A** and **B**, and 5 days for **C** and **D**

3 · Regulation of L Channels by Phosphorylation

As introduced earlier, L-type Ca channels are *regulated* by receptor-dependent processes [3,6,7]. A direct demonstration that peptides associated with L-type Ca channels are phosphorylated by cAMP-dependent protein kinase *in intact cells* has not yet been achieved. However, we have found that the skeletal muscle α_1 peptide is an excellent substrate in vitro for cAMP-dependent and other protein kinases. The 165 kD component can be phosphorylated in its native state in rabbit skeletal muscle T-tubule membranes by cAMP-dependent protein kinase and a multifunctional Ca/CaM-dependent protein kinase in vitro (Fig. 4). The initial rates of phosphorylation of the α_1 peptide by both protein kinases compared favorably to the rates of phosphorylation of established substrates of the enzymes (Fig. 4). Phosphorylation of the α_1 peptide by both protein kinases was additive [49]. cAMP-dependent protein kinase phosphorylated the α_1 peptide preferentially at serine residues, while the Ca/CaM-dependent protein kinase induced the formation of slightly more phosphothreonine than phosphoserine [49]. The phosphorylation could be reversed by the Ca-dependent phosphatase calcineurin, which dephosphorylated the membrane-bound α_1 peptide that had been previously phosphorylated by either protein kinase [50].

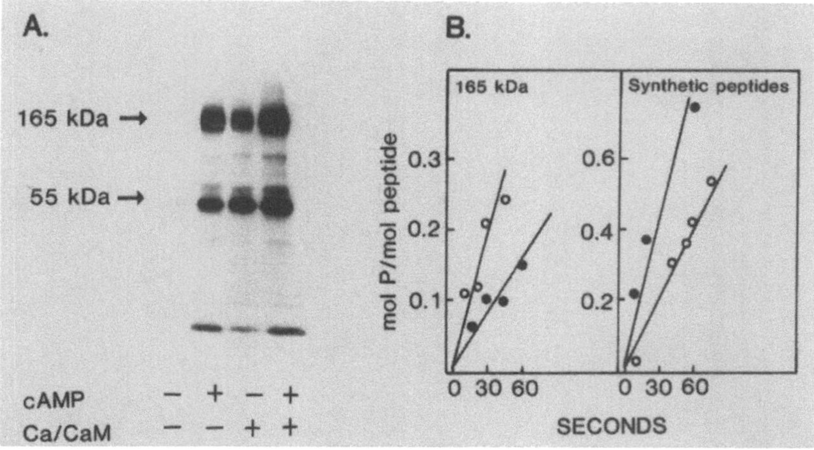

Fig. 4A,B. Phosphorylation of the 165 kD DHP receptor in skeletal muscle transverse tubule membranes by cAMP-dependent and Ca/calmodulin-dependent protein kinases. **A** Autoradiogram depicting phosphorylation of the 165 kD DHP receptor by each protein kinase alone and together. Transverse tubule membranes were phosphorylated with the purified catalytic subunit of cAMP-dependent protein kinase or with partially purified preparations of a Ca/calmodulin-dependent protein kinase. DHP receptors were purified from the phosphorylated membranes and electrophoresed on 5–15% gels. Shown is an autoradiogram of the resulting gel. **B** Initial rates of phosphorylation of the 165 kD peptide and two synthetic peptides by cAMP- and Ca/CaM-dependent protein kinases. The *left side* of **B** shows the rate of incorporation of ^{32}P into the membrane-bound 165 kD peptide in the presence of either 0.1 μM catalytic subunit of cAMP-dependent protein kinase (•), or 6 μg/ml (approximately 1-2 nM) of Ca/CaM-dependent protein kinase (○). The *right side* of **B** shows the initial rates of phosphorylation of the synthetic peptide "kemptide" (25 nM) by cAMP-dependent protein kinase (0.1 μM) (•), and the Ca/CaM-dependent protein kinase substrate analog (400 μM) by the Ca/CaM-dependent protein kinase (6 μg/ml) (○)

The Ca and phospholipid-dependent protein kinase, protein kinase C, has been shown to modulate the activity of Ca channels in a variety of cells [51–54]. We have found that the α_1 peptide is an efficient substrate for protein kinase C when the peptide is phosphorylated in its membrane-bound state; 1.5–2.0 mol phosphate/mol peptide was incorporated within 2 min in incubations carried out at 37°C (Fig. 5). In contrast, the purified protein was a poorer substrate for the kinase; this may reflect that the protein is not in the proper conformation to serve as a substrate in detergent solution [55]. Prior phosphorylation of the α_1 peptide by cAMP-dependent protein kinase prevented subsequent phosphorylation of the peptide by protein kinase C [55]. Phosphoamino acid analysis indicated that protein kinase C phosphorylated the α_1 peptide at both serine and threonine residues [55]. Phosphopeptide mapping experiments showed that cAMP- and Ca/CaM-dependent protein kinases and protein kinase C phosphorylated unique, as well as shared sites, in the α_1 peptide (Fig. 6). The unique site for protein kinase C was peptide 1, while those for cAMP- and Ca/CaM-dependent protein kinases were peptides 2 and 3, and 4 and 5, respectively (Fig. 6). As the α_1 peptide may be the major functional unit of DHP-sensitive Ca channels, the results suggest that the phosphorylation-dependent modulation of Ca channel activity by neurotransmitters may involve phosphorylation of the α_1 peptide at multiple sites. The results are in agreement with the suggestion from the predicted amino acid sequence for

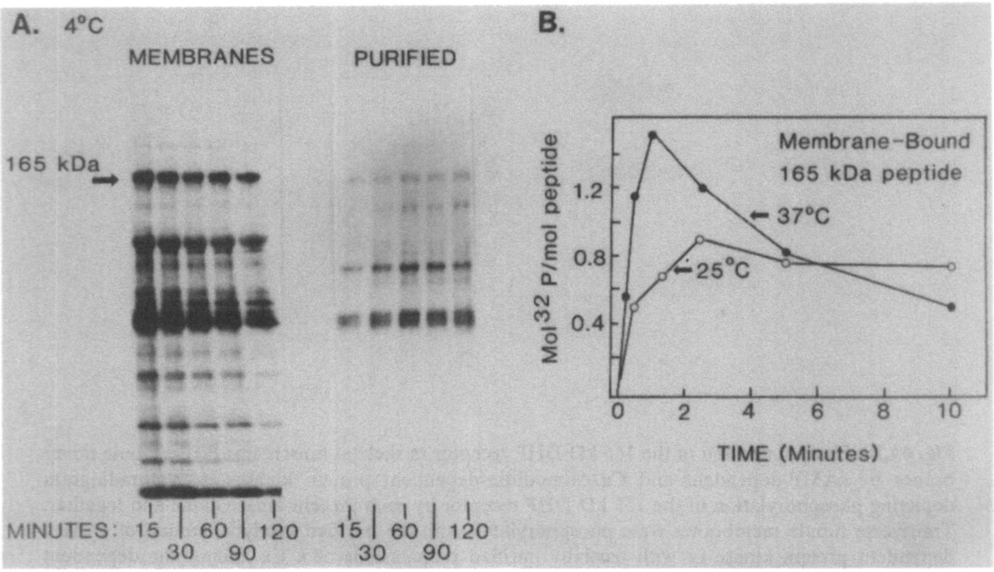

Fig. 5A,B. Time courses of phosphorylation of the membrane-bound and purified skeletal muscle 165 kD DHP receptor by protein kinase C. Transverse tubule membranes (**A**, *left panel*, and **B**) or partially purified DHP receptors (**A**, *right panel*) were phosphorylated by protein kinase C (10 nM) at 4°C (**A**), 25°C (O, **B**), or 37°C (•, **B**). After electrophoresis of the samples on SDS gels, autoradiograms were produced and the 165 kD phosphopeptide localized. **A** Autoradiograms comparing the phosphorylation of the membrane-bound and purified 165 kD peptide by protein kinase C at 4°C. **B** The 165 kD peptide was excised from the dried gels and the ^{32}P incorporation was quantitated by scintillation counting

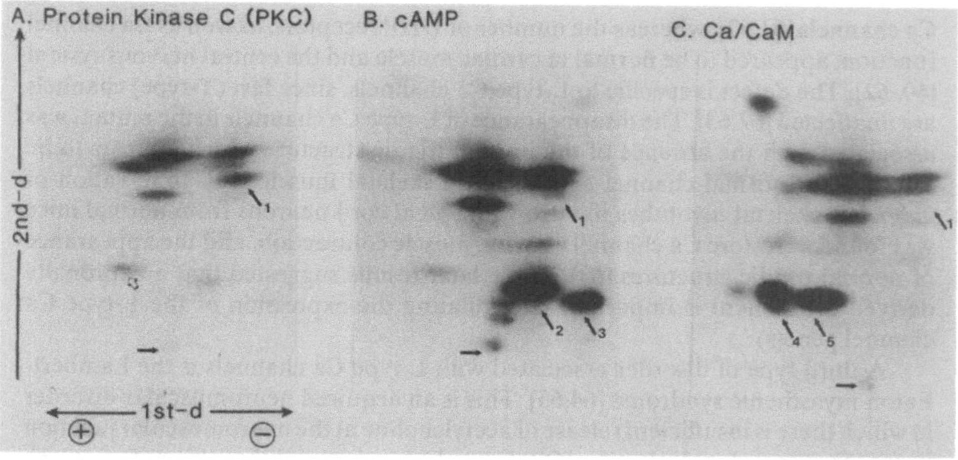

Fig. 6. Two-dimensional phosphopeptide maps obtained after phosphorylation of the 165 kD skeletal muscle DHP receptor by (**A**) protein kinase C, (**B**) cAMP-dependent protein kinase, and/or (**C**) a multifunctional Ca/calmodulin-dependent protein kinase. The 165 kD peptide was phosphorylated in transverse tubule membranes for 5 min at 25°C by the protein kinases indicated. After electrophoresis on SDS gels, the bands corresponding to the 165 kD phosphopeptide were excised and processed for phosphopeptide mapping

the α_1 peptide which indicates there are multiple potential phosphorylation sites in the peptide [40]. All sites are predicted to be located cytoplasmically.

Phosphorylation of the 55 kD, putative β subunit, has also been observed in in vitro studies with cAMP- and Ca/CaM-dependent protein kinases, as well as by protein kinase C [27,49,50,55,56]. Future studies are needed to elucidate which of the putative channel components may be phosphorylated in the intact cell and contribute to activation of the channels.

4 L-type Ca Channels and Disease

Abnormal regulation of L-type Ca channel function and/or synthesis has been demonstrated to occur in several disease states. In one model of cardiac hypertrophy, the hereditary cardiomyopathy observed in hamsters, cardiac myocytes have elevated Ca concentrations [57], and inhibitors of L-type Ca channels have been found to be effective in treating the disease [58]. The number of binding sites for Ca channel inhibitors, as well as the activity of related voltage-dependent Ca channels, were found to be markedly increased in cardiomyopathic hamsters [59]. These results make it tempting to speculate that these increases may be involved in the pathogenesis of this cardiomyopathy.

A second case of Ca channel associated pathology corresponds to a hereditary skeletal muscle disease. In the genetically dysgenic mouse, *mdg/mdg*, skeletal muscles do not contract. Therefore, birth cannot occur and the disease must be studied at the fetal stage. Skeletal muscles from fetal *mdg/mdg* mice have been shown to lack receptors for Ca channel effectors [60], as well as functional L-type

Ca channels [61,62], whereas the number of DHP receptors, as well as Ca channel function, appeared to be normal in cardiac muscle and the central nervous system [60–62]. The defect is specific to L-type Ca channels, since fast (T-type) channels are unaffected [62,63]. The disappearance of L-type Ca channels in the mutant was associated with the absence of the normal triadic structure which appears to be essential for normal channel expression in skeletal muscle [60]. Innervation of *mdg/mdg* skeletal myotubes in vitro with spinal cord neurons from normal mice was found to restore Ca channel activity, muscle contraction, and the appearance of normal triadic structures [63]. These later results suggested that a neuronally derived component is important in regulating the expression of the L-type Ca channel gene(s).

A third type of disorder associated with L-type Ca channels is the Lambert-Eaton myasthenic syndrome [64,65]. This is an acquired neuromuscular disorder in which there is insufficient release of acetylcholine at the neuromuscular junction in response to a depolarization [66]. It has been proposed that this autoimmune disease is due to circulating antibodies against Ca channels [64,65]. Autoantibodies from patients with this disease were tested for their effects on Ca channels in cultured bovine adrenal chromaffin cells. The IgG from affected patients caused a reduction in the number of functional Ca channels without altering the properties of the unaffected Ca channels [65]. Serum from control patients had no effects. As the antibodies tested were active toward L-type Ca channels, these results may suggest a role for L-type Ca channels in neurotransmitter release.

Other instances of abnormal Ca channel activity have been observed. Since abnormally high levels of cholesterol are well known to represent a risk factor in cardiovascular diseases, it is interesting that a dramatic decrease in the cholesterol content of cardiac cells was found to be associated with a drastic diminution of L-type Ca channel activity [67]. This altered channel activity was associated with a major decrease, and eventual total suppression, of contractile activity. The results show that the membrane environment of the L-type Ca channels is crucial for their proper functioning.

Future studies of voltage-dependent Ca channels should provide more information about their structures and regulation in normal and abnormal states. It is likely that different types of voltage-dependent Ca channels will be isolated, and will provide the basis for the elucidation of the molecular events involved in regulation of channel function.

Acknowledgments. The work from the authors' laboratories cited in this review was supported by NIH grant HL23306 (MMH), the CNRS and the Association des Myopathes (ML). MMH was an Established Investigator of the American Heart Association.

References

1. Nowycky MC, Fox AP, Tsien RW (1985) Three types of neuronal calcium channel with different calcium agonist sensitivity. Nature (London) 316:440–443
2. McCleskey EW, Fox AP, Feldman D, Tsien RW (1986) Different types of calcium channels. J Exp Biol 124:177–190
3. Hosey MM, Lazdunski M (1988) Calcium channels: molecular pharmacology, structure and regulation. J Membrane Biol 104:81–105

4. Miller RJ (1987) Multiple calcium channels and neuronal function. Science 235:46–52
5. Chen C, Corbley MJ, Roberts TM, Hess P (1988) Voltage-sensitive calcium channels in normal and transformed 3T3 fibroblasts. Science 239:1024–1026
6. Reuter H (1983) Calcium channel modulation by neurotransmitters, enzymes and drugs. Nature (London) 301:569–574
7. Tsien RW, Bean BP, Hess P, Lansman JB, Nilius B, Nowycky MC (1986) Mechanisms of calcium channel modulation by β-adrenergic agents and dihydropyridine calcium agonists. J Mol Cell Cardiol 18:691–710
8. Schmid A, Renaud J-F, Lazdunski M (1985) Short term and long term effects of β-adrenergic effectors and cyclic AMP on nitrendipine-sensitive voltage-dependent Ca^{2+} channels of skeletal muscle. J Biol Chem 260:13041–13046
9. Arreola J, Calvo J, Garcia MC, Sanchez JA (1987) Modulation of calcium channels of twitch skeletal muscle fibres of the frog by adrenaline and cyclic adenosine monophosphate. J Physiol 393:307–330
10. Osterrieder W, Brum G, Hescheler J, Trautwein W, Flockerzi V, Hofmann F (1982) Injection of subunits of cyclic AMP-dependent protein kinase into cardiac myocytes modulates Ca^{2+} current. Nature (London) 298:576–578
11. Brum G, Flockerzi V, Hofmann F, Osterrieder W, Trautwein W (1983) Injection of catalytic subunit of cAMP-dependent protein kinase into isolated cardiac myocytes. Pflüger's Arch 398:147–154
12. Bkaily G, Sperelakis N (1984) Injection of protein kinase inhibitor into cultured heart cells blocks slow calcium channels. Am J Physiol 246:H630–H634
13. Chad JE, Eckert R (1986) An enzymatic mechanism for calcium current inactivation in dialyzed *helix* neurons. J Physiol 378:31–51
14. Armstrong D, Eckert R (1987) Voltage-activated calcium channels that must be phosphorylated to respond to membrane depolarization. Proc Natl Acad Sci USA 84:2518–2522
15. Bean BP (1984) Nitrendipine block of cardiac calcium channels: high affinity binding to the inactivated state. Proc Natl Acad Sci USA 81:6388–6392
16. Sanguinetti MC, Kass RS (1984) Voltage-dependent block of calcium channel current in the calf cardiac Purkinje fiber by dihydropyridine calcium channel antagonists. Circ Res 55:336–348
17. Cognard C, Romey G, Galizzi J-P, Fosset M, Lazdunski M (1986) Dihydropyridine-sensitive Ca channels in mammalian skeletal muscle in culture: electrophysical properties and interactions with Ca channel activator (Bay K 8644) and inhibitor (PN 200–100). Proc Natl Acad Sci USA 83:1518–1522
18. Kunze DL, Hamilton SL, Hawkes MJ, Brown AM (1987) Dihydropyridine binding and calcium channel function in clonal rat adrenal medullary tumor cells. Mol Pharmacol 31:401–409
19. Fosset M, Jaimovich E, Delpont E, Lazdunski M (1983) [^{3}H]Nitrendipine receptors in skeletal muscle. J Biol Chem 258:6086–6092
20. Glossmann H, Ferry DR, Boschek CB (1983) Purification of the putative calcium channel from skeletal muscle with the aid of [^{3}H]-nimodipine binding. Naunyn-Schmiedeberg's Arch Pharmacol 323:1–11
21. Nicola-Siri L, Sanchez JA, Stefani E (1980) Effect of glycerol treatment on the calcium current of frog skeletal muscle. J Physiol (London) 305:87–96
22. Changeux J-P, Revaḥ F (1987) The acetylcholine receptor molecule: allosteric sites and the ion channel. Trends Neurosci 10:245–250
23. Catterall WA (1984) The molecular basis of neuronal excitability. Science 223:653–661
24. Striessnig J, Knaus H-G, Grabner M, Moosburger K, Seitz W, Lietz H, Glossmann H (1987) Photoaffinity labelling of phenylalkylamine receptor of the skeletal muscle transverse-tubule calcium channel. FEBS Lett 212:247–253
25. Leung AT, Imagawa T, Campbell KP (1987) Structural characterization of the 1,4-dihydropyridine receptor of the voltage-dependent Ca^{2+} channel from rabbit skeletal muscle. J Biol Chem 262:7943–7946
26. Sieber M, Nastainczyk W, Zubor V, Wernet W, Hofmann F (1987) The 165-kD peptide of the purified skeletal muscle dihydropyridine receptor contains the known regulatory sites of the calcium channel. Eur J Biochem 167:117–122
27. Takahashi M, Seagar MJ, Jones JF, Reber BFX, Catterall WA (1987) Subunit structure of dihydropyridine-sensitive calcium channels from skeletal muscle. Proc Natl Acad Sci USA 84:5478–5482

28. Hosey MM, Barhanin J, Schmid A, Vandaele S, Ptasienski J, O'Callahan C, Cooper C, Lazdunski M (1987) Photoaffinity labelling and phosphorylation of a 165 kilodalton peptide associated with dihydropyridine and phenylalkylamine-sensitive calcium channels. Biochem Biophys Res Commun 147:1137–1145
29. Vaghy PL, Striessnig J, Miwa K, Knaus H-G, Itagaki K, McKenna E, Glossmann H, Schwartz A (1987) Identification of a novel 1,4-dihydropyridine- and phenylalkylamine-binding polypeptide in calcium channel preparations. J Biol Chem 262:14337–14342
30. Galizzi J-P, Borsotto M, Barhanin J, Fosset M, Lazdunski M (1986) Characterization and photoaffinity labeling of receptor sites for the Ca^{2+} channel inhibitors *d-cis*-diltiazem, (±)-bepridil, desmethoxyverapamil, and (+)-PN 200–110 in skeletal muscle transverse tubule membranes. J Biol Chem 261:1393–1397
31. Sharp AH, Imagawa T, Leung AT, Campbell KP (1987) Identification and characterization of the dihydropyridine-binding subunit of the skeletal muscle dihydropyridine receptor. J Biol Chem 262:12309–12315
32. Borsotto M, Barhanin J, Fosset M, Lazdunski M (1985) The 1,4-dihydropyridine receptor associated with the skeletal muscle voltage-dependent Ca^{2+} channel. J Biol Chem 26:14255–14263
33. Curtis BM, Catterall WA (1984) Purification of the calcium antagonist receptor of the voltage-sensitive calcium channel from skeletal muscle transverse tubules. Biochemistry 23:2113–2118
34. Flockerzi V, Oekem H-J, Hofmann F, Pelzer D, Cavalié A, Trautwein W (1986) Purified dihydropyridine-binding site from skeletal muscle t-tubules is a functional calcium channel. Nature (London) 323:66–68
35. Smith JS, McKenna EJ, Ma J, Vilven J, Vaghy PL, Schwartz A, Coronado R (1987) Calcium channel activity in a purified dihydropyridine-receptor preparation of skeletal muscle. Biochemistry 26:7182–7188
36. Talvenheimo JA, Worley JF, Nelson MT (1987) Heterogeneity of calcium channels from a purified dihydropyridine receptor preparation. Biophys J 52:891–899
37. Barhanin J, Coppola T, Schmid A, Borsotto M, Lazdunski M (1987) The calcium channel antagonists receptor from rabbit skeletal muscle. Eur J Biochem 164:525–531
38. Curtis BM, Catterall WA (1986) Reconstitution of the voltage-sensitive calcium channel purified from skeletal muscle transverse tubules. Biochemistry 25:3077–3083
39. Horne WA, Ghany MA, Racker E, Weiland GA, Oswald RE, Cerione RA (1988) Functional reconstitution of skeletal muscle Ca^{2+} channels: separation of regulatory and channel components. Proc Natl Acad Sci USA 85:3718–3722
40. Tanabe T, Takeshima H, Mikami A, Flockerzi V, Takahashi H, Kangawa K, Kojima M, Matsuo H, Hirose T, Numa S (1987) Primary structure of the receptor for calcium channel blockers from skeletal muscle. Nature (London) 328:313–318
41. Noda M, Shimizu S, Tanabe T, Takai T, Kayano T, Ikeda T, Takahashi H, Nakayama H, Kanaoka Y, Minamino N, Kangawa K, Matsuo H, Raftery MA, Hirose T, Inayama S, Hayashida H, Miyata T, Numa S (1984) Primary structure of *electrophorus electricus* sodium channel deduced from cDNA sequence. Nature (London) 312:121–127
42. Noda M, Takayuki I, Kayano T, Suzuki H, Takeshima H, Kurasaki M, Takahashi H, Numa S (1986) Existence of distinct sodium channel messenger RNAs in rat brain. Nature (London) 320:188–192
43. Leung AT, Imagawa T, Block B, Franzini-Armstrong C, Campbell KP (1988) Biochemical and ultrastructural characterization of the 1,4-dihydropyridine receptor from rabbit skeletal muscle. J Biol Chem 263:994–1001
44. Morton ME, Froehner SC (1987) Monoclonal antibody identifies a 200-kDa subunit of the dihydropyridine-sensitive calcium channel. J Biol Chem 262:11904–11907
45. Cognard C, Lazdunski M, Romey G (1986) Different types of Ca^{2+} channels in mammalian skeletal muscle cells in culture. Proc Natl Acad Sci USA 83:517–521
46. Rosenberg RL, Hess P, Reeves JP, Smilowitz H, Tsien RW (1986) Calcium channels in planar lipid bilayers: insights into mechanisms of ion permeation and gating. Science 231:1564–1566
47. Cooper CL, Vandaele S, Barhanin J, Fosset M, Lazdunski M, Hosey MM (1987) Purification and characterization of the dihydropyridine-sensitive voltage-dependent calcium channel from cardiac tissue. J Biol Chem 262:509–512
48. Chang FC, Hosey MM (1988) Dihydropyridine and phenylalkylamine receptors associated with cardiac and skeletal muscle calcium channels are structurally different. J Biol Chem 263:18929–18937

49. O'Callahan CM, Hosey MM (1988) Multiple phosphorylation sites in the 165-kilodalton peptide associated with dihydrophyridine-sensitive calcium channels. Biochemistry 27:6071–6077
50. Hosey MM, Borsotto M, Lazdunski M (1986) Phosphorylation and dephosphorylation of dihydropyridine-sensitive voltage-dependent Ca^{2+} channel in skeletal muscle membranes by cAMP- and Ca^{2+}-dependent processes. Proc Natl Acad Sci USA 83:3733–3737
51. Marchetti C, Brown AM (1988) Protein kinase activator 1-oleoyl-2-acetyl-*sn*-glycerol inhibits two types of calcium currents in GH_3 cells. Am J Physiol 254:C206–C210
52. Dosemeci A, Dhallan RS, Cohen NM, Lederer WJ, Rogers TB (1988) Phorbol ester increases calcium current and stimulates the effects of angiotensin II on cultured and neonatal rat heart myocytes. Circ Res 62:347–357
53. Rane SG, Dunlap K (1986) Kinase C activator 1,2-oleoylacetylglycerol attenuates voltage-dependent calcium current in sensory neurons. Proc Natl Acad Sci USA 83:184–188
54. DiVirgilio F, Pozzan T, Wolheim CB, Vincentini LM, Meldolesi J (1986) Tumor promoter phorbol myristate acetate inhibits Ca^{2+} influx through voltage-gated Ca^{2+} channels in two secretory cell lines PC12 and RINm5F. J Biol Chem 261:32–35
55. O'Callahan CM, Ptasienski J, Hosey MM (1988) Phosphorylation of the 165 kDa dihydropyridine/phenylalkylamine receptor from skeletal muscle by protein kinase. C J Biol Chem 263:17342–17349
56. Curtiz BM, Catterall WA (1985) Phosphorylation of the calcium antagonist receptor of the voltage-sensitive calcium channel by cAMP-dependent protein kinase. Proc Natl Acad Sci USA 82:2528–2532
57. Lossnitzer K (1975) Genetic induction of a cardiomyopathy. Handbook of experimental pharmacology, vol 16, Springer Berlin Heidelberg New York, pp 309–344
58. Rouleau JL, Chuck LHS, Hollosi G, Kidd P, Sievers RE, Wikman-Coffelt J, Parmley WW (1982) Verapamil preserves myocardial contractility in the hereditary cardiomyopathy of the Syrian hamster. Circ Res 50:405–412
59. Wagner JA, Reynolds IJ, Weisman HF, Dudeck P, Weisfeldt ML, Snyder SH (1986) Calcium antagonist receptors in cardiomyopathic hamster: selective increases in heart, muscle, brain. Science 232:515–518
60. Pincon-Raymond M, Rieger F, Fosset M, Lazdunski M (1985) Abnormal transverse tubule system and abnormal amount of receptors for Ca^{2+} channel inhibitors of the dihydropyridine family in skeletal muscle from mice with embryonic muscular dysgenesis. Dev Biol 112:458–466
61. Romey G, Rieger F, Renaud JF, Pincon-Raymond M, Lazdunski M (1986) The electrophysical expression of Ca^{2+} channels and of apamin sensitive Ca^{2+} activated K^+ channels is abolished in skeletal muscle cells from mice with muscular dysgenesis. Biochem Biophys Res Commun 136:935–940
62. Beam KG, Knudson CM, Powell JA (1986) A lethal mutation in mice eliminates the slow calcium current in skeletal muscle cells. Nature (London) 320:168–170
63. Rieger F, Bournaud R, Shimahara T, Garcia L, Pincon-Raymond M, Romey G, Lazdunski M (1987) Restoration of dysgenic muscle contraction and calcium channel function by co-culture with normal spinal cord neurons. Nature (London) 330:563–566
64. Fukunoya H, Engel AG, Lang B, Newson-Davis J, Vincent A (1983) Passive transfer of Lambert-Eaton myasthenic syndrome with IgG from man to mouse depletes the presynaptic membrane active zones. Proc Natl Acad Sci USA 80:7636–7640
65. Kim YI, Neher E (1988) IgG from patients with Lambert-Eaton syndrome blocks voltage-dependent calcium channels. Science 239:405–408
66. Lambert EH, Elmqvist D (1971) Quantal components of end-plate potentials in the myasthenic syndrome. Ann NY Acad Sci 183:183–199
67. Renaud J-F, Schmid A, Romey G, Nana JL, Lazdunski M (1986) Melvinolin, an inhibitor of cholesterol biosynthesis, drastically depresses Ca^{2+} channel activity and uncouples excitation from contraction in cardiac cells in culture. Proc Natl Acad Sci USA 83:8007–8011

Cell Membrane Abnormalities in Genetic Hypertension

G. BIANCHI[1], P. FERRARI, S. SALARDI, P. SALVATI, M. FERRANDI, L. TORIELLI and B. R. BARBER

1 Introduction

Alterations of cell function and structure have been described both in man with essential hypertension (Garay and Meyer 1975; Canessa et al. 1980; Cusi et al. 1981) and rat models with genetic types of hypertension (De Mendoca et al. 1982; Duhm et al. 1983). Possible cause-effect relationships between these cell abnormalities and the development of hypertension have therefore been proposed. The polygenic character of the human disease, the influence of environmental factors and methodological limitations to applying the best experimental design in humans, however, make it difficult to distinguish between those alterations that may be direct phenotypic expressions of the primary genetic defect and those which are secondary readjustments of the cell homeostasis.

An animal model with genetic hypertension could represent an ideal tool for studying this problem and furnishing the methodological background for approaching it in the human disease. However, even in an animal model some considerations must be taken into account when looking for a cause-effect relationship between alterations in cell cation transports and hypertension. In fact, due to the complex feedback mechanisms regulating cell function, a primary genetic abnormality affecting a determined biochemical pathway could trigger secondary changes aimed at establishing a new functional equilibrium of the cell (Albert et al. 1983). Moreover, a cell dysfunction, such as an ion transport alteration, affects the overall organ function with a consequent involvement, at the whole body level, of other districts and hormonal or humoral factors which try to compensate and mask the primary defect (Guyton et al. 1974). Therefore, a cell functional alteration may represent: (1) a direct phenotypic expression of the primary genetic abnormality causing hypertension; (2) a secondary change due to the complex network of feedback mechanisms regulating cell function, and caused by a genetic abnormality affecting some other biochemical pathway; (3) the effect of a linkage between genes affecting ion transports and blood pressure, but not reciprocally related; (4) an epiphenomenon not relevant to blood pressure control.

To distinguish among these possibilities we have studied the Milan hypertensive strain of rats (MHS) and its normotensive control (MNS) at different biological levels of organization (whole body, organ and cell function and

Istituto Patologia Medica Università di Sassari and Istituto Ricerche Farmitalia Carlo Erba, Nerviano, Milano, Italy
[1]To whom correspondence should be addressed

A. Azzi, Z. Drahota and S. Papa
Molecular Basis of Membrane-Associated Diseases

molecular cell structure) to establish whether: (a) a difference between MHS and MNS in a given organ function might be responsible also for the difference in blood pressure and whether it is or not caused by secondary hormonal, hemodynamic or neural factors; (b) the organ alteration is consistent with a given cell function abnormality; (c) the cell abnormality is primarily determined in the stem cell or secondarily affected by a humoral factor and genetically associated to blood pressure levels in a segregating F2 population obtained by mating F1 (MHS × MNS) hybrids. (d) The cell abnormality may be due to a given biochemical or structural alteration more directly related to the primary gene defect. This chapter reviews the principal results obtained along these lines in Milan rats.

2 Organ Dysfunction

Experiments of kidney cross-transplantation between MHS and MNS rats demonstrated that hypertension can be transplanted with the kidney (Bianchi et al. 1974; Fox and Bianchi 1976). Special attention was therefore addressed to the study of renal function and it was shown that during the prehypertensive stage the kidney of MHS, as compared to that of MNS, had a faster glomerular filtration rate (GFR) (Ferrari et al. 1982), increased urinary output and decreased urine osmolality and renin secretion (Bianchi et al. 1975). Moreover, studies of sodium balance showed renal sodium retention in MHS (Bianchi 1975). These differences were found not only in vivo but also when renal function was studied in isolated perfused kidney (Salvati et al. 1984), thus supporting the hypothesis of an intrinsic defect of MHS kidney not related to some extrarenal cause.

Moreover, studying in vitro sodium transports in vesicle preparations from different segments of renal tubuli we found, in MHS, an increased amiloride-sensitive Na^+ uptake across brush border membranes from kidney cortex (Hanozet et al. 1985) and a faster Na-K cotransport in luminal vesicles from thick ascending limb (Ferrandi, in press). In our view the hypothesis that best reconciles the increased GFR, the decreased renin secretion and the reduced concentrating ability of kidneys with the transplantation and the in vitro sodium transport results in MHS is that postulated by a faster sodium reabsorption across the tubular cell (Bianchi and Ferrari 1983; Bianchi and Barlassina 1983; Bianchi et al. 1984). This hypothesis, implying a cell defect in ion transports, may also be tested by studying red blood cells (RBC) according to both the indications given in points (c) and (d) and on the assumption that the genetic abnormality in sodium transport across tubular cells may also be found in a cell model such as RBC.

3 Cell Dysfunction

RBC of MHS rats, when compared to MNS, showed peculiar characteristics: some, such as a smaller cell volume and a reduced intracellular Na^+ content, were present both before and after hypertension development and others, such as a faster Na-K cotransport (Ferrari et al. 1987a), were detectable only during the prehypertensive age up to 3–5 months. The Na-K pump activity did not show any difference between the two strains until hypertension was fully developed, then it became lower in MNS (Ferrari et al. 1987a).

To test whether these abnormalities are genetically determined within the stem cell or due to some extracellular factor, we transplanted bone marrow from MHS and MNS rats into previously irradiated F1 (MHS × MNS) hybrids (Bianchi et al. 1985). The control F1 hybrids which did not receive exogenous bone marrow died within 15 days after irradiation. The F1 rats that received bone marrow from MHS or MNS rats survived in good health and 5 months after transplantation their RBC showed the same characteristics as those of the respective bone marrow donors. This result demonstrated that the biochemical mechanisms responsible for the functional differences between MHS and MNS RBC are genetically determined within the stem cells. Moreover, a genetic association between RBC dysfunction and hypertension was demonstrated in an F2 segregating generation obtained by crossing the F1 hybrids (Bianchi et al. 1985). In particular, we measured blood pressure and RBC cotransport in individual F2 rats and we demonstrated a positive, statistically significant correlation ($2+0.52, p < 0.01$) between these two phenotypic traits (Bianchi et al. 1985). This correlation may be due either to a pleiotropic effect or to a linkage of single genes or groups of genes regulating RBC functions and blood pressure. At the moment we cannot distinguish between these two possibilities and genetic studies are being conducted to answer this question.

4 Analogies Between RBC and Renal Tubular Cell Dysfunctions

The results of a faster Na^+ transport across the cell membrane both in RBC and kidney tubular cells of MHS, as described in the previous sections, are consistent with the hypothesis that a primary change in tubular Na^+ reabsorption may be the cause of the kidney dysfunction and hypertension in MHS. Moreover, direct measurements of cell volume and intracellular Na^+ content in proximal tubular cells showed, as in RBC, a reduction of these two parameters in MHS, while no differences were observed between the two strains when distal tubular cells or liver cells were compared (Ferrari et al. 1987b; Beck et al. 1983). This functional and physiological link between RBC and tubular cells favours the hypothesis that the genetic association between ion transport abnormality and blood pressure is due to a pleiotropic effect of the same gene and not to a linkage. However, the molecular mechanisms underlying the functional characteristics of MHS cells remain to be elucidated.

5 Possible Molecular, Biochemical Mechanisms Underlying the Cell Functional Changes

According to point (d) we are now approaching the problem of identifying the molecular abnormality causing the ion transport and cell volume alterations of MHS by studying the biochemical and kinetic characteristics of Na^+ fluxes in RBC and trying to correlate them with a structural difference between MHS and MNS cells. In fact, among the many biochemical differences between the cells of the two strains, those that are linked to molecular alterations are most likely to be primary and genetically determined. In particular, we addressed our research to the study of the protein composition of the cell membrane, since a relationship between cell membrane (in particular, the membrane skeleton) and the functional alterations of MHS cells was suggested by the following observations:

1. The smaller RBC volume of MHS rats was maintained in resealed ghosts, even in the absence of transportable ions on both sides of the membrane (Ferrari et al. 1986), indicating that a structural alteration of the cell membrane and not a cytoplasmic factor is most probably responsible for this defect;
2. When the differences in RBC volume between MHS and MNS were abolished by hypotonic swelling, Na-K cotransport differences also disappeared (Ferrari et al. 1987a);
3. Kinetic analysis of the RBC Na-K cotransport showed that its faster rate in MHS is due to a greater activation of this system at physiological, intracellular Na concentrations, because of increased affinity for internal Na_i (Ferrari et al. 1988). This suggests a conformational alteration of this carrier in MHS more than a difference in the total number of cotransporter sites.
4. The Na-K cotransport measured in erythrocyte, inside-out vesicles (IOV), which are deprived of membrane skeleton, did not show any rate difference between MHS and MNS (Ferrari et al. 1986). Furthermore, the kinetic analysis of IOV cotransport as a function of Na showed that the removal of the membrane skeleton from RBC membranes abolished also the differences in affinity for Na between the two strains (Ferrari et al. 1988). All these findings therefore supported the hypothesis that a different or absent membrane skeleton protein is involved in the cell volume and ion transport abnormalities of MHS. Using immunochemical and molecular biology techniques, we started a systematic study of the cell membrane protein composition of MHS and MNS, obtaining some interesting results by performing a cross-immunization of MHS with ghosts of MNS and vice versa. In this experiment, only MHS rats, when immunized with MNS membrane skeletons, were able to raise antibodies against a 105 kD protein located in RBC membrane skeleton (Ferrandi et al. 1987). This protein, which is also present in brain and renal tissue, has been recognized to be adducin (Gardner and Bennett 1986), a peripheral protein able to favour the actin-spectrin network assembly (Gardner and Bennett 1987), to bind calmodulin and phosphatidylserine and to be phospharylated by protein kinase C (Wolf and Sahyounss 1986; Mische et al. 1987). All these functions and characteristics may be in keeping with a regulatory role, of adducin, on the integral membrane proteins participating in the regulation of cell volume and ion transports. We are now performing studies aimed at identifying functional and structural differences in adducin between MHS and MNS. In particular, using the antibody raised by MHS, we have screened a mouse cDNA library identifying a positive clone for a 31 kD protein which cross-reacts with the MHS Ab and is able to induce the production of the anti 105 kD Ab when injected in rabbit. The sequence of this clone has been obtained and studies are now in progress for isolating the whole gene of adducin from MHS and MNS cDNA libraries (Salardi et al. in preparation).

Several investigators have studied cation fluxes in RBC of other strains of genetically hypertensives rats (Duhm et al. 1983; Ferrari et al. 1987b; Rosati et al. 1988). Unfortunately, different methodologies were used in different laboratories and therefore comparison is difficult. However, in all the genetically hypertensive strains, an alteration of the cell ion handling has been observed even before the development of hypertension, as in SHR and MNS rats.

Since in hypertensive humans, subpopulations with different alterations of RBC cation transports have been recognized (Canessa et al. 1980; Cusi et al. 1981), it is conceivable that different hypertensive strains may represent models for these subpopulations.

In conclusion, we believe that a fruitful approach for understanding the role of particular cell function alterations in arterial hypertension is to consider them as part of an overall cell abnormality. Taking advantage of the fact that most strains of rats with genetic hypertension are fully inbred and therefore homozygous at practically all loci, a combined approach of genetic analysis, cell physiology and immunochemical techniques should allow us to demonstrate the existence of even minor differences in protein structure. Thus, the sequence of events going from a molecular structural abnormality to the overall cell and organ dysfunction might be clarified. Finally, the identification of a specific protein defect would aid detection through molecular biology techniques of the gene or genes responsible for that molecular alteration.

References

Albert B, Bry D, Lewis J, Raff M, Roberts K, Watson JD (1983) Molecular biology of the cell. Garland, New York, pp 199–250

Beck F, Bianchi G, Dorge A, Rick R, Schramm M, Thurau K (1983) Sodium and potassium concentrations of renal cortical cells in two animal models of primary arterial hypertension. J Hypertens Suppl 2:38–39

Bianchi G, Barlassina C (1983) Renal function in essential hypertension. In: Genest J, Kuchel O, Hamet P, Cantin M (eds) Hypertension: physiopathology and treatment, 2nd edn. McGraw-Hill, New York, pp 54–73

Bianchi G, Ferrari P (1983) Animal models for arterial hypertension. In: Genest J, Kuchel O, Hamet P, Cantin M (eds) Hypertension: physiopathology and treatment, 2nd edn. McGraw-Hill, New York, pp 534–555

Bianchi G, Fox U, Di Francesco GF, Giovannetti AM, Pagetti D (1974) Blood pressure changes produced by kidney cross-transplantation between spontaneously hypertensive rats and normotensive rats. Clin Sci Mol Med 47:435–448

Bianchi G, Baer PG, Fox U, Duzzi L, Pagetti D, Giovannetti AM (1975) Changes in renin, water balance, and sodium balance during development of high blood pressure in genetically hypertensive rats. Circ Res 36; 37 (Suppl 1):I-153–I-161

Bianchi G, Ferrari P, Barber BR (1984) The Milan hypertensive strain. In: de Jong W (ed) Experimental and genetic models of hypertension. Elsevier, Amsterdam New York [Birkenhager WH, Reid JL (eds) Handbook of hypertension, vol 4] pp 328–349

Bianchi G, Ferrari P, Trizio D, Ferrandi M, Torielli L, Barber BR, Polli E (1985) Red blood cell abnormalities and spontaneous hypertension in the rat, a genetically determined link. Hypertension 7:319–325

Canessa M, Adragna N, Solomon HS, Connolly TM, Tosteson D (1980) Increased sodium-lithium countertransport in red cells of patients with essential hypertension. N Engl J Med 302:772–776

Cusi D, Barlassina C, Ferrandi M, Lupi GP, Ferrari P, Bianchi G (1981) Familial aggregation of cation transport abnormalities and essential hypertension. Clin Exp Hypertens 3:871–884

De Mendoca M, Knorr A, Grichois ML, Ben Ishay D, Garay RP, Meyer P (1982) Erythrocytic sodium ion transport systems in primary and secondary hypertension of the rat. Kidney Int 21 (Suppl 2): S-69-S-75

Duhm J, Gobel BO, Beck FX (1983) Sodium and potassium ion transport accelerations in erythrocytes of DOC, DOC-salt two-kidney, one clip, and spontaneously hypertensive rats: role of hypokalemia and cell volume. Hypertension 5:642-652

Ferrandi M, Salardi S, Modica R, Bianchi G (1987) Antigenic properties of erythrocyte membrane from a hypertensive strain of rats. Endocitobiology 3, 503:584-587

Ferrandi M, Parenti P, Ferrari P, Bianchi G, Braw R, Karlish SJB (1988) Na, K, Cl cotransporter per mediated Rb fluxes in membrane vesicles from kidneys of normotensive and hypertensive rats. (in press)

Ferrari P, Cusi D, Barber BR, Barlassina C, Vezzoli G, Duzzi L, Minotti E, Bianchi G (1982) Erythrocyte membrane and renal function in relation to hypertension in rats of the Milan hypertensive strain. Clin Sci 63:61s-64s

Ferrari P, Ferrandi M, Torielli L, Canessa M, Bianchi G (1987a) Relationship between erythrocyte volume and sodium transport in the Milan hypertensive rat and age-dependent changes. J Hypertens 5:199-206

Ferrari P, Torielli L, Ferrandi M, Bianchi G (1986) Volumes and Na transports in intact red blood cells, resealed ghosts, and inside out vesicles of Milan hypertensive rats. Ann NY Acad Sci 488:561-563

Ferrari P, Nussdorfer G, Torielli L, Salvati P, Tripodi G, Niutta E, Bianchi G (1987b) Renal function and cell characteristics during the development of essential and genetic hypertension. In: Hofman A, Grobbee DE, Schalekamp MAD (eds) Proceeding of the early pathogenesis of primary hypertension. Elsevier, Amsterdam, pp 116-9

Ferrari P, Torielli L, Ferrandi M, Bianchi G (1988) The role of membrane skeleton in the alteration of Na-K cotransport (CO) in Milan hypertensive rats (MHS). 12th Sci Meet Int Soc Hypertension, Abstr

Fox U, Bianchi G (1976) The primary role of the kidney in causing the blood pressure difference between the Milan hypertensive strain (MHS) and normotensive rats. Clin Exp Pharmacol Physiol Suppl 3:71-74

Garay RP, Meyer P (1975) A new test showing abnormal net Na^+ and K^+ fluxes in erythrocytes of essential hypertensive patients. Lancet 1:349-353

Gardner K, Bennett V (1986) A new erythrocyte membrane associated protein with calmodulin binding activity. J Biol Chem 261:1339-1348

Gardner K, Bennett V (1987) Modulation of spectrin-actin assembly by erythrocyte adducin. Nature (London) 328:359-362

Guyton AC, Coleman TG, Cowley AW, Manning RD, Norman A, Ferguson JD (1974) A systems analysis approach to understanding long-range arterial blood pressure control and hypertension. Circ Res 35:159-176

Hanozet GM, Parenti P, Salvati P (1985) Presence of a potential sensitive Na^+ transport across renal brush border membrane vesicles from rat of the Milan hypertensive strain. Biochem Biophys Acta 819:179-186

Mische SM, Mooseker MS, Morrow JS (1987) Erythrocyte adducin: a calmodulin-regulated actin-bundling protein that stimulates spectrin-actin binding. J Cell Biol 105:2837-2845

Rosati C, Meyer P, Garay R (1988) Sodium transport kinetics in erythrocytes from spontaneously hypertensive rats. Hypertension 11:41-48

Salardi S, Saccardo B, Borsani G, Modica R, Ferrandi M, Tripodi MG, Soria M, Ferrari P, Baralle FE, Sidoli A, Bianchi G (1988) Erythrocyte adducin differential properties in the normotensive and hypertensive rats of the Milan strain. Characterization of spleen adducin m-RNA. (in preparation)

Salvati P, Pinciroli GP, Bianchi G (1984) Renal function of isolated perfused kidneys from hypertensive (MHS) and normotensive (MNS) rats of the Milan strain at different ages. J Hypertens 2(Suppl 3):351-354

Wolf M, Sahyounss N (1986) Protein kinase C and phosphatidylserine bind to M_r 110,000/115,000 polypeptides enriched in cytoskeletal and postsynaptic density preparations. J Biol Chem 261:13327-13332

Transmembrane Cation Transport: An Approach to the Study of the Molecular Basis of Hypertension

L. MAZZANTI[1], R.A. RABINI[1], N. CESTER[2], C. ROMANINI[2] and E. BERTOLI[1]

1 Introduction

The cause and pathophysiological events of pregnancy-induced hypertension (PIH) remain controversial. The outstanding clinical features of this condition seem to be an increased vascular reactivity, vasoconstriction, decrement of plasma volume, and expansion of the interstitial fluid (Lindheimer and Katz 1985). Such alterations may be due to an abnormal sodium distribution between the interstitial and intravascular compartments.

Multiple defects in sodium transport have also been described in essential hypertension, with controversial results of the studies investigating transmembrane ion fluxes (Epstein 1986). However, general agreement exists on the fact that an increase in intracellular Na^+ content may be a major factor in essential hypertension. Such an alteration might enhance the vascular sensitivity to pressor substances such as angiotensin II, through a secondary increase in intracellular Ca^{2+} (Blaustein 1984).

By the same mechanism hypothesized in essential hypertension, a derangement of cellular sodium homeostasis is likely to play a considerable role in the pathophysiological mechanisms of PIH. However, the abnormalities of cation transport have never been systematically explored through normal and hypertensive pregnancies and discrepancies exist among the studies presently available which examine sodium handling under such conditions (Rubython and Morgan 1983). The purpose of our study was to investigate cellular sodium homeostasis during PIH, which may be considered as a model of transient or reversible hypertension. Moreover, we examined the cellular Na^+ activity and transport in the erythrocytes and in the syncitiothrophoblast placental membranes from normal and hypertensive pregnant women and in cord erythrocytes of infants from normal and PIH mothers.

We also studied the Ca^{2+}-ATPase activity in a smaller group of patients.

2 Material and Methods

We determined the intraerythrocyte sodium activity and the Na^+, K^+ ATPase activity in the erythrocyte membranes from 77 women affected by PIH and 133 normotensive pregnant women. Umbilical cord blood of infants from 28 normal

[1]Institute of Biochemistry, University of Ancona, School of Medicine, Ancona, Italy
[2]Obstetrics and Gynecology, University of Ancona, School of Medicine, Ancona, Italy

A. Azzi, Z. Drahota and S. Papa
Molecular Basis of Membrane-Associated Diseases

and 21 hypertensive mothers was also analyzed. The Ca^{2+} ATPase activity was determined in erythrocytes from ten PIH and ten normal women. Moreover, we analyzed 26 placentas from physiological pregnancies used as controls and 16 placentas from women affected by PIH.

Criteria for diagnosis and inclusion in this study required the diastolic BP readings to be higher than 90 mmHg on two or more determinations in women previously normotensive and without proteinuria. The mean age of patients was 27.4 ± 1.5 for PIH and 28.2 ± 1.8 for controls. The degree of parity was not different between the two groups.

Erythrocyte Membrane Preparation. Heparinized blood samples were centrifuged (4500 g) to remove plasma. Red cells were washed twice with isotonic NaCl, lysed hypotonically in 5 mmol/1 phosphate buffer solution (pH 8) and centrifuged in Kontron at 20 000 g for 20 min at 4°C. Resultant membranes were washed with phosphate buffer of decreasing molarity in order to completely remove the haemoglobin (Burton 1981).

Preparation of Microvillus Plasma Membranes. The preparation of microvillus plasma membranes (MV) was accomplished according to the method of Whitsett and Wallick (Whitsett and Wallick 1980).

Term placenta (39–41 weeks of gestational age) was obtained within 15 min after delivery and placed on ice. The tissue was cut into fragments and washed in Krebs-Henseleit buffer. About 30 g tissue was then placed in ice-cold 150 mM NaCl and stirred gently for 30 min. After gauze filtration, the filtrate was centrifuged for 15 min at 20,000 × g in Centricon H-401 (rotor A80.1). The supernatant was centrifuged at 100,000 × g in Centricon T-2070 (rotor TFT 80.13), resuspended and centrifuged at 7000 × g for 10 min. The pellet, representing the microvillus plasma membrane, was then used for biochemical determinations.

Intraerythrocyte Sodium Activity. Heparinized blood samples were centrifuged at 3000 g for 10 min at 4°C. After removal of plasma, the RBCs were rinsed three times with 110 mmol/1 MgC12 solution. The haematocrit of the washed erythrocyte suspension was measured.

After ultrasonic lysis (Braunsonic 1510), intracellular sodium activity was determined in the hemolysates by an ion-selective electrode analyzer (Beckman E4A) after proper calibration. Corrections were made for extracellular trapped volume. The intra-assay coefficient of variation as calculated from values obtained from 20 determinations on the same sample in a single assay was 0.9% for Na^+.

Na^+, K^+ ATPase Activity. The Na^+, K^+ activated Mg^{2+} dependent ATPase activity was determined on erythrocyte membranes by the Kitao method (Kitao and Hattori 1983). The ATPase activity was assayed by incubating ghosts at 37°C in 1 ml medium containing MgC12 (5 mmol/1), NaCl (140 mmol/l), KCl (14 mmol/l) in Tris HCl 40 mmol/l, pH 7.7. The ATPase reaction was started by the addition of 3 mmol/l NaATP and stopped 20 min later by the addition of 1 ml of 15% trichloracetic acid. Inorganic phosphate (Pi) hydrolyzed from the reaction was measured according to Fiske and Subbarow (Fiske and Subbarow 1925).

Enzyme activity was expressed as the difference in inorganic phosphate released in the presence and absence of 10 mmol/l ouabain. The ATPase activity assayed in the presence of ouabain was subtracted from the total Mg^{2+} dependent ATPase activity to calculate the activity of the ouabain-sensitive Na^+, K^+ ATPase. The results are expressed as μmol Pi/(mg membrane proteins × 60 min).

Protein concentration was determined by the Lowry method (Lowry et al. 1951), using albumin as a standard.

Ca^{2+} ATPase Activity. The Ca^{2+} ATPase activity was determined by the method of Davis (Davis 1985).

3 Results

The results are shown in Tables 1, 2, 3.

Table 1. Intraerythrocyte Na^+ activity (mean ± SD) expressed as mmol/l

	Controls	PIH[a]
Maternal erythrocytes	5.8 ± 1.2	7.0 ± 1.9**
Cord blood erythrocytes	8.4 ± 1.8	10.6 ± 2.5**

[a]** = $p < 0.01$.

Table 2. Na^+, K^+ ATPase activity (mean ± SD), expressed as μmol Pi/(mg protein × 60 min)

	Controls	PIH[a]
Maternal erythrocytes	0.83 ± 0.06	0.69 ± 0.07**
Cord blood erythrocytes	0.84 ± 0.07	0.79 ± 0.08*
Microvillus membranes	1.35 ± 0.09	0.96 ± 0.05**

[a]* = $p < 0.05$; ** = $P < 0.01$.

Table 3. Ca^{2+} ATPase activity (mean ± SD), expressed as μmol Pi/(mg protein × 90 min)

	Controls	PIH[a]
Maternal erythrocytes	0.123 ± 0.045	0.235 ± 0.056**
Microvillus membranes	0.323 ± 0.050	0.425 ± 0.060**

[a]** = $p < 0.01$.

4 Discussion

The present study indicates an abnormality in sodium handling by the erythrocyte membrane and the syncitiothrophoblast plasma membrane during PIH, which is similar to the alterations previously described in essential hypertensive patients. The increased cellular sodium activity observed in the hypertensive women compared with gestational age-matched controls might account for the pathogenesis of the disease, according to Blaustein's hypothesis (Blaustein 1984). The observed higher RBC Na^+ activity in PIH might be due to the contemporary decreased activity of the sodium, potassium-ATPase. The data obtained from the umbilical cord blood samples raise the possibility that either a genetic defect is expressed in the erythrocyte membrane of infants from PIH mothers, or a circulating inhibitor of the sodium pump is present in PIH.

The sodium and calcium transport in the placental microvillus membranes reflects the same abnormal pattern present in the erythrocytes, demonstrating that in PIH the cation flux alterations involve different cellular types. Moreover, the modifications of the cation transport in syncitiothrophoblast plasma membrane are of outstanding clinical and biochemical interest for the primary function of placental microvillus as a barrier between maternal and fetal blood. However, the involvement of the electrolytic regulation mechanisms as determinants of PIH remains presently incompletely understood and further studies are under way to clarify the role of sodium, potassium-ATPase under this pathological condition.

Acknowledgement. The present work was supported by a grant of Regione Marche.

References

Blaustein MP (1984) Sodium transport and hypertension. Where are we going? Hypertension 6:445–53

Burton GW, Ingold KU, Thompson KE (1981) An improved procedure for the isolation of ghost membranes from human red blood cells. Lipids 16:946

Davis FB (1985) Erythrocyte membrane Ca^{2+} ATPase responsiveness to in vivo glucose administration. Diabetes 34:639–43

Epstein FH (1986) Cellular sodium transport in essential hypertension. N Engl J Med 314:222–9

Fiske C, Subbarow Y (1925) The colorimetric determination of phosphorus. J Biol Chem 66:375–400

Kitao T, Hattori K (1983) Inhibition of erythrocyte ATPase activity by aclacynomycin and reverse effects of ascorbate on ATPase activity. Experientia 39:1362–4

Lindheimer MD, Katz AI (1985) Hypertension in pregnancy. N Engl J Med 313:675–80

Lowry OH, Rosenburg MY, Farr AL, Randall RT (1951) Protein measurement with the Folin phenol reagent. J Biol Chem 193:265–75

Rubython J, Morgan DB (1983) The effects of pregnancy and pregnancy induced hypertension on active sodium transport in the erythrocyte. Clin Chim Acta 132:91–9

Whitsett JA, Wallick ET (1980) ^3H ouabain binding and Na^+/K^+-ATPase activity in human placenta. Am J Physiol 238:E38–E45

Transmembrane Ionic Fluxes in the Brain in Energy-Depleted Cells

B. K. Siesjö

1 Introduction

The membranes of brain cells, particularly neurons, have a high resting permeability to ions. This "leakiness" is reflected in the high consumption of energy which, to a very large extent, seems to be utilized to pump ions against constant leak fluxes along dissipative pathways. Thus, when energy production is arrested, as occurs in anoxia or in hypoglycemic coma, ion gradients in the brain are quickly dissipated.

The brain normally uses glucose as its main or even sole substrate, and the glucose extracted is usually completely oxidized to CO_2 and water. In the rat, the cortical metabolic rate for oxygen ($CMRO_2$) is around 4.5 $\mu mol\ g^{-1}\ min^{-1}$, and for glucose (CMRgl) about 0.8 $\mu mol\ g^{-1}\ min^{-1}$ (Siesjö 1978; Sokoloff 1981). In man, the values are lower (about half those cited). Nonetheless, the human brain, which accounts for about 2% of the body weight, utilizes about 20% of the oxygen consumed by the body. It has been estimated that close to 50% of the energy generated by oxidative phosphorylation is utilized to pump ions against their thermodynamic gradients, particularly to drive the Na^+, K^+-dependent ATPase (for reviews, see Siesjö 1978; Astrup 1982; Hansen 1985). In support, when sufficient amounts of anesthetics are given to abolish all spontaneous (or evoked) electrical activity, $CMRO_2$ and CMRgl are reduced to 40–50% of control. The remaining fraction, which is sometimes called "basal" energy metabolism, represents energy consumed by processes which are not directly coupled to electrical activity. The nature of the processes involved is obscure.

The cerebral metabolic rate can be increased 2.5- to 3.0-fold during sustained seizures (for literature, see Siesjö 1978). Thus, at normal body temperature there is a six- to eight-fold difference in metabolic rate between the deeply anesthethized, electrically silent brain, and the brain which is involved in sustained seizure activity. In hypothermia, the cerebral metabolic rate is even further reduced and, if one combines barbiturate anesthesia with a lowering of body temperature by 10°C, $CMRO_2$ is reduced to 15% of control (see Astrup et al. 1980). Obviously, there is a 20-fold difference in metabolic rate between this drug-depressed, hypothermic brain and that driven (at normal temperature) to generate epileptic discharges.

The question arises whether these extreme differences in metabolic rate are accompanied by corresponding differences in the capacity of the tissue to maintain ion homeostasis. This does not seem to be the case since cerebral extracellular ion

Laboratory for Experimental Brain Research, University of Lund, Sweden

A. Azzi, Z. Drahota and S. Papa
Molecular Basis of Membrane-Associated Diseases

concentrations (and cellular energy state) remain unperturbed in all functional states examined except seizures. Even in the latter condition, the cerebral energy state is only marginally affected (for references, see Siesjö et al. 1986; Siesjö and Wieloch 1986) and perturbation of ion homeostasis is moderate (e.g. Heinemann et al. 1977; Astrup et al. 1979; Nicholson 1980). Thus, even with intense seizure activity cerebral extracellular K^+ activity (K_e^+), which is normally 3 mM, does not rise above a ceiling value of 10 mM, and Ca_e^{2+} usually falls moderately (10–30%) from the normal value of about 1.3 mM.

The situation is different in energy-depleted cells, viz. under the conditions of ischemia/hypoxia and severe hypoglycemia. Under these conditions, failure of ATP synthesis is accompanied by loss of ion homeostasis, and of cell volume regulation. It is the purpose of the present chapter to discuss the association between a failing energy source and dissipative ion fluxes. It seems justified that this area is intensely explored, particularly since loss of ion homeostasis has been linked to the brain damage which is often incurred in ischemia and hypoglycemia (for reviews, see Siesjö 1981, 1984, 1988a; Raichle 1983; Meldrum 1983; Siesjö and Wieloch 1985). Although this is particularly true of the deranged calcium metabolism, there is circumstantial evidence that cellular influx of Na^+ and Cl^-, with osmotically obliged water, can lead to osmolytic damage of cellular domains endowed with channels for Na^+ and Cl^- entry (e.g. Rothman and Olney 1986).

The question arises whether ion homeostasis is extensively deranged only when energy sources dwindle. Although this is generally the case, solid evidence exists that an equally extensive, albeit transient, disturbance of ion homeostasis occurs in spreading depression (SD), a condition which is a propagated disturbance of cell membrane functions accompanied by relatively moderate energy failure (Nicholson 1979, 1980; Miess and Paschen 1984; Hansen 1985; Lauritzen 1987). I will return to the problems related to ionic fluxes during spreading depression in a subsequent section.

Most of the results and concepts discussed in this chapter are derived from data obtained with ion-specific microelectrodes (ISMs) which measure cerebral extracellular concentrations (activities) of ions and/or extracellular fluid (ECF) volume. More comprehensive descriptions of ion metabolism of the brain, normal or perturbed, are found in previous or recent review articles (Somjen 1979; Nicholson 1979, 1980; Hansen 1985; Siesjö 1988a). In the present work, the emphasis is on loss of ion homeostasis in energy-depleted cells, and on its likely association with cell dysfunction and cell death.

2 Ischemia

2.1 Failure of Energy Metabolism and Loss of Ion Homeostasis

I will consider two types of ischemic insults. One represents a situation in which cerebral blood flow (CBF) is gradually reduced until oxygen consumption is insufficient to cover the metabolic needs of the tissue. With this type of ischemia, our question is: At what CBF value is ion homeostasis deranged, and what is the sequence of ionic changes observed? The other type of ischemia is one in which

CBF is suddenly and completely arrested. Now, another question is posed: What is the time sequence of events elicited, and how do the ionic changes relate to the rate at which the endogenous energy stores are tapped off?

Progressive Reduction in CBF. Figure 1A illustrates how the cerebral energy state is affected by a gradual lowering of the cerebral perfusion pressure, in this case elicited by a reduction of mean arterial blood pressure (MABP). Clearly, the tissue contents of phosphocreatine (PCr) and ATP are maintained until massive energy failure sets in at an MABP of about 30 mmHg (Siesjö and Zwetnow 1970). Other data demonstrate that two compensatory mechanisms delay the final energy

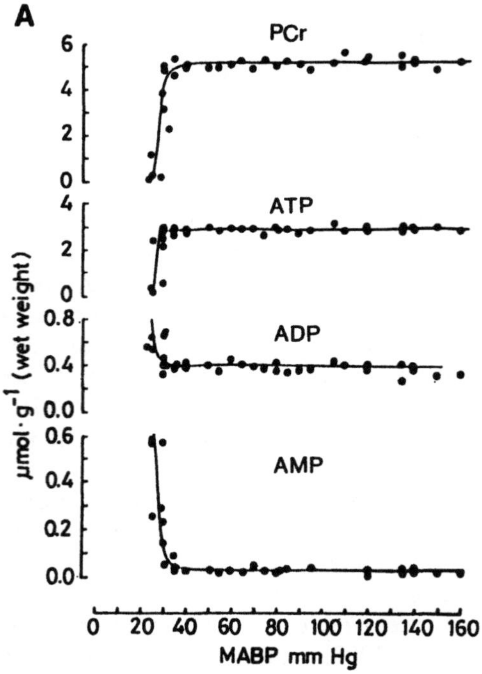

Fig. 1A,B. Ischemic thresholds for energy failure and loss of ion homeostasis. **A** Cerebral cortical concentrations of PCr, ATP, ADP, and AMP as functions of mean arterial blood pressure, which was gradually lowered by bleeding (data from Siesjö and Zwetnow 1970). **B** CBF is related to the extracellular K^+ and Ca^{2+} concentration during progressive lowering of cerebral perfusion pressure. Note increase in extracellular K^+ at CBF values below 15 ml/100 g/min (Harris and Symon 1984)

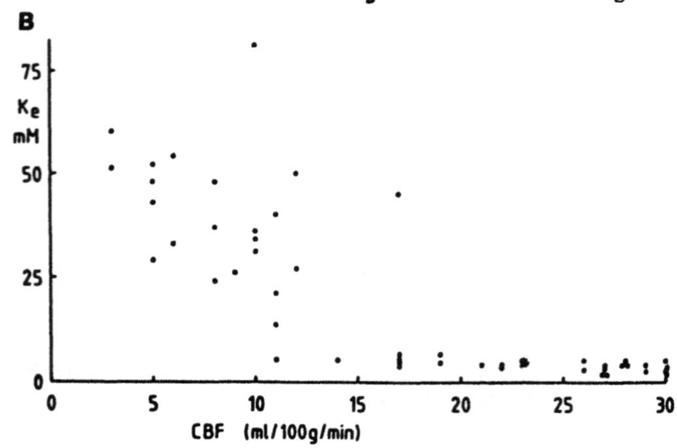

failure. One is vascular autoregulation, i.e. the dilatation of resistance vessels in response to a lowering of perfusion pressure, and the other is increased extraction of oxygen. When the latter mechanism has been exhausted, the avalanche character of catabolic events sets in.

The corresponding changes in extracellular ion concentrations have been most clearly enunciated in experiments in which blood pressure was manipulated in primates with an occluded middle cerebral artery (MCA) (Harris et al. 1981), or in rats which had their carotid arteries clamped (Harris and Symon 1984). Figure 1B, which gives data for the rat, illustrate the rise in extracellular K^+ when CBF fell below 0.15 ml g^{-1} min^{-1}, i.e. below about 20% of control. Similar experiments in MCA-occluded baboons have defined two ischemic thresholds: one representing a CBF value below which spontaneous and evoked electrical activity ceases and another one, at a lower CBF value, below which loss of ion homeostasis occurs (see Astrup et al. 1981). It seems very likely that transgression of the first threshold involves functional inactivation of neurons at a marginally altered energy state, while the second threshold, characterized by rapid dissipative ion fluxes, is reached when the cerebral energy state is suddenly deranged. Usually, CBF in the focus of an ischemic lesion caused by a stroke is so low that energy failure and loss of ion homeostasis are at hand. However, the periphery of the lesion ("the penumbra zone") is considered to contain functionally inactive but viable cells which maintain their structural integrity above a certain critical CBF value. In these regions, the energy sources suffice to maintain membrane potential and ionic gradients. However, any extra stress imposed on the cells can tip the balance in favor of a catabolic cascade (see Nedergaard 1988).

Experiments of this type clearly showed that degrees of ischemia exist in which K_e^+ is moderately increased but Ca_e^{2+} is at normal values (Harris et al. 1981; Harris and Symon 1984). In fact, the results defined a critical concentration of K_e^+ (10–15 mM) above which a depolarizing shift occurred of the DC potential between brain and extracranial sites, accompanied by an accelerated release of K^+ and a rapid reduction of Ca_e^{2+}. As will be discussed below, this "terminal depolarization" probably heralds the point at which neuronal depolarization is sufficiently severe to drastically increase membrane permeability to ions, this in turn enhancing energy failure in a mutually reinforcing cascade of events.

Suddenly Occurring Anoxia. Our next concern is the sequence of events following induction of sudden and complete ischemia. Figure 2A shows changes in cerebral cortical concentrations of PCr, ATP, and lactate (Nordström and Siesjö 1978), and Fig. 2B the extracellular concentrations of K^+, Ca^{2+}, Na^+, and Cl^-, as well as the size of the ECS volume (for original references and further documentation, see Hansen 1985). Anoxia of only 10-s duration significantly reduces tissue PCr and ATP contents (Nilsson et al. 1975), and the tissue ATP store approaches zero after 2 min. At that time, tissue lactate content has increased to about 10 mM kg^{-1}.

The corresponding changes in extracellular ion activities (Fig. 2B) demonstrate two major phases of events. In the first (phase 1 according to Hansen 1985) K_e^+ increases slowly towards 10 mM with little or no changes in Ca_e^{2+}, Na_e^+ and Cl_e^-. Since measurements indicate that ECF volume shrinks in that period (towards 80% of control), the unchanged activities of Na^+ and Cl^- suggest that a fraction of the Na^+ and Cl^- contained in ECF is translocated into intracellular domains, with

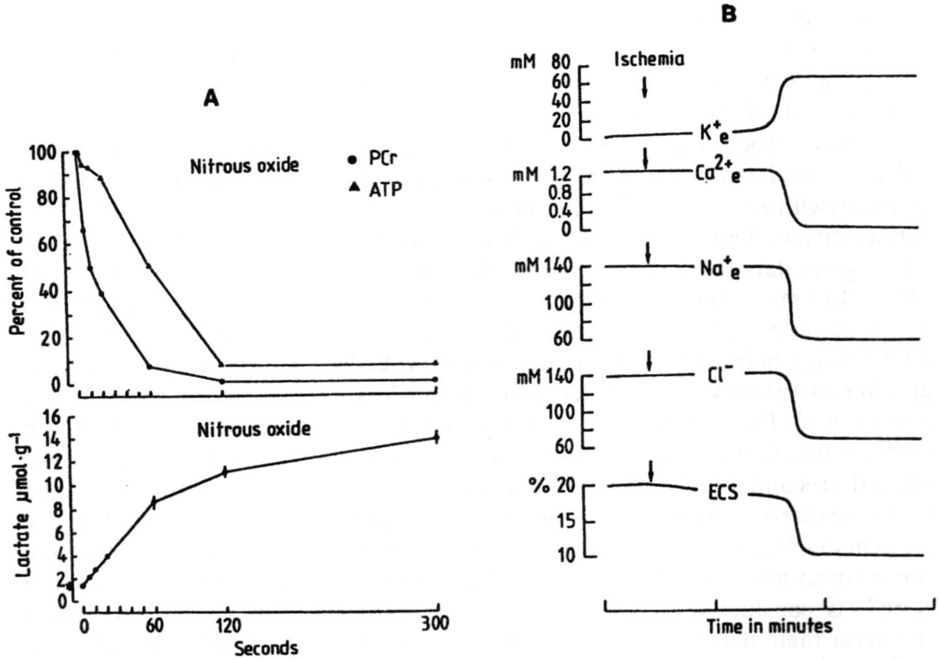

Fig. 2A,B. Sequence of changes in cerebral energy metabolism and ion homeostasis following induction of complete anoxia (ischemia). **A** Cerebral cortical concentrations of PCr, ATP, and lactate during the first 300 s of ischemia (data from Nordström and Siesjö 1978). **B** Corresponding changes in the extracellular concentrations of K^+, Ca^{2+}, Na^+, and Cl^-, as well as in the extracellular fluid space volume (ECS). Data from Hansen (1985) (Siesjö 1988b)

osmotically obligated water. Since Ca^{2+}_e *increases* somewhat in phase 1, it is questionable if calcium is at all translocated from extra- to intracellular fluids. One could, in fact, envisage that cells extrude calcium released from endogenous stores. At least, cellular uptake of Ca^{2+} from ECF in this period seems unlikely.

During phase 2, K^+ is released at a rapid rate, and the extracellular concentration subsequently reaches values of 50–70 mM. At the point of this sudden K^+ release, the DC potential shifts in the negative direction, the ECF concentrations of Ca^{2+}, Na^+ and Cl^- fall precipituously, and the ECS volume is reduced below 50% of control. The logical explanation is that depolarization of a critical degree dramatically increases membrane permeability to all ions under consideration.

What is the relationship between changes in cellular energy state and alterations in extracellular ion concentrations? Tentatively, one may regard the gradual increase in K^+_e to a critical value (or rather, the accompanying depolarization) as the trigger of the "shock" opening of ionic gates, and the sudden loss of ion homeostasis. We may further assume that the loss of K^+ from cells during phase 1 occurs because relative energy failure slows down ATP-driven Na^+/K^+ exchange, while the transition to phase 2 signals overt ATP shortage. Clearly, if this is so the time to terminal depolarization should be delayed and hastened by measures with retarded and accelerated energy failure, respectively. Furthermore, this time should vary indirectly with the state of opening of ion channels during phase 1. Figure 3 shows that this is the case. Thus, induction of preischemic hypoglycemia shortens and hyperglycemia increases the time to terminal depolarization (Hansen 1978) probably because glycolytic production of ATP delays energy failure during anoxia (Hansen and Nordström 1978). The data in Fig. 4 demonstrate that, at similar glucose concentrations, the time to terminal depolarization varies inversely with preischemic metabolic rate (Astrup et al. 1980). This is an intriguing finding since it suggests that the preischemic rate of energy metabolism leaves its imprint on energy expenditure during anoxia/ischemia, either because the state of opening of ion gates is preserved for some time in the anoxic brain, or because some other process coupled to ATP expenditure continues to deplete available energy stores at a rate which is set by the preischemic conditions.

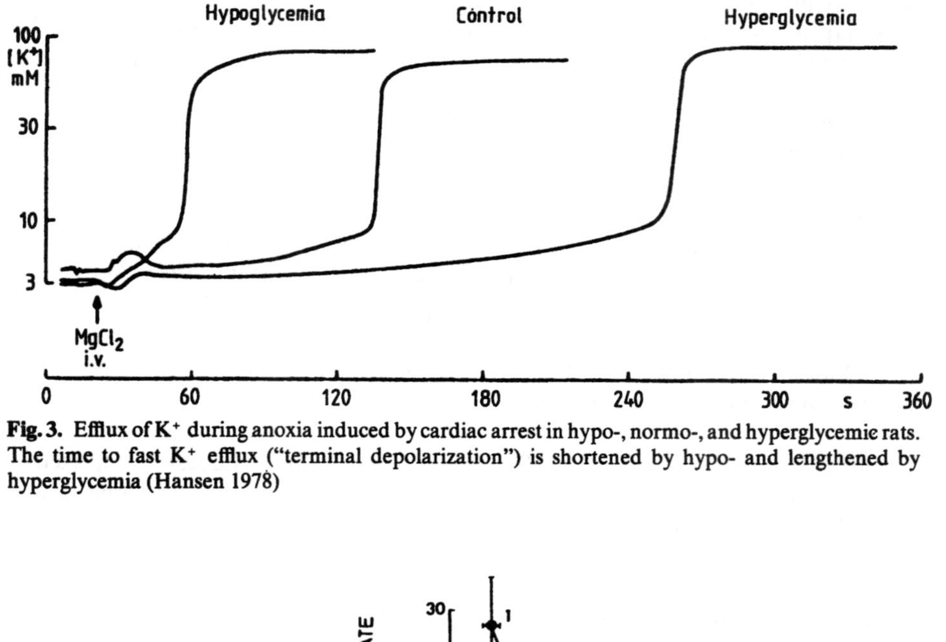

Fig. 3. Efflux of K^+ during anoxia induced by cardiac arrest in hypo-, normo-, and hyperglycemic rats. The time to fast K^+ efflux ("terminal depolarization") is shortened by hypo- and lengthened by hyperglycemia (Hansen 1978)

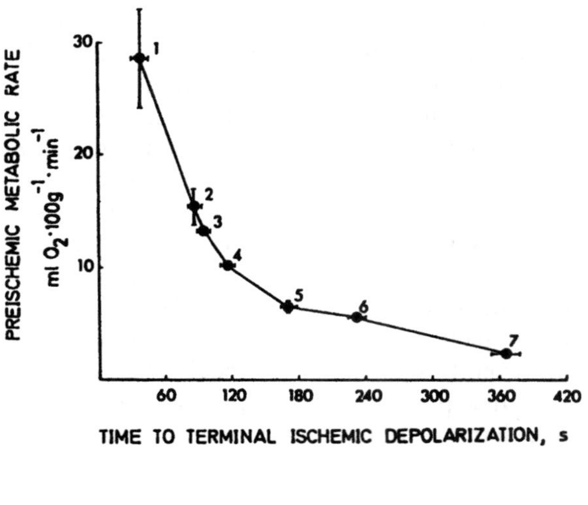

Fig. 4. Effect of the preischemic cerebral metabolic rate for oxygen ($CMRO_2$; ml O_2/(100 g)/min) on the time to the terminal ischemic depolarization. The numbers *1–7* refer to experimental groups in which $CMRO_2$ was varied between the extreme values seen during status epilepticus (*1*) and during combined hypothermia-barbiturate anesthesia (*7*). The values are means ±SEM (Astrup et al. 1980)

2.2 Ionic Gates Involved: Facts and Speculation

The results quoted lead to two questions. First, what is the ionic basis for the electrical silence in the period preceding the terminal depolarization? Second, which are the ionic gates opened during this terminal depolarization and what is the trigger?

Ionic Changes Preceding Massive Depolarization. As discussed in some detail by Hansen (1985), EEG activity ceases at a time (15–20 s) when K_e^+ is only very moderately increased, and Ca_e^{2+}, Na_e^+ and Cl_e^- are hardly altered at all. The reason for the block of synaptic activity is not known, but speculations encompass presynaptic depolarization with inhibition of transmitter release caused by relative energy failure, and hyperpolarization of postsynaptic elements. The first possibility appears rather unlikely since the overall ATP content is reduced by only about 10%; furthermore, neurophysiological data fail to support the assumption of a presynaptic block (see Hansen 1985). The second hypothesis was proposed by Krnjevic (1975), who suggested that both anesthesia and hypoxia hyperpolarize cells by a mechanism involving Ca^{2+}-triggered K^+ efflux. There is circumstantial evidence for such a mechanism since hypoxia in tissue slices in vitro leads to hyperpolarization coupled to a decreased membrane resistance due to enhanced K^+ conductance (Hansen et al. 1982). Thus, enhanced K^+ efflux during phase 1 may hyperpolarize neurons and arrest neuronal activity and, at the same time, contribute to the slow rise in K_e^+. It is not known, however, whether such a hypothetical opening of K^+ gates is triggered by calcium. If it is, the required rise in Ca_i^{2+} may be caused by the release of sequestered or bound Ca^{2+} as a result of the intracellular acidosis. This contention is supported by results reported following microinjection of acid into axons (Abercrombie and Hart 1986; however, also see Busa and Nuchitello 1984).

Terminal Depolarization: Voltage-Dependent Ionic Events. The results depicted in Fig. 2B suggest that channels are suddenly opened, or transporters activated, which allow K^+ to rapidly leave cells, and Ca^{2+}, Na^+, and Cl^- to enter, once massive depolarization sets in. One possible explanation for this is an increase in ion conductances caused by opening of voltage-sensitive ion channels. A likely sequence of events starts with a sudden increase in K^+ conductance and an ensuing flooding of the ECF with K^+, resulting in depolarization of pre- and postsynaptic elements, and of glial cells. Theoretically, K^+ conductance can increase either because depolarization causes voltage-sensitive K^+ channels to open, or because a rise in Ca_i^{2+} has an equivalent effect on calcium-activated ones. It is not known, however, whether such channels have the necessary conductance to transmit large amounts of K^+ into ECF, or whether they remain open at the high Ca^{2+} and low pH_i encountered (see above). The demonstration by Noma (1983) of ATP-dependent K^+ channels in muscle leaves a third possibility since one can envisage that depletion of ATP triggers a massive efflux of K^+ (see Hansen 1985). At present, however, this possibility is speculative.

Once pre- and postsynaptic endings (or glial cells) depolarize, voltage-sensitive channels for Ca^{2+}, Na^+, and Cl^- would open, allowing influx of all these ions.

We now know that voltage-sensitive calcium channels (VSCCs) are activated by relatively strong depolarization, and that they mediate calcium influx and raise Ca_i^{2+} in neurons in primary culture (Kudo and Oguro 1986; Murphy et al. 1988; see also Miller 1987). Figure 5A summarizes current knowledge on VSCCs in pre- and postsynaptic neuronal elements. It also illustrates the now well-known fact that VSCCs are of at least three types, only one of which (the L-type) is sensitive to the dihydropyridine type of calcium antagonist (Nowycky and Tsien 1985). It is not known where the VSCCs are localized but it seems clear that the presynaptic channel causing transmitter release is not of the L-type, and it is likely that some VSCCs, including L channels, are found in the cell soma membrane. Again,

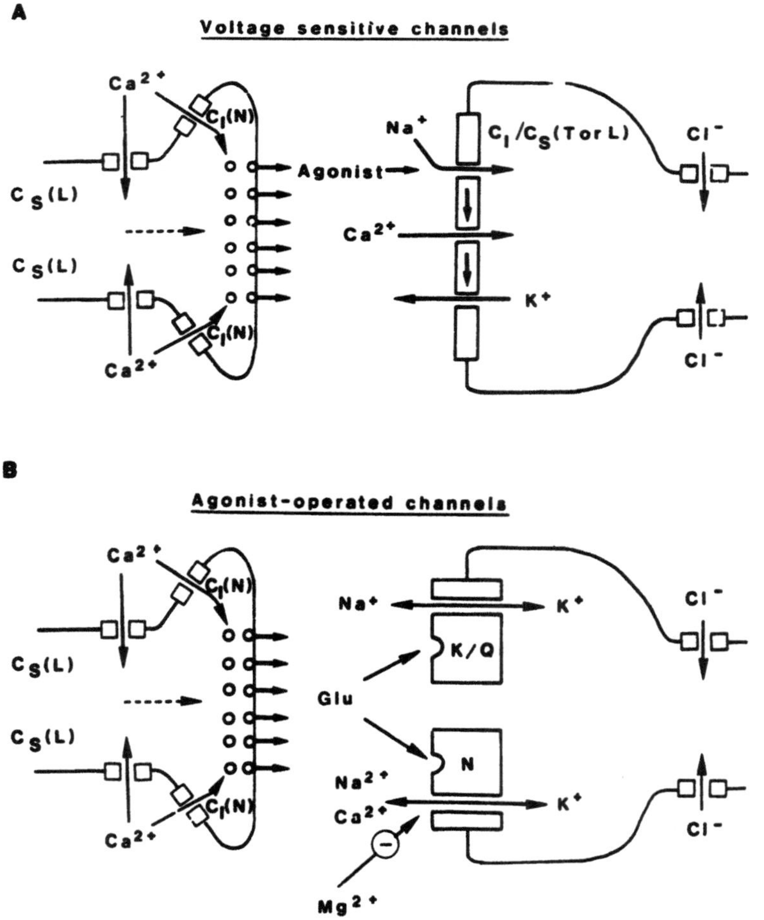

Fig. 5 A,B. Voltage-sensitive channels in pre- and postsynaptic structures (**A**) and postsynaptic agonist-operated channels (**B**). The voltage-sensitive channels have been assumed to be of the L, T, and N types, the L type being sensitive (C_s) and the other types insensitive (C_i) to calcium antagonists of the dihydropyridine class. Agonist-operated channels, which are activated by either glutamate or a related amino acid, have been assumed to be gated by kainate/quisqualate (K/Q) or by NMDA (*N*) receptors. It is assumed that only the latter opens a conductance for Ca^{2+}

one must ask whether the conductances of these channels are large enough and whether the channel opening time is sufficient to account for the rapid influx of calcium during anoxia. The question also remains how Na^+ is transmitted into the cells.

Terminal Depolarization: Transmitter-Dependent Ionic Events. The second possible sequence of events is one in which depolarization acts primarily to hasten release of transmitters which then secondarily increase membrane conductances to ions. The trigger of these events may still be the depolarization caused by a sudden increase in K^+ conductance with an ensuing release of K^+, e.g. from postsynaptic elements. If this is so, the sequence proposed differs from the first only in the sense that it provides for yet another type of Na^+ and Ca^{2+} conductances, viz. those activated by agonist-receptor interactions. It is now clear that anoxia leads to massive release of excitatory amino acids (EAAs), mainly glutamate and aspartate to ECF (Benveniste et al. 1984; Hagberg et al. 1985).

It has been suspected for a long time that release of glutamate can activate postsynaptic Na^+ receptors, thus leading to Na^+ influx and depolarization. In fact, van Harreveld (1966, 1970) assumed that the rapid depolarization during asphyxia was due to release of glutamate from depolarized presynaptic endings, with opening of glutamate-activated Na^+ channels in postsynaptic elements, mostly dendrites; this author also assumed that Na^+ influx, with forced entry of Cl^-, was responsible for a fluid shift from extra- to intracellular fluids and for swelling of apical dendrites.

Following these remarkably foresighted deductions, much has been learned about EAA-operated Na^+ and Ca^{2+} channels in the brain (see Fagg 1985; Collingridge 1985; Fagg et al. 1986; Mayer 1987; Mayer and Westbrook 1987; Foster and Fagg 1988; Miller 1987). At least three types of glutamate receptors exist, selectively activated by kainate (K), quisqualate (Q), and N-methyl-D-aspartate (N or NMDA) (Fig. 5B). When glutamate (or aspartate?) activates the K/Q type of receptor a channel for monovalent ions is opened (see Collingridge 1985; Foster and Fagg 1988). This channel would, by allowing Na^+ influx, depolarize the postsynaptic membrane and set the stage for forced entry of Cl^-, with osmotically obliged water and resulting cell swelling. Since this channel is unselectively permeable to monovalent cations, it could, at least in theory, contribute to the rapid loss of K^+ during phase 2. When, on the other hand, glutamate activates the NMDA type of receptor, a larger conductance channel is opened which allows calcium entry. There is some uncertainty whether these two main types of channels are separate entities, or whether the conductance mechanisms discussed represent different functional states of one and the same channel (Jahr and Stevens 1987; Mayer 1987; Cull-Candy and Usowicz 1987). To facilitate discussion, I will consider the channels separately.

The NMDA-gated channels show some unusual characteristics since they are blocked by Mg^{2+} in a voltage-dependent manner (meaning that the block is relieved when the membrane depolarizes) (Nowak et al. 1984) and since the receptor action of NMDA is strongly reinforced by small amounts of glycine (Nowak et al. 1987). Probably, glycine acts at an allosteric site, separate from that which binds the agonists. There are no specific antagonists of the K/Q type of receptor which, however, can be blocked by some peptides, including γ-glutamylglycine (DGG) (Fagg 1985; Collingridge 1985). Specific antagonists of the NMDA channels exist, the most potent being the ω-phosphonic acid derivates D-2-amino-phosphonovalerate (AP5) and D-2-amino-phosphonoheptanoate (AP7). Since these poorly penetrate the blood-brain barrier, it is of even larger interest that non-competitive NMDA antagonists exist, which block the NMDA gated channel. Of these, the dissociative anesthetics, ketamine and phencyclidine, as well as MK801, readily pass the blood-brain barrier.

Experiments on dissociated neuronal cultures in vitro have clearly shown that not only VSCCs (see above) but also AOCCs, the latter activated by glutamate or NMDA, mediate influx of calcium and allow Ca_i^{2+} to rise in response to agonist-receptor interaction (Kudo and Oguro 1986; Murphy et al. 1988; Miller 1987). The question arises whether one or both of these types of calcium channels have a dominating influence on calcium influx during anoxia. Recent results demonstrate that neither calcium antagonists of the dihydropyridine type (Peters 1986; Höller et al. 1986) nor competitive (AP7) or non-competitive (MK801) NMDA antagonists (Hansen et al. 1988; Zhang et al. 1989) delay or curtail calcium influx during anoxia. Interestingly, however, the non-dihydropyridine calcium antagonist flunnarizine increased the K^+ threshold associated with terminal depolarization and Ca^{2+} influx, thus considerably delaying anoxia-induced cellular uptake of calcium (Peters 1986; Höller et al. 1986).

These results suggest that the major portion of the calcium taken up by cells during ischemia does not enter via the L-type calcium channel, or via NMDA-gated agonist-operated channels. How do such results fit with those showing that, in vitro, depolarization induced by K^+ and stimulation of NMDA receptors with either glutamate or NMDA raises Ca_i^{2+} by mechanisms which are counteracted by classical calcium antagonists and NMDA antagonists, respectively? My tentative interpretation is that although these channels play important functional roles in transmitting calcium signals across excitable membranes, depolarization during anoxia/ischemia opens a more indiscriminate type of channel, one allowing influx of both Ca^{2+} and Na^+, perhaps also accelerating K^+ efflux. It seems likely that drugs of the flunnarizine type, although not acting on conventional Na^+ and Ca^{2+} channels, act by delaying this "shock" opening of cation channels of low selectivity (see below).

2.3 Acid Base Events

Ischemia causes H^+ to accumulate in both intra- and extracellular fluids. The source of the acid generated is ICF. In a formal sense, anaerobic glycolysis does not lead to lactic acid production since the H^+ ions are formed in other reactions than that catalyzed by lactate dehydrogenase, mainly during ATPase activity (Alberti and Cuthbert 1982; Hochachka and Mommsen 1983). However, if the extra ATP formed during glycolysis is hydrolyzed to ADP and P_i so that the ATP concentration remains constant, the conversion of 1 mol glucose to 2 mol lactate will always be associated with the formation of 2 mol H^+. Clearly, if ATP is broken down during anoxia/ischemia there will be an additional release of H^+. However, less than one H^+ is formed per ATP hydrolyzed. Furthermore, since the ATP content is about 3 mmol kg^{-1}, while the lactate content during ischemia usually rises to values of around 15 mM kg^{-1}, the dominant source of H^+ is, for all practical purposes, glucose (see Siesjö 1985, 1988b).

Since the buffering capacity of cerebral intracellular fluids is at least approximately known, it is possible to calculate changes in pH and PCO_2 when increasing amounts of lactate plus H^+ accumulate (Siesjö 1985, 1988b). Changes in pH_i can be compared to, and validated by, those calculated from measured total CO_2 content, tissue PCO_2, and pH_e (von Hanwehr et al. 1986; Smith et al. 1986). The changes derived in these ways are realistic estimates of intracellular events provided that the various intracellular fluids can be considered homogeneous. Although this is a controversial issue (see Kraig et al. 1986; Boris-Möller et al. 1988) I will assume, for the sake of this discussion, that major compartmentation between cells does not exist. One can calculate that when the lactate content increases by 5,10,15,20, and 25 mM kg^{-1} during complete ischemia, the pH should fall from a value of 7.0 to about 6.8, 6.5, 6.3, 6.0, and 5.8, respectively, and PCO_2 increases from a normal value of 45 to 75, 115, 160, 210, and 285 mm Hg, respectively. These values are reasonably close to those derived from experiments in which the total CO_2 content, $PtCO_2$, and pH_e were measured (von Hanwehr et al. 1986; Smith et al. 1986).

How is this intracellular acid load distributed between intra- and extracellular fluids? An efflux of acid will, of course, curtail the decrease in pH_i. Such redistribution of acid is probably of only marginal importance since the size of the ECF is, during anoxia/ischemia, not more than about 10% of the ICF volume. However, changes in pH_e give valuable hints to mechanisms of transmembrane flux of H^+ and HCO_3^-.

Two mechanisms exist for acid transport, neither of which require specialized conductance channels or ion exchangers. The first mechanism of acid-base equilibration is obvious: CO_2 will diffuse across cell membranes until the PCO_2 is the same in all compartments. The increase in PCO_2 will itself lower pH_e substantially (see Siesjö 1988b). Another source of H^+ is lactic acid which could diffuse from cells to ECF in the unionized form, dissociating in H^+ and lactate following its egress (e.g. Roos and Boron 1981; Boron 1983).

These mechanisms would favour equilibration of H^+ between intra- and extracellular fluids. Two findings suggest that also other mechanisms must exist for H^+ and HCO_3^- transport across membranes of the anoxic/ischemic brain. First, following induction of anoxia, the smooth fall in pH_i is interrupted by a transient alkalinization, which is coincident in time with the terminal depolarization (Kraig et al. 1983, 1985; Mutch and Hansen 1984) (Fig. 6). During this alkaline shift, H^+ must be transmitted into cells, or HCO_3^- must leave. Second, measurements of pH_e in ischemia with progressively increasing lactate contents have given the surprising result that pH_e shows two plateau levels, with a rapid transition zone between the two at tissue lactate levels of 17–18 mM kg^{-1} (Kraig et al. 1985, 1986). This implies that appreciable extra- to intracellular H^+ gradients could be upheld at moderate and severe degrees of lactate accumulation.

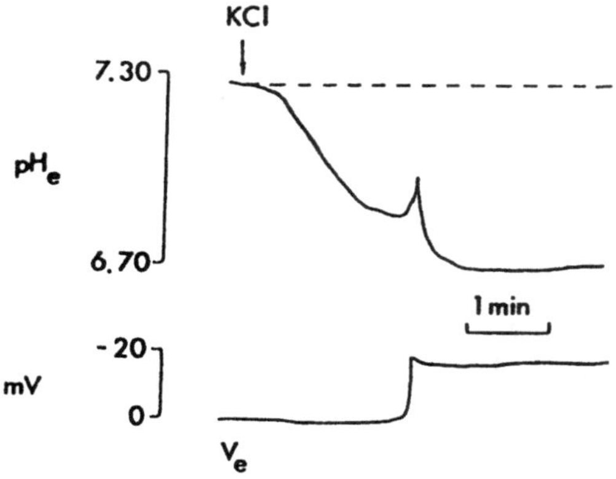

Fig. 6. Changes in pH_e of parietal cortex, and of the DC potential between the cortex and an extracerebral site, following anoxia induced by cardiac arrest (i.v. KCl injection). Note acidification interrupted by a transient alkalinization when the DC potential shifts in the negative direction (Mutch and Hansen 1984)

Theoretically, the mechanisms involved can be of two kinds: electrically neutral Na^+/H^+ and/or Cl^-/HCO_3^- antiporters, and conductance channels for H^+ and HCO_3^-. It is tempting to assume that any net H^+ or HCO_3^- flux across membranes occurs by Na^+/H^+ or Cl^-/HCO_3^- exchange (e.g. Roos and Boron 1981; Boron 1983; Grinstein and Rothstein 1986). However, since the Na^+ and Cl^- gradients are quickly dissipated during anoxic depolarization these antiporters may not influence, in any important way, the steady state distribution of H^+. It is also difficult to envisage that electroneutral Cl^-/HCO_3^- exchange is activated by depolarization, explaining the transient alkaline shift. In support, the results published by Mutch and Hansen (1984) demonstrate that blockers of the cation and anion antiporters had little effect on the extracellular acidification during ischemia.

Could conductance channels be involved? As recognized already by Caldwell (1956), the H^+ gradient across membranes is so small that H^+ transport can influence transmembrane pH only if the conductance is markedly increased. There is now evidence that such conductance changes are biologically feasible (Thomas and Meech 1982). In theory, H^+ could enter or leave cells (depending on the electrochemical gradient) through glutamate-activated cation channels. There is precedence for that since Endres et al. (1986) showed that stimulation of dorsal root cells in vitro with glutamate or its structural analogues increased pH_e, at the same time decreasing pH_i. However, since the HCO_3^- gradient is larger, diffusion of HCO_3^- could provide a mechanism for H^+ shuttling. It has recently been shown that HCO_3^- can be transported via voltage-dependent Cl^- channels at a rate sufficient to significantly alter pH_i during activation of GABA receptors (Kaila and Voipio 1987). Clearly, it seems justified to explore to what extent cation and anion antiporters and conductance channels influence the steady state distribution of H^+ and HCO_3^- during ischemia.

3 Hypoglycemia

Hypoglycemia is another condition leading to energy deprivation. As discussed in detail elsewhere (Siesjö 1978, 1988c; Siesjö and Agardh 1983) energy failure develops around plasma glucose concentrations of 1 mM. However, the loss of energy homeostasis is less related to the plasma glucose level than to the functional state, as this is reflected in the EEG. As Fig. 7 shows, the tissue energy state deteriorates when spontaneous EEG activity is lost ("isoelectric EEG"). This, in turn, correlates to the truly critical event, viz. massive depolarization of membranes. Thus, the cerebral energy state is upheld just prior to depolarization, as defined by the pronounced negative shift in DC potential, and, rapidly deteriorates once depolarization occurs (Harris et al. 1984; Wieloch et al. 1984).

The data of Fig. 7 illustrate one of the principal differences between ischemia and hypoglycemia. In hypoglycemic coma of even extended duration, significant amounts of ATP and other nucleotide triphosphates persist (about 20–30%). This suggests that a new steady state has been achieved in which energy expenditure matches the reduced ATP levels. It seems likely that the ATP levels, at the reduced levels encountered during hypoglycemic coma, vary somewhat with small variations in plasma glucose concentrations. Marked reductions of blood pressure and, thus, of CBF, are known to further reduce the cellular energy state (Auer et al. 1986). Probably, such variations in energy supply critically affect many of the metabolic functions which persist during coma. For example, the free fatty acid content of the tissue varies inversely with the adenylate energy charge, suggesting that the rate of reacylation of lysophospholipids is a function of ATP concentration (Wieloch et al. 1984).

3.1 Ion Homeostasis

Astrup and Norberg (1976) demonstrated that K_e^+ dramatically increased when hypoglycemia was severe enough to abolish EEG activity. Interestingly, no increase in K^+ was noted in the predepolarization period, even though the EEG was

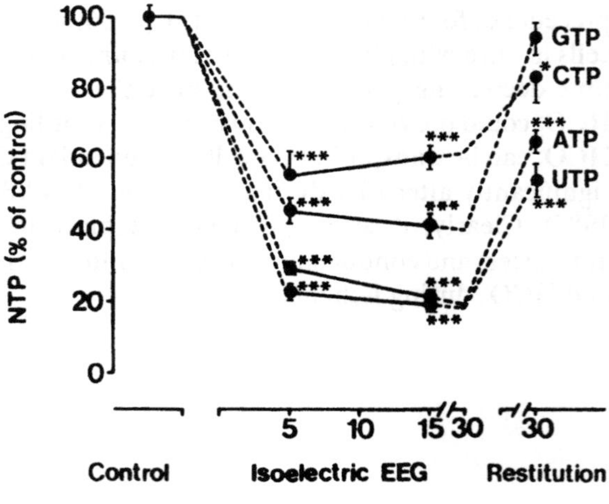

Fig. 7. Changes in cortical nucleoside triphosphate concentrations during insulin-induced hypoglycemic coma and restitution following glucose administration. Statistical analysis by Student's t-test: * $p < 0.05$; *** $p < 0.001$ (Chapman et al. 1981)

depressed, or showed interspersed polyspike activity. Since the efflux of K^+ could precede or follow the cessation of EEG activity by a few minutes, it seemed clear that such cessation did not correlate tightly to the depolarization accompanying loss of K^+ homeostasis. The results presented, and those subsequently published (Pelligrino et al. 1981; Harris et al. 1984), demonstrated that the final depolarization was usually preceded by an initial, and partly reversible event which was interpreted as a spreading depression. Initially, the depolarization was usually confined to one hemisphere; the delay in EEG flattening of the second hemisphere was usually 2–4 min. Thus, an isoelectric EEG seems to be a reliable indicator of bilateral membrane depolarization first after the initial 5-min isoelectric period.

Figure 8 shows that at the high blood pressures maintained during coma, K_e^+ reaches values of about 40 mM, i.e. lower than during severe ischemia (Pelligrino et al. 1981). In all probability, K^+ is being pumped back into the cell at the expense of available ATP. In support of this contention, a lowering of blood pressure raises K_e^+. Furthermore, when the comparison is made at similar blood glucose values, there is a tight, inverse correlation between blood pressure and K_e^+ (see Fig. 8, lower panel).

As Fig. 9 illustrates, efflux of K^+ is coupled to uptake of Ca^{2+}, both during the first depolarization, which is followed by partial recovery, and during the second ("final") depolarizaton (Harris et al. 1984; Wieloch et al. 1984). This coupling of K^+ and Ca^{2+} fluxes supports the hypothesis that whenever K_e^+ reaches a certain, critical value, calcium channels are opened. However, hypoglycemia differs from ischemia in that the increase in K_e^+ is abrupt, similar to that observed in spreading depression (see below).

Changes in Na_e^+ and Cl_e^- have not been reported in hypoglycemia. However, since depolarization during hypoglycemic coma is accompanied by an impedance change, similar to that observed in ischemia (Pelligrino et al. 1981), it appears likely that also hypoglycemia leads to cellular uptake of Na^+ and Cl^-, with osmotically obliged water.

3.2 Spreading Depression and Ion Homeostasis

In view of the occurrence of SD-like alterations preceding the final depolarization in hypoglycemia, it is justified to recall ionic shifts in SD (see Nicholson 1980; Hansen 1985; Lauritzen 1987). SD is a propagated electrical disturbance which can be elicited by KCl application, strong electrical stimulation, or stab wounds, and which spreads across the cortical surface at a rate of 3–5 mm min^{-1}. The propagated electrical disturbance is accompanied by efflux of K^+, with K_e^+ values usually exceeding 20 mM, and by influx of Ca^{2+}, Na^+, and Cl^-, by a negative DC shift, and by changes suggesting a decrease in ECF volume to about 50% of control.

Although SD leads to small changes in high energy phosphates, SD is by itself not caused by energy failure and the presence of an energy source allows rapid restoration of a normal ionic distribution (Miess and Paschen 1984). It has been proposed that one of two changes can account for the SD: a localized, marked increase in K_e^+, and the release of sufficient amounts of glutamate to cause excitation and depolarization. Theoretically, any of these changes could propagate over large distances if the K^+ (or glutamate) released from one neuron is not

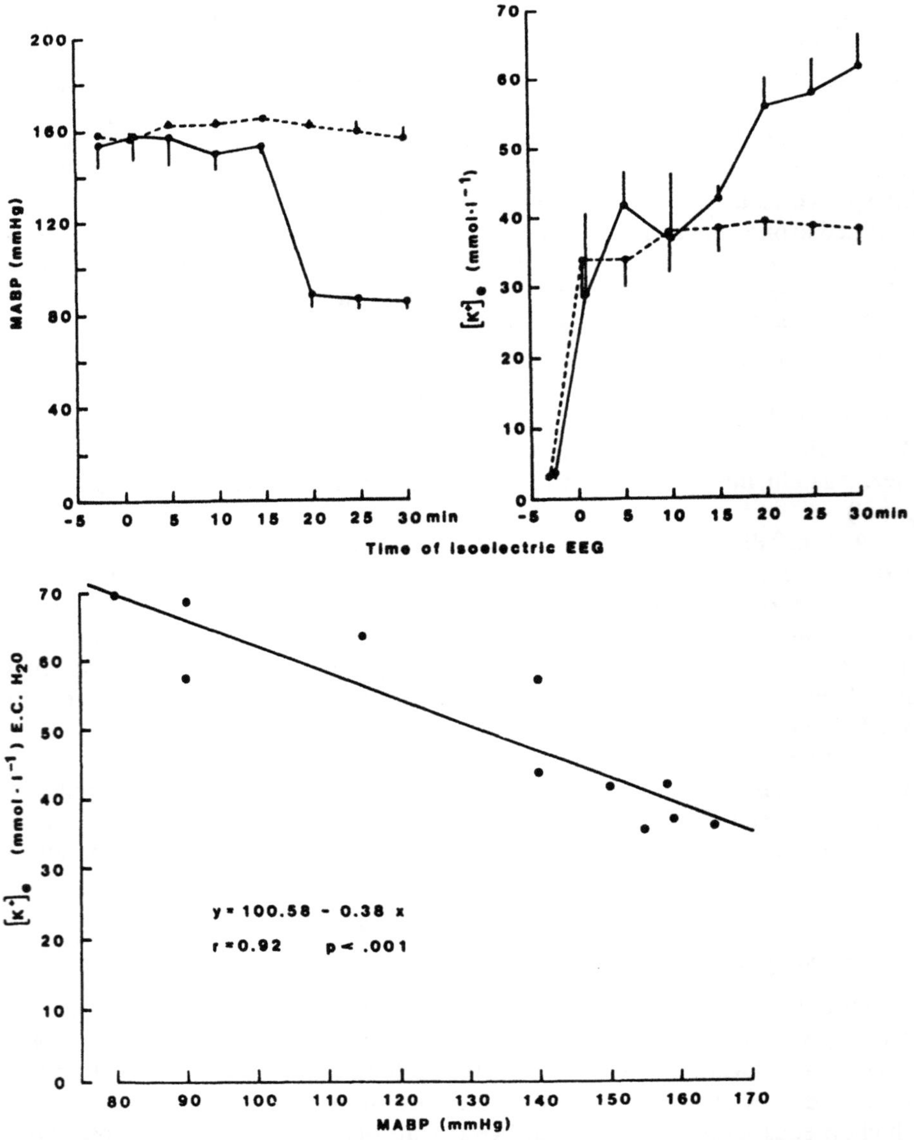

Fig. 8. *Upper panel:* Extracellular K^+ activity in rat cerebral cortex as a function of mean arterial blood pressure in rats during severe, insulin-induced hypoglycemia. *Left panel:* Blood pressure in "normotensive" animals (*broken line*) as well as in those in which blood pressure was reduced to 80 mmHg after an "isoelectric" period of 15 min (*continuous line*). Corresponding values for extracellular K^+ activity are shown in the *right panel*. Values are means ± SE. *Lower panel:* Relationship between mean arterial blood pressure and extracellular K^+ activity in cerebral cortex of rats in which blood glucose concentrations were reduced to very similar values (0.41–0.53 μmol g^{-1}) by means of insulin injection (Pelligrino et al. 1981)

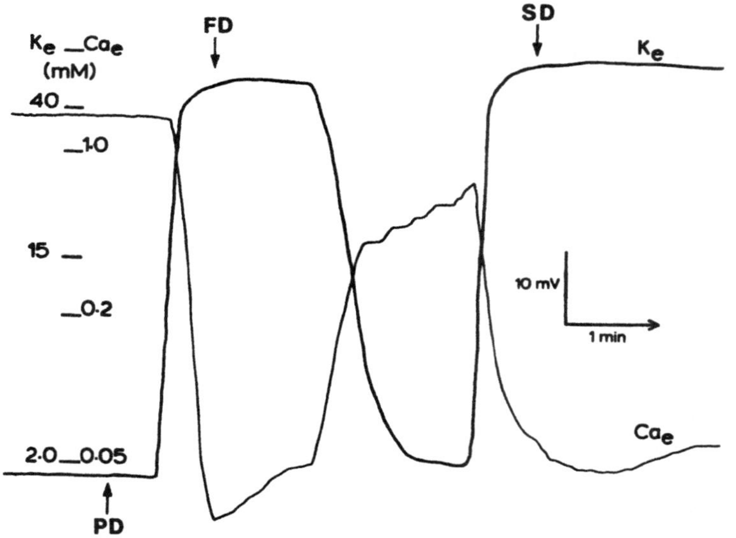

Fig. 9. Influence of severe hypoglycemia on brain extracellular calcium and potassium activities. *FD* First depolarization; *SD* second depolarization; *PD* partial depolarization (Wieloch et al. 1984)

prevented from acting on the next by efficient reuptake mechanisms. This may explain why SD can only be elicited in brain tissues in which neurons are tightly packed.

3.3 Hypoglycemia and SD: Similarities and Dissimilarities

It has previously been suggested that ischemia and SD share many features in common, the major difference being the reversibility of the ionic changes in SD, as contrasted to ischemia (Hansen 1985). In view of the changes observed in hypoglycemia, it would appear that hypoglycemia and SD are even more similar. Thus, one could speculate that the only major difference between the two conditions is that the energy sources in the brain during hypoglycemia are too feeble to allow more than an aborted attempt towards recovery.

It has recently been shown that ketamine, a noncompetitive NMDA receptor antagonist, which blocks the NMDA gated Ca^{2+} channel, also blocks cortical spreading depression in rats (Goroleva et al. 1987). This strongly indicates that receptor activation by glutamate and Ca^{2+} influx are involved in SD. Supportive data were reported by Hansen et al. (1988, see also Zhang et al. 1989), who obtained similar results using MK801. These authors also showed that, following administration of MK801, the usual stimulus elicited a transient, moderate rise in K_e^+, but it caused neither a SD, nor a decrease in Ca_e^{2+}.

As discussed above, MK801 did not delay or abort changes in Ca_e^{2+} following induction of anoxia. The results of Hansen et al. (1988) also showed that the drug failed to alter the Ca_e^{2+} response to hypoglycemia. Do these results show that the ionic channels activated during the two conditions are different, as implied by Hansen et al. (1988)? My tentative conclusion would be another one. Thus, the results from the two laboratories demonstrate that the calcium influx associated with glutamate activation of calcium channels is required to *elicit* SD. However, once SD is elicited, the same type of channels may be activated as in hypoglycemia. The difference may then simply be that hypoglycemia bereaves the cells of the major part of their energy source, thus preventing ion fluxes which promote repolarization. For example, provided that ATP is available, repolarization may occur by the operation of an electrogenic Na^+/K^+ pump, or by efflux of K^+, triggered by Ca^{2+} or other activators of K^+ channels.

4 Concluding Remarks

The results discussed are, at least at first glance, difficult to conceptualize. What is particularly hard to understand is that although dihydropyridine antagonists and NMDA antagonists block or curtail the calcium influx into isolated neurons, which is elicited by depolarization and NMDA agonists, respectively, neither of these drugs seem to affect calcium influx during anoxia or hypoglycemia in vivo. Furthermore, it appears odd that NMDA antagonists hinder the initiation of spreading depression although that condition is accompanied by ionic shifts similar to those observed in anoxia or hypoglycemia.

It is my suggestion that the seemingly discordant results can be explained on the basis of the following two assumptions. First, the rapid ionic shifts occurring during anoxia, hypoglycemia and SD are caused by the opening of a large, unspecific ion conductance. Second, this conductance is activated when Ca_i^{2+} is increased, and reaches a certain limiting value.

There are precedences for both assumptions. Thus, Phillips and Nicholson (1979, see also Nicholson and Kraig 1981) concluded that the depolarization associated with SD allowed the indiscriminate passage into cells of anions below a certain critical size. Furthermore, it is now clear that many cells contain a non-specific cation channel, activated by Ca^{2+} (Partridge and Swandulla 1988).

Clearly, if these assumptions are valid one can predict that a very strong stimulus, such as electrical stimulation, application of KCl, or a stab wound, elicits an SD simply because it opens calcium channels, allowing Ca_i^{2+} to rise to the threshold value. One can envisage that NMDA antagonists block the initiation of SD by preventing Ca^{2+} influx through NMDA-gated channels. Presumably, activation of VSCCs does not suffice to raise Ca_i^{2+} to the limiting value in a situation with a preserved energy source for calcium extrusion.

In hypoglycemia, the situation is different since the impediment of energy production, when sufficiently severe, cause dissipative ion fluxes which quickly exhaust the cellular energy stores. In this situation, therefore, Ca_i^{2+} is sooner or later apt to rise to the limiting value, even if the NMDA-gated calcium channels are blocked. One can envisage that since the brain during hypoglycemia continues to produce some ATP, partial recovery of the perturbation of ion homeostasis is possible. However, once the large, unspecific ion conductance is activated, the ionic leak may be large enough to waste available energy stores, creating a vicious cycle. It is also easy to understand that since anoxia completely interrupts energy production and causes massive depolarisation, and since calcium influx occurs by multiple pathways, NMDA antagonists alone do not affect calcium influx.

References

Abercrombie RF, Hart CE (1986) Calcium and proton buffering and diffusion in isolated cytoplasm from Myxicola axons. Am J Physiol 250 (Cell Physiol 19):C391–C405

Alberti KGMM, Cuthbert C (1982) The hydrogen ion in normal metabolism: a review. In: Porter R, Lawrenson G (eds) Metabolic acidosis, CIBA Found Symp '87. Pitman, London, pp 1–19

Astrup J (1982) Energy-requiring cell functions in the ischemic brain. J Neurosurg 56:482–497

Astrup J, Norberg K (1976) Potassium activity in cerebral cortex in rats during progressive severe hypoglycemia. Brain Res 103:418–423

Astrup J, Blennow G, Nilsson B (1979) Effects of reduced cerebral blood flow upon EEG pattern, cerebral extracellular potassium, and energy metabolism in the rat cortex during bicuculline-induced seizures. Brain Res 177:115–126

Astrup J, Rehncrona S, Siesjö BK (1980) The increase in extracellular potassium concentration in the ischemic brain in relation to the preischemic functional activity and cerebral metabolic rate. Brain Res 199:161–174

Astrup J, Siesjö BK, Symon L (1981) Thresholds in cerebral ischemia – the ischemic penumbra. Stroke 12(6):723–725

Auer R, Hall P, Ingvar M, Siesjö BK (1986) Hypotension as a complication of hypoglycemia leads to enhanced energy failure but no increase in neuronal necrosis. Stroke 17(3):442–449

Benveniste H, Brejer J, Schousboe A, Diemer NH (1984) Elevation of the extracellular concentration of glutamate and aspartate in rat hippocampus during transient cerebral ischemia monitored by intracerebral microdialysis. J Neurochem 43:1369–1374

Boris-Möller F, Drakenberg T, Elmdén K, Forsén S, Siesjö BK (1988) Evidence against major compartmentalization of H^+ in ischemic rat brain tissue. Neurosci Lett 85:113–118

Boron WF (1983) Topical review: transport of H^+ and ionic weak acids and bases. J Membrane Biol 72:1–16

Busa WB, Nuchitello R (1984) Metabolic regulation via intracellular pH. Am J Physiol 246 (Regul Integr Comp Physiol 15):R409–R438

Caldwell PC (1956) Intracellular pH. In: Bourne GH, Danielli JF (eds) International review of cytology, vol 5. Academic Press, New York London, pp 229–277

Chapman AG, Westerberg E, Siesjö BK (1981) The metabolism of purine and pyrimidine nucleotides in rat cortex during insulin-induced hypoglycemia and recovery. J Neurochem 36:179–189

Collingridge GJ (1985) Long-term potentiation in the hippocampus: mechanisms of initiation and modulation by neurotransmitters. TIPS 6:407–411

Cull-Candy SG, Usowicz MM (1987) Multiple-conductance channels activated by excitatory amino acids in cerebellar neurons. Nature (London) 325:525 ff

Endres W, Ballanyi K, Serve G, Grafe P (1986) Excitatory amino acids and intracellular pH in motoneurons of the isolated frog spinal cord. Neurosci Lett 72:54–58

Fagg GE (1985) L-glutamate, excitatory amino acid receptors and brain function. TINS 8:207–210

Fagg GE, Foster AC, Ganong AH (1986) Excitatory amino acid synaptic mechanisms and neurological function. TIPS 9:357–363

Foster AC, Fagg GE (1988) Taking apart NMDA receptors. Nature (London) 329:395–396

Goroleva NA, Koroleva VI, Amemori T, Pavlik V and Bures J (1987) Ketamine blockade of cortical spreading depression in rats. Electroencephalogr Clin Neurophysiol 66:440–447

Grinstein S and Rothstein A (1986) Mechanisms of regulation of the Na^+/H^+ exchanger: topical review. J Membrane Biol 90:1–12

Hansen AJ (1978) The extracellular potassium concentration in brain cortex following ischemia in hypo- and hyperglycemic rats. Acta Physiol Scand 102:324 ff

Hansen AJ (1985) Effect of anoxia on ion distribution in the brain. Physiol Rev 65:101–147

Hansen AJ, Nordström C-H (1979) Brain extracellular potassium and energy metabolism during ischemia in juvenile rats after exposure to hypoxia for 24 h. J Neurochem 32:915–920

Hansen AJ, Hounsgaard J and Jahnsen H (1982) Anoxia increases potassium conductance in hippocampal nerve cells. Acta Physiol Scand 115:301–310

Hansen AJ, Lauritzen M, Wieloch T (1988) NMDA antagonists inhibit cortical spreading depression, but not anoxic depolarization. Acta Physiol Scand 132:6

Harris RJ, Symon L (1984) Extracellular pH, potassium, and calcium activities in progressive ischaemia or rat cortex. J Cerebr Blood Flow Metabol 4:178–186

Harris RJ, Symon L, Branston NM, Bayhan M (1981) Changes in extracellular calcium activity in cerebral ischaemia. J Cerebr Blood Flow Metabol 1:203–209
Harris RJ, Wieloch TW, Symon L, Siesjö BK (1984) Cerebral extracellular calcium activity in severe hypoglycemia: relation to extracellular potassium and energy state. J Cerebr Blood Flow Metabol 4:187–193
Heinemann U, Lux HD, Gutnick MJ (1977) Extracellular free calcium and potassium during paroxysmal activity in the cerebral cortex of the cat. Exp Brain Res 27:237–243
Hochachka PW, Mommsen TP (1983) Protons and anaerobiosis. Science 219:1391–1397
Höller M, Dierking H, Dengler K, Tegtmeier F, Peter T (1986) Effect of flunarizine on extracellular ion concentration in the rat brain under hypoxia and ischemia. In: Battistini N, Fiorani P, Courbier R, Plum F, Fieschi C (eds) Acute brain ischemia medical and surgical therapy, vol 32. Raven, New York, pp 229–236
Jahr CE, Stevens CF (1987) Glutamate activities multiple single channel conductances in hippocampal neurons. Nature (London) 325:522–525
Kaila R, Voipio J (1987) Postsynaptic fall in intracellular pH induced by GABA-activated bicarbonate conductance. Nature (London) 330:163–165
Kraig RP, Ferreira-Filho CH, Nicholson C (1983) Alkaline and acid transients in cerebellar microenvironment. J Neurophys 49:831–850
Kraig RP, Pulsinelli WA, Plum F (1985) Hydrogen ion buffering during complete brain ischemia. Brain Res 342:281–290
Kraig RP, Pulsinelli WA, Plum F (1986) Carbonic acid buffer changes during complete brain ischemia. Am J Physiol 250 (Regul Integr Comp Physiol 19(:R348–R357
Krnjevic K (1975) Coupling of neuronal metabolism and electrical activity. In: Ingvar DH, Lassen NA (eds) Brain work. Alfred Benzon Symp VIII. Munksgaard, Copenhagen
Kudo Y, Oguro A (1986) Glutamate-induced increase in intracellular Ca^{2+} concentration in isolated hippocampal neurones. Br J Pharmacol 89:191–198
Lauritzen M (1987) Cortical spreading depression as a putative migraine mechanism. TINS 1:8–13
Mayer M (1987) Two channels reduced to one. Nature 325:480–481
Mayer ML, Westbrook GL (1987) Cellular mechanisms underlying excitotoxicity. TINS 2(10):59–61
Meldrum BS (1983) Metabolic factors during prolonged seizures and their relation to nerve cell death. In: Delgado-Escueta AV, Waterlain CG, Treiman DM, Porter RJ (eds) Advances in neurology, vol 34: Status epilepticus. Raven, New York, pp 261–275
Miess G, Paschen W (1984) Regional changes of blood flow, glucose, and ATP content determined on brain sections during a single passage of spreading depression in rat brain cortex. Exp Neurol 84:249–258
Miller RJ (1987) Multiple calcium channels and neuronal function. Science 235:46–52
Murphy SN, Thayer SA, Miller RJ (1988) The effects of excitatory amino acids on intracellular calcium in single mouse striatal neurons in vitro. J Neurosci (in press)
Mutch WA, Hansen AJ (1984) Extracellular pH changes during spreading depression and cerebral ischemia: mechanisms of brain pH regulation. J Cerebr Blood Flow Metabol 4:17–27
Nedergaard M (1988) Mechanisms of brain damage in focal cerebral ischemia. Acta Neurol Scand 77:3–24 (Suppl)
Nicholson C (1979) Brain-cell microenvironment as a communication channel. In: Schmitt FO, Worden FG (eds) The neurosciences: 4th study program. MIT, Cambridge, pp 457–471
Nicholson C (1980) Measurement of extracellular ions in the brain. Trends Neurosci 3:216–218
Nicholson C, Kraig RP (1981) The behaviour of extracellular ions during spreading depression. In: Zeuthen T (ed) The application of ion-selective microelectrodes. Elsevier/North Holland, Biomedical Press, Amsterdam, pp 217–238
Nilsson B, Norberg K, Nordström C-H, Siesjö BK (1975) Rate of energy utilization in the cerebral cortex of rats. Acta Physiol Scand 93:569–571
Nordström C-H, Siesjö BK (1978) Effects of phenobarbital in cerebral ischemia. Pt I: Cerebral energy metabolism during pronounced incomplete ischemia. Stroke 9:327–335
Nowak L, Bregestovski P, Ascher P, Herbert A, Prochiantz A (1984) Magnesium gates glutamate-activated channels in mouse central neurones. Nature (London) 307:462–465
Nowak L, Ascher P, Berwald-Netter Y (1987) Ionic channels in mouse astrocytes in culture. J Neurosci 7:101–109
Nowycky MCAP, Tsien RW (1985) Three types of neuronal calcium channel with different calcium agonist sensitivity. Nature (London) 316:440–443

Noma A (1983) ATP-regulated K^+ channels in cardiac muscle. Nature (London) 305:147–148
Partridge LD, Swandulla D (1988) Calcium-activated non-specific cation channels. TINS 11(2):69–72
Pelligrino D, Almquist L-O, Siesjö BK (1981) Effects of insulin-induced hypoglycemia on intracellular pH and impedance in the cerebral cortex of the rat. Brain Res 221:129–147
Peters T (1986) Calcium in physiological and pathological cell function. Eur Neurol 25 (Suppl 1):27–44
Phillips JM, Nicholson C (1979) Anion permeability in spreading depression investigated with ion-sensitive microelectrodes. Brain Res 173:567–571
Pumain R, Heinemann U (1985) Stimulus- and amino acid-induced calcium and potassium changes in rat neocortex. J Neurophysiol 53(1):1–16
Raichle M (1983) The pathophysiology of brain ischemia. Ann Neurol 13:2–10
Ross A, Boron WF (1981) Intracellular pH. Physiol Rev 61(2):296–434
Rothman SM, Olney JW (1986) Glutamate and the pathophysiology of hypoxic-ischemic brain damage. Ann Neurol 19:105–111
Siesjö BK (1978) Brain energy metabolism. John Wiley & Sons, Chichester New York
Siesjö BK (1981) Cell damage in the brain: a speculative synthesis. J Cerebr Blood Flow Metabol 1:155–185
Siesjö BK (1984) Cerebral circulation and metabolism (Review Article) J Neurosurg 60:883–908
Siesjö BK (1985) Acid-base homeostasis in the brain: physiology, chemistry, and neurochemical pathology. In: Kogure K, Hossmann K-A, Siesjö BK, Welsh FA (eds) Progress in brain research, vol 63. Elsevier, New York, pp 121–154
Siesjö BK (1988a) Historical overview. Calcium, ischemia, and death of brain cells. Ann NY Acad Sci 522:638–661
Siesjö BK (1988b) Acidosis and ischemic brain damage. Neurochem Pathol (in press)
Siesjö BK (1988c) Hypoglycemia, brain metabolism, and brain damage. Diabetes/Metabol Rev 4(2):113–144
Siesjö BK, Agardh C-D (1983) Hypoglycemia. In: Lajtha A (ed) Handbook of neurochemistry, vol 3. Metabolism in the nervous system, 2nd edn. Plenum Press, New York London, pp 353–379
Siesjö BK, Wieloch T (1985) Cerebral metabolism in ischaemia: neurochemical basis for therapy. Br J Anaesth 57:47–62
Siesjö BK, Wieloch T (1986) Epileptic brain damage: pathophysiology and neurochemical pathology. In: Delgado-Escueta AV, Ward AA Jr., Woodbury DM, Porter RJ (eds) Advances in neurology, vol 44. Raven, New York, pp 813–847
Siesjö BK, Zwetnow NN (1970) The effect of hypovolemic hypotension on extra- and intracellular acid-base parameters and energy metabolites in the rat brain. Acta Physiol Scand 79:114–124
Siesjö BK, Ingvar M, Wieloch T (1986) Cellular and molecular events underlying epileptic brain damage. Ann NY Acad Sci 462:207–223
Smith M-L, von Hanwehr R, Siesjö BK (1986) Changes in extra- and intracellular pH in the brain during and following ischemia in hyperglycemic and in moderately hypoglycemic rats. J Cerebr Blood Flow Metabol 6:574–583
Sokoloff L (1981) Relationships among local functional activity, energy metabolism, and blood flow in the central nervous system. Fed Proc 40:2311–2316
Somjen GG (1979) Extracellular potassium in the mammalian central nervous system. Annu Rev Physiol 41:159–177
Thomas RC, Meech RW (1982) Hydrogen ion currents and intracellular pH in depolarized voltage-clamped snail neurones. Nature (London) 299:826–827
van Harreveld A (1966) Brain tissue electrolytes. Butterworths, London
van Harreveld A (1970) A mechanism for fluid shifts specific for the central nervous system. In: Wycis H (ed) Current research in neurosciences, vol 10. Topical problems of psychiatry and neurology. Karger, Basel New York, pp 62–70
von Hanwehr R, Smith M-L, Siesjö BK (1986) Extra- and intracellular pH during near-complete forebrain ischemia in the rat. J Neurochem 46:331–339
Wieloch T, Harris RJ, Symon L, Siesjö BK (1984) Influence of severe hypoglycemia on brain extracellular calcium and potassium activities, energy, and phospholipid metabolism. J Neurochem 43:160–168
Zhang E, Lauritzen M, Wieloch T, Hansen AJ (1989) Calcium movements in brain during failure of energy metabolism. In: Hartmann A, Kuschinsky W (eds) Cerebral ischemia and calcium. (in press)

3 Mitochondrial Genetic Diseases

Mitochondrial Diseases

S. DIMAURO[1], A. LOMBES, H. NAKASE, M. ZEVIANI, R. RIZZUTO, C. T. MORAES and E. A. SCHON

1 Historical Considerations

The very concept of a mitochondrial disease was introduced in 1962 when Luft, Ernster, and workers described a young Swedish woman with severe hypermetabolism not due to thyroid dysfunction but rather to "loose coupling" of muscle mitochondria (Luft et al. 1962). Although the work of Luft et al. had shown that elegant biochemical investigations could be performed in human muscle biopsies, in the decade that followed the attention of clinical researchers was directed to morphological changes of muscle mitochondria. Ultrastructural studies of muscle, pioneered by N. K. Gonatas and G. M. Shy (Shy and Gonatas 1964; Shy et al. 1966), led to the description of "pleoconial myopathy" (excessive number of normal-looking mitochondria) and "megaconial myopathy" (accumulation of giant mitochondria), and refinement of pre-existing histochemical techniques, especially a modification of the Gomori trichrome stain (Engel and Cunningham 1963), allowed the recognition of abnormal proliferation of mitochondria in affected fibers ("ragged red fibers", RRF). Useful as they were, and still are, in the detection of patients, morphological criteria are not adequate for a rational classification of "mitochondrial myopathies" for several reasons: (1) mitochondrial changes are not specific for primary mitochondrial diseases but are also seen in other neuromuscular disorders in which they presumably represent secondary changes; (2) The different ultrastructural changes of mitochondria (excessive number; increased size; bizarre shape; abnormal cristae; paracrystalline or osmophilic inclusions) do not distinguish different mitochondrial diseases; (3) morphological changes are not necessarily present in primary mitochondrial diseases [for instance, they are not seen in carnitine palmitoyltransferase (CPT) deficiency or in Leigh syndrome].

Starting in the early 1970s, biochemical studies of mitochondria isolated from human muscle biopsies or from postmortem tissues were performed in an increasing number of patients with suspected mitochondrial diseases (Spiro et al. 1970; French et al. 1972; DiMauro et al. 1973). Various techniques were applied, including studies of oxidation and phosphorylation of different substrates by polarography, analysis of cytochromes by reduced-minus-oxidized spectra, and measurement of individual mitochondrial enzymes or portions of the respiratory

H. Houston Merritt Clinical Research Center for Muscular Dystrophy and Related Diseases, College of Physicians and Surgeons, Columbia University, New York, NY 10032, USA
[1]To whom correspondence should be addressed

A. Azzi, Z. Drahota and S. Papa
Molecular Basis of Membrane-Associated Diseases

chain. These investigations helped to identify the site of the defect in mitochondrial metabolism. Recently, purification of individual enzymes or enzyme complexes and the availability of antibodies against holoenzymes and enzyme subunits have allowed the exploration of biochemical defects at the molecular level through various immunochemical (ELISA, Western blot) and immunocytochemical techniques. As a result of these studies, a rational biochemical classification of the mitochondrial diseases is finally taking shape (see below).

The functional consequences of mitochondrial defects in muscle have been explored by conventional standardized exercise tests and by ^{31}P nuclear magnetic resonance (NMR) spectroscopy. This non-invasive technique allows us to follow the changes in pH and in the concentration of crucial energy compounds (ATP, AMP, Pi) during exercise (Argov et al. 1987; Duboc et al. 1987). Characteristic abnormal patterns have been identified in mitochondrial myopathies, which differ from those seen in other metabolic myopathies with impaired energy metabolism, such as some glycogenoses (see Argov, this Vol.).

Molecular genetic techniques are now being applied to the study of mitochondrial diseases. The human mitochondrial DNA (mtDNA) was sequenced in 1981 (Anderson et al. 1981) and molecular probes for the mitochondrial genes are available. Southern analysis of muscle mtDNA has recently revealed major deletions in some patients with mitochondrial myopathies (Holt et al. 1988), a finding confirmed in our laboratory (Zeviani et al. 1988b). An increasing number of nuclear genes encoding mitochondrial proteins are being isolated and sequenced, and cDNA probes are being used to identify molecular genetic defects in patients through Southern and Northern analysis and in situ hybridization of affected tissues.

While the astounding progress in our understanding of the biochemical causes of mitochondrial diseases has been largely due to the application of concepts and techniques developed by basic scientists, conversely, human pathology offers unique biological "models" in which basic concepts can be verified. This is the reason for the convergence of interests between "basic" and "clinical" scientists which has taken place in recent years and which has fostered joint meetings such as the present one.

2 Clinical Considerations

The distinct symptomatology of Luft disease, so neatly explained by the loose coupling found in muscle mitochondria, set a deceiving precedent. The many patients with mitochondrial myopathy (defined by the presence of morphologically abnormal mitochondria in the muscle biopsy) who were described after Luft's report, had heterogenous and non-specific clinical presentations. Weakness was the predominant feature, but this could be congenital or late in onset, mild or severe, static or progressive, and with varied distribution. The only clinical feature that was frequently associated with the presence of RRF in limb muscles was progressive external opthalmoplegia (PEO).

It was soon realized that RRF were often seen in patients in whom myopathy was neither the only nor the predominant clinical feature: these patients had multisystem disorders often dominated by brain disease and accompanied by lactic acidosis. The term "mitochondrial encephalomyopathy" was introduced to distinguish these disorders from the pure myopathies (Shapira et al. 1977). The "splitters" among clinicians thought that three of these disorders were distinctive enough to be identified as separate syndromes (Berenberg et al. 1977; Rowland et al. 1983; Fukuhara 1983; Pavlakis et al. 1984; DiMauro et al. 1985). The first, Kearns-Sayre syndrome (KSS) is characterized by the invariant triad: onset before age 20, PEO, pigmentary retinal degeneration, and at least one of the following: complete heart block, CSF protein above 100 mg/dl, cerebellar syndrome. The second syndrome, myoclonus epilepsy with ragged red fibers (MERRF), includes myoclonus, ataxia, and epilepsy. The third syndrome, bearing the acronym MELAS (mitochondrial encephalomyopathy with lactic acidosis, and stroke-like episodes), is characterized by onset in childhood, intermittent vomiting, proximal limb weakness, and episodes resembling vascular accidents. This classification did not sit well with the "lumpers", who regarded all mitochondrial encephalomyopathies as manifestations of a continuous spectrum of symptoms and signs, and stressed common rather than distinctive features, such as short stature, dementia, hearing loss, lactic acidosis (Petty et al. 1986). Although clinical classifications are not precise and occasional patients do share characteristics of two syndromes, recent biochemical and genetic data bolster the position of the "splitters". From a genetic point of view, all patients with KSS are "simplex" cases, with no other affected individuals in the family. On the other hand, MERRF seems to be transmitted by non-Mendelian, maternal inheritance, suggesting a genetic defect of mtDNA (Rosing et al. 1985). Recent findings of mtDNA deletions in patients with KSS but not in patients with MERRF or MELAS further strengthens the concept that KSS is a separate entity (Zeviani et al. 1988b). From the biochemical point of view, there is increasing evidence that MELAS is due to complex I deficiency (Kobayashi et al. 1987; Nishizawa et al. 1987; Koga et al. 1989), which would differentiate this syndrome from the other two not only clinically but also biochemically.

3 Biochemical Considerations

Mitochondrial diseases can be classified into five broad categories based on the affected area of mitochondrial metabolism (Morgan-Hughes 1986; DiMauro et al. 1987a). *Defects of mitochondrial transport* are best exemplified by CPT deficiency, which impairs transfer of long-chain fatty acids from the cytosol into the mitochondrial matrix. In muscle, CPT deficiency probably results in acute energy shortage during prolonged exercise and fasting, situations in which long-chain fatty acid oxidation is a crucial source of energy. In fact, patients experience muscle breakdown and myoglobinuria with prolonged exertion, especially if associated with fasting (DiMauro and Papadimitriou 1986).

Defects of substrate utilization are exemplified by pyruvate dehydrogenase complex (PDHC) deficiency and by defects in the β-oxidation pathway. The PDHC is composed of five enzymes, two of which, a kinase and a phosphatase, regulate the activity of the complex by phosphorylation (deactivation) and dephosphorylation (activation) of the first enzyme component (El). Syndromes of varying severity are associated with PDHC deficiency, from intermittent ataxia to congenital lactic acidosis, failure to thrive, developmental delay, hypotonia, and death in infancy (Stansbie et al. 1986).

Defects of the Krebs cycle are best illustrated by fumarase deficiency, a recently described disease causing severe infantile encephalomyopathy and early death (Zinn et al. 1986). The laboratory hallmark of the disease is the excretion of fumaric and succinic acid in the urine. Both mitochondrial and cytosolic isozymes of fumarase are decreased and the defect appears to be generalized.

The best example of *defects of oxidation-phosphorylation coupling* remains Luft disease, an exceedingly rare condition (only two patients have been described) characterized by severe hypermetabolism of non-thyroid origin (Luft et al. 1962; DiMauro et al. 1976). In both patients, isolated muscle mitochondria showed a maximal respiratory rate even in the absence of ADP, an indication that respiratory control was lost. Studies of calcium transport by isolated mitochondria suggested a vicious cycling of calcium between mitochondria and cytosol, but the molecular abnormality remains unknown.

Defects of the respiratory chain have been the subject of many studies in recent years.

Defects of complex I (NADH-coenzyme Q reductase) cause two main syndromes, a pure myopathy with exercise intolerance and myalgia, or an encephalomyopathy. The encephalomyopathy can occur in infancy with severe lactic acidosis, hypotonia, seizures, respiratory insufficiency and early death, or it may have later onset (childhood or adulthood) and a less severe course. As mentioned above, complex I deficiency has been documented in numerous patients with MELAS and it appears to be the most important if not the exclusive cause of this syndrome.

In an infant with fatal infantile generalized complex I deficiency, the biochemical defect was further localized between FMN and CoQ, and electron paramagnetic resonance (EPR) of liver submitochondrial particles showed almost complete absence of the iron-sulphur clusters of complex I (Moreadith et al. 1984). Immunological studies using antibodies against the whole complex or against the four main subunits of the iron-protein fragment showed selective absence of two polypeptides (Moreadith et al. 1987).

Defects of complex III also cause two main syndromes, a myopathy and an encephalomyopathy. Immunological studies showed that antibodies against complex III purified from beef heart cross-reacted with four subunits of human muscle complex III, representing core protein 1 and 2, peptide VI, and cytochrome *c*. Using these antibodies in Western blots of muscle mitochondrial extracts from a patient with complex III deficiency myopathy, Darley-Usmar et al. showed that core proteins and peptide VI were decreased, whereas cytochrome *c* was present in a normal amount (Darley-Usmar et al. 1983).

4 Cytochrome *c* Oxidase (COX) Deficiency

We are considering the defects of complex IV (cytochrome *c* oxidase) separately and in more detail because these disorders have been the subject of much work in our laboratory (DiMauro et al. 1988a,b). COX, the last component of the respiratory chain, catalyzes the transfer of reducing equivalents from cytochrome *c* to molecular oxygen, utilizing the energy liberated in the reaction to sustain a transmembrane proton-pumping activity (Kadenbach et al. 1987). The holoenzyme contains as redox centers two copper atoms and two unique heme A iron porphyrins bound to a multisubunit protein frame embedded in the mitochondrial inner membrane. In mammals, the apoprotein is composed of 13 subunits, the three larger of which (subunits I to III) are associated with the prosthetic group and perform both catalytic and proton-pumping activities. They are encoded by mtDNA and are synthesized in mitochondria. The ten smaller subunits (IV, Va, Vb, VIa, VIb, VIc, VIIa, VIIb, VIIc, VIII, nomenclature of Kadenbach et al. 1983) are encoded by nuclear DNA and synthesized on cytoplasmic ribosomes. The functions of the nuclear-encoded subunits have not been fully clarified, but a regulatory role has been proposed by Kadenbach and co-workers, who have also shown that some of these subunits occur as tissue-specific isoforms (Kuhn-Nentwig and Kadenbach 1985; Kadenbach et al. 1987).

As seen with defects of complex I and complex III, the clinical phenotypes of COX deficiency also fall into two main categories, one characterized by myopathy, the other dominated by brain disease. In the first group, the most common disorder is *fatal infantile myopathy* which causes generalized weakness, respiratory insufficiency, and death before 1 year of age. There is lactic acidosis and, often, renal dysfunction with glycosuria, phosphaturia and aminoaciduria. In these patients, COX deficiency is confined to skeletal muscle, sparing heart, liver and brain (Bresolin et al. 1985): this finding supports the notion of tissue-specific isozymes. The amount of immunologically reactive enzyme protein was markedly decreased in muscle extracts or isolated mitochondria when tested by ELISA or by immunocytochemistry of frozen sections. However, the subunit composition of the mutant COX appeared normal when examined by SDS-PAGE after immunoprecipitation (Bresolin et al. 1985). In contrast to this invariably fatal condition, some children with severe myopathy and lactic acidosis at birth improve spontaneously and are virtually normal by age 2 or 3 years. Serum lactate and muscle biopsy also return to normal in this *benign infantile mitochondrial myopathy*, which is due to a reversible defect of COX (DiMauro et al. 1983). The enzyme activity is markedly decreased in muscle biopsies taken soon after birth, but returns to normal in the first year of life. The reversibility of the enzyme defect in these patients suggests that the genetic defect may affect a COX subunit which is not only tissue-specific but also developmentally regulated. A mutation of a fetal or neonatal isozyme would correct spontaneously when the mature isozyme begins to be expressed.

There are two encephalomyopathies associated with COX deficiency: Leigh syndrome and MERRF. *Leigh syndrome* (subacute necrotizing encephalomyelopathy) is characterized by psychomotor regression, brainstem abnormalities, and seizures. Neuropathology consists of focal, symmetrical, necrotic lesions extending from thalamus to pons and involving the inferior olives and the posterior columns of the spinal cord. COX deficiency has been reported in about 15 patients with Leigh syndrome, including two sets of siblings (Miyabayashi et al. 1983; Arts et al. 1987b). We have seen two more sets of affected siblings, both from consanguineous parents, strongly suggesting autosomal recessive transmission. The enzyme defect is generalized but partial, with residual COX activities ranging from about 15% of the normal mean in muscle to 35% in brain (DiMauro et al. 1987). The residual COX shows normal kinetics and normal subunit composition by Western blot.

COX deficiency was also generalized and partial in two unrelated patients with *MERRF syndrome* (Mendell et al. 1987; Berkovic et al. 1987). In both cases, family history was compatible with maternal transmission. We studied many postmortem tissues from one of these patients. Residual COX activity varied in different tissues and even in different samples from the same tissue. This is in keeping with a biochemical error due to a mutation of mtDNA, which ought to be generalized but is variably expressed in different tissues, depending on the relative proportion of normal and mutant mtDNA in each tissue. The Km for cytochrome *c* was abnormally low, and immunological studies showed a decrease of the holenzyme but a disproportionately greater decrease of the mtDNA-encoded subunit II than of the nuclear-encoded subunit IV. However, Northern and Southern analyses showed that the gene for subunit II had a normal size and was normally transcribed. Taken together, these data suggest a point mutation or a very small deletion of the mitochondrial gene encoding COX II (Lombes et al. 1989).

5 COX-Deficiencies: Genetic Considerations

The large number of subunits, some encoded by nuclear DNA, others by mtDNA, and the complexity of post-translational events required for the import and correct assembly of nuclear-encoded subunits explain how inherited COX deficiencies may be due to many different genetic errors. At least three mechanisms can be considered: (1) Alterations of transcription or translation of mtDNA-encoded polypeptides (I to III); (2) alterations of transcription or translation of nuclear-encoded polypeptides (IV to VIII); (3) alterations of the post-translational processing of the cytoplasmically synthesized precursors of subunits IV to VIII. Interaction with mitochondrial receptors, translocation, and proteolytic processing of these precursors are indispensable for correct assembly in the inner mitochondrial membrane (Pfanner et al. 1988). Diseases due to mechanisms 2 or 3 ought to be transmitted by Mendelian inheritance, while diseases due to mechanism 1 should be transmitted by maternal inheritance.

To define the molecular genetic basis for the COX-deficiency disorders, we are in the process of cloning the cDNAs encoding the ten nuclear-encoded subunits. We have already isolated and characterized four full-length cDNAs of human COX-encoding subunits IV, Va, Vb, and VIII (Zeviani et al. 1987, 1988a; Rizzuto et al. 1988). We have begun to use these cDNAs, together with those already available, encoding COX I to III, as probes to analyze the DNA and RNA of patients with the two generalized forms of COX deficiency, Leigh and MERRF syndromes. In a patient with Leigh syndrome, we performed a transcriptional analysis of RNA derived from brain tissue. Relative to β-actin (used as an internal standard), the steady-state transcriptional level of the patient's nuclear genes was similar to control, but there was a significant decrease in the patient's steady-state mitochondrial transcription (not only for COX I to III, but also for a non-COX gene, ATPase 6). Thus, the ratio of mitochondrial: nuclear COX transcription was markedly reduced in the patient while the total amount of mtDNA estimated by Southern analysis in the patient's tissue was the same as in the control. These results point to a regulatory error in overall mitochondrial transcription rather than to a specific error in the structural genes of COX (Nakase et al. 1988).

An unexpected development has occurred in our understanding of KSS, because we have found major deletions of mtDNA in muscle biopsies of seven patients with KSS, but not in patients with other mitochondrial encephalomyopathies, including MERRF and MELAS (Zeviani et al. 1988b). The deletions varied in size from 2.0 to 7.0 Kb and did not localize to any single region of the mitochondrial genome. The proportion of deleted genomes in each KSS patient ranged from 45 to 75% of total mtDNA. There was no direct correlation between the size or site of the deletions, biochemical abnormality (partial COX deficiency is often seen in KSS), or clinical severity. These data bolster arguments that KSS is a unique disorder and genetic in origin, despite its sporadic occurrence. These findings raise a number of interesting questions both at a basic and clinical level: (1) Are there "hot spots" in the mitochondrial genome that are more susceptible to deletion?; (2) What is the actual mechanism of deletion?; (3) Are mtDNA deletions expressed in tissues other than muscle?; (4) If deletions are expressed in non-muscle tissues, is the relative proportion of deleted vs non-deleted mtDNAs different in clinically unaffected tissues such as liver?

In their seminal paper, Holt et al. (1988) found no deletions in mtDNA from leukocytes of patients with deletions in muscle mtDNA, and they attributed this pattern to the frequent cell division of leukocyte precursors, allowing selection against the survival of cells with deleted mtDNAs. If deletions are expressed in fibroblasts from patients with KSS, then we will investigate the "selection hypothesis" experimentally by following the relative number of deleted mtDNAs in subsequent cell generations in vitro.

The deletions could arise from either of two mechanisms: (1) inheritance of a pre-existing deletion in the mother, or (2) spontaneous mutation of mtDNA in the oocyte or in the zygote. Against the first possibility, and in support of the second, is the absence of affected mothers or siblings. We will obtain muscle biopsies from mothers of patients with typical KSS and investigate the presence or absence of mtDNA deletions. Lack of deletions would further strengthen the hypothesis that mtDNA deletions arose by new mutation in the oocyte or in the zygote.

Acknowledgements. Some of the work discussed was supported by Center Grants NS 11766 from the National Institute of Neurological and Communicative Disorders and Stroke, and from the Muscular Dystrophy Association, and by a donation from Dr. and Mrs. Libero Danesi.

Drs. Lombes and Rizzuto are supported by Fellowships from the Muscular Dystrophy Association, and Dr. Moraes by a Fellowship from the Brazilian Research Council (CNPq). We thank Mrs. Shirley Susarchick for typing the manuscript.

References

Anderson S, Bankier AT, Barrell BG, DeBruijn MHL, Coulson AR, Drouin J, Eperson IC, Nierlich DP, Roe BA, Sariger F, Schreier PH, Smith AJH, Staden R, Young IG (1981) Sequence and organization of the human mitochondrial genome. Nature (London) 290:457–465

Argov Z, Bank WJ, Maris J, Peterson P, Chance B (1987) Bioenergetic heterogeneity of human mitochondrial myopathies: phosphorus magnetic resonance spectroscopy study. Neurology 37:257–262

Arts WFM, Scholte HR, Loonen MCB, Przyrembel J, Fernandes J, Trijbels JMF, Luft-Houwen SEM (1987) Cytochrome *c* oxidase deficiency in subacute necrotizing encephalomyelopathy. J Neurol Sci 77:103–115

Berenberg RA, Pellock JM, DiMauro S, Schotland DL, Bonilla E, Eastwood A, Hays AP, Vicale CT, Behrens M, Chutorian A, Rowland LP (1977) Lumping or splitting? "Ophthalmoplegia-plus" or Kearns-Sayre syndrome? Ann Neurol 1:37–54

Berkovic SF, Carpenter S, Karpati G, Andermann F, Andermann E, Shoubridge E, Arnold D (1987) Cytochrome *c* oxidase deficiency: a remarkable spectrum of clinical and neuropathological findings in a single family. Neurology 37:223 (Abstr)

Bresolin N, Zeviani M, Bonilla E, Miller RH, Leech RW, Shanske S, Nakagawa M, DiMauro S (1985) Fatal infantile cytochrome *c* oxidase deficiency: decrease of immunologically detectable enzyme in muscle. Neurology 35:802–812

Darley-Usmar VM, Kennaway NG, Buist NRM, Capaldi RA (1983) Deficiency in ubiquinone-cytochrome *c* reductase in a patient with mitochondrial myopathy and lactic acidosis. Proc Natl Acad Sci USA 80:5103–5106

DiMauro S, Papadimitriou A (1986) Carnitine palmitoyltransferase deficiency. In: Engel AG, Banker BQ (eds) Myology, vol 2. McGraw-Hill, New York, pp 1697–1708

DiMauro S, Schotland DL, Bonilla E, Lee CP, Gambetti PL, Rowland LP (1973) Progressive opthalmoplegia, glycogen storage, and abnormal mitochondria. Arch Neurol 29: 170–179

DiMauro S, Bonilla E, Lee CP, Schotland DL, Scarpa A, Conn H, Chance B (1976) Luft's disease: further biochemical and ultrastructural studies of skeletal muscle in the second case. J Neurol Sci 27: 217–232

DiMauro S, Nicholson JF, Hays AP, Eastwood AB, Papadimitriou A, Koenigsberger R, DeVivo DC (1983) Benign infantile mitochondrial myopathy due to reversible cytochrome *c* oxidase deficiency. Ann Neurol 14:226–234

DiMauro S, Bonilla E, Zeviani M, Nakagawa M, DeVivo DC (1985) Mitochondrial myopathies. Ann Neurol 17:521–538

DiMauro S, Bonilla E, Zeviani M, Servidei S, DeVivo DC, Schon EA (1987a) Mitochondrial myopathies. J Inherit Metab Dis 10 (Suppl 1):113–128

DiMauro S, Servidei S, Zeviani M, DiRocco M, DeVivo DC, DiDonato S, Uziel G, Berry K, Hoganson G, Johnsen SD, Johnson PC (1987b) Cytochrome *c* oxidase deficiency in Leigh syndrome. Ann Neurol 22:498–506

DiMauro S, Zeviani M, Servidei S, Prelle A, Miranda AF, Bonilla E, Schon EA (1988a) Biochemical and molecular aspects of cytochrome *c* oxidase deficiency. In: DiDonato S et al. (eds) Advances in neurology, vol 48. Raven, New York, pp 93–106

DiMauro S, Zeviani M, Rizzuto R, Lombes A, Nakase H, Bonilla E, Miranda A, Schon EA (1988b) Molecular defects in cytochrome oxidase in mitochondrial diseases. J Bioenerg Biomembr 20:353–364

Duboc D, Jehenson P, Tran Dinh S, Marsac C, Syrota A, Fardeau M (1987) Phosphorus NMR spectroscopy study of muscular enzyme deficiencies involving glycogenolysis and glycolysis. Neurology 37:664–671

Engel WK, Cunningham CC (1963) Rapid examination of muscle tissue: an improved trichrome stain method for fresh-frozen biopsy sections. Neurology 13:919–963

French JH, Sherard ES, Lubell H, Brotz M, Moore CL (1972) Trichopoliodystrophy. I. Report of a case and biochemical studies. Arch Neurol 26:229–244

Fukuhara N (1983) Myoclonus epilepsy and mitochondrial myopathy. In: Scarlato G, Cerri C (eds) Mitochondrial pathology in muscle diseases. Piccin Medical Books, Padova, pp 88–110

Holt IJ, Harding AE, Morgan-Hughes JA (1988) Deletions of mitochondrial DNA in patients with mitochondrial myopathies. Nature (London) 331:717–719

Kadenbach BL, Ungibauer M, Jarausch J, Buge U, Kuhn-Nentwig L (1983) The complexity of respiratory complexes. Trends Biochem Sci 8:398–400

Kadenbach BL, Kuhn-Nentwig L, Buge U (1987) Evolution of a regulatory enzyme: cytochrome *c* oxidase (complex IV). Curr Top Bioenerg 15:114–161

Kobayashi M, Morishita H, Sugiyama N, Yokochi K, Nakano M, Wada Y, Hotta Y, Terauchi A, Nonaka I (1987) Two cases of NADH-coenzyme Q reductase deficiency: relationship to MELAS syndrome. J Pediat 110:223–227

Koga Y, Nonaka I, Kobayashi M, Tojyo M, Nihei K (1988) Muscle pathology in complex I (NADH CoQ reductase) deficiency. Ann Neurol 24:749–756

Kuhn-Nentwig L, Kadenbach B (1985) Isolation and properties of cytochrome *c* oxidase from rat liver and quantification of immunological differences between isozymes from various rat tissues with subunit-specific antisera. Eur J Biochem 149:147–158

Lombes A, Mendell JR, Nakase H, Barohn RJ, Bonilla E, Zeviani M, Yates AJ, Omerza J, Gales TL, Nakahara K, Rizzuto R, Engel WK, DiMauro S (1989) Myoclonic epilepsy and ragged-red fibers with cytochrome oxidase deficiency: neuropathology, biochemistry, and molecular genetics. Ann Neurol (in press)

Luft R, Ikkos D, Palmieri G, Ernster L, Afzelius B (1962) A case of severe hypermetabolism of nonthyroidal origin with a defect in the maintenance of mitochondrial respiratory control: a correlated clinical, biochemical and morphological study. J Clin Invest 41:1776–1804

Mendell JR, Barohn RJ, Yates AJ, Omerza J, Gales TL, Lombes A, Bonilla E, Zeviani M, Engel WK, DiMauro S (1987) Autopsy case of myoclonic epilepsy and ragged red fibers with cytochrome *c* oxidase deficiency: evidence for a maternally inherited biochemical defect. Ann Neurol 22:128 (Abstr)

Miyabayashi S, Narisawa K, Tada K, Sakai K, Kobayashi K, Kobayashi Y (1983) Two siblings with cytochrome *c* oxidase deficiency. J Inherit Metab Dis 6:121–122

Moreadith RW, Batshaw ML, Ohnishi T, Kerr D, Knox B, Jackson D, Hruban R, Olson J, Reynafarje B, Lehninger AL (1984) Deficiency of the iron-sulfur clusters of mitochondrial reduced nicotinamide-adenine-dinucleotide-ubiquinone oxidoreductase (complex I) in an infant with congenital lactic acidosis. J Clin Invest 74:685–697

Moreadith RW, Cleeter MWJ, Ragan CI, Batshaw ML, Lehninger AL (1987) Congenital deficiency of two polypeptide subunits of the iron-protein fragment of mitochondrial complex I. J Clin Invest 79:463–467

Morgan-Hughes JA (1986) The mitochondrial myopathies. In: Engel AG, Banker BQ (eds) Myology, vol 2. McGrawHill, New York, pp 1709–1743

Nakagawa M, Miranda AF (1987) A monoclonal antibody against cytochrome *c* oxidase distinguishes cardiac and skeletal muscle mitochondria. Exp Cell Res 168:44–52

Nakase H, Zeviani M, Rizzuto R, DiMauro S, Schon EA (1988) Molecular genetic analysis of a patient with Leigh syndrome. Neurology 38:174 (Abstr)

Nishizawa M, Tanaka K, Shinozawa K, Kuwabana T, Atsumi T, Miyatake T, Ohama E (1987) A mitochondrial encephalomyopathy with cardiomyopathy. A case revealing a defect of complex I in the respiratory chain. J Neurol Sci 78:189–201

Pavlakis SG, Philipps PC, DiMauro S, DeVivo DC, Rowland LP (1984) Mitochondrial myopathy, encephalopathy, lactic acidosis and stroke-like episodes: a distinctive clinical syndrome. Ann Neurol 16:481–488

Petty RKH, Harding AE, Morgan-Hughes JA (1986) The clinical features of mitochondrial myopathy. Brain 109:915–938

Pfanner N, Hartl F, Neupert W (1988) Import of proteins into mitochondria: a multi-step process. Eur J Biochem 175:205–212

Rizzuto R, Nakase H, Zeviani M, DiMauro S, Schon EA (1988) Subunit Va of cytochrome *c* oxidase is highly conserved between human and bovine. Gene 69:245–256

Rosing HS, Hopkins LC, Wallace DC, Epstein CM, Weidenheim K (1985) Maternally inherited mitochondrial myopathy and myoclonic epilepsy. Ann Neurol 17:228–237

Rowland LP, Hays AP, DiMauro S, DeVivo DC, Behrens M (1983) Diverse clinical disorders associated with morphological abnormalities of mitochondria. In: Scarlato G, Cerri C (eds) Mitochondrial pathology in muscle diseases. Piccin Medical Books, Padova, pp 143–158

Shapira Y, Harel S, Russell A (1977) Mitochondrial encephalomyopathy: a group of neuromuscular disorders with defects in oxidative metabolism. Isr J Med Sci 13:161–164

Shy CM, Gonatas NK (1964) Human myopathy with giant abnormal mitochondria. Science 145:493–496

Shy CM, Gonatas NK, Perez M (1966) Two childhood myopathies with abnormal mitochondria: I. Megaconial myopathy. II. Pleoconial myopathy. Brain 89:133–158

Spiro AJ, Moore CL, Prineas JW, Strasberg PM, Rapin I (1970) A cytochrome-related inherited disorder of the nervous system and muscle. Arch Neurol 23:103–112

Stansbie D, Wallace SJ, Marsac C (1986) Disorders of the pyruvate dehydrogenase complex. J Inherit Metab Dis 9:105–119

Zeviani M, Nakagawa M, Herbert J, Lomax MI, Grossman LI, Sherbany AA, Miranda AF, DiMauro S, Schon EA (1987) Isolation of a cDNA clone encoding subunit IV of human cytochrome *c* oxidase. Gene 55:207–217

Zeviani M, Sakoda S, Sherbany AA, Nakase H, Rizzuto R, Samitt CE, DiMauro S, Schon EA (1988a) Sequence of cDNAs encoding subunit Vb of human and bovine cytochrome *c* oxidase. Gene 65:1–11

Zeviani M, Moraes CT, DiMauro S, Nakase H, Bonilla E, Schon EA, Rowland LP (1988b) Deletions of mitochondrial DNA in Kearns-Sayre syndrome. Neurology 38:1339–1346

Zinn AB, Kerr DS, Hoppel CL (1986) Fumarase deficiency: a new cause of mitochondrial encephalomyopathy. New Engl J Med 315:469–475

Mitochondrial Myopathy: Biochemical Approaches to Respiratory Chain-Linked Electron Transfer and Energy Coupling

C.P. LEE[1,3], M.E. MARTENS[2], P.L. PETERSON[2], S.H. TSANG[1] and M. SCIAMANNA[1]

1 Introduction

In 1959, Ernster and associates [1] pioneered biochemical studies of isolated mitochondrial preparations from skeletal muscle of patients with neuromuscular diseases. They correlated the clinical, ultrastructural and biochemical studies of a patient with hypermetabolism of non-thyroid origin, and clearly demonstrated the critical role of respiratory control in living cells [1,2]. This disease was later named Luft syndrome. These studies marked the beginning of a new era of research concerning biochemical studies of mitochondrial metabolism on biopsies derived from patients with mitochondrial metabolic disorders. Since then there have been many reports of altered mitochondrial structure and function in patients with neuromuscular diseases. The literature is large and diverse. A number of review articles have appeared in recent years [3–7].

Figure 1 shows a literature survey on the impairments in respiratory chain-linked electron transfer reported during the period from 1980 to date. The number of reports published since 1980 is plotted vs the respiratory chain (NADH oxidase). There have been 26 reports of defects in Complex I (NADH-CoQ reductase) [8–33], 20 in Complex III (QH_2-cyt.*c* reductase) [12,13,18,19,22,31,34–47] and 28 in Complex IV (cyt. *c* oxidase) [12,26,34,35,37,41,45,48–68]. In most cases the site of abnormality was associated with deficiency of one or more of the enzyme(s) indicated. In three cases, an impairment on the path of electron flow from cytochrome *c* to cytochrome oxidase and an excess of cytochrome *c* have been demonstrated [61,63,67]. This point will be discussed in more detail later. In many of the studies shown in Fig. 1, the respiratory chain defects were identified in a single patient and were not and/or could not be repeated because of the invasive nature of open muscle biopsy. Furthermore, different and sometimes conflicting conclusions on the same disease have been reported by different investigators. A good example is the mitochondrial encephalomyopathy, Kearns-Sayre syndrome. Four kinds of impairments in mitochondrial respiratory chain-linked functions have been reported. These are: (1) cytochrome *c* oxidase deficiency [50,58], (2) generalized deficiency of respiratory chain complexes and/or coenzyme Q [22,47,68,69], (3) an undefined lesion without cytochrome *c* oxidase deficiency [70], and (4) impaired electron transfer between cytochrome *c* and cytochrome oxidase

[1]Department of Biochemistry, Wayne State University School of Medicine, Detroit, MI 48201, USA
[2]Department of Neurology, Wayne State University School of Medicine, Detroit, MI 48201, USA
[3]To whom correspondence should be addressed

A. Azzi, Z. Drahota and S. Papa
Molecular Basis of Membrane-Associated Diseases

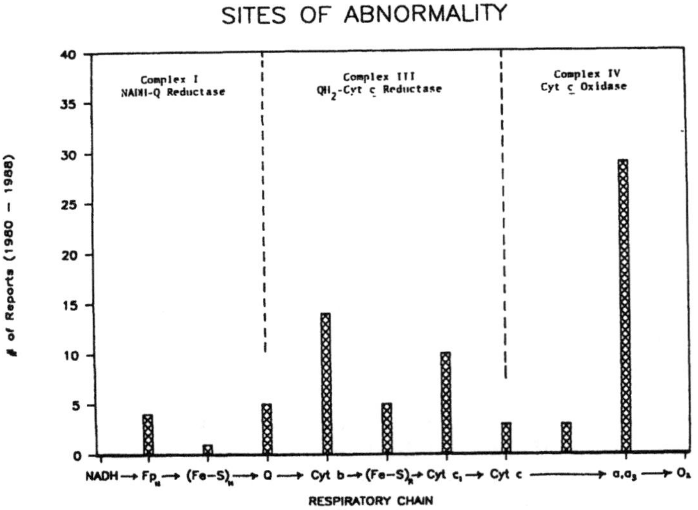

Fig. 1. Impairments in respiratory chain-linked electron transfer reported from 1980–1988

with cytochrome *c* oxidase deficiency and excess cytochrome *c* [61,63,66,67]. It is apparent that (1) and (3) are directly contradictory. The conclusion in (4) was based on experimental findings in our laboratory which will be discussed in more detail later. The diversity of biochemical findings in Kearns-Sayre syndrome may be explained by differences in experimental techniques for preparation and assay of mitochondrial preparations, the choice of parameters measured in the face of very small samples sizes, patient selection (a number of defects that result in systemic impairment of mitochondrial function may be characterized by very similar clinical features), or a true biochemical heterogeneity among Kearns-Sayre patients. It is essential, therefore, to establish what the limitations of the type of preparation and assay procedures are and what, if any, artifactual problems may arise during mitochondrial isolation and characterization. Only then can the scientific merits of such studies be assessed.

In this chapter, the approaches to mitochondrial preparation and characterization reported in the literature will first be summarized. Then, a systematic approach to the study of mitochondrial disease will be presented. Experimental data obtained from patients with the mitochondrial encephalomyopathy, Kearns-Sayre syndrome, will be used to illustrate the importance of the choice of preparation and parameters assayed in small biopsy specimens.

2 Literature Survey

2.1 Preparations of Intact Mitochondria and Mitochondrial Fragments

Mitochondrial preparations have been isolated from both fresh [71–75] and frozen [73] skeletal muscle. However, intact mitochondrial preparations can only be isolated from fresh tissue. Those isolated from frozen tissues are mitochondrial membrane fragments, although some investigators have assumed that they were intact mitochondria. Crude mitochondrial preparations with undefined composition have also been made from both fresh and frozen tissues [cf. Fig. 2 and refs. 71,73]. A large number of biochemical studies reported in the literature were done with crude mitochondrial preparations, or the post-nuclear supernatant, as referred to by some investigators [21,76].

Another important difference in the isolation procedures for skeletal muscle mitochondrial preparations among various laboratories is the isolation media. Both electrolyte and non-electrolyte isotonic media have been used. However, as Ernster and Nordenbrand [71] pointed out nearly 30 years ago, for skeletal muscle, the separation of mitochondria from other components is very difficult in non-electrolyte medium, in contrast to other types of tissues, such as heart and liver. The myofibrils tend to form a gelatinous consistency in non-electrolyte medium, making their disintegration very difficult. The isotonic electrolyte medium developed by Chappell and Perry [77] has been found to be very successful for the isolation of intact skeletal muscle mitochondria [71,72,75].

A flow diagram of the procedures used to isolate crude and intact mitochondrial preparations from fresh skeletal muscle is shown in Fig. 2. There are two major steps: disruption of the tissue and isolation of mitochondria by differential centrifugation. The muscle cells can be disrupted in two ways: (1) homogenization with a glass homogenizer; or (2) initial digestion with proteinase (Nagarse [72] or trypsin [76]), followed by disruption with either a glass homogenizer or ultra-turrax. In our experience, the yield and quality of the mitochondrial preparation are determined by the homogenization step and, if a proteinase is used, the amount of enzyme used and the time period of enzyme digestion. Prolonged incubation with proteinase will cause inactivation of mitochondrial enzyme(s), such as pal-

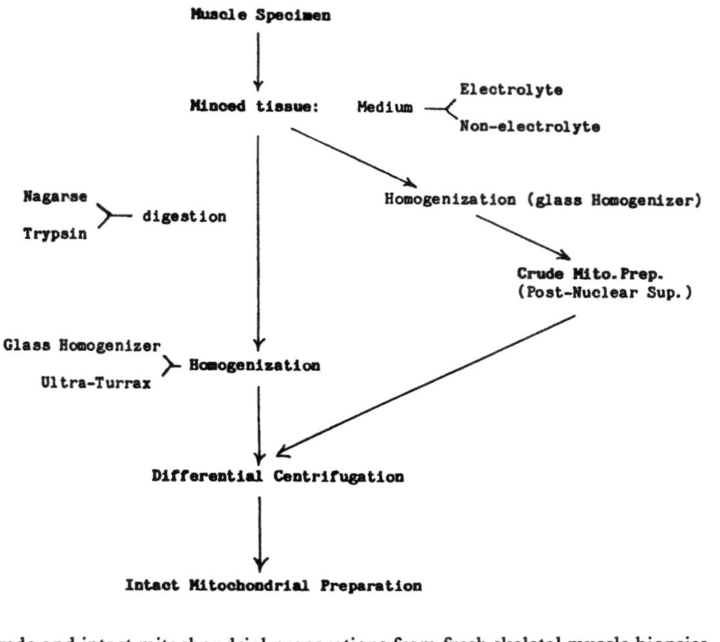

Fig. 2. Isolation of crude and intact mitochondrial preparations from fresh skeletal muscle biopsies

mityl CoA dehydrogenase in the case of Nagarse [78], and NADH and succinate oxidases in the case of trypsin [79]. Too short an incubation will, on the other hand, result in a low yield of mitochondria contaminated by other cellular components (e.g. myosin).

2.2 *Assays of Mitochondrial Preparations*

The assay parameters used to study mitochondrial preparations from patients with mitochondrial disease that have been reported in the literature are summarized in Table 1. In the case of isolated and purified intact mitochondrial preparations, one or more of the parameters listed in Table 1 have been used to characterize both the electron transfer and/or energy coupling capacities of the mitochondrial preparations. In the case of crude mitochondrial preparations from either fresh or frozen muscle, only the rate(s) of electron transfer of partial reactions of the respiratory chain has been measured. The energy coupling capacity of the crude preparations cannot be measured because of the complexity and undefined composition of the preparations. In the case of fresh biopsy specimens, studies of partial reactions in some cases cannot provide quantitative data on the individual enzyme (or enzyme complex) measured. Since some mitochondria may remain intact, substrates such as NADH are not permeable; thus, the activities will be severely underestimated. In some cases, quantitative information regarding individual enzymes (or enzyme complexes) can be derived from studies of partial reactions (e.g. in sonicated suspensions), but these results will not reveal how well the enzyme complexes interact with each other. Measurements of the rates of partial reactions together

Table 1. Assay parameters for mitochondrial preparations

A. Isolated and purified mitochondrial preparations
1. *Respiratory chain-linked oxidative phosphorylation*
2. *Respiration with* ^{14}*C-labeled substrates*
3. *Partial reactions of the respiratory chain:*
   - NADH oxidase
   - NADH dehydrogenase
   - Succinate dehydrogenase
   - NADH-CoQ reductase
   - NADH-cytochrome *c* reductase
   - Succinate-cytochrome *c* reductase
   - Cytochrome *c* oxidase
4. *Content of the respiratory chain carriers*
5. *Steady state kinetics of the respiratory chain carriers*

B. Crude mitochondrial preparations
1. *Partial reactions of the respiratory chain:*
   - NADH dehydrogenase
   - Succinate dehydrogenase
   - NADH-cytochrome *c* reductase
   - Succinate-cytochrome *c* reductase
   - Cytochrome *c* oxidase
2. *Respiration with* ^{14}*C-labeled substrates*
3. *NADH oxidase*
4. *Succinate oxidase*

with those of succinate oxidase and the antimycin A-sensitive NADH oxidase would resolve this problem [cf. ref.6]. Unfortunately, there have been only a few cases in which NADH and/or succinate oxidase activities were also measured. The diversity and complexity of biochemical studies in this field and the difficulties in comparing the results reported from different laboratories are apparent.

3 Systematic Approaches to Respiratory Chain-Linked Electron Transfer and Energy Coupling

In this section the systematic approach developed in our laboratory, using experimental data obtained from patients with Kearns-Sayre syndrome, is presented. The techniques employed are straightforward and can be readily reproduced in other laboratories. The complete analyses (i.e. mitochondrial isolation and characterization) can be accomplished with 2 g wet weight of skeletal muscle.

3.1 Isolation and Purification of Intact and Tightly Coupled Skeletal Muscle Mitochondrial Preparations

Figure 3 shows a schematic diagram of the isolation of intact mitochondria from small human biopsy specimens (0.8–3 g wet weight). This method was developed in our laboratory 20 years ago [72], and has since been refined. It has now been successfully applied to less than 1 g (wet weight) of tissue [67]. The critical steps, as we indicated earlier, are: the amount of Nagarse, the time of enzyme digestion, and the method of homogenization. The best conditions are indicated in the figure. Using this procedure, we have isolated tightly coupled mitochondrial preparations from as little as 0.8 g (wet weight) of human skeletal muscle. The yields are 3–5 mg mitochondrial protein per g tissue. The yields can be higher in some patients with accumulation of mitochondria in their skeletal muscle.

3.2 Biochemical Studies of Isolated Mitochondrial Preparations from Kearns-Sayre Syndrome Patients

Kearns-Sayre syndrome (KSS), a mitochondrial encephalomyopathy, was first described by Kearns and Sayre in 1958 [80]. Cardiac abnormalities and problems with the eye (retinal abnormalities and involvement of ocular muscles) are typical clinical features of this illness [cf. ref.7]. As described above, four kinds of conclusions regarding the impairment of mitochondrial electron transfer were made (cf. Sect. 1).

Figure 4 shows two typical experiments measuring the respiratory chain-linked oxidative phosphorylation and calcium transport in mitochondria isolated from a skeletal muscle biopsy specimen of a patient with KSS. Pyruvate + malate were used as substrate in the top trace. Succinate was used as substrate in the bottom trace. The ADP-induced respiratory control was clearly seen in both cases. In addition, respiration-dependent Ca^{2+} transport was also demonstrated (cf. bottom trace). Table 2 summarizes the respiratory and phosphorylating activities of isolated mitochondrial preparations derived from four KSS patients. The

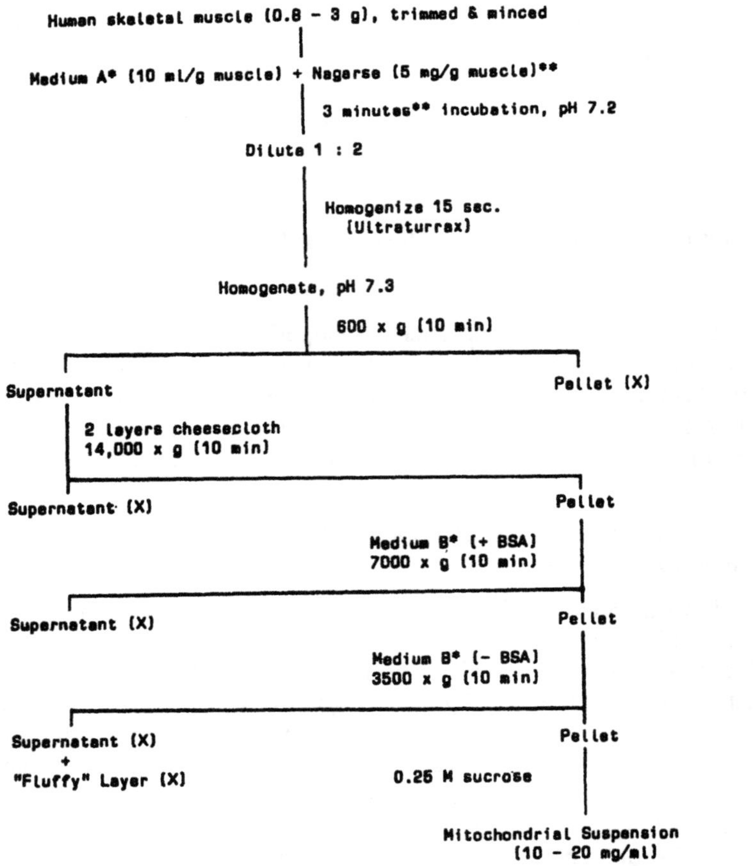

*: Medium A: 0.1 M KCl, 50 mM Tris-HCl, pH 7.5, 5 mM $MgCl_2$, 1 mM EDTA, 1 mM ATP.
Medium B: 0.1 M KCl, 50 mM Tris-HCl, pH 7.5, 1 mM $MgCl_2$, 0.2 mM EDTA, 0.2 mM ATP.

**: Dependent on the potency of the batch of Nagarse

Fig. 3. Isolation of intact mitochondria from small human skeletal muscle biopsies

mitochondrial preparations were tightly coupled with both succinate and NAD-linked substrates. The accompanying phosphorylating efficiencies (ADP/O ratios) with both kinds of substrates were in the normal range. However, the rates of respiration were considerably lower than controls with these substrates and with ascorbate + PMS (cytochrome oxidase). In analyzing these rates, it was apparent that cytochrome oxidase could not have been the rate-limiting step of the impaired respiratory rates seen with succinate and NAD-linked substrates.

In order to identify the site of impairment, the steady state kinetics of the respiratory chain carriers were investigated. Figure 5 shows the substrate-reduced *minus* oxidized difference spectra of skeletal muscle mitochondria isolated from a KSS patient (Fig. 5 bottom) and a control (Fig. 5 top). The steady state spectra (I) were recorded under state 3 conditions immediately after the addition of substrates

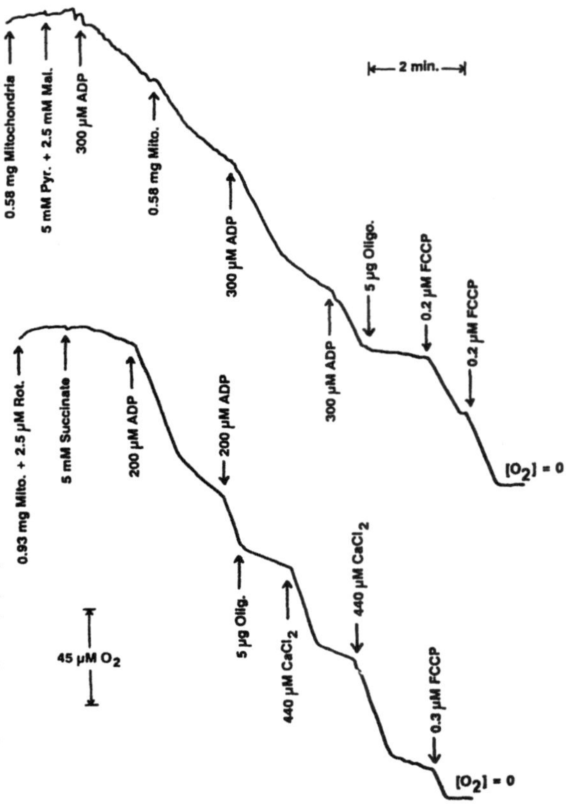

Fig. 4. Polarographic traces of mitochondria isolated from the skeletal muscle biopsy of a KSS patient oxidizing pyruvate + malate (*top*) and succinate + rotenone (*bottom*). The reaction medium consisted of 150 mM sucrose, 25 mM Tris-HCl, 10 mM phosphate, pH 7.5, in a volume of 1.0 ml. Other additions were as indicated; temp. = 30°C

Table 2. Respiratory and phosphorylating activities of isolated skeletal muscle mitochondria from Kearns-Sayre syndrome patients and controls

	Respiratory rate (n-atoms O/min/mg protein)[a]			
Substrates	State 3	State 3U[b]	RCI	ADP/O
Pyruvate + malate				
Patients (n = 4)	37 ± 8[d]	63 ± 17*	3.1 ± 0.1***	3.0 ± 0.3
Controls (n = 7)	164 ± 7	167 ± 12	4.6 ± 0.5	3.5 ± 0.2
Glutamate + malate				
Patients (n = 2)	43 ± 1*	52 ± 20*	2.2 ± 0.6**	2.6 ± 0.05*
Controls (n = 5)	169 ± 8	206 ± 11	7.0 ± 1.1	3.3 ± 0.1
Succinate + rotenone				
Patients (n = 4)	60 ± 11*	70 ± 15*	2.5 ± 0.5	1.9 ± 0.1
Controls (n = 7)	205 ± 10	202 ± 15	3.0 ± 0.2	1.8 ± 0.1
Ascorbate + PMS				
Patient No. 1-A	62			
-B[c]	163			
Patient No. 2	172			
Patient No. 3	1655			
Patient No. 4	2262			
Controls (n = 3)	2158 ± 706			

[a] Values are the mean ± SE.
[b] State 3U = uncoupled rate.
[c] Fresh autopsy.
[d] Statistical significance (Student's t-test): $*P < 0.01$; $**P < 0.05$; $***P < 0.10$.

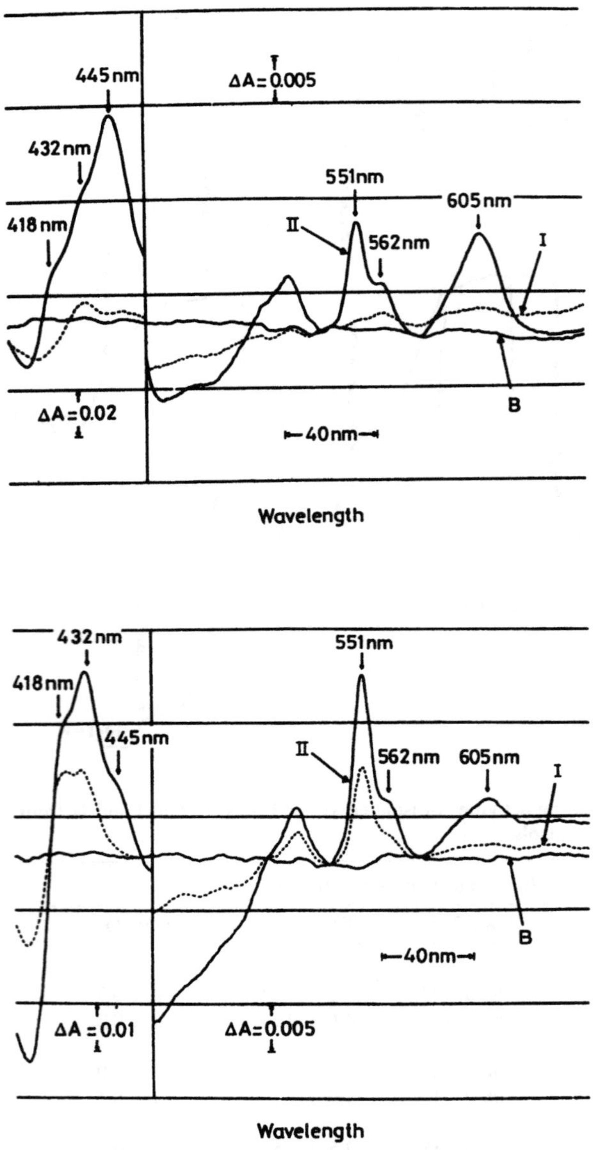

Fig. 5. Reduced *minus* oxidized difference spectra of mitochondria isolated from skeletal muscle biopsies of a control (*top*) and a KSS patient (*bottom*). The reaction medium was as described in Fig. 4. Substrates were 5 mM succinate, 5 mM pyruvate, 5 mM malate and 0.6 mM ADP. Mitochondrial protein = 1.1 mg/ml. Volume = 1.0 ml; temp. = room temperature. *B* Baseline (oxidized *minus* oxidized); *I* (*dotted line*) = reduced *minus* oxidized at steady state; *II* (*solid line*) = reduced *minus* oxidized at anaerobiosis. The *small arrows* indicate the α and γ peaks of cytochromes $a + a_3$, *b*, and $c + c_1$ [67]

into the sample cuvette. Spectra II were recorded following anaerobiosis. Three distinctive differences can be clearly seen. (1) At both the steady state and anaerobic state, the reduced levels of the *c* cytochromes in the patient's mitochondria were significantly higher than those in the control; (2) the cytochrome oxidase content in the KSS patient was significantly lower than that of the control; and (3) a cross-over point between cytochrome *c* and cytochrome oxidase is clearly shown in the steady state spectrum. These findings were later confirmed in two additional KSS patients. Table 3 summarizes the reduction levels of respiratory chain pigments in isolated human skeletal muscle mitochondria by substrates at steady state and anaerobiosis, and by dithionite. With the exception of patient No. 1 (cf. Table 2) whose mitochondria were from a fresh autopsy specimen, all were from fresh biopsy specimens. These data clearly demonstrated that the site of impairment in electron transfer of NADH and succinate oxidase was on the path from cytochrome *c* to cytochrome oxidase. The following two possible explanations are considered: (1) alteration(s) in the structure of cytochrome *c* and/or cytochrome oxidase; and/or (2) alterations in the insertion of cytochrome *c* and/or cytochrome oxidase into the membrane.

The immediate question is: Is the remaining cytochrome oxidase fully active? Based on the rate of respiration and the content of cytochrome oxidase heme *a*, the turnover number (TN) of cytochrome oxidase was calculated (Table 4). The TN gives an indication of the functional state of the remaining enzyme in the mitochondria. It is apparent that the TN of patients No. 3 and No. 4 were within the normal range as compared with those of controls, indicating that the cytochrome oxidase remaining in these two patients' mitochondria was not defective. In the case of patient No. 1, the TN of cytochrome oxidase was approximately 50% of those seen in the controls. This is not surprising, since the disease in patient No. 1 was by far the severest among the four. She died at age 13. At this advanced stage there may be secondary changes that caused additional injury.

Table 3. Reduction of respiratory chain pigments in isolated skeletal muscle mitochondria from Kearns-Sayre syndrome patients and controls

Pigment		Respiratory pigment content (n mol/mg protein)[a,b]		
		Steady state	Anaerobic state	Dithionite
a	Patients	0.01 ± 0.01	0.08 ± 0.01*	0.07 ± 0.01*
	Controls	0.01 ± 0.01	0.30 ± 0.05	0.34 ± 0.04
a_3	Patients	0.01 ± 0.004	0.14 ± 0.03**	0.20 ± 0.05*
	Controls	0.03 ± 0.02	0.40 ± 0.09	0.52 ± 0.06
b	Patients	0.05 ± 0.02	0.19 ± 0.04	0.25 ± 0.05
	Controls	0.07 ± 0.03	0.25 ± 0.02	0.29 ± 0.04
$c + c_1$	Patients	0.32 ± 0.07**	0.65 ± 0.08	0.68 ± 0.08
	Controls	0.08 ± 0.04	0.50 ± 0.02	0.53 ± 0.03
FP	Patients	0.17 ± 0.10	0.63 ± 0.24	0.83 ± 0.27
	Controls	0.15 ± 0.07	0.92 ± 0.18	1.03 ± 0.16

[a] Values are the mean ± SE for patients (n = 4) and controls (n = 3).
[b] Statistical significance (Student's t-test): *$P < 0.01$; **$P < 0.05$.

Table 4. Turnover number of cytochrome oxidase

Mitochondrial preparation	Turnover number (s^{-1})
Patient (No. 1)	50
Patient (No. 3)	200
Patient (No. 4)	300
Controls (n = 3)	110
Literature[a]	150–250

[a]Data from beef heart and rat liver mitochondria.

We were very fortunate to be able to receive autopsy material of patient No. 1 and to perform further studies [66]. Table 5 presents a summary of the rates of respiration and their accompanying phosphorylation efficiencies with various substrates of the mitochondrial preparations isolated from fresh skeletal muscle, heart, liver and brain. The data obtained from skeletal muscle confirmed our earlier results. The respiratory rates of brain, liver and heart mitochondrial preparations were also considerably lower than expected based on results derived from rat tissues. These autopsy specimens were taken approximately 3 h after the patient's death. It is remarkable that tightly-coupled mitochondrial preparations could be isolated from fresh autopsy specimens even after a prolonged period of time. These data indicate that one can derive pertinent data on mitochondrial respiratory chain-linked oxidative phosphorylation from fresh autopsy specimens. The cytochrome contents, relative to controls, of mitochondrial preparations from the fresh autopsy tissues are summarized in Table 6. Rats were the source of controls for brain, liver and heart mitochondria. The data show that the degree of deficiency of cytochrome oxidase was the highest in skeletal muscle and the lowest in heart mitochondria. Partial deficiencies of cytochromes *b* and *c* in brain and liver mitochondria and an excess of cytochrome *c* in heart and skeletal muscle mitochondria were also seen. Steady state kinetic analysis of the respiratory chain pigments (not shown) revealed a cross-over point between cytochrome *c* and cytochrome oxidase in all four tissues similar to that found in skeletal muscle biopsies [61,63,67].

It is well established that cytochrome *c* is a peripheral protein, whereas cytochrome c_1 an integral protein of the mitochondrial inner membrane. The contents of cytochromes *a*, *b* and $c+c_1$ of mitochondrial preparations isolated from both fresh and frozen skeletal muscle autopsy specimens of patient No. 1 have been measured. The contents of cytochromes *a* (0.10 vs 0.14 nmol/mg protein) and *b* (0.20 vs 0.24 nmol/mg protein) were comparable between mitochondrial preparations from fresh or frozen autopsy tissues, respectively, whereas the content of *c*-cytochrome in the mitochondrial preparation from frozen tissue (0.32 nmol/mg) was less than half of that seen in the preparation from fresh autopsy tissue (0.69 nmol/mg). These data indicate that more than half of the *c*-cytochromes seen in the mitochondrial preparations from the fresh specimen became dissociated from the membrane during the freezing and thawing and/or isolation process, strongly supporting the conclusion that the excess *c*-cytochromes seen in mitochondrial preparations from fresh tissue of KSS patients is cytochrome

Table 5. Respiratory chain-linked oxidative phosphorylation of mitochondrial preparations from fresh autopsy specimens of a Kearns-Sayre patient

Tissue	Substrate	Respiratory rate (n-atoms 0/min/mg protein) State 3	RCI[a]	ADP/O
Skeletal muscle	Pyr + Mal	32.3	2.8	2.3
	Succ + Rot	41.0	ND	ND
	Asc + PMS	163.0	ND	ND
Heart	Pyr + Mal	55.5	2.0	2.4
	Succ + Rot	121.6	1.3	0.8
Liver	Pyr + Mal	16.6	1.7	1.9
	Succ + Rot	40.2	1.7	1.3
Brain	Pyr + Mal	25.6	1.6	3.1

[a]RCI: respiratory control index.

Table 6. Cytochrome content of mitochondrial preparations derived from fresh autopsy specimens of a Kearns-Sayre patient (percent of controls)[a]

Tissue	Cyt. *a* (%)	Cyt. *b* (%)	Cyt. *c* (%)
Brain	40	65	40
Liver	40	60	50
Heart	60	100	120
Skeletal muscle	15	80	140

[a]Source of controls: rats for brain, liver and heart; human biopsy for skeletal muscle.

c rather than cytochrome c_1. Dissociation of cytochrome *c* from the mitochondrial membrane was also seen in mitochondrial preparations from frozen skeletal muscle of normal rats. These findings raise a very significant question. That is, does freezing and thawing of the tissue alter the binding of membrane-bound enzyme(s)? The answer appears to be yes; if frozen specimens had been used, the excess cytochrome *c* associated with the mitochondrial membrane of KSS patients would have been missed.

One would predict that a defect in the pathway of electron transfer along the respiratory chain would have both extensive and complex effects on mitochondrial and cellular metabolism. The most obvious effect would, of course, be an impaired rate of ATP production resulting from a decreased rate of mitochondrial substrate utilization (Fig. 6). Additionally, a block in electron transfer in the region between cytochrome *c* and cytochrome oxidase could also cause serious secondary damage to the mitochondrial membrane and the cell as a whole as a result of the buildup of toxic peroxides and other oxygen-containing free radicals [81].

Based on years of experience, the protocol for biochemical studies on respiratory chain-linked functions of human skeletal muscle specimens described above

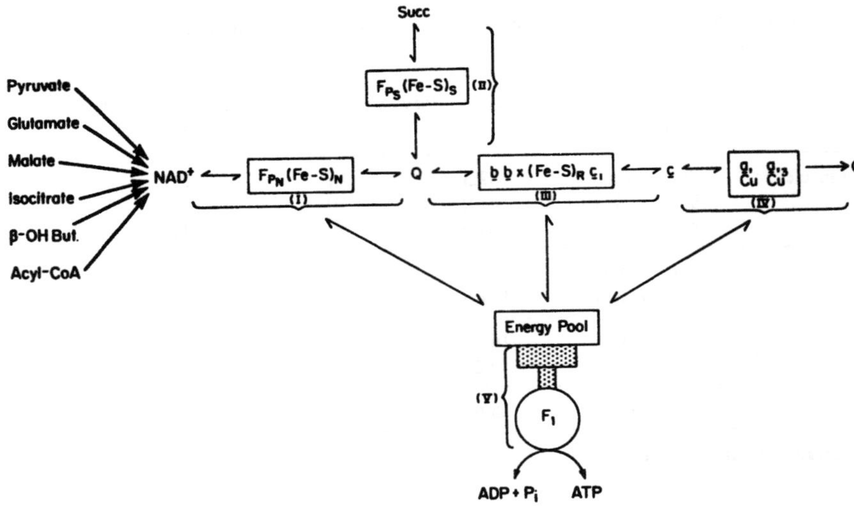

Fig. 6. Respiratory chain-linked oxidative phosphorylation

was developed. The two main features of this protocol are: (A) preparation of tightly-coupled skeletal muscle mitochondria with high quality and yield requires fresh tissue specimens, electrolyte medium and disruption of the tissue (± proteinase) under strictly controlled conditions; and (B) the mitochondrial preparations are characterized with respect to their respiratory rates with various substrates and the accompanying energy-coupling capacity, and the redox states of the respiratory chain carriers at steady state, anaerobic and chemically reduced states. Additional assays, such as those for the partial reactions of the respiratory chain, the partial reactions of ATP synthase, ion transport, etc., are based on the results of (A) and (B), above. Results derived solely from partial reactions of the respiratory chain do not provide any information on either the interaction between the enzyme complexes or the energy-coupling capacity. For example, the rate-limiting step of the impaired rate of respiration could be located on the path of electron flow between the two enzymes rather than in the enzyme itself, as we have demonstrated in the case of Kearns-Sayre syndrome [61,63,66,67], or there could be a defect in energy coupling, as was seen in Luft syndrome [1,2].

The availability of semimicro-techniques for isolation and characterization of intact mitochondrial preparations provides us with a valuable tool for examining the functional activities of respiratory chain enzymes. Analyses of these studies in conjunction with the clinical assessment will provide a means for identifying the underlying biochemical lesion in these disorders.

Acknowledgement. This research is supported by grants from NIH and MDA of the United States.

References

1. Ernster L, Ikkos D, Luft R (1959) Enzymatic activities of human skeletal muscle mitochondria: a tool in clinical metabolic research. Nature (London) 184:1851–1854
2. Luft R, Ikkos D, Palmieri G, Ernster L, Afzelius B (1962) A case of severe hypermetabolism of nonthyroid origin with a defect in the maintenance of mitochondrial respiratory control – a correlated clinical, biochemical and morphological study. J Clin Invest 41:1776–1804
3. Carafoli E, Roman I (1980) Mitochondria and disease. Mol Aspects Med 3:295–429
4. DiMauro S, Bonilla E, Zeviani M, Nakagawa M, DeVivo DC (1985) Mitochondrial myopathies. Ann Neurol 17:521–538
5. Morgan-Hughes JA (1986) The mitochondrial myopathies. In: Engel AG, Banker BQ (eds) Myology, vol 2. McGrawHill, New York, pp 1709–1743
6. Scholte HR (1988) The biochemical basis of mitochondrial diseases. J Bioenerg Biomembr 20:161–191
7. Peterson PL, Martens ME, Lee CP (1988) Mitochondrial encephalomyopathies. In: Riggs J (ed) Neurologic clinics, vol 6. Saunders, Philadelphia, pp 529–544
8. Busch BFM, Scholte HR, Arts WF, Luyt-Houwen IEM (1981) A mitochondrial myopathy with a respiratory chain defect and carnitine deficiency. In: Busch HFM, Jennekens FGI, Scholte HR (eds) Mitochondria and muscular diseases. Mefar, Beetsterzwaag, NL, pp 207–211
9. Land JM, Morgan-Hughes JA, Clark JB (1981) Mitochondrial myopathy – biochemical studies revealing a deficiency of NADH cytochrome *b* reductase activity. J Neurol Sci 50:1–13
10. Prick MJJ, Gabreels FJM, Renier WO, Trijbels JMF, Sengers RCA, Sloof JL (1981) Progressive infantile poliodystrophy: association with disturbed pyruvate oxidation in muscle and liver. Arch Neurol 38:767–772
11. Arts WFM, Scholte HR, Bogaard JM, Kerrebijn KF, Luyt-Houwen IEM (1983) NADH-CoQ reductase deficient myopathy: successful treatment with riboflavin. Lancet 2:581–582
12. Barth PG, Scholte HR, Berden JA, Van der Klei-Van Moorsel JM, Luyt-Houwen IEM, Van't Veer-Korthof ETh, Van der Harten JJ, Sobotka-Plojhar MA (1983) An X-linked mitochondrial disease affecting cardiac muscle, skeletal muscle and neutrophil leucocytes. J Neurol Sci 62:327–355
13. Scholte HR, Busch HFM, Barth PG, Beekman RP, Berden JA, Duran M, Luyt-Houwen IEM, Przyrembel H, Roth B, DeVries S (1983) Carnitine deficiency and mitochondrial respiratory chain blockade. In: Scarlato G, Cerri C (eds) Mitochondrial pathology in muscle diseases. Piccin Med Books, Padua, pp 215–228
14. Moreadith RW, Batshaw ML, Ohnishi T, Kerr D, Knox B, Jackson D, Hruban R, Olson J, Reynafarje B, Lehninger AL (1984) Deficiency of the iron-sulfur clusters of mitochondrial reduced nicotinamide adenine dinucleotide-ubiquinone oxidoreductase (complex III) in an infant with congenital lactic acidosis. J Clin Invest 47:685–697
15. Morgan-Hughes JA, Hayes DJ, Clark JB, Landon DN, Swash M, Stark RJ, Rudge P (1984) Mitochondrial encephalomyopathies: biochemical studies in two cases revealing defects in the respiratory chain. Brain 105:553–582
16. Papadimitriou P, Neustein HB, DiMauro S, Stanton R, Bresolin N (1984) Histocytoid cardiomyopathy of infancy: a deficiency of reducible cytochrome *b* in heart mitochondria. Pediatr Res 18:1023–1028
17. Sherratt HSA, Johnson MA, Turnbull DM (1986) Defects of complex I and complex IV in skeletal muscle from patients with chronic progressive external ophthalmoplegia. Ann NY Acad Sci 488:508–510
18. Folkers K, Wolaniuk J, Simonsen R, Morishita M, Vadhanavikit S (1985) Biochemical rationale and the cardiac response of patients with muscle disease to therapy with coenzyme Q. Proc Natl Acad Sci USA 82:4513–4516
19. Morgan-Hughes JA, Hayes DJ, Cooper M, Clark JB (1985) Mitochondrial myopathies: deficiencies localized to complex I and complex III of the mitochondrial respiratory chain. Biochem Soc Trans 13:648–650
20. Scholte HR, Luyt-Houwen IEM, Busch HFM, Vaandrager-Verduin MH (1986) Biochemical criteria for NADH-CoQ reductase deficiency. Ann NY Acad Sci 488:513–516
21. Fischer JC, Ruitenbeck W, Gabreels FJM, Janssen AJM, Renier WO, Sengers RCA, Stadhouders AM, Ter Laak HJ, Trijbels JMF, Veerkamp JH (1986) A mitochondrial encephalomyopathy: the first case with an established defect at the level of coenzyme Q. Eur J Pediatr 144:441–444

22. Ogasahara S, Nishikawa Y, Yorifuji S, Soga F, Nakamura Y, Takahashi M, Hashimoto S, Kono N, Tarui S (1986) Treatment of Kearns-Sayre syndrome with coenzyme Q_{10}. Neurology 36:45–53
23. Kobayashi M, Morishita H, Sugiyama N, Yokochi K, Nakano M, Wada Y, Hotta Y, Rerauchi A, Nonaka I (1986) Mitochondrial myopathy, encephalopathy, lactic acidosis and stroke-like episodes syndrome and NADH-CoQ reductase deficiency. J Inherit Metab Dis 9:301–304
24. Peterson PL, Martens ME, Lee CP, Wolfe D, Hatfield J, Gilroy J (1986) MELAS – a disorder of mitochondrial metabolism. Neurology 36(Suppl 1):94
25. Robinson BH, Ward J, Goodyear P, Baudet A (1986) Respiratory chain defects in the mitochondria of cultured skin fibroblasts from three patients with lactacidemia. J Clin Invest 77:1422–1427
26. Roodhooft AM, Van Acker KJ, Martin JJ, Ceuterick C, Scholte HR, Luyt-Houwen IEM (1986) Benign mitochondrial myopathy with deficiency of NADH-CoQ reductase and cytochrome *c* oxidase. Neuropediatrics 17:221–226
27. Scholte HR, Luyt-Houwen IEM, Busch HFM (1985) Difficulties in assessing biochemical properties of abnormal muscle mitochondria. J Inherit Metab Dis 8(Suppl 2):149–150
28. Tanaka K, Nishikimi M, Suzuki S, Ozawa T, Okino E, Takahashi H (1986) Multiple cytochrome deficiency and deteriorated mitochondrial polypeptide composition in fatal infantile mitochondrial myopathy and renal dysfunction. Biochem Biophys Res Commun 137:911–916
29. Van Erven PM, Ruitenbeek W, Gabreels FJM, Renier WO, Fischer JC, Janssen AJM (1986) Subacute necrotizing encephalomyelopathy (Leigh syndrome) associated with disturbed oxidation of pyruvate, malate and 2-oxoglutarate in muscle and liver. Neuropediatrics 17:28–32
30. Nishizawa M, Tanaka K, Shinozawa K, Kuwabara T, Atsumi T, Miyatake T, Ohama E (1987) A mitochondrial encephalomyopathy with cardiomyopathy. A case revealing a defect of complex I in the respiratory chain. J Neurol Sci 78:189–201
31. Scholte HR, Busch HFM, Luyt-Houwen IEM, Vaandrager-Verduin MHM, Przyrembel H, Arts WFM (1987) Defects in oxidative phosphorylation. Biochemical investigations in skeletal muscle and expression of the lesion in other cells. J Inherit Metab Dis 10(Suppl 1):81–97
32. Peterson PL, Martens ME, Lee CP, Nigro MA, Kotagal S, McGillivary D, MacGregor DL (1988) MELAS – a disorder in NAD-linked substrate oxidation. Neurology 38(Suppl 1):268
33. Kotagal S, Peterson PL, Martens ME, Lee CP, Nigro M, Archer CR (1988) Impaired NADH-CoQ reductase activity in a child with moyamoya syndrome. Pediatr Neurol 4:241–244
34. Stansbie D, Dormer RL, Hughes IA, Minchom PE, Hendry GAF, Jones OTG, Cross AR, Sherratt HSA, Turnbull DM, Johnson MA (1982) Mitochondrial myopathy with skeletal muscle cytochrome oxidase deficiency. J Inherit Metab Dis 5(Suppl 1):27–28
35. Boustany RN, Aprille JR, Halperin J, Levy H, DeLong GR (1983) Mitochondrial cytochrome deficiency presenting as a myopathy with hypotonia, external ophthalmoplegia, and lactic acidosis in an infant and as fatal hepatopathy in a second cousin. Ann Neurol 14:462–470
36. Darley-Usmar VM, Kennaway NG, Buist NRM, Capaldi RA (1983) Deficiency in ubiquinone cytochrome *c* reductase in a patient with mitochondrial myopathy and lactic acidosis. Proc Natl Acad Sci USA 80:5103–5106
37. Minchom PE, Dormer RL, Hughes IA, Stansbie D, Cross AR, Hendry GAF, Jones OTG, Johnson MA, Sherratt HSA, Turnbull DM (1983) Fatal infantile mitochondrial myopathy due to cytochrome *c* oxidase deficiency. J Neurol Sci 60:453–463
38. Przyrembel H, Scholte HR, Busch HFM, Blom W, Arts WFM, Luyt-Houwen IEM (1983) Chronic lactic acidemia and lipid myopathy due to cytochrome c_1 deficiency in liver and skeletal muscle: treatment with vitamin C. In Scarlato G, Cerri C (eds) Mitochondrial pathology in muscle diseases
39. Sengers RCA, Fischer JC, Trijbels JMF, Ruitenbeek W, Stadhouders AM, Ter Laak HJ, Jaspar HHJ (1983) A mitochondrial myopathy with a defective respiratory chain and carnitine deficiency. Eur J Pediatr 140:332–337
40. Behbehani AW, Goebel H, Osse G, Gabriel M, Langenbeck U, Berden J, Berger R, Schutgens RBH (1984) Mitochondrial myopathy with lactic acidosis and deficient activity of muscle succinate cytochrome *c* oxidoreductase. Eur J Pediatr 143:67–71
41. Hayes DJ, Lecky BRF, Landon DN, Morgan-Hughes JA, Clark JB (1984) A new mitochondrial myopathy: biochemical studies revealing a deficiency in the cytochrome-bc_1 complex (complex III) of the respiratory chain. Brain 107:1165–1177
42. Kennaway NG, Buist NRM, Darley-Usmar VM, Papadimitriou A, DiMauro S, Kelley RI, Capaldi RA, Blank NK, D'Agostino A (1984) Lactic acidosis and mitochondrial myopathy associated with deficiency of several components of complex III of the respiratory chain. Pediatr Res 18:991–999

43. Riggs JE, Schochet S, Fakedej AV, Papadimitriou A, DiMauro S, Crosby TW, Gutmann L, Moxley RT (1984) Mitochondrial encephalomyopathy with decreased succinate-cytochrome *c* reductase activity. Neurology 34:48–53
44. Argov Z, Bank WJ, Maris J, Elef S, Kennaway NG, Olson RE, Chance B (1986) Treatment of mitochondrial myopathy due to complex III deficiency with vitamins K_3 and C: a ^{31}P-NMR follow-up study. Ann Neurol 19:598–602
45. Sengers RCA, Trijbels JMF, Bakkeren JAJM, Ruitenbeek W, Fisher JC, Jansen AJM, Stadhouders AM, Ter Laak HJ (1984) Deficiency of cytochromes *b* and aa_3 in muscle from a floppy infant with cytochrome oxidase deficiency. Eur J Pediatr 141:178–180
46. Reichmann H, Rohkamm R, Zeviani M, Servidei S, Ricker K, DiMauro S (1986) Mitochondrial myopathy due to complex III deficiency with normal reducible cytochrome *b* concentration. Arch Neurol 43:957–961
47. Bresolin N, Moggio M, Bet L, Gallanti A, Prelle A, Nobile-Orazio E, Adobbati L, Ferrante C, Pellegrini G, Scarlato G (1987) Progressive cytochrome *c* oxidase deficiency in a case of Kearns-Sayre syndrome: morphological, immunological, and biochemical studies in muscle biopsies and autopsy tissues. Ann Neurol 21:564–572
48. Rimoldi M, Bottacchi E, Rossi L, Cornelio F, Uziel G, DiDonato S (1982) Cytochrome *c* oxidase deficiency in muscles of a floppy infant without mitochondrial myopathy. J Neurol 227:201–207
49. DiMauro S, Nicholson JF, Hays AP, Eastwood AB, Papadimitriou A, Koenigsberger R, DeVivo DC (1983) Benign infantile mitochondrial myopathy due to reversible cytochrome *c* oxidase deficiency. Ann Neurol 14:226–234
50. Johnson MA, Turnbull DM, Dick DJ, Sherratt HSA (1983) A partial deficiency of cytochrome *c* oxidase in chronic progressive external ophthalmoplegia. J Neurol Sci 60:31–53
51. Maehara M, Ogasawara N, Mizutani N, Watanabe K, Suzuki S (1983) Cytochrome *c* oxidase deficiency in Menkes kinky hair disease. Brain Dev 5:533–540
52. Miyabayashi S, Narisawa K, Tada K, Sakai K, Kobayashi K, Kobayashi Y (1983) Two siblings with cytochrome *c* oxidase deficiency. J Inherit Metab Dis 6:121–122
53. Prick MJJ, Gabreels FJM, Trijbels JMF, Janssen AJ, LeCoultre R, Van Dam K, Jaspar HH, Ebels EJ, Op de Coul AA (1983) Progressive poliodystrophy (Alper's disease) with a defect in cytochrome aa_3 in muscle: a report of two unrelated patients. Clin Neurol Neurosurg 85:57–70
54. Trijbels F, Sengers R, Monnens L, Janssen A, Willems J, Ter Laak H, Stadhouders A (1983) A patient with lactic acidemia and cytochrome oxidase deficiency. J Inherit Metab Dis 6(Suppl 2):127–128
55. Bresolin N, Zeviani M, Bonilla E, Miller RH, Leech RW, Shanske S, Nakagawa M, DiMauro S (1985) Fatal infantile cytochrome *c* oxidase deficiency: decrease of immunologically detectable enzyme in muscle. Neurology 35:802–812
56. Byrne E, Trounce I (1985) Oxygen electrode studies with human skeletal muscle mitochondria in vitro. A re-appraisal. J Neurol Sci 69:319–333
57. DiMauro S, Zeviani M, Bonilla E, Bresolin N, Nakagawa M, Miranda AF, Moggio M (1985) Cytochrome *c* oxidase deficiency. Biochem Soc Trans 13:651–653
58. Turnbull DM, Johnson MA, Dick DJ, Cartlidge NEF, Sherratt HSA (1985) Partial cytochrome oxidase deficiency without subsarcolemmal accumulation of mitochondria in chronic progressive ophthalmoplegia. J Neurol Sci 70:93–100
59. Zeviani M, Nonaka I, Bonilla E, Okino E, Moggio M, Jones S, DiMauro S (1985) Fatal infantile mitochondrial myopathy and renal dysfunction caused by cytochrome *c* oxidase deficiency. Immunological studies in a new patient. Ann Neurol 17:414–417
60. Angelini C, Bresolin N, Pegolo G, Bet L, Rinaldo P, Trevisan C, Vergani L (1986) Childhood encephalomyopathy with cytochrome *c* oxidase deficiency, ataxia, muscle wasting, and mental impairment. Neurology 36:1048–1052
61. Peterson PL, Martens ME, Lee CP (1986) Kearns-Sayre syndrome: cytochrome oxidase deficiency, and impairment in electron transfer between cytochrome *c* and cytochrome oxidase. EBEC Rep 4:81
62. Arts WFM, Scholte HR, Loonen MCB, Przyrembel H, Fernandez J, Trijbels JMF, Luyt-Houwen IEM (1987) Cytochrome *c* oxidase deficiency in subacute necrotizing encephalomyopathy. J Neurol Sci 77:103–115
63. Lee CP, Martens ME, Peterson PL, Hatfield JS (1987) Cytochrome oxidase and neuromuscular disease. In: Papa S, Chance B, Ernster L (eds) Cytochrome systems: molecular biology and bioenergetics. Plenum, New York, pp 407–414

64. Servidei S, Lazaro RP, Bonilla E, Barron KD, Zeviani M, DiMauro S (1987) Mitochondrial encephalomyopathy and partial cytochrome *c* oxidase deficiency. Neurology 37:58–63
65. Zeviani M, Peterson PL, Bonilla E, DiMauro S (1987) Benign reversible muscle cytochrome *c* oxidase (COX) deficiency: a second case. Neurology 37:64–67
66. Lee CP, Martens ME, Peterson PL, Hart Z, Nigro MA, Chang CH (1988) Kearns-Sayre syndrome: tissue-specific cytochrome deficiency. Abstr TU-407, 14th Int Congr Biochemistry, July 1988, Prague, CSSR Piccin Med Books, Padua, pp 193–195
67. Martens ME, Peterson PL, Lee CP, Nigro MA, Hart Z, Glasberg M, Hatfield JS, Chang CH (1988) Kearns-Sayre syndrome: biochemical studies of mitochondrial metabolism. Ann Neurol 24:630–637
68. Allen RJ, DiMauro S, Coulter DL, Papadimitriou A, Rottenberg SP (1983) Kearns-Sayre syndrome with reduced plasma and cerebrospinal fluid folate. Ann Neurol 13:679–682
69. Yorifuji S, Ogasahara S, Takahashi M, Tarui S (1985) Decreased activities in mitchondrial inner membrane electron transport system in muscle from patients with Kearns-Sayre syndrome. J Neurol Sci 71:65–75
70. Mitsumoto H, Aprille JR, Wray SH, Nemni R, Bradley WG (1983) Chronic progressive external ophthalmoplegia (CPEO): clinical, morphologic, and biochemical studies. Neurology 33:452–461
71. Ernster L, Nordenbrand K (1967) Skeletal muscle mitochondria. Meth Enzymol 10:86–94
72. Makinen M, Lee CP (1968) Biochemical studies of skeletal muscle mitochondria: I. Microanalysis of cytochrome content, oxidative and phosphorylative activities of mammalian skeletal muscle mitochondria. Arch Biochem Biophys 126:75–82
73. Bookelman H, Trijbels JMF, Sengers RCA, Janssen AJM (1978) Measurement of cytochromes in human skeletal muscle mitochondria isolated from fresh and frozen stored muscle specimens. Biochem Med 19:366–373
74. Lee CP, Martens ME, Jankulovaska L, Neymark M (1979) Defective oxidative metabolism of myodystrophic skeletal muscle mitochondria. Muscle Nerve 2:340–348
75. Nedergaard J, Cannon B (1979) Overview – Preparation and properties of mitochondria from different sources. Meth Enzymol 55:3–28
76. Fischer JC, Ruitenbeck W, Stadhouders AM, Trijbels JMF, Sengers RCA, Ganssen AJM, VeerKamp JH (1985) Investigation of mitochondrial metabolism in small human skeletal muscle biopsy specimens. Improvement of preparation procedure. Clin Chim Acta 145:89–100
77. Chappell JP, Perry SV (1954) Biochemical and osmotic properties of skeletal muscle mitochondria. Nature (London) 173:1094–1095
78. Pande SV, Blanchaer MC (1970) Preferential loss of ATP-dependent long-chain fatty acid activating enzyme in mitochondria prepared using Nagarse. Biochim Biophys Acta 202:43–48
79. Luzikov VN, Romashina LV (1972) Studies on stabilization of the oxidative phosphorylation system. II. Electron transfer-dependent resistance of succinate oxidase and NADH oxidase systems of submitochondrial particles to proteinases and cobra venom phospholipase. Biochim Biophys Acta 267:37–47
80. Kearns TP, Sayre GP (1958) Retinitis pigmentosa, external ophthalmoplegia and complete heart block. Arch Ophthalmol 60:280–283
81. Chance B, Sies H, Boveris a (1979) Hydroperoxide metabolism in mammalian organs. Physiol Rev 59:527–605

Phosphorus Magnetic Resonance Spectroscopy (^{31}P NMR) as a Tool for in Vivo Monitoring of Mitochrondrial Muscle Disorders

Z. ARGOV

1 Introduction

Magnetic resonance spectroscopy is a relatively new technique for the study of in vivo biochemistry. It offers a look into several biochemical reactions in normally functioning intact tissues. This chapter will not describe the basic NMR experiment and the various atoms that can be studied by it. For an introduction to the NMR field the reader is referred to monographs on the subject (e.g. Gadian [1]). The chapter will also only briefly describe the basics of phosphorus nuclear magnetic resonance spectroscopy (^{31}P NMR). I will concentrate on the use of ^{31}P NMR in muscle mitochondrial disorders and related conditions. Findings in other disorders of energy metabolism (e.g. glycolytic defects) will not be described. Data on normal functioning muscle will be presented only if they are relevant to mitochondrial diseases.

2 ^{31}P NMR

Because of the basic physical behavior of phosphorus atoms, ^{31}P NMR is an inherently insensitive analytical technique. Using magnetic field strength of 2 Tesla, in vivo ^{31}P NMR of muscle can record only metabolites that are present in concentrations higher than about 0.5 mM. ^{31}P NMR spectra from muscle are easy to interpret and quantify since only a small number of phosphorus-containing compounds are present in high enough concentrations that enable detection in vivo.

The typical ^{31}P NMR spectrum of muscle at rest is shown in Fig. 1. There are five major peaks: three of the phosphate atoms in the ATP molecule (in positions alpha, beta and gamma), one of phosphocreatine (PCr) and one of inorganic phosphates (Pi). The latter peak contains both $H_2PO_4^-$ and HPO_4^{--}. The chemical shift (the spectral "distance") between PCr and Pi peaks is determined by the intracellular pH. Thus, if properly calibrated, the ^{31}P NMR spectrum can accurately determine the intracellular pH of muscle cells [2].

Can ^{31}P NMR measure the various compounds quantitatively? The area under the spectral peak is determined by the amount of this metabolite present in the sample recorded. The amount of muscle seen depends on the size of the external coil used to record the NMR signals. In studies of human muscle the sample volume

Department of Neurology, Hadassah University Hospital, Jerusalem, Israel

A. Azzi, Z. Drahota and S. Papa
Molecular Basis of Membrane-Associated Diseases

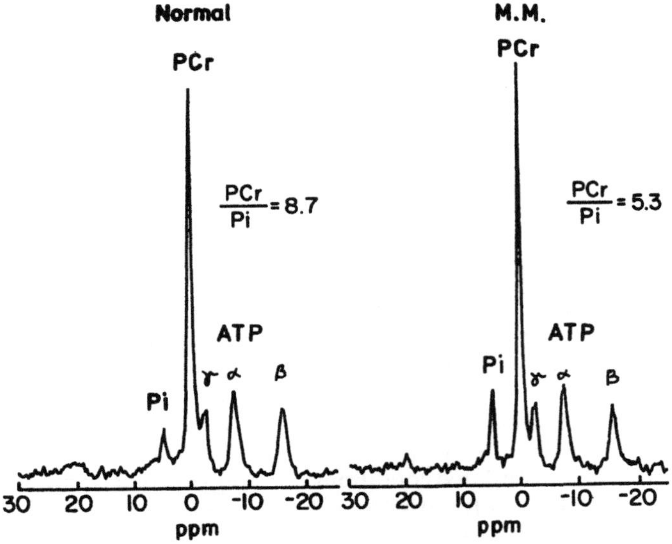

Fig. 1. Two 5-min ^{31}P NMR spectra (60 averaged scans) of normal arm muscle (*left*) and a case of moderate mitochondrial myopathy (M.M.; *right*) due to a complex 1 defect. The various peaks of the spectrum are indicated and the relevant PCr/Pi ratios are given

is a few cm^3 and the signals are averaged from this volume. However, the tissue in this volume is not only muscle (also recorded are skin and subcutaneous tissue) and calibration to absolute amounts and concentrations requires an internal standard with known concentration. Most studies have used the ATP peak, assuming an unchanged average ATP concentration of about 8 mM[3]. This concentration may be a "crude" average in normal persons, but it remains to be seen whether diseased muscle cells retain the same concentration of ATP. Thus, for quantitative measurements in diseased muscle, ratios between compounds are used more frequently than single peak amounts. Luckily, the ratios between some of these peaks have a bioenergetic significance as will be described below.

A high resolution spectrum, such as that shown in Fig. 1, takes a few minutes to record. When studying a muscle at rest this presents no problem. However, in studying muscle metabolism it is important to make measurements during exercise. The changes that occur in the concentrations of PCr and Pi in working muscle are very rapid and a method with a time resolution of 0.5–1 min (as e.g. ^{31}P NMR) will fail to accurately represent those changes. To solve this problem, the steady-state type of exercise protocol was developed.

Before recounting the steady-state protocol, the basic changes that occur in the ^{31}P NMR spectrum will be described. During exercise, the levels of PCr fall and those of Pi rise stoichiometrically [4]. A spectrum of a moderately working muscle is shown in Fig. 2. It can be seen that during such exercise, the levels of ATP do not change. Once exercise is stopped, the ^{31}P NMR rapidly returns to its resting state characteristics: PCr is high again and Pi is low. During this post-exercise recovery there is some loss of signal and the whole spectrum is "noisier" with smaller peaks.

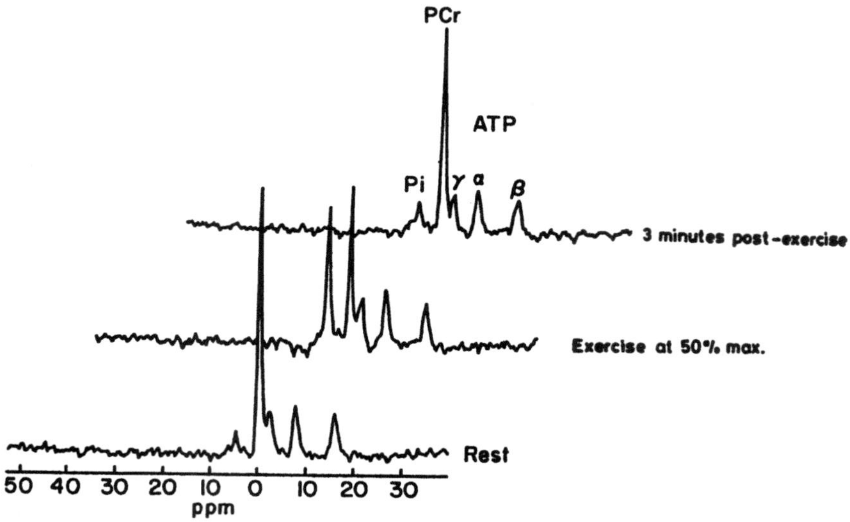

Fig. 2. Two-min spectra of normal arm muscle during rest, work at 50% of personal maximum and after recovery from exercise. Note the reversible changes in PCr/Pi ratios and the constant level of ATP during the exercise. The post-exercise spectrum has smaller peaks because of increased blood flow to the arm (see text)

This may be due to increased blood volume and flow to muscle that is maximal during this period. Thus, to evaluate recovery it is better again to look at ratios; we quantify the recovery by the change in PCr/Pi units per time (see, for example, Fig. 3 discussed below).

The basic principle of the steady-state protocol was demonstrated by a ^{31}P NMR study of isolated heart mitochondria in vitro [5]. In this work, mitochondria were adjusted to give a series of steady states. Each state had a distinct ATPase rate (corresponding to the ATP synthesis by oxidative metabolism) and a ^{31}P NMR measured PCr/Pi value (corresponding to the intracellular ADP concentration). When the values of ATPase were plotted against the inverse ratios of PCr/Pi, a rectangular hyperbola was obtained. Such a profile is expected when the entire oxidative metabolism is regarded as a single process that follows Michaelis-Menten kinetics of single enzyme systems [6,7]. The human exercise protocols used in our laboratory follow a very similar approach.

Graded levels of work rate were performed by the wrist flexors of the arm, monitored by an ergometer. After a few minutes of constant work performance, the PCr/Pi ratio was stabilized and measured and the subject increased the work rate until a new steady state was achieved. It should be noted that these were pseudo steady states, since not all determinants remained constant (e.g. the pH was not constant when high levels of work rates were reached). However, when work rate was plotted against Pi/PCr, a profile very reminiscent of a rectangular hyperbola was obtained [6,7]. The kinetic meanings of these findings are discussed elsewhere [7]. If a steady-state protocol can be followed, the slope of the graph will give a comparable look at oxidative metabolism in different subjects. Patients with

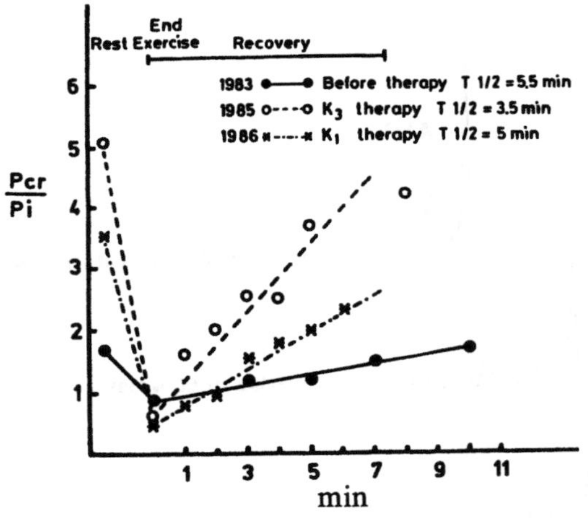

Fig. 3. Post-exercise recovery in the patient with complex 3 deficiency before and during two therapeutic approaches. The points of end exercise are very similar, thus, the recovery slopes are comparable (see text)

muscle disease find it difficult to follow several steady states. Thus, we sometimes used a modified version of the above protocol in which only a single level of work rate had to be maintained. This was 50% of personal maximal power output, as measured by the ergometer, kept until a 2–3 min steady state was achieved. Figure 2 was in fact taken from such a protocol in a normal control. The importance of following the changes of PCr/Pi ratios during transitions in muscle metabolism relate to the bioenergetic basis of the reactions recorded by ^{31}P NMR.

3 The Bioenergetic Basis for the ^{31}P NMR Studies

Cytosolic phosphorylation potential (PP) is characteristic of the energy balance between ATP utilization and production and, thus, represents the energy state of muscle cells and their ability to perform work at any given time [5,8]. This ratio also plays a direct role in the control of mitochondrial respiration. It is defined as:

$$PP = ATP/ADP \times Pi. \quad (1)$$

The creatine kinase (CK) reaction which rapidly controls the cytosolic ATP concentration and its availability to mitochondria, is:

$$PCr + ADP + H^{+} \leftrightarrows ATP + Cr \quad \text{(Cr is free creatine).} \quad (2)$$

This reaction is normally at or near equilibrium. The CK reaction equilibrium constant (k) is:

$$K = [ATP] \times [Cr]/[PCr] \times [ADP] \times [H].$$

Derived from above is $[ATP] = K \times [PCr] \times H \times [ADP]/[Cr]$,
which when substituted in the PP equation gives:

$$PP = [PCr]/[Pi] \times K \times [H] \times 1/[Cr]. \quad (3)$$

From Eq. (3) it can be seen that the tissue energetic state is directly proportional to the ^{31}P NMR measurable ratio of PCr/Pi. Thus, PCr/Pi is extensively used in in vivo metabolic observation of tissues and organs. Knowing the tissue concentration of total creatine and ATP, the PP can be directly calculated by in vivo ^{31}P NMR of muscle. This is done by further evaluation of Eq. (3). Free creatine [Cr] is determined as:

$$[\mathrm{Cr}] = [\mathrm{Cr}]\,\mathrm{total} - [\mathrm{PCr}],$$

where [PCr] is calculated from PCr/ATP × [ATP]. Thus, the PP equation is now:

$$\frac{\mathrm{PCr} \times [\mathrm{H}] \times \mathrm{K}}{\mathrm{Pi} \times \{[\mathrm{Cr}]\,\mathrm{total} - (\mathrm{PCr/ATP}) \times [\mathrm{ATP}]\}}. \quad (4)$$

PCr/ATP is calculated from the ^{31}P NMR spectrum and [H] is also directly determined from the ^{31}P NMR spectrum, since intracellular pH is measurable. However, to make these calculations fully relaxed spectra have to be used because the ratios of PCr/ATP and less PCr/Pi are affected by the mode of the ^{31}P NMR spectral collection [1]. The general relationship between PP and PCr/Pi remains valid regardless of the type of spectrum followed.

The above calculations are, of course, accurate only when a homogeneous tissue or cell collection is studied by ^{31}P NMR. The estimation of PP becomes more complicated when an inhomogeneous tissue is studied. In animal muscle Meyer et al. [9] have shown that slow twitch, oxidative red muscles have significantly lower PCr/Pi values than fast twitch, glycolytic white muscles. Thus, during all the studies reported here one has to remember that we are recording an average value of probably several PCr/Pi ratios.

4 Early ^{31}P NMR Studies in Mitochondrial Myopathies (MM)

The clinical and biochemical characteristics of MM are reviewed in this volume by S. DiMauro and C.P. Lee. This is a group of disorders characterized by muscle weakness and exercise intolerance. The mitochondria are structurally abnormal as seen by electron microscopy. It is assumed that the mitochondria are also functionally impaired in MM. Indeed, in many patients the biochemical defect in oxidative phosphorylation has been demonstrated in vitro in muscle samples or isolated mitochondria. Since ^{31}P NMR monitors some aspects of oxidative metabolism, it is not surprising that some of the initial studies with this new technique were performed using patients with MM.

Radda and Taylor [3] tried to summarize these studies by dividing the observations into three groups:

1. Patients with reduction in post-exercise recovery of PCr, but with normal ^{31}P NMR at rest. This was first described in two sisters with MM due to NADH-coenzyme Q reductase deficiency [10]. Also observed in these patients was an early fall in PCr during mild exercise.
2. A significantly elevated level of Pi at rest (resulting in low PCr/Pi), but with normal post-exercise recovery. This was first described by Gadian et al. [11] in one patient with a mitochondrial disorder, in which the defective site was not characterized. It was probably the first reported case of MM studied by ^{31}P

NMR. It is interesting to note that in these patients, only a modest reduction of intracellular pH with rapid pH recovery after exercise was found.
3. No ^{31}P NMR abnormalities, a situation present in eight presumed cases of MM. It is not clear if the latter patients had clinical involvement of skeletal muscle.

These initial observations have suggested that the ^{31}P NMR findings in MM may be very variable, as are the clinical and biochemical profiles. Another very important early report was that of Eleff et al. [12] who studied a patient with a well-documented complex 3 deficiency. This patient had ^{31}P NMR abnormalities at rest, during exercise and following muscle work. Thus, this patient first demonstrated the possibility that in cases with marked bioenergetic impairment, the ^{31}P NMR will show abnormalities at each tested step of the oxidative metabolism. This report was also revolutionary because it showed that ^{31}P NMR can be used as an aid in designing new therapy for MM (the details will be described below).

There was a clear need for a study of a large group of these rare disorders, using consistent protocols, to further characterize the ^{31}P NMR changes in MM. This was first done by Arnold et al. [13]. Twelve patients with MM, well characterized clinically and histologically, were reported. The biochemical lesions were, however, less characterized. In three patients, the defective site was found to be in NADH-coenzyme Q reductase and in another three it was probably at coenzyme Q. The other six cases were not defined biochemically. At rest, 8 of the 12 patients had abnormally low PCr levels, resulting in low PCr/Pi values. Assuming unchanged ATP concentration, the phosphorylation potential was calculated and found to be reduced in 10 of the 12. During exercise, the patients performed less "work" than the normal control subjects (work was not measured directly, but only estimated through pressures achieved against a manometer), yet their PCr/Pi values dropped to lower values compared with controls (0.45 vs 0.66). Recovery of PCr concentrations after exercise was slow in only 3 of 10 tested patients. The authors also had the impression that intracellular pH reductions during exercise were less marked in MM compared with normals.

The problems this study raised were:

1. Only a limited scope of mitochondrial diseases was tested. Was the pattern similar in patients with other defective sites?
2. Of the 12 patients selected for evaluation of muscle ^{31}P NMR in MM, 3 or 4 had minimal or no clinical involvement of skeletal muscle. Could that alter the homogeneity of the results?
3. There were problems in evaluation of exercise in very weak MM patients. Are the findings really comparable to a normal group of subjects?
4. Resynthesis of PCr relies exclusively on oxidative metabolism, as the same authors showed [14]. Why then did only 30% of the patients have abnormal recovery rates?

Our work tried to re-examine the ^{31}P NMR profile in MM and relate the findings to the above and other problems. Some of the results have already been published and presented here will be a larger group of patients. I will try to elucidate the ^{31}P NMR results in our MM group in view of recent observations made in other neuromuscular diseases. The study was performed in the Departments of Biochemistry/Biophysics and Neurology at the University of Pennsylvania, Philadelphia. Several researchers took part in the various studies, but the major contribution was made by B. Chance and W.J. Bank. Other physicians who contributed to the studies are given in the Acknowledgements.

5 Aims of the Study

The basic study of 18 well-documented MM cases was carried out in the attempt to answer the following questions:

1. Do all mitochondrial disorders have abnormal ^{31}P NMR?
2. What are the most sensitive ^{31}P NMR indicators of mitochondrial disease?
3. Are the abnormal findings at rest and during exercise always in agreement?
4. Do similar metabolic disorders show similar ^{31}P NMR profiles?

The methods used in our studies will not be described here, and the reader is referred to some of the original reports for technical details [4,12,15].

6 Criteria for Entry of Patients to the Study

In order to study only clear-cut cases of mitochondrial disorders, patients were selected according to the following criteria:

1. Only patients with clinical evidence of skeletal muscle involvement were studied. Mitochondrial disorders may remain limited to the ocular muscles or may not involve striated muscle at all. Patients with such localized disease may not show any ^{31}P NMR changes in their skeletal muscles. Four patients with ophthalmoplegia and abnormal mitochondria in ocular muscles only were in fact studied separately. Three of them did not demonstrate any abnormalities (Argov and Bank unpublished), comparable to the normal findings obtained by Arnold et al. [13].
2. Histological evidence of abnormal mitochondria in skeletal muscle was obtained in all subjects. This included both the presence of ragged-red fibers and abundant structurally abnormal mitochondria by electron microscopy.
3. Infants and young children were not included. This was because they could not perform exercise in the magnet and thus, the dynamic part of the ^{31}P NMR protocol could not be evaluated. Thus, the ages of the patients studied were 14 to 52 years.
4. An attempt to locate the biochemical site of the mitochondrial disease was performed in all patients. This was regarded as an important criterion for patient selection. There were many other patients who were diagnosed as having MM on clinical and histological evidence only. Such patients were not included in the study as we attempted to correlate the defect type with the ^{31}P NMR features. This criterion selected more clearly defined MM with more variability in the biochemical defects. The biochemical studies were performed using either fresh

muscle mitochondria or frozen muscle samples (or both) in the laboratories of S. DiMauro (Columbia-Presbyterian, New York) and C.P. Lee (Wayne State, Detroit). Several patients with only histologically diagnosed MM were later studied separately and it should be mentioned that they showed a similar pattern of ^{31}P NMR abnormalities (Argov and Bank unpublished).

By such selection we were able to reach several patients with unique mitochondrial disorders that represented the wide gamut of MM. Included were patients with complex 1 defects (three cases), complex 3 deficiency (one case), cytochrome c oxidase impairment (five cases), impaired ATP associated metabolism (two cases) and presumed disturbed substrate availability or metabolism (two cases). In some patients the site could not be accurately determined even after the biochemical studies.

The clinical picture in the selected patients was also varied, as the spectrum of mitochondrial diseases is wide. While in some subjects the disease affected skeletal muscle only, others had associated involvement of brain and retina. However, on clinical grounds we separated them into two groups: severe disease (7 cases) and mild to moderate illness (11 cases). All patients were studied during a metabolically stable phase in their disease, avoiding mainly the study of patients with acute lactic acidosis.

7 Rest PCr/Pi

During the resting state, the muscle has a very low metabolic rate and the mitochondria are in state 4 [4]. That probably means that the PP is at its highest. In normal human muscles we found the PCr/Pi to be 8.5 ± 2.1 ($n = 20$). Assuming [ATP] of 8.0 mM, total muscle creatine of about 40 mM, intracellular pH at rest of 7.07 and $K_{ck} = 1.66 \times 10^{-9}$ M^{-1}, this corresponds to a PP of $100\,000 \pm 30\,000$ M^{-1}.

In 15/18 (83%) of the patients, PCr/Pi was low. It varied from 1.7 to 7.8, averaging 5.0 ± 1.75 ($p < 0.01$). We are not sure to what levels of PP such PCr/Pi correspond because the muscle ATP concentration-[ATP] is unknown in many disease states. It seems reasonable to assume that in mitochondrial disorders, the [ATP] in muscle will not remain normal. This, however, was never studied systematically. If [ATP] is indeed lower in MM than in normal controls then, according to Eq. (4) above, the PP calculated will have even lower levels.

The finding of low PCr/Pi at rest in MM is not surprising. In fact, it was thought to be a specific finding in mitochondrial diseases, which affects the energy state of muscle. When more ^{31}P NMR studies were performed in other muscular disorders it became clear that low PCr/Pi at rest is not specific to primary MM. Table 1 lists those disorders with low PCr/Pi at rest, already reported.

In carnitine deficiency and muscle cramp the energy pathways are directly affected. It is thus logical to assume that the energy state of muscle cells was altered leading to low PCr/Pi in these diseases. There is a variety of findings suggesting that the lack of thyroid hormones affects mitochondrial metabolism [18]. In malignant hyperthermia, a drug-induced syndrome with dominantly-inherited susceptibility,

Table 1. Disorders with low PCr/Pi at rest

1.	Mitochondrial myopathies
2.	Primary carnitine deficiency [16]
3.	State after muscle cramp in myophosphorylase deficiency [17]
4.	Hypothyroid myopathy [18]
5.	Muscular dystrophy [19]
6.	Myotonic dystrophy [20]
7.	Motor neuron and other denervating disorders [21]
8.	Polymyositis [20]
9.	Muscle injury [20]
10.	Recurrent myoglobinuria of undetermined origin [20]
11.	Malignant hyperthermia susceptibles [22]
12.	Several animal models of muscle diseases [23]

mitochondrial uncoupling during attacks was suspected [22]. But all the other disorders are not known to result from a primary disturbance in energy metabolism. In the process of muscle cell death, intracellular calcium accumulates and this may interfere with mitochondrial metabolism. Such a possible secondary mitochondrial malfunction remains to be characterized in any destructive myopathy. Yet, by losing specificity, we have gained wider usage of rest PCr/Pi, as will be described in the section dealing with therapy monitoring.

8 Changes During Exercise

Limited exercise capacity is one of the major features of mitochondrial disorders. When energy metabolism is stressed, a rapid decline in PP, expressed as a premature decrease of the PCr/Pi, is expected in mitochondrial myopathies. Indeed, this was seen in all our patients; after 3–6 min of moderate muscle work the PCr/Pi values dropped below 1. The problem was how to quantify this feature. If the patients tried to maintain a steady-state exercise for a few minutes, they usually reached levels of PCr/Pi which were nonstable (usually, even normal subjects cannot maintain steady-state exercise at PCr/Pi ratios below 1 [4,6,7]). Thus, we could obtain an exercise performance of two to three steady states, the amount needed to accurately calculate the slope of work rate to Pi/PCr function, in only five patients. The slopes were found to be very low: 3.6–20 J/min/Pi/PCr (normal range in sedentary persons 30–50 J/min/Pi/PCr). It can be argued that this was partly due to the fact that the maximal work performance of these muscles was reduced. That means that we were looking at that part of the rectangular hyperbola where the slope was already diminished (i.e. near the maximal work rate [6,7]). This is probably partially true. However, in two patients we recorded the PCr/Pi values reached while they performed steady-state exercise at 50% of their individual maximal capacity. It was found to be 0.65 and 0.7. The normal range in 20 controls was 1.1 ± 0.3 [16], as demonstrated in one case of Fig. 2. This is one possible way to compare exercises between diseased and nondiseased muscles. Even at such a single, moderate level of exercise, other patients with MM found it difficult to maintain even a few minutes of steady-state exercise. Thus, although our results and those of other groups indicate early and rapid reduction of PCr levels during exercise in MM we are still in search of a good way to quantify these abnormalities.

One of the interesting findings of exercising muscles with mitochondrial malfunction was their relative resistance to intracellular acidosis. It was thought that since in such muscle disorders oxidative metabolism is impaired, the cells will rely on enhanced glycolytic activity during exercise, resulting in intracellular lactic acid accumulation. Clinically, many of the mitochondrial disorders present with lactic acidemia and low serum pH, presumably due to the above mechanism. Surprisingly, a different picture was observed when intracellular pH was recorded in MM by ^{31}P NMR during exercise. In normal subjects, exercise at levels of PCr/Pi < 1 usually led to a rapid reduction in pH. Presumably at this stage mitochondrial metabolism was fully activated and normal muscle used glycolysis for further production of ATP [6,7]. In 12 controls exercising for more than 3 min at PCr/Pi < 1, intracellular pH was reduced to 6.65–6.7. In 13 of 18 patients intracellular pH dropped only to 6.78–7.0 (the intracellular pH at rest in normal muscle was 7.0 $\pm$ 0.05 and was similar in the MM group). Four other patients had a pH reduction in the normal range and one patient had an early, prompt reduction in intracellular pH, before reaching PCr/Pi values of 1. The reason for the reduced acidosis observed in several patients is not clear. We postulated that the chronic overproduction of lactic acid in muscle cell with mitochondrial impairment leads to stimulation of intracellular buffering mechanisms that partially prevented acidification of the intracellular compartment. This proposed mechanism has yet to be characterized.

9 Post-Exercise Recovery

The return of the reduced PCr/Pi at the end of exercise to its resting levels is a process that is primarily controlled by oxidative metabolism [14]. The enhanced glycolytic metabolism, activated by muscle contraction, stops and only the fully activated mitochondria (being at state 3) resupply the PCr levels and are responsible for the return of PP (and the PCr/Pi) to its pre-exercise levels. Thus, monitoring the post-exercise recovery of the PCr/Pi should be a good indicator for mitochondrial function. This can be easily tested even in very weak patients with marked exercise intolerance. That is because the only thing the patient has to do is exercise for few minutes to reach a depleted level of PCr/Pi. Hopefully, this can be done when the last 2–3 min of exercise are at a steady state of PCr/Pi. Indeed, all our patients were able to meet these requirements. However, before evaluating the results of post-exercise recovery of PCr/Pi in MM an important fact has to be mentioned: the rate (slope) of recovery is normally dependent on the end-exercise state [14]. Specifically, the lower the end-exercise PCr/Pi, the lower is the rate of recovery. This finding was attributed to the level of acidosis that was reached at the end of exercise [14]. The more acidotic the normal muscle cell became during exercise, the slower was the recovery process. While this mechanism is probably important in many conditions it may not be the only one. We have recorded differential recovery rates, depending on the end-exercise PCr/Pi, in myophosphorylase deficiency (McArdle's syndrome) [17]. In this condition there is a full block of glycolysis with no intracellular acidification even in maximally depleting exercise. Thus, other mechanisms may be responsible for the different rates of post-exercise recovery. However, if post-exercise recovery rates from similar end-exercise PCr/Pi levels are compared, accurate appreciation of mitochondrial function can be obtained.

In 17 of 18 patients (94%) the rate of PCr/Pi recovery after exercise was abnormal. It ranged from 0.2 to 2.02 units per minute with an average of 0.93 ± 0.54 units per minute (normal rate from a similar end-exercise PCr/Pi range of 0.6–0.9 was 1.8 ± 0.3 units per minute, $p < 0.001$). Thus, post-exercise recovery of PCr/Pi was the most sensitive ^{31}P NMR indicator of mitochondrial malfunction and probable the most "pure" of all.

10 Conclusions from the Basic ^{31}P NMR Study of MM

The main findings in the group of 18 patients with varied mitochondrial defects were:

1. The most sensitive ^{31}P NMR indicator of mitochondrial impairment was the reduced rate of PCr/Pi recovery after exercise.
2. PCr/Pi at rest was also a sensitive indicator of mitochondrial disease. However, as further studies have shown, this was not a specific marker for primary mitochondrial disease.
3. Exercise was so impaired in these patients that accurate evaluation of it by ^{31}P NMR was very limited.

We believe that the triad of the three features: low PCr/Pi at rest, reduced slope of work rate to Pi/PCr and slow post-exercise recovery is diagnostic of mitochondrial disease and we have not found this combination in many other muscular diseases that we tested. This combination was, however, recorded in two hypothyroid myopathy patients [18]. Thyroidectomized rats also showed excessive PCr/Pi reduction during exercise and slowed PCr/Pi recovery when compared to normal rats [18]. This led us to conclude that mitochondrial impairment plays an important role in hypothyroid myopathy and is most likely responsible for the exercise intolerance in this condition.

What was the correlation between the ^{31}P NMR findings and the clinical disorders that were present in the 18 patients? First, we found no relationship between the site of defect and the degree of ^{31}P NMR impairment. In three of our patients with a defect in complex 1 we found the diagnostic triad in one and only a moderate reduction of post-exercise recovery in the others. Second, when we scored the degree of ^{31}P NMR impairment, we found that in seven patients the findings were severe (PCr/Pi at rest < 5.0, recovery rate < 1 unit/min) and in the others, moderate or mild. This ^{31}P NMR score was similar to the subdivision of patients into severe and moderate-mild cases by clinical evaluation. Thus, the ^{31}P NMR findings best correlated with the degree of in vivo bioenergetic impairment as expressed in the clinical picture.

Another interesting observation was the discrepancy between reduced PCr/Pi at rest and slowed post-exercise recovery. In one of the MM cases PCr/Pi at rest was very low (3.8), but recovery was normal at 2.0 units/min. In another patient, the resting state was normal (PCr/Pi = 8.1) but recovery was markedly slowed (0.62 units/min). The second observation can be explained by the fact that the energy metabolism was more stressed during recovery and, thus, the impairment was "unmasked". However, such an explanation does not apply for the first patient. While this was only an observation in a single case, data from previous works support this discrepancy [3,13]. It may be suggested that the mechanisms determining the two ^{31}P NMR features: PCr/Pi at rest (a long-term change) and post-exercise recovery of PCr/Pi (a rapid change), are different.

11 ^{31}P NMR Monitoring of Therapy

One of the first ^{31}P NMR reports of MM was that of a patient with complex 3 deficiency who was treated with vitamins K_3 (menadione) and C[12]. This patient had myopathy, exercise intolerance and lactic acidosis. Since the site of the biochemical defect was well characterized (thanks to the work of Kennaway and Buist, University of Oregon, Eugene; [24]), a treatment with electron acceptors that could "bridge" the defect was designed by B. Chance. It was thought that vitamin K_3 would accept electrons from coenzyme Q, and transfer them to ascorbate, which would then transfer them further to cytochrome c. This mechanism was based on the known relative redox potentials of these compounds relative to each other and was supposed to restore another site of ATP production along the electron transport chain [12]. Since these substances were recognized drugs, they were given in very large doses. Their administration resulted in improvement of two ^{31}P NMR features: the rest PCr/Pi rose from 1.7 to 3.5 (Fig. 4) and the recovery rate was markedly increased (Fig. 3). These changes occurred within days, thus treatment was continued for 1 year and was associated with clinical improvement. Although the mechanism suggested by Chance was later questioned: the "bridge" might not be operational, but the vitamins were working as electron scavengers, the clinical and ^{31}P NMR improvement was well maintained, as a follow-up study has shown [25]. Furthermore, when the dose of K_3 was doubled, PCr/Pi values rose to 6 (Fig. 4) [25]. If proof was needed that the K_3 was indeed operative, it came after we tried to replace K_3 with K_1 because of a possible, better cell penetration by the lipid soluble K_1. After a short period of K_1 treatment, the PCr/Pi dropped again to 3.5 (Fig. 4) and the recovery rate was somewhat slower (Fig. 3). Original treatment was resumed. This very unique patient suggested that PCr/Pi values at rest may serve as a monitor of therapy (conventional and experimental).

The above notion was strengthened by ^{31}P NMR observations in other neuromuscular diseases. In muscular dystrophy of the Duchenne type, it was shown that there was an age-related reduction in PCr/Pi [19]. Since this was clearly parallel to the changes in disease severity, it was suggested that PCr/Pi values represent the deterioration in the disease state. Support for that came from our observation in a patient with chronic polymyositis, an inflammatory-destructive disease of muscle, who during an acute exacerbation showed rest PCr/Pi values which dropped from 4.5 to 3.3 [21].

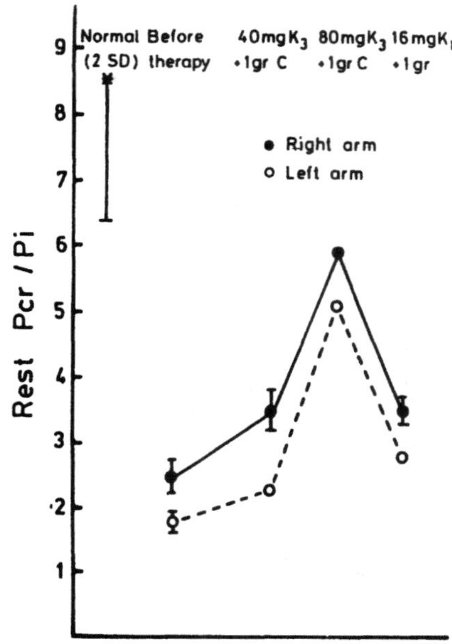

Fig. 4. The changes in the PCr/Pi ratios at rest in a patient with complex 3 deficiency following various modes of therapy (see text). *Bars* = 2 SD

The studies of McCully et al. [20] on muscle injury have clearly demonstrated that resting muscle PCr/Pi can change quickly even in acute disorders. After maximal eccentric exercise of the wrist flexors, rest PCr/Pi fell to abnormal levels within 24 h (4.9 ± 1.0, n = 5), associated with pain and slight edema. During the following 3–5 days the PCr/Pi levels rose back to the initial values in all the subjects tested. It could be concluded that ^{31}P NMR has the potential to monitor the disease state in very chronic as well as acute muscle disorders.

The above findings led to more ^{31}P NMR studies of muscle diseases with potential therapy (Fig. 5). Primary muscle carnitine deficiency probably leads to impaired mitochondrial function due to lack of substrates (long-chain fatty acids). In one patient, the rest PCr/Pi values before therapy were 5.5 and they rose to normal values of 8.1 after 18 months of carnitine supplementation [16]. In this patient the improvement in the bioenergetic state of muscle cells was probably due to enhanced mitochondrial function. Less clear was the same effect in polymyositis, in which conventional immunosuppressive therapy (steroids) showed clinical improvement, associated with a rise of PCr/Pi from 6.1 to 8.0 [21]. In hypothyroid myopathy, the abnormally low values of PCr/Pi at rest in two patients rose to normal levels after 3 months of hormonal supplementation [18].

We have used this information to monitor another experimental therapy of mitochondrial diseases, steroids. In one family with several female members of three generations affected with weakness and exercise intolerance, steroids were shown to have some beneficial effects (this family was studied by T. Heimann-Patterson, Hahnemann Hospital, Philadelphia [26]. Five members of this family were given 46 mg/d of methylprednisolone for 3 months and their ^{31}P NMR pattern was recorded before and after therapy. In four members the PCr/Pi values rose in association with better exercise tolerance (only three are shown in Fig. 5

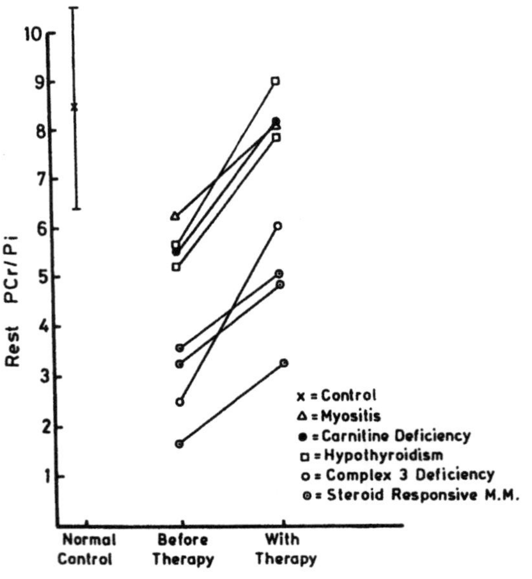

Fig. 5. The effects of therapy on PCr/Pi values at rest in various neuromuscular diseases. The mode of therapy in each case is described in the text. The normal range ± 2 SD is also shown

because of technical reasons). The elder grandmother, who was severely affected, did not improve. Two patients of this family also showed improvement in their post-exercise recovery rate by 118 and 150% (the others were too weak to exercise; [26]).

It can be concluded that although reduced rest PCr/Pi values are not specific for primary mitochondrial diseases, they may be of use in monitoring therapy of neuromuscular diseases and designing experimental therapeutic tests. Since ^{31}P NMR recording of PCr/Pi at rest is simple even in young children and babies, who do not cooperate during exercise, the wider use of this technique in monitoring therapy is expected.

12 Animal Models

There is no well-characterized, inherited animal model of MM. Thus, Hayes et al. [27] used a drug-induced model. They injected diphenyleneiodonium, an NADH-coenzyme Q reductase inhibitor, in rats and studied with ^{31}P NMR the stimulated gastrocnemius muscle in vivo. No change was observed at rest. Enhanced PCr depletion during exercise and slowed recovery afterwards were found. These data are comparable to those observed in the human condition with the defect being at a similar site. In the rat model, however, unlike the human condition, the intracellular acidosis was more marked than in the control group. This may be due to the fact that the model is acute and compensatory changes in buffering systems could not occur, as they do in the chronic human diseases.

As mentioned, we studied hypothyroid rats and found similar changes, which we attributed to mitochondrial impairment [18]. In the hypothyroid rats pH changes were comparable to those of normal animals. However, we tried to avoid excessive stimulation and this may explain the lack of marked acidosis in both normal and diseased groups. It is obvious, however, that the hypothyroid rat is not a good model of primary mitochondrial disease and the changes are secondary.

In dystrophic hamsters, several biochemical studies have demonstrated mitochondrial abnormalities [23]. Again, the mitochondrial impairment is thought to be secondary, probably to intracellular calcium accumulation. The ^{31}P NMR pattern in these animals was also indicative of mitochondrial malfunction showing the triad of low PCr/Pi at rest, enhanced reduction of PCr/Pi during muscle work and very reduced post-exercise recovery. This is a destructive myopathy that affects the heart muscle too, thus our findings of disturbed muscle oxidative metabolism may be due to several factors. It should be remembered that in marked heart failure, the oxygen supply to muscle is also reduced and this may affect the muscle ^{31}P NMR profile, especially during work. This should be taken into account when studying patients in whom MM is associated with cardiac involvement.

The field of ^{31}P NMR still awaits the finding of a "pure" primary chronic mitochondrial disease in an animal. Until then, the heterogeneity of bioenergetic profiles in mitochondrial diseases has to be regarded as multifactorial.

13 Comments

The studies performed by us and by other groups have clearly demonstrated the bioenergetic heterogeneity of MM as judged by ^{31}P NMR. It is clear that ^{31}P NMR can be a sensitive tool for detecting mitochondrial malfunction in vivo (primary and secondary). This is easier in muscle because of the relatively simple methods needed to record ^{31}P NMR spectra of muscle. Theoretically, similar changes may be detected in other affected tissues, such as heart and brain. The PCr/Pi values in these tissues are much lower (1.5–3.0), partly because a true resting state is nonexistent. Also, there are numerous technical problems in recording from these organs in vivo ^{31}P NMR. The development of methods for depth recording will certainly increase the usage of ^{31}P NMR for heart and especially brain metabolic studies. The first report of mitochondrial encephalomyopathy in which ^{31}P NMR of brain was studied showed an abnormal brain spectrum with low PCr/Pi [28].

^{31}P NMR will mainly remain a research tool in myology. It will not replace the in vitro biochemical studies in determining the site of defects. However, it will probably find clinical usage for screening and detecting possible mitochondrial impairments, especially in cases with exercise intolerance [21]. Another very promising use for ^{31}P NMR in MM will be the monitoring of therapy, both proven and experimental. This is the only noninvasive method that can repeatedly evaluate oxidative metabolism. The potential use for therapy monitoring may not be limited only to primary diseases of the mitochondrial metabolism.

Acknowledgements. Many of the observations reported here were performed while I was a Muscular Dystrophy Association Post-Doctoral Research Fellow in the Department of Biochemistry/Biophysics, University of Pennsylvania, Philadelphia. I was fortunate to work with Dr. Britton Chance of this Department and Dr. William Bank from the Neurology Department and thank them for their many contributions and support during the 3-year period.

Grants from the Muscular Dystrophy Association and the National Institute of Health (NS 08075) supported the human and animal research.

Many physicians have given me permission to use their patients and were responsible for the follow-up of patients and their therapies. I thank especially Dr. P. Peterson (Wayne State University, Detroit); Drs. R. Rowland, S, DiMauro, A. Penn (Columbia-Presbyterian, New York); Drs. N.R.M. Buist, N.G. Kennaway (University of Oregon, Eugene); Dr. Heiman-Patterson (Hahnemann University, Philadelphia) and Dr. Schotland (University of Pennsylvania, Philadelphia) for their help and involvement in the studies.

References

1. Gadian DG (1982) Nuclear magnetic resonance and its applications to living systems. Clarendon, Oxford
2. Taylor DJ, Bore PJ, Styles P, Gadian DG, Radda GK (1983) Bioenergetics of intact human muscle: a ^{31}P NMR study. Mol Biol Med 1:77–94
3. Radda GK, Taylor DJ (1985) Application of nuclear magnetic resonance spectroscopy in pathology. Int Rev Exp Pathol 2:1–58
4. Chance B, Eleff S, Leigh JS, Jr., Sokolow D, Sapega A (1981) Mitochondrial regulation of phosphocreatine/inorganic phosphate ratios in exercising human muscle: a gated ^{31}P NMR study. Proc Natl Acad Sci USA 78:6714–6718
5. Gyulai L, Roth Z, Leigh JS, Jr., Chance B (1985) Bioenergetic studies of mitochondrial oxidative phosphorylation using ^{31}P NMR. J Biol Chem 260:3947–3954
6. Chance B, Leigh JS, Jr., Clark BJ, Maris J, Kent J, Nioka S, Smith D (1985) Control of oxidative metabolism and oxygen delivery in human skeletal muscle: a steady state analysis of the work/energy cost transfer function. Proc Natl Acad Sci USA 82:8384–8388
7. Chance B, Clark BJ, Nioka S, Subramanian H, Maris JM, Argov Z, Bode H (1985) Phosphorus nuclear magnetic resonance spectroscopy in vivo. Circulation 72 (Suppl 4):103–110
8. Gibbs C (1985) The cytoplasmic phosphorylation potential. J Mol Cardiol 17:727–731
9. Meyer RA, Brown TR, Kushmerick J (1985) Phosphorus nuclear magnetic resonance of fast and slow twitch muscle. Am J Physiol 248:C279–C287
10. Radda GK, Bore PJ, Gadian DG, Ross BD, Styles P, Taylor DJ, Morgan-Hughes J (1982) ^{31}P NMR examination of two patients with NADH-CoQ reductase deficiency. Nature (London) 295:608–609
11. Gadian DG, Radda GK, Ross BD, Hockaday J, Bore P, Taylor D, Styles P (1981) Examination of a myopathy by phosphorus nuclear magnetic resonance. Lancet ii:774–775
12. Eleff S, Kennaway NG, Buist NRM, Darley-Usmar VM, Capaldi RA, Bank WJ, Chance B (1984) ^{31}P NMR study of improvement in oxidative phosphorylation by vitamins K3 and C in a patient with a defect in electron transport at complex 3 in skeletal muscle. Proc Natl Acad Sci USA 81:3529–3533
13. Arnold DL, Taylor DJ, Radda GK (1985) Investigation of human mitochondrial myopathies by phosphorus magnetic resonance spectroscopy. Ann Neurol 18:189–196
14. Arnold DL, Matthews PM, Radda GK (1984) Metabolic recovery after exercise and the assessment of mitochondrial function in human skeletal muscle in vivo by means of ^{31}P NMR. Magn Resonance Med 1:307–315
15. Argov Z, Bank WJ, Maris J, Peterson P, Chance B (1987) Bioenergetic heterogeneity of human mitochondrial myopathies: phosphorus magnetic resonance spectroscopy study. Neurology 37:257–262
16. Argov Z, Maris J, Fishbeck K, Bank W, Chance B (1985) In vivo study of human lipid myopathies by ^{31}P NMR spectroscopy. Ann Neurol 18:119 (Abstr)

17. Argov Z, Bank WJ, Maris J, Chance B (1987) Muscle energy metabolism in McArdle's syndrome by in vivo phosphorus magnetic resonance spectroscopy (^{31}P NMR). Neurology 37:1720–1724
18. Argov Z, Renshaw P, Boden B, Winokur A, Bank WJ (1988) The effects of thyroid hormones on skeletal muscle bioenergetics: an in vivo ^{31}P NMR study of humans and rats. J Clin Invest 81 (in press)
19. Younkin DP, Berman P, Sladky J, Bank WJ, Chance B (1987) ^{31}P NMR studies in Duchenne muscular dystrophy: age related metabolic changes. Neurology 37:257–262
20. McCully KK, Argov Z, Boden B, Brown R, Bank WJ, Chance B (1988) Detection of muscle injury in humans with ^{31}P NMR magnetic resonance spectroscopy. Muscle Nerve 11:212–216
21. Bank WJ, Argov Z (1987) The value of ^{31}P NMR in the diagnosis and monitoring the course of human myopathies. Ann NY Acad Sci 508:448–450
22. Olgin J, Argov Z, Rosenberg H, Chance B (1988) Non-invasive evaluation of malignant hyperthermia susceptibility with phosphorus nuclear magnetic resonance spectroscopy. Anesthesiology 68:507–513
23. Argov Z, Maris J, Damico L, Chance B (1988) Mitochondrial malfunction of dystrophic hamster muscle: in vivo ^{31}P NMR study. J Neurol Sci 86:185–193
24. Kenaway NG, Buist NRM, Darley-Usmar VM, Papadimitriou A, DiMauro S, Kelley RI, Capaldi RA, Blank NK, D'Agostino A (1984) Lactic acidosis and mitochondrial myopathy associated with deficiency of several components of complex 3 of the respiratory chain. Pediatr Res 18:991–999
25. Argov Z, Chance B, Maris J, Eleff S, Kennaway NG, Chance B, Olson R (1986) Treatment of mitochondrial myopathy due to complex 3 deficiency with vitamins K3 and C: a follow-up study. Ann Neurol 19:598–602
26. Heiman-Patterson TP, Argov Z, DiMauro S, Bonilla E, Tahmoush AJ, Bank W (1989) ^{31}P NMR studies during methylprednisolone treatment of a familial steroid-responsive mitochondrial disorder. Neurology 39 (Suppl 1): 337 (Abstr)
27. Hayes DJ, Byrne E, Shoubridge EA, Morgan-Hughes JA, Clark JB (1985) Experimentally induced defects of mitochondrial metabolism in rat skeletal muscle. Biochem J 229:109–117
28. Hayes DJ, Hilton-Jones D, Arnold DL, Galloway G, Styles P, Duncan J, Radda GK (1985) A mitochondrial encephalomyopathy: a combined ^{31}P NMR magnetic resonance and biochemical investigation. J Neurol Sci 71:105–118

Cytochemical and Immunocytochemical Studies of Human Mitochondrial Myopathies

M. A. JOHNSON

1 Introduction

The traditional role of cytochemical techniques in the investigation of mitochondrial myopathies is that of providing a preliminary diagnosis, and the presence of a respiratory chain disorder is often detected initially by means of mitochondrial 'marker' techniques used on tissue sections. Subsequent identification of the precise site of the defect within the respiratory chain is usually, and more appropriately, the province of biochemical investigation.

Virtually all current biochemical methodologies for the investigation of mitochondrial disorders involve the use of mitochondrial fractions isolated from tissue homogenates. In the majority of situations this practice causes no apparent problems since many respiratory chain disorders appear to affect the muscle fibre population in a relatively homogeneous manner. An example is the fatal infantile form of cytochrome *c* oxidase (Complex IV) deficiency (van Biervliet et al. 1977) and the same appears to be true of Complex IV deficiency associated with some cases of Leigh's disease (subacute necrotizing encephalomyelopathy (Willems et al. 1977). Little is known as yet of the distribution within muscle fibre populations of defects of Complex I and Complex III (Morgan-Hughes et al. 1977, 1979) associated with human myopathies.

There are two main approaches which may be adopted if it is necessary to investigate metabolic disorders at the level of individual muscle fibres. The first is the methodology pioneered by Lowry et al. (1978) involving microdissection of single fibres and the adaptation of biochemical assay techniques for use on very small muscle samples. The second is the development of microphotometric assays of enzyme activity in individual fibres in situ in tissue sections, an approach used by Pette and co-workers (Pette 1981; Reichmann and Pette 1982).

The partial cytochrome oxidase deficiency found in association with chronic progressive external ophthalmoplegia (CPEO) affords a striking example of a mitochondrial myopathy where this type of technique is essential for the effective investigation of the underlying defect. Initial qualitative cytochemical studies of biopsies from CPEO patients showed that a variable percentage of individual muscle fibres contained no demonstrable cytochrome oxidase activity (Johnson et al. 1983) whereas some fibres showed normal activity. If the proportion of enzyme-deficient fibres was low, assays of cytochrome oxidase in mitochondrial fractions frequently showed no significant decrease. The conflicting results of some

Department of Neurology, University of Newcastle upon Tyne, UK

A. Azzi, Z. Drahota and S. Papa
Molecular Basis of Membrane-Associated Diseases

previous biochemical studies of cytochrome oxidase activity in CPEO patients (DiMauro et al. 1980; Mitsumoto et al. 1983) are largely explicable in these terms.

The aim of this communication is to review what is currently known of the underlying defect(s) in patients with CPEO and similar syndromes, with particular reference to studies of respiratory chain abnormalities at a cellular level. These investigations include qualitative cytochemical studies, microphotometric assays of enzyme activity in single muscle fibres and immunocytochemical labelling of components of individual respiratory complexes.

2 Clinical Features

The syndrome of CPEO is characterized by progressive gaze restriction due to dysfunction of extraocular muscles, and is often associated with proximal limb weakness (Drachman 1968). The condition is frequently regarded as a more limited form of the Kearns-Sayre syndrome (KSS) in which patients show, in addition, varying combinations of the following features: cardiac conduction defects, cerebellar ataxia, neurosensory deafness, pigmentary retinopathy and abnormal shortness of stature (Kearns and Sayre 1958). The age of onset of clinical symptoms of KSS is generally before 15 years whereas CPEO may not become apparent until the patient is much older; in extreme cases this may be at over 60 years.

3 Routine Metabolic Studies

Investigations of blood metabolites in the two groups of patients may or may not be informative. Abnormally high levels of lactate, pyruvate and alanine have been recorded (Sulaiman et al. 1974; Reske-Nielsen et al. 1976; Johnson et al. 1983), either at rest or after exercise or fasting. It is probable that the presence or absence of such findings is largely dependent upon the proportion of the skeletal muscle mass affected by the primary defect(s).

4 Histopathology

Morphological mitochondrial abnormalities in skeletal muscle in oculocraniosomatic disorders were first described by Olson and co-workers (1972). Characteristically these consist of dense, peripheral aggregations of mitochondria which assume a granular appearance in tissue sections. In Gomori trichrome-stained material they appear as irregular red areas, often distorting the fibre periphery. This appearance has led to the common usage of the term 'ragged-red fibres' (RRF) to describe the phenomenon. The abnormality is, however, more appropriately demonstrated using specific cytochemical 'marker' enzyme techniques which allow the distribution of mitochondria to be visualized (Pearse 1972).

Ragged-red fibres are not confined to the muscle of KSS and CPEO patients. They have been described in syndromes characterized by lactic acidaemia and exercise intolerance associated with defects of Complex I and Complex III

(Morgan-Hughes et al. 1977, 1979). They are also an integral abnormality of the MERFF syndrome (myoclonus epilepsy with ragged-red fibres) which usually has an onset before 20 years (Fukuhara et al. 1980). It is also important to point out that absence of RRF can in no way by taken to indicate normal mitochondrial function in CPEO patients since occasional affected individuals show no RRF on light microscopy and only minor ultrastructural abnormalities (Turnbull et al. 1985).

5 Quantitative Cytochemistry

In order to investigate the mitochondrial defect in CPEO and KSS further we screened a series of muscle biopsies using a cytochemical technique for the demonstration of cytochrome oxidase activity (Seligman et al. 1968). This technique uses 3,3′diaminobenzidine (DAB) as electron donor in the respiratory chain at the level of cytochrome *c* (cyt. *c*). As a result of electron transfer catalyzed by cytochrome oxidase, DAB undergoes oxidative polymerization and cyclization to form an indamine polymer. This insoluble brown reaction product shows a particulate (mitochondrial) distribution within muscle fibres in tissue sections.

In our initial series of seven CPEO and KSS patients (Johnson et al. 1983) we found that ca. 5–30% muscle fibres showed no demonstrable cytochrome oxidase activity. We have since screened > 30 patients with CPEO or KSS and have found that cytochrome oxidase-negative fibres are an invariable feature of the muscle defect. On occasions $< 5\%$ fibres may be affected but only in exceptional circumstances have we found $> 50\%$ fibres to be cytochrome oxidase-negative. These observations have been corroborated by other groups (Müller-Höcker et al. 1983; Byrne et al. 1985). The recent studies of Bresolin et al. (1987) are of particular interest since they document a large increase in the proportion of affected fibres in a patient who was biopsied and subsequently studied 7 years later at autopsy. Such 'follow-up' studies are rare but indicate that the enzyme defect appears to be progressive.

Whereas the cytochemical properties of normal muscle fibres are uniform along the length of the fibres, it has been shown that morphological mitochondrial abnormalities may be confined to segments of muscle fibres (Sengers et al. 1986). We have found this to apply also to the cytochrome oxidase defect in CPEO and, since the defect is progressive, this may indicate fibres in the process of transition. Nevertheless, the majority of enzyme-deficient fibres could be traced for distances exceeding 1000 μm in tissue sections.

We found that the majority of fibres showing mitochondrial clumping (RRF) in SDH preparations also had severely decreased cytochrome oxidase activity (Fig. 1) or no activity. However, the proportion of cytochrome oxidase-negative fibres almost always exceeded that of RRF in individual biopsies, suggesting that the enzyme abnormalities may precede the morphological changes. In occasional patients up to 30% fibres were cytochrome oxidase-negative although no RRF were seen.

Estimation of the proportion of fibres which are cytochrome oxidase-negative is an unsatisfactory method of quantifying the defect, since it takes no account of the proportion of fibres in which enzyme activity is decreased though still detect-

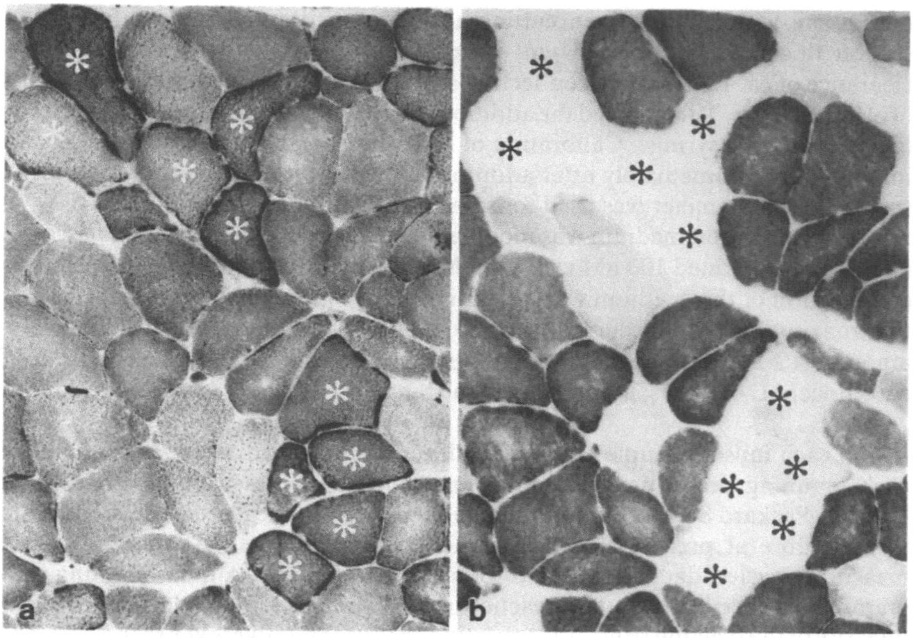

Fig. 1a,b. Serial sections from muscle biopsy from a CPEO patient showing: **a** peripheral mitochondrial clumping (ragged-red fibres) demonstrated using a cytochemical method for succinate dehydrogenase; **b** corresponding fibres (*asterisks*) have no cytochemically-demonstrable cytochrome oxidase activity (Seligman technique)

able. In normal skeletal muscle cytochrome oxidase activity is highest in Type 1 (slow-twitch oxidative) fibres, followed by Type 2A (fast-twitch oxidative/glycolytic) fibres, and is lowest in Type 2B (fast-twitch glycolytic) fibres. In order to detect an increase or decrease in enzyme activity in individual fibres, activity must be related to fibre type; this is done by reference to serial tissue sections in which the activity of myofibrillar ATPase is demonstrated (Brooke and Kaiser 1970).

In our initial studies we used end-point microphotometric estimations of enzyme activity in tissue sections, but measurement of reaction rates by means of a series of readings taken as the reaction proceeds is a more accurate method of cytochemical assay (Pette 1981).

6 Microphotometric Enzyme Determination

Increase in the amount of final reaction product (FRP) in cytochemical reactions may be measured by monitoring the increase in absorbance (Δ E) at the absorption maximum of the FRP. For the indamine polymer derived from DAB this is ca. 450 nm. We used a Zeiss UMSP 30 microphotometer fitted with a computer-controlled scanning stage allowing absorbance measurements to be made in multiple fibres simultaneously. Frozen sections were cut using a Reichert Frigocut cryostat microtome with motor-driven cutting action to ensure maximal uniformity of section thickness. Sections (8 μm) were mounted on standard glass slides. A clearance of ca. 1.0 mm between section and cover-glass was maintained by strips of electrical tape. This allowed the addition of ca. 0.2 ml reaction medium by means of a hypodermic syringe. Calibration of zero absorbance was made by taking an initial reading immediately after addition of the reaction medium. A measuring spot of 25 μm diameter was used for samples of adult muscle.

The incubation medium was modified from that described by Seligman et al. (1968) and contained 100 μM cyt. *c*, 4 mM DAB in 0.1 M phosphate, pH 7.0. The composition of this medium was determined by preliminary work which showed that maximal activity occurred using > 80 μM cyt. *c* in this assay system. The amount of FRP produced showed a linear relationship to the concentration of chromogen (DAB) present throughout the range 1.0–6.0 mM. Assays were done at 25°C.

In each muscle sample at least 100 fibres were assayed, using 20–25 fibres in each scanning procedure. The co-ordinates of the fibres were stored using a Hewlett-Packard Series 200 computer and ten cycles of measurements were made in each fibre at predetermined intervals (normally 100 s). At the end of ten measuring cycles, data were stored for statistical analysis. A data-handling programme was used to calculate reaction rate (slope) and intercept and degree of linearity of the reaction (Fig. 2). Correlation coefficients (r) were generally > 0.995, though in fibres with very low activity, instrument error assumed greater importance and lower values of r were found. There was no departure from linearity in later stages of the measuring cycles (up to 900 s) even in fibres with very high activity.

The distribution of cytochrome oxidase activities in subpopulations (Types 1, 2A and 2B) of normal muscle fibres is shown in Fig. 3; reaction rates are given as

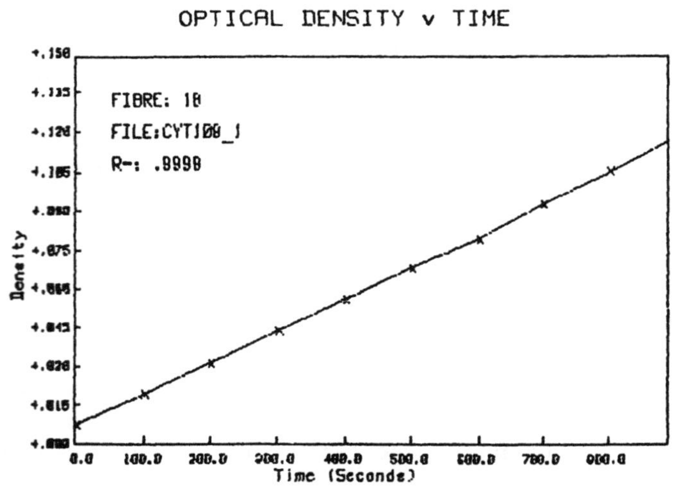

Fig. 2. Computer graph of the increase in optical density (Δ E) due to formation of reaction product in the cytochemical assay of cytochrome oxidase activity in a single muscle fibre

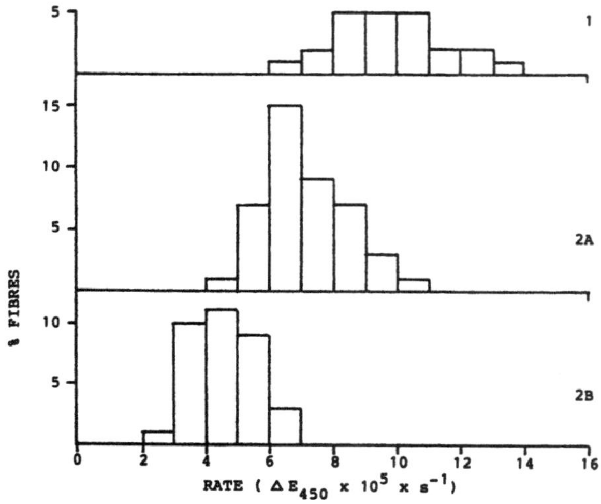

Fig. 3. Histogram showing cytochrome oxidase activities in single muscle fibres of normal human m.quadriceps (patient B.K.). Note the gradient of activities: Type *1* fibres > Type *2A* fibres > Type *2B* fibres

$\Delta E_{450} \times s^{-1}$. Samples of m.quadriceps (n = 10) from patients in whom there was no evidence of a neuromuscular disorder served as controls. Analysis of individual histograms of activity distribution allowed the following upper and lower limits to be set:

Type 1 fibres	$5–15 \times 10^{-5} \times s^{-1}$
Type 2A fibres	$3–10 \times 10^{-5} \times s^{-1}$
Type 2B fibres	$2–6 \times 10^{-5} \times s^{-1}$

Since reaction rates in individual fibres of CPEO patients were to be compared with normal, this was more meaningful than the calculation of mean rates. Very few fibres of any type showed reaction rates $< 3.0 \times 10^{-5} \times s^{-1}$ and none $< 2.0 + 10^{-5} \times s^{-1}$. Fibres with activities below this level correspond to the 'cytochrome oxidase-negative' fibres seen on qualitative inspection of reacted tissue sections. Absolute zero in this assay is ca. $0.3 \times 10^{-5} \times s^{-1}$, the value obtained in the presence of 5 mM azide.

Activity histograms from two patients with CPEO are shown in Figs. 4 and 5. In Fig. 4 it is seen that the cytochrome oxidase activity of the muscle is only moderately decreased. Many fibres show normal activity and the proportion of fibres with activities less than the lower limit of normal for that fibre type is not great. However, it is also seen that some fibres of *all* fibre types show activities $< 2.0 \times 10^{-5} \times s^{-1}$, i.e. the decrease in cytochrome oxidase activity is independent of fibre type. Figure 5 shows the distribution of activities in a muscle sample which is more severely affected; nevertheless, some fibres still retain normal activity. There is sometimes a clearly bimodal distribution of activities within one fibre type, indicative of separate populations of normal and enzyme-deficient fibres. This finding is totally different from that in muscle from Leigh's disease patients where fibre populations show a homogeneous decrease in cytochrome oxidase activity.

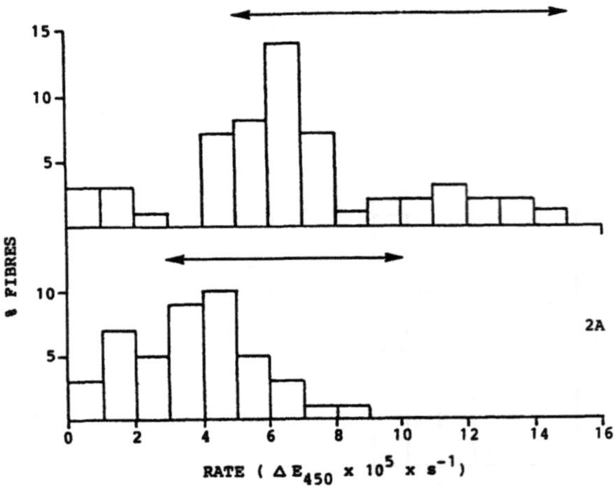

Fig. 4. Cytochrome oxidase activities in single muscle fibres of m.quadriceps from CPEO patient (J.G.). *Arrows* show normal range of activities in Type *1* and Type *2A* fibres; no Type 2B fibres were present in this biopsy. Note preservation of normal activities in the majority of fibres in this patient

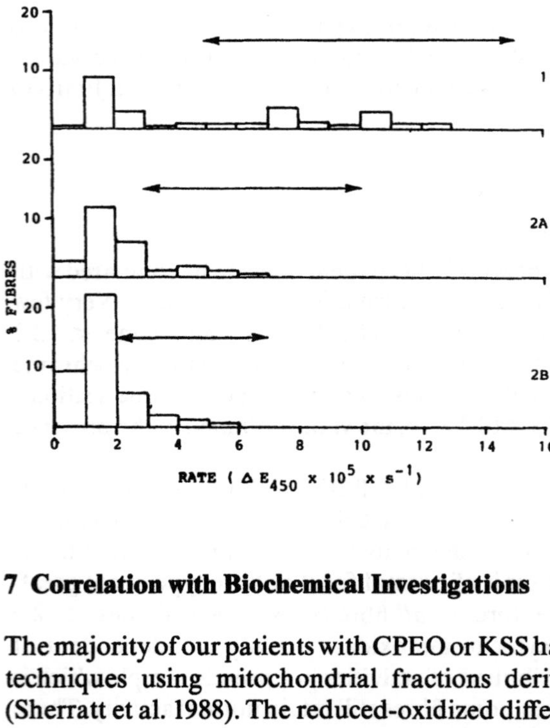

Fig. 5. Cytochrome oxidase activities in single muscle fibres of m.quadriceps from CPEO patient (D.B.). This patient is more severely affected with the majority of fibres of all fibre types showing activities less than the lower limit of normal (*arrows*)

7 Correlation with Biochemical Investigations

The majority of our patients with CPEO or KSS have also been studied by standard techniques using mitochondrial fractions derived from muscle homogenates (Sherratt et al. 1988). The reduced-oxidized difference spectra of the cytochromes were determined at -196°C; cytochrome concentrations of the muscle mitochondrial fractions were measured using the extinction coefficients given by Bookelman et al. (1978). Substrate oxidations were determined spectrophotometrically at 425–475 nm at 30°C with succinate or glutamate plus malate as substrates, using ferricyanide as final electron acceptor (Turnbull et al. 1982). Cytochrome oxidase activity was determined using the spectrophotometric method described by Smith (1955).

In severely cytochrome oxidase-deficient patients, cytochrome spectra revealed a corresponding decrease in cyt. aa_3 (Johnson et al. 1983; Turnbull et al. 1985) but this was not apparent in patients in whom comparatively few enzyme-deficient fibres were found.

A good correlation between conventional spectrophotometric assays of cytochrome oxidase and that using microphotometry of tissue sections (Seligman method) was reported in experimental studies of copper-deficient rat skeletal muscle (Turnbull et al. 1983). Subsequent studies of human myopathies have shown a broad measure of agreement between results of the two techniques. However, results obtained on three of our patients, using a different microphotometric method (Ballantyne and Bright 1979), conflict with biochemical findings. Although the latter showed the patients to have decreases of only 25–50% of normal activity, microphotometric estimations of cytochrome oxidase activity showed that *all* muscle fibres in the patients' biopsies had activities below the lower limit of normal (Reichmann et al. 1985). The probable reason for this anomaly is that the assay method used cannot be made totally independent of endogenous cyt. *c* concentration (Ballantyne and Bright 1979). Any decrease in endogenous cyt. *c* as compared with the normal control muscles will exert an additional effect and will be reflected in a spuriously large decrease in oxidase activity.

8 Immunocytochemical Studies of Cytochrome Oxidase Subunits

Antisera to individual subunits of cytochrome oxidase (Kuhn-Nentwig and Kadenbach 1986) purified from human heart have been used in immunocytochemical investigations of muscle biopsies from 17 CPEO patients (Johnson et al., 1988). The object of this study was to see whether any individual enzyme subunits were selectively affected in this disorder. Polyclonal antisera were raised in rabbits to subunits II/III (mitochondrially-coded) and to subunits IV, Vab, VIa, VIbc, VIIa, VIIbc and VIII (nuclear-coded). The antisera were characterized by immunoblotting (Kuhn-Nentwig and Kadenbach 1985) and ELISA, and were found to react only with their corresponding subunits except for the antiserum to subunit VIIa which also reacted with VIIbc.

In normal human skeletal muscle (Fig. 6a,b) particulate (mitochondrial) immunoreactivity was seen with all antisera except that raised against subunit VIa. This may be due to reactive epitopes being located in the lipid domain of the inner mitochondrial membrane and inaccessible to the antisera even after procedures designed to achieve maximal penetration of antisera to intramitochondrial sites. The frozen sections used for all immunolabelling studies were post-fixed in formol-calcium (4% formaldehyde, 0.1 M calcium chloride, pH 7.0) and transferred through a graded ethanol series prior to immunostaining. Primary antisera were used at the highest dilution to give consistent immunolabelling in all fibres of six normal muscle samples. Visualization was achieved using unlabelled swine anti-rabbit immunoglobulins (Dako, Z196) followed by rabbit peroxidase-anti-peroxidase complex (PAP; Dako, Z113) as described by Sternberger et al. (1970). Controls included substitution of non-immune rabbit serum as the first-layer reagent, and preabsorption of antisera with excess purified antigen (18h/4°C). These control procedures gave negative reactions in tissue sections.

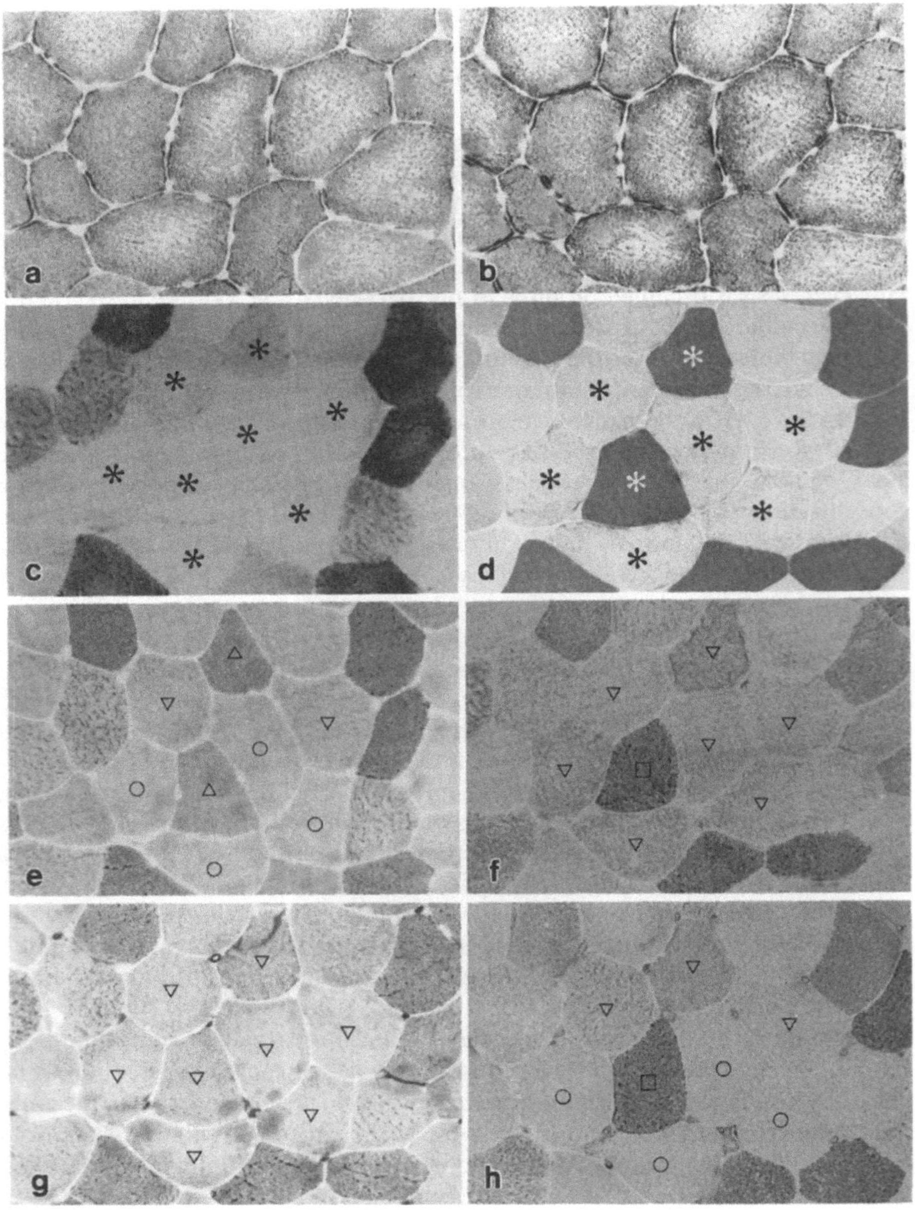

Fig. 6a-h.

Serial sections of CPEO biopsies were used to demonstrate the catalytic activity of cytochrome oxidase (Seligman technique) and the immunolocalization of enzyme subunits (PAP technique) in individual muscle fibres. In all CPEO biopsies, a proportion of fibres was found to show abnormal immunolabelling. Some fibres were devoid of any immunoreactivity; in others this was severely decreased and sometimes, although discrete, mitochondrial labelling was absent, diffuse reactivity was seen throughout the muscle fibres (Fig. 6c-h). Samples of at least 100 muscle fibres were used and, depending on the proportion of cytochrome oxidase-negative fibres included, subunit immunoreactivity was analyzed in 25–75 enzyme-deficient fibres per biopsy. Representative results from three patients are shown in Fig. 7.

Regression analysis of the numbers of fibres deficient in (1) cytochrome oxidase catalytic activity and (2) immunoreactivity towards each of the subunit

Patient AJ							Patient RL							Patient DB						
CYTOX	II/III	IV	Vab	VIbc	VIIa	VIIbc	CYTOX	II/III	IV	Vab	VIbc	VIIa	VIIbc	CYTOX	II/III	IV	Vab	VIbc	VIIa	VIIbc
o	o	o	o	o	o	o	*	o						o	o		o		o	o
*			o	o	o	o	o	o				o	o	o					o	
o	o	o		o	o		o	o	o	o	o	o	o	*					o	o
*		o	o	o	o	o	o	o	o	o	o	o	o	o	o		o	o	o	o
o	o	o	o	o	o	o	*	o	o			o	o	o						
o	o	o	o	o	o		o	o				o	o	o		o		o	o	o
o				o		o	o	o		o		o	o	o	o		o		o	o
o	o	o	o	o	o	o	o	o					o	o	o		o	o		o
*		o	o		o	o	*	o					o	o	o		o	o	o	o
o	o	o	o		o	o	*	o		o				o	o			o	o	o
o	o	o	o		o	o	o	o	o	o		o	o	o	o			o	o	o
o	o	o	o		o	o	*	o	o			o	o	o	o	o	o	o	o	o
o		o	o		o	o	o	o			o	o		o	o	o	o	o	o	o
o	o	o	o	o	o	o	o	o	o	o				o	o			o	o	o
o	o	o	o	o	o	o	o	o		o	o	o	o	o	o		o	o	o	o
o	o	o	o	o	o	o	o	o	o	o	o	o	o	*				o	o	o
o	o	o	o	o	o	o	*	o	o		o	o	o	o	o		o	o	o	o
o	o		o			o	*	o			o	o	o	o	o		o	o	o	o
o	o	o	o	o	o	o	o	o						*	o				o	
o	o						o	o	o		o		o	o	o	o		o	o	o
*		o	o			o	o	o	o	o		o	o	o	o	o	o	o	o	o
o	o		o				*	o	o	o				o	o		o	o	o	o
o							*	o		o				o	o		o	o	o	o
o	o	o	o	o	o	o	o	o	o	o	o	o		o	o		o		o	o
o	o	o	o	o	o	o	*	o	o	o				o	o		o		o	o

Fig. 7. Correlation of subunit immunolabelling and cytochrome oxidase activity in samples of 25 fibres from three CPEO patients. Fibres with negative or severely decreased immunostaining are designated (○). Absence of cytochrome oxidase activity is also shown as (○) and severely decreased enzyme activity as (*)

Fig. 6. a Normal human skeletal muscle showing immunocytochemical localization of cytochrome oxidase subunits II/III (mitochondrially-coded) using subunit-specific antisera and peroxidase labelling (PAP technique). **b** Serial section showing immunolocalization of cytochrome oxidase subunit IV (nuclear-coded). Note particulate (mitochondrial) reaction in all fibres. **c** Muscle from patient with CPEO showing a cluster of eight fibres (*asterisks*); six of these show no demonstrable cytochrome oxidase activity and two show severely decreased activity (Seligman technique). **d** Section serial to (**c**) showing that two fibres are Type 1 (*dark*) and six Type 2A (*pale*). Myofibrillar ATPase technique after pH 4.6 preincubation. **e** Subunit II/III immunoreactivity showing negative reaction (○), decreased reaction (▽) and diffuse reaction (△) in these fibres in serial section. **f** Subunit IV immunoreactivity in serial section. Decreased reaction is seen in seven fibres (▽) but normal immunolabelling (□) is retained in one. **g** Subunit VIbc immunolabelling shows moderate to severe decrease in all eight fibres (▽). **h** Subunit VIIa immunolabelling shows negative reaction (○) in four fibres, moderate to severe decrease in three and preservation of normal reaction (□) in one antisera was carried out. There was close correlation ($r = 0.84, p < 0.001$) between loss of catalytic activity and absentce or severely decreased immunolabelling of subunits II/III. Nuclear-coded regulatory subunits showed variable correlation with loss of catalytic activity. In some instances this was high ($r > 0.725, p < 0.001$); subunit Vab 0.773, subunit VIbc 0.745, subunit VIIa 0.836, subunit VIIbc 0.833. However, many cytochrome oxidase-deficient fibres retained immunoreactivity of subunit IV ($r = 0.441, p > 0.05$).

The mean percentages of cytochrome oxidase-deficient fibres showing abnormal immunostaining with each of the subunit antisera were calculated from percentages of affected fibres in individual patients ($n = 17$). Results are given as % ± SD: subunit II/III 81 ± 18, subunit IV 51 ± 20, subunit Vab 61 ± 17, subunit VIbc 53 ± 18, subunit VIIa 59 ± 21, subunit VIIbc 70 ± 20 (reactions with anti-VIII sera were not strong enough to permit evaluation in many biopsies).

All the patients studied showed a fairly constant pattern of severely and less severely depleted subunits. It was also clear that no one subunit was exclusively affected. Using anti-subunit II/III sera, 11/17 patients showed decreased immunostaining in 80–100% fibres in which catalytic activity was impaired; this is not surprising since subunits II/III are involved in the catalytic function of the enzyme complex.

Studies of 'nearest-neighbour' relationships of subunits from rat liver using cross-linking agents (Jarausch and Kadenbach 1985) have shown that many of the nuclear-coded polypeptides are linked spatially to the large mitochondrially coded subunits. It is therefore possible that some nuclear-coded subunits cannot be properly incorporated into the inner mitochondrial membrane if the large catalytic subunits are absent. Diffuse immunoreactivity suggestive of extra-mitochondrial localization could be due to unincorporated subunits or, alternatively, to subunit proteins or breakdown products thereof liberated from abnormal mitochondria.

Antisera raised against cytochrome oxidase holoenzyme have also been used in immunocytochemical studies of KSS and CPEO patients. Significant decreases in immunoreactivity were reported in enzyme-deficient fibres by Müller-Höcker et al. (1985) but similar studies by Johnson et al. (1985) and Bresolin et al. (1987) failed to find any appreciable reduction in immunolabelling which corresponded to loss of catalytic function.

9 Other Respiratory Chain Abnormalities in CPEO Patients

There is considerable evidence that other sites in the respiratory chain are involved in the mitochondrial defect in CPEO patients, in addition to the partial deficiency of Complex IV described above. For example, in the series of patients studied by Sherratt et al. (1984, 1986) and Johnson et al. (1988) 5/9 patients showed associated Complex I abnormalities and Complex III deficiency was found in one additional patient. These defects were identified in polarographic studies and by spectrophotometric assays of the individual respiratory complexes made on mitochondrial fractions.

There is as yet no cytochemical assay which is specific for Complex I activity. The technique for the demonstration of NADH-tetrazolium reductase activity

(Pearse 1972), which is extensively used in muscle histopathology, demonstrates not only NADH-ubiquinone reductase activity but also that of NADH-cyt. b_5 reductase which is present in the outer mitochondrial membrane and the sarcoplasmic reticulum. Until a specific assay for Complex I is developed for use on tissue sections, several questions remain unanswered: (1) is the partial defect of Complex I in these patients homogeneously distributed in the muscle fibre population? (2) if not, are the same fibres affected by both Complex I and Complex IV defects? In addition, it is possible that mild Complex I or Complex III deficiencies in CPEO patients might remain undetected for the same reasons that apply to Complex IV defects investigated solely using muscle homogenates.

An alternative method of studying Complex I deficiencies at the cellular level is by immunocytochemical labelling. In our laboratories, two patients with severe Complex I defects associated with exercise intolerance and muscle weakness have been investigated using antisera to Complex I holoenzyme (Cleeter and Ragan 1985). This has shown a large decrease in immunoreactivity in the muscle fibre populations as a whole as compared with normal control tissue. However, anti-holoenzyme sera are not ideally suited for the detection of less widespread defects and have given no clear-cut results when applied to muscle from CPEO patients known to have partial Complex I defects. In immunocytochemical procedures involving anti-holoenzyme sera, it is likely that high reactivity towards some epitopes may mask significant decreases elsewhere.

Abnormalities of mitochondrial ATPase (mtATPase) in the muscle of CPEO patients have been reported by Müller-Höcker and co-workers (1985). Previous biochemical studies have demonstrated that loose coupling of oxidative phosphorylation is accompanied by an increased endogenous activity of this enzyme (Ernster et al. 1959). Cytochemical techniques for the demonstration of mtATPase activity in situations of close and loosely coupled oxidative phosphorylation (Meijer 1972) have shown an apparent difference in cytochemical latency in the two conditions. In normal muscle an uncoupler must be used to overcome the latency of mtATPase activity in tissue sections; in cases of loosely coupled oxidative phosphorylation, activity is demonstrable without the prior use of an uncoupler. This phenomenon has been described in a variety of mitochondrial myopathies (Sengers et al. 1984; Müller-Höcker et al. 1986) and has been considered to be a non-specific indication of mitochondrial dysfunction. Although the condition occurs in CPEO patients in association with cytochrome oxidase deficiency, the fibre populations affected by the two defects are not identical (Müller-Höcker et al. 1985). In the series of patients investigated by Johnson et al. (1988), 12/17 biopsies showed similar abnormalities in varying degrees of severity, but no study of the correlation of the two defects was made.

10 Genetic Considerations

The nature of the genetic defect in KSS and CPEO has been the subject of speculation for some time. In addition to a partial deficiency of Complex IV, defects involving Complexes I and III and mitochondrial ATPase may occur; frequently more than one defect is present in individual patients. Since the enzyme proteins involved are all coded in part by the mitochondrial genome (Anderson et al. 1981), this might suggest primary abnormalities of this genome. However, the major proportion of subunits in the mitochondrial respiratory complexes are nuclear-coded and imported into mitochondria; thus any genetic error which disrupts the sequence of events in their post-translational processing (Schatz and Butow 1983) could also give rise to dysfunction of the corresponding complexes (DiMauro et al. 1985).

Since mitochondrial DNA (mtDNA) is maternally inherited, evidence has been sought of maternal transmission in mitochondrial myopathies. With the exception of the syndrome associated with myoclonus epilepsy (MERFF syndrome) investigated by Rosing et al. (1985), this evidence is singularly lacking. A recent survey by Harding et al. (1985) showed that most cases of CPEO or KSS were apparently sporadic.

The occurrence of tissue-specific isoforms of respiratory chain enzymes is governed by the nuclear genome (Kadenbach et al. 1982) and it has been suggested that tissue-specific disorders will tend to be associated with nuclear-coded proteins (Minchom et al. 1983). The recent studies of Holt et al. (1988a, 1988b) have shown, nevertheless, that deletions of the mitochondrial genome may be detectable in skeletal muscle but not in leucocytes. One of the reasons for this may be that single leucocytes containing predominantly deleted mtDNA may be eliminated from the cell population. However, a skeletal muscle fibre is a stable syncitium in which defective mitochondria may be able to co-exist within a predominantly normal mitochondrial population. It has been shown that phenotypic expression in individual cells will depend on the relative proportion of normal and abnormal mtDNA present (Oliver and Wallace 1982). In situations favouring the replication of abnormal mitochondria in skeletal muscle, myopathy may become apparent though its onset may be delayed for many years.

The particular deletions of mtDNA reported by Holt and co-workers (1988a) involved 16 patients, of whom 8 had deficiencies of Complex I, 4 had Complex III deficiency and in the remaining 4 patients no respiratory chain defect had been identified. Another 28 patients with various forms of mitochondrial myopathy showed no apparent mtDNA deletions. It is of interest that one of the deletions illustrated in their communication (1988b) includes subunits II/III of cytochrome oxidase and two subunits of mtATPase as well as elements of Complex I. Data on the distribution of the cytochrome oxidase defect in individual muscle fibres in CPEO is consistent with the heterogeneity of mitochondrial populations which would result from such deletions. It would also be expected that where more than one respiratory complex is rendered abnormal by the same deletion, the resulting defects would show the same tissue distribution. The cytochemical and immunocytochemical methodology discussed in this communication is of direct relevance to the investigation of phenotypic expression at the cellular level in such situations.

Acknowledgements. I acknowledge with thanks the help and collaboration of Prof. B. Kadenbach and Dr. M. Droste (University of Marburg, FRG) who provided the subunit-specific antisera. I am also grateful to my colleagues in the University of Newcastle upon Tyne, UK: Dr. D. M. Turnbull (Department of Neurology), Dr. H. S. A. Sherratt (Department of Pharmacological Sciences) and Dr. A. J. Clarke (Department of Human Genetics) for their contributions to these studies.

References

Anderson S, Bankier AT, Barrell BG, de Bruijn MHL, Coulson AR, Drouin J, Eperon IC, Nierlich DP, Roe BA, Sanger F, Schreier PH, Smith AJH, Staden R, Young IG (1981) Sequence and organisation of the human mitochondrial genome. Nature (London) 290:457–464

Ballantyne B, Bright JE (1979) Comparison of kinetic and end-point microdensitometry for the direct quantitative histochemical assessment of cytochrome oxidase activity. Histochem J 11:173–186

Bookelman H, Trijbels JMF, Sengers RAC, Jansen AJM (1978) Measurement of cytochromes in human skeletal muscle mitochondria isolated from fresh and frozen stored muscle specimens. Biochem Med 19:366–373

Bresolin N, Moggio M, Bet L, Gallanti A, Prelle A, Nobile-Orazio E, Adobbati L, Ferrante C, Pellegrini G, Scarlato G (1987) Progressive cytochrome *c* oxidase deficiency in a case of Kearns-Sayre syndrome: morphological, immunological and biochemical studies in muscle biopsies and autopsy tissues. Ann Neurol 21:564–572

Brooke MH, Kaiser KK (1970) Three "myosin adenosine triphosphatase" systems. The nature of their pH lability and sulfhydryl dependence. J Histochem Cytochem 18:670–672

Byrne E, Dennet X, Trounce I, Burdon J (1985) Partial cytochrome oxidase (aa_3) deficiency in chronic progressive external ophthalmoplegia. Histochemical and biochemical studies. J Neurol Sci 71:257–271

Cleeter MWT, Ragan CI (1985) The polypeptide composition of the mitochondrial NADH: ubiquinone reductase complex from several mammalian species. Biochem J 230:739–746

DiMauro S, Mendell JR, Sahenk Z, Bachman D, Scarpa A, Schofield R, Reinier C (1980) Fatal infantile mitochondrial myopathy and renal dysfunction due to cytochrome *c* oxidase deficiency. Neurology 30:795–804

DiMauro S, Bonilla E, Zeviani M, Nakagawa M, DeVivo DC (1985) Mitochondrial myopathies. Ann Neurol 17:521–538

Drachman DA (1968) Ophthalmoplegia plus, the neurodegenerative disorders associated with progressive external ophthalmoloplegia. Arch Neurol 18:654–674

Ernster L, Ikhos D, Luft R (1959) Enzymic activities of human skeletal muscle mitochondria: a tool in clinical metabolic research. Nature (London) 184:1851–1854

Fukuhara N, Takiguchi S, Shirakawa K, Tsubaki T (1980) Myoclonus epilepsy associated with ragged-red fibres (mitochondrial abnormalities): disease entity or syndrome? Light and electron microscopic studies of two cases and review of literature. J Neurol Sci 47:117–133

Harding AE, Petty RKH, Morgan-Hughes JA (1985) Mitochondrial myopathy: a genetic study. J Med Genet 22:138

Holt IJ, Cooper JM, Morgan-Hughes JA, Harding AE (1988a) Deletions of muscle mitochondrial DNA. Lancet i:1462

Holt IJ, Harding AE, Morgan-Hughes JA (1988b) Deletions of muscle mitochondrial DNA in patients with mitochondrial myopathies. Nature (London) 331:717–719

Jarausch J, Kadenbach B (1985) Structure of the cytochrome *c* oxidase complex of rat liver. 1. Studies on nearest-neighbour relationship of polypeptides with cross-linking reagents. Eur J Biochem 146:211–217

Johnson MA, Turnbull DM, Dick DJ, Sherratt HSA (1983) A partial deficiency of cytochrome *c* oxidase in chronic progressive external ophthalmoplegia. J Neurol Sci 60:31–35

Johnson MA, Kadenbach B, Kuhn-Nentwig L, Fulthorpe JJ, Sherratt HSA, Turnbull DM (1985) Immunocytochemical studies of cytochrome oxidase deficiencies in human skeletal muscle. Biochem Soc Trans 13:729

Johnson MA, Kadenbach B, Droste M, Old SL, Turnbull DM (1988) Immunocytochemical studies of cytochrome oxidase subunits in skeletal muscle of patients with partial cytochrome oxidase deficiencies. J Neurol Sci 87:75–90

Kadenbach B, Hartmann R, Glanville R, Buse G (1982) Tissue-specific genes code for polypeptide VIa of bovine liver and heart cytochrome *c* oxidase. FEBS Lett 138:236–238

Kearns TP, Sayre GP (1958) Retinitis pigmentosa, external ophthalmoplegia and complete heart block. Arch Ophthalmol 60:280–289

Kuhn-Nentwig L, Kadenbach B (1985) Orientation of rat liver cytochrome *c* oxidase subunits investigated with subunit-specific antisera. Eur J Biochem 153:101–104

Kuhn-Nentwig L, Kadenbach B (1986) Isolation and characterisation of human heart cytochrome *c* oxidase. J Bioenerg Biomembr 18:307–314
Lowry CV, Kimmey JS, Felder S, Chi MM-Y, Kaiser KK, Passoneau PN, Kirk KA, Lowry OH (1978) Enzyme patterns in single human muscle fibres. J Biol Chem 253:8269–8277
Meijer AEFH (1972) Mitochondria with defective respiratory control of oxidative phosphorylation isolated from muscle tissues of thyroidectomised rabbits. J Neurol Sci 16:445–453
Minchom PE, Dormer RL, Hughes RA, Stansbie D, Cross AR, Hendry GAF, Jones OTG, Johnson MA, Sherratt HSA, Turnbull DM (1983) Fatal infantile mitochondrial myopathy due to cytochrome *c* oxidase deficiency. J Neurol Sci 60:454–463
Mitsumoto H, Aprille JR, Wray SH, Nemni S, Bradley WG (1983) Chronic progressive external ophthalmoplegia (CPEO): clinical, morphologic and biochemical studies. Neurology 33:452–461
Morgan-Hughes JA, Darveniza P, Kahn SN, Landon DN, Sherratt RM, Land JM, Clark JB (1977) A mitochondrial myopathy characterised by a deficiency in reducible cytochrome *b*. Brain 100:617–640
Morgan-Hughes JA, Darveniza P, Landon DN, Lan JM, Clark JB (1979) A mitochondrial myopathy with a deficiency of respiratory chain NADH-CoQ reductase activity. J Neurol Sci 43:27–46
Müller-Höcker J, Pongratz D, Hübner G (1983) Focal deficiency of cytochrome *c* oxidase in skeletal muscle of patients with progressive external ophthalmoplegia. Virchows Arch (Cell Pathol) 402:61–71
Müller-Höcker J, Stünkel S, Pongratz D, Hübner G (1985) Focal deficiency of cytochrome *c* oxidase and of mitochondrial ATPase with histochemical evidence of loosely coupled oxidative phosphorylation in a mitochondrial myopathy of a patient with bilateral ptosis – an enzyme histochemical, immunocytochemical and fine structural study. J Neurol Sci 69:27–36
Müller-Höcker J, Pongratz D, Hübner G (1986) Activation of mitochondrial ATPase as evidence of loosely coupled oxidative phosphorylation in various skeletal muscle disorders. J Neurol Sci 74:199–213
Oliver NO, Wallace DC (1982) Assignment of two mitochondrially synthesised polypeptides to human mitochondrial DNA and their use in the study of intracellular mitochondrial interaction. Mol Cell Biol 2:30–41
Olson W, Engel WK, Walsh GO, Einaugher R (1972) Oculocraniosomatic neuromuscular disease with "ragged-red" fibers. Histochemical and ultrastructural changes in limb muscles of a group of patients with idiopathic progressive external ophthalmoplegia. Arch Neurol 26:193–211
Pearse AGE (1972) Histochemistry, theoretical and applied, 3rd edn. Churchill-Livingstone, London
Pette D (1981) Microphotometric measurement of initial maximum reaction rates in quantitative enzyme histochemistry in situ. Histochem J 13:319–327
Reichmann H, Pette D (1982) A comparative microphotometric study of succinate dehydrogenase activity levels in type I, IIA and IIB fibres of mammalian and human muscles. Histochemistry 74:27–42
Reichmann H, Johnson MA, Turnbull DM, Sherratt HSA (1985) The cytochemical determination of enzyme activities in single skeletal muscle fibres from patients with a partial deficiency of cytochrome oxidase. Biochem Soc Trans 13:730
Reske-Nielsen E, Lou HC, Lowes M (1976) Progressive external ophthalmoplegia. Evidence for a generalised mitochondrial disease with a defect in pyruvate metabolism. Acta Ophthalmol 54:37–44
Rosing HS, Hopkins LC, Wallace DC, Epstein CM, Weidenheim K (1985) Maternally inherited mitochondrial myopathy and myoclonic epilepsy. Ann Neurol 17:228–237
Schatz G, Butow RA (1983) How are proteins imported into mitochondria? Cell 32:316–318
Seligman AM, Karnovsky MJ, Wasserkrug HL, Hanker JS (1968) Non-droplet ultrastructural demonstration of cytochrome oxidase activity with a polymerising osmophilic reagent, diaminobenzidine (DAB). J Cell Biol 38:1–14
Sengers RCA, Stadhouders AM, Trijbels JMF (1984) Mitochondrial myopathies – clinical, morphological and biochemical aspects. Eur Pediat 141:192–207
Sengers R, Stadhouders A, Bastiaensen L (1986) 'Ragged-red' fiber pathology due to micro-angiopathy? Muscle Nerve 9 (Suppl):184
Sherratt HSA, Cartlidge NEF, Johnson MA, Turnbull DM (1984) Mitochondrial myopathy with partial cytochrome oxidase deficiency and impaired oxidation of NADH-linked substrates. J Inherit Metab Dis 7 (Suppl 2):107–108

Sherratt HSA, Johnson MA, Turnbull DM (1986) Defects of complex I and complex IV in skeletal muscle from patients with chronic progressive external ophthalmoplegia. Ann NY Acad Sci 488:508–519

Sherratt HSA, Whatmough NJ, Johnson MA, Turnbull DM (1988) Methods for study of normal and abnormal skeletal muscle mitochondria. Methods Biochem Anal 33:243–336

Smith L (1955) Spectrophotometric assay of cytochrome *c* oxidase. Methods Biochem Anal 2:427–434

Sternberger LA, Hardy PH, Cuculis JJ, Meyer HG (1970) The unlabelled antibody-enzyme method of immunohistochemistry. Preparation and properties of soluble antigen-antibody complex (horseradish peroxidase-antihorseradish peroxidase) and its use in identification of spirochaetes. J Histochem Cytochem 18:315–333

Sulaiman WR, Doyle D, Johnson RH, Jennett S (1974) Myopathy with mitochondrial inclusion bodies: histological and metabolic studies. J Neurol Neurosurg Psychiat 37:1236–1246

Turnbull DM, Sherratt HSA, Davies DM, Sykes AG (1982) Tetra-cyano-2,2-bipyridineiron (III), an improved electron acceptor for the spectrophotometric assay of β oxidation and succinate dehydrogenase in intact mitochondria. Biochem J 206:511–516

Turnbull DM, Spoors TC, Cullen MJ, Johnson MA, Sherratt HSA (1983) The effect of copper deficiency on skeletal and cardiac muscle mitochondria. In: Scarlato G, Cerri C (eds) Mitochondrial pathology in muscle diseases. Piccin, Padua pp 251–260

Turnbull DM, Johnson MA, Dick DJ, Cartlidge NEF, Sherratt HSA (1985) Partial cytochrome oxidase deficiency without subsarcolemmal accumulation of mitochondria in chronic progressive external ophthalmoplegia. J Neurol Sci 70:93–100

Van Biervliet JPGM, Bruinvis L, Ketting D, DeBree PK, Vander Heiden C, Wadman SK, Willems JL, Bookelman H, Van Haelst V, Monnens LAH (1977) Hereditary mitochondrial myopathy with lactic acidaemia, a de Toni-Fanconi-Debré syndrome and a defective respiratory chain in voluntary striated muscles. Pediat Res 11:1088–1092

Willems JL, Monnens LAH, Trijbels JMF, Veerkamp JH, Meijer AEFH, Van Dam K, Van Haelst V (1977) Leigh's encephalomyelopathy in a patient with cytochrome oxidase deficiency in muscle tissue. Pediatrics 60:850–857

Regulatory Complexity of Cytochrome *c* Oxidase and Its Defective Manifestation in Mitochondrial Diseases

B. KADENBACH[1], F.-J. HÜTHER[1], U. BÜGE[1], A. SCHLERF[1] and M.A. JOHNSON[2]

1 Introduction

The expenditure of energy in different cells and tissues of mammalian organisms varies, depending on the type of tissue, its developmental stage, and various extracellular signals, like growth factors, hormones, neurotransmitters and metabolites. Because about 95% of the energy (ATP) is synthesized in mitochondria by oxidative phosphorylation, the rate of cellular energy synthesis is roughly proportional to the rate of cell respiration. The reduction of oxygen to water is catalyzed by cytochrome *c* oxidase, the terminal enzyme of the mitochondrial respiratory chain (complex IV). Like two other enzyme complexes of the respiratory chain, the NADH-dehydrogenase (complex I) and the ubiquinone-cytochrome *c* oxidoreductase (complex III), cytochrome *c* oxidase converts the redox energy into a proton electrochemical gradient ($\Delta\mu H^+$) across the inner mitochondrial membrane, which is used by the ATP-synthase to drive the endergonic synthesis of ATP from ADP and phosphate (Fig. 1). The activity of cytochrome *c* oxidase can be principally regulated by three different mechanisms:

1. By isosteric effectors, i.e. the concentration of substrates, namely the concentrations of reduced cytochrome *c* and oxygen [1], and the $\Delta\mu H^+$ across the inner mitochondrial membrane [2,3].
2. By allosteric effectors (for review see [4]). Recently ATP and ADP could be identified as allosteric effectors of cytochrome *c* oxidase (see below).
3. By the cellular enzyme concentration, which is varied by the rate of gene expression and the rate of enzyme degradation.

In addition, the degree of coupling between respiration and phosphorylation in mitochondria was suggested to be variable [5–7]. In particular, free fatty acids were found to modify the degree of coupling in mitochondria [8–11]. For cytochrome *c* oxidase, in intact mitochondria, a variable H^+/e^- stoichiometry was found, which was suggested to depend on intracellular conditions [12]. In proteoliposomes the H^+/e^- stoichiometry was also found to vary depending on the pH and the presence of divalent cations [13].

[1]Biochemie, Fachbereich Chemie der Philipps-Universität Hans-Meerwein-Strasse, D-3550 Marburg, FRG
[2]Department of Neurology, University of Newcastle upon Tyne, England

A. Azzi, Z. Drahota and S. Papa
Molecular Basis of Membrane-Associated Diseases

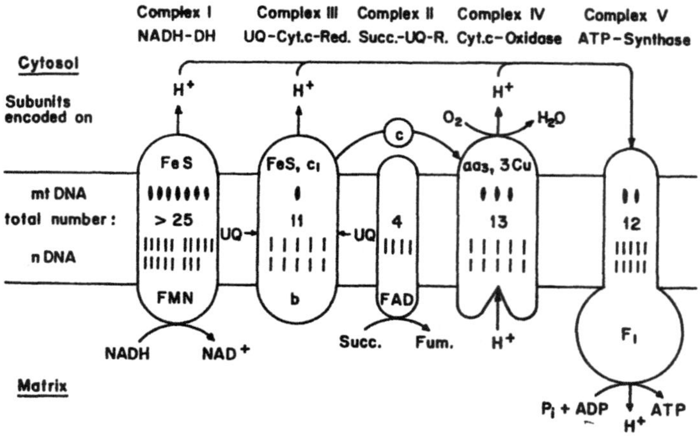

Fig. 1. Enzyme complexes of the inner mitochondrial membrane involved in oxidative phosphorylation. The scheme indicates the total number of subunits in each complex and the number of subunits encoded on mtDNA and nuclear DNA. The probable sidedness of the redox center is also indicated

2 Subunit Composition and Regulatory Complexity

Cytochrome *c* oxidase from eukaryotic organisms is composed of multiple subunits encoded on two different genomes: the mitochondrial and the nuclear genome. The three catalytic subunits, which carry the five redox centers [14] are encoded on mitochondrial DNA, whereas additional subunits are encoded on nuclear DNA. Since these additional subunits are absent in bacterial cytochrome *c* oxidase, they have been ascribed a regulatory function [4,14,15].

It was postulated that the nuclear-encoded subunits regulate the rate of respiration and the H^+/e^- stoichiometry in cytochrome *c* oxidase by specific binding of effectors like nucleotides, ions, substrates etc. The binding of an effector was assumed to change the conformation of the subunit, which is transmitted to the catalytic subunits via tight contact sites within the hydrophobic domain of the enzyme complex [4]. A schematic drawing of these structural features of the enzyme complex is shown in Fig. 2.

The following experimental observations support the regulatory model of mammalian cytochrome *c* oxidase:

1. The number of subunits increase with the evolutionary stage of the organism. In addition to the three catalytic subunits, no regulatory subunits are found in cytochrome *c* oxidase from bacteria, five in *Dictyostelium discoideum* (a slime mold), six in yeast, eight to nine in fish and ten in mammals [14].
2. Most nuclear-encoded subunits have a transmembranous hydrophobic domain (except Va, Vb and VIb) [14], and are in close contact to the mitochondrial-encoded subunits [16].
3. Various anions, including ATP, ADP and phosphate change the conformation of the enzyme complex, each in a different specific manner, as indicated by modified visible spectra [17].

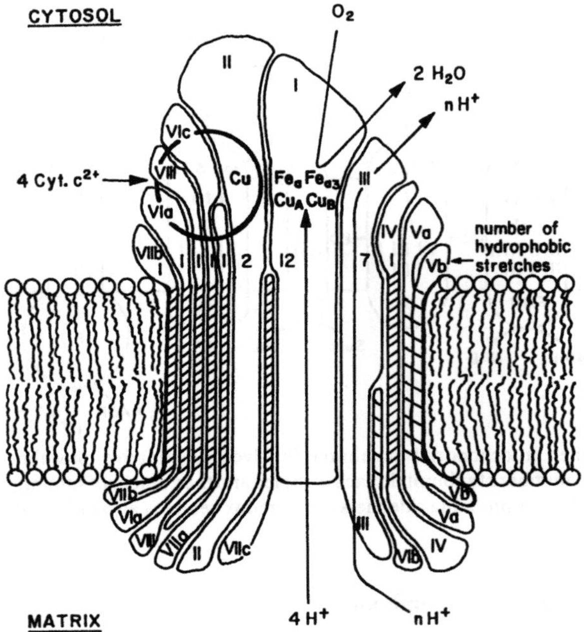

Fig. 2. Regulatory model of mammalian cytochrome *c* oxidase. The model takes into consideration the participation of some nuclear-encoded subunits in the binding domain for cytochrome *c*, the transmembranous orientation of most nuclear-encoded subunits, and the presence of five redox centers in the complex [14]. It is suggested that hydrophilic domains of nuclear-encoded subunits act as receptors for intracellular signals, like nucleotides, ions, substrates etc. [4]

4. The kinetics of ferrocytochrome *c* oxidation in reconstituted bovine heart cytochrome *c* oxidase is influenced by extraliposomal [17–19] and intraliposomal nucleotides [20]. The increase of the K_m for cytochrome *c* by intraliposomal ATP and its decrease by intraliposomal ADP must be due to interaction with nuclear-encoded subunits, because no effect of intraliposomal nucleotides is obtained with the reconstituted enzyme from *Paracoccus denitrificans* [21], which lacks these subunits (see Fig. 3).
5. A specific photoaffinity labelling by 2-azido [γ-^{32}P] ATP of subunit VIa of bovine heart cytochrome *c* oxidase was demonstrated [22]. In contrast, 8-azido-[γ-^{32}P] ATP was found to bind unspecifically to most subunits of the bovine heart enzyme [17,19], probably due to the unnatural conformation of this ATP derivative [23].
6. The respiratory control ratio of reconstituted bovine heart cytochrome *c* oxidase could be specifically decreased by low concentrations of free fatty acids [24].

3 Multiple Tissue- and Developmental-Specific Isozymes Occur in Mammals

By comparing the SDS-gel electrophoretic subunit pattern of isolated cytochrome *c* oxidase from liver and heart, different apparent molecular weights were found for subunits VIa, VIIa and VIII [25,26]. The different structures of the corresponding

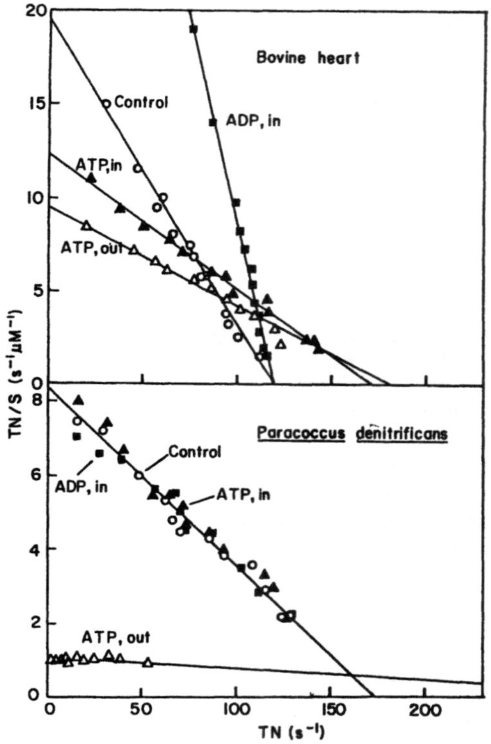

Fig. 3. Comparison of the effects of intraliposomal and extraliposomal nucleotides on the kinetics of cytochrome *c* oxidation by reconstituted cytochrome *c* oxidase from bovine heart and *Paracoccus denitrificans*. The data, presented for the low-affinity phase of cytochrome *c* oxidation in reversed Eadie-Hofstee plots, were measured by the photometric method at increasing concentrations of ferrocytochrome *c* with proteoliposomes containing the enzyme from bovine heart [20] and *P. denitrificans* [21]. The concentrations of intraliposomal (*in*) nucleotides were 10 mM (bovine) or 15 mM (*P. denitrificans*), and of extraliposomal (*out*) ATP 10 mM

subunits from liver and heart were verified by different N-terminal amino acid sequences [14,27,28], different immunological relationships [29,30], and more recently by different gene structures of the liver-type and heart-type subunit VIa [31]. Different immunological relationships were also found between corresponding subunits of the enzyme from adult and fetal tissues (liver, heart and skeletal muscle) [30].

The functional relevance of tissue-specific subunits was demonstrated by different kinetic properties of the enzyme from bovine liver and heart [32], and from different kinetic responses of the reconstituted liver and heart enzymes upon different liposomal lipids and phosphates [33]. In Fig. 4 the opposite effect of intraliposomal phosphate on the V_{max} of the two reconstituted enzymes is shown. The V_{max} of the heart enzyme decreases, that of the liver enzyme increases, but for both enzymes the K_m for cytochrome *c* increases when intraliposomal Hepes is exchanged by 20 mM phosphate.

The basic principle of isozyme variability in the mammalian cytochrome *c* oxidase complex was recently enlightened by analysis of the subunit structure of

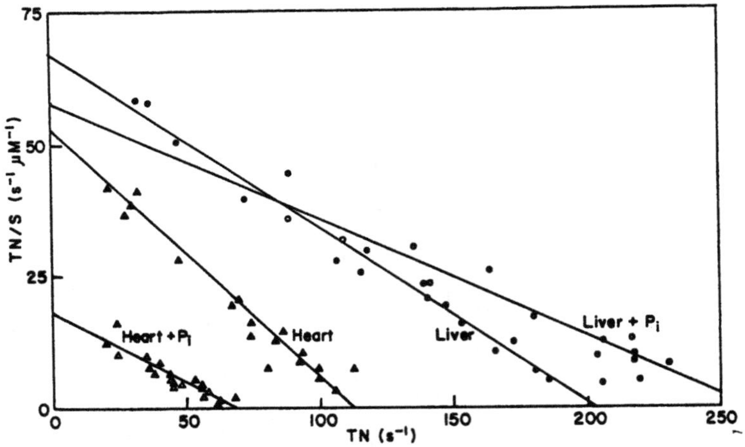

Fig. 4. Influence of intraliposomal phosphate on the kinetics of cytochrome *c* oxidation by reconstituted cytochrome *c* oxidase from bovine heart and liver. Reconstitution of the enzymes was performed in the presence of 100 mM K/Hepes, pH 7.4 (*closed symbols*) or 20 mM KP_i, pH 7.4 200 mM sucrose (*open symbols*). The photometric assay of activity was performed in 10 mM K/Hepes, pH 7.4, 40 mM KCl as described [33]

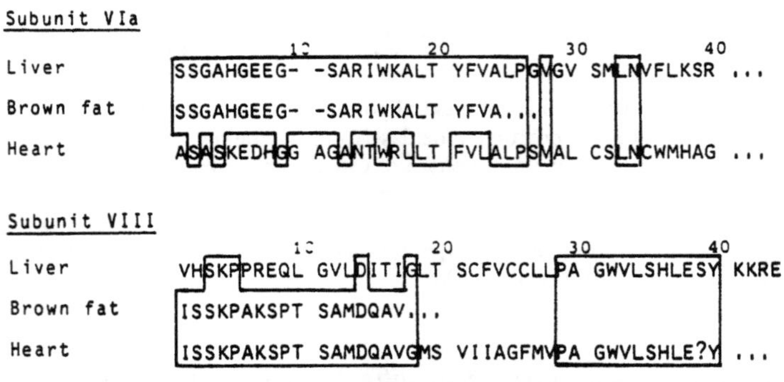

Fig. 5. N-terminal amino acid sequences of cytochrome *c* oxidase subunits VIa and VIII from rat liver, heart and brown fat tissue. The amino acid sequences of subunits VIa from liver and heart [31] and the C-terminal part of subunit VIII from liver [38] were deduced from cDNA sequences. The other N-terminal amino acids were determined by sequencing of gel bands [34]

the isolated enzyme complex from rat brown fat tissue [34]. In the liver and heart enzyme of the rat subunits VIa and VIII are different (50 and 41% amino acid sequence homology, respectively). Surprisingly, in brown fat cytochrome *c* oxidase, the liver-type subunit VIa and the heart-type subunit VIII (see Fig. 5) are expressed. Thus, it appears that the enzyme complex is organized like a modular construction system: In each isozyme three identical catalytic subunits are assembled together with variable forms of ten different nuclear-encoded regulatory subunits. In mammalian organisms the enzyme always contains 13 subunits. In lower eukaryotes, where signals like hormones and neurotransmitters do not occur, the enzyme complex contains correspondingly less regulatory subunits. The large intrinsic regulatory complexity of the mammalian enzyme complex emerges from the calculation of theoretically possible isozymes. Assuming two different forms for each of the ten nuclear-encoded subunits, $2^{10} = 1024$ different isozymes are possible. From immunological relationships between corresponding subunits of the enzyme from various adult and fetal tissues, two or more different forms for each of the ten nuclear-encoded subunits could be deduced [30].

4 Isogenes, Gene Families and Retroposons of Cytochrome *c* Oxidase

The mammalian mitochondrial genome encodes for only 13 different proteins, all of which are components of membrane-bound energy-transducing enzyme complexes (see Fig. 1). Because of its maternal inheritance, mitochondrial DNA is assumed to be identical in all tissues of an organism [35]. Therefore, the basic catalytic property of cytochrome *c* oxidase, associated with the three mitochondrial-encoded subunits, is identical in all isozymes.

The variation of cytochrome *c* oxidase regulation in different tissues is apparently due to a tissue-, developmental- and probably hormone-specific expression of different forms of the ten nuclear-encoded subunits. Isogenes of cytochrome *c* oxidase, i.e. different genes coding for the same subunit, could be recently identified for the first time for subunit VIa of the rat liver and heart enzyme [31]. The two cDNAs show no homology in the 3′-noncoding region, and the 60% homology of the coding region results in only 59% amino acid sequence homology (Fig. 6). This result supports the assumption that the two genes evolved independently after liver and heart occurred as separate tissues. Whereas the heart-type gene for subunit VIa is expressed in heart and skeletal muscle, but not in liver, brain and kidney, the liver-type gene is expressed in all tissues, but only to a small extent in skeletal muscle, as shown from Northern blot analysis [30]. Thus, two different subunits VIa are expressed in the heart, the extent of each appears to depend on a yet unknown regulatory signal. This conclusion confirms the well-known observation that in SDS-polyacrylamide gels of the heart enzyme occasionally four different bands can be seen in the region of subunit VI (two subunits VIa, subunits VIb and VIc). Comparison of the rat heart subunit VIa, with the amino acid sequence of bovine heart subunit VIa, as determined from the isolated protein [36], shows 82% homology [37].

The genes for nuclear-encoded subunits of cytochrome *c* oxidase occur in multiple forms (gene families). This was demonstrated for subunits VIc and VIII from rat liver, where five to seven and four to five different genes were suggested to occur in the rat genome by Southern blot analysis [38]. The multiple genes, identified by multiple bands in Southern blots, do not necessarily include the tissue-specific isogenes, because they may not show enough sequence homology for hybridization with each other, as shown for the two isogenes of subunit VIa. In fact, different gene families might exist for both the liver- and the heart-type gene, as followed from three to five different bands obtained with each gene probe in a Southern blot [37]. Therefore, the number of different functional genes coding for each of the ten nuclear-encoded subunits of cytochrome *c* oxidase remains unknown.

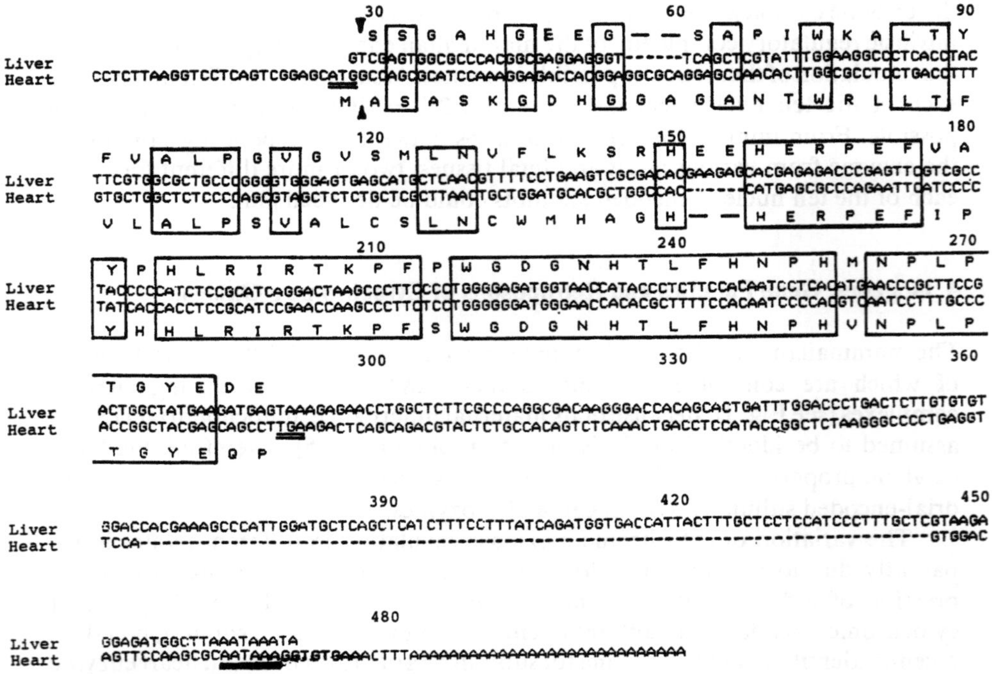

Fig. 6. cDNA and deduced amino acid sequences of two isogenes coding for subunit VIa of rat heart and rat liver cytochrome *c* oxidase. *Dashes* indicate insertions in both nucleotide and protein sequences. Start and termination codons as well as polyadenylation signals are *double underlined*. Amino acid sequence homologies are in *large boxes*. The first amino acid of the mature protein precedes an *arrow head*. Corrected version of published sequence [31]

From the analysis of genomic clones, which have been screened with cDNA probes for subunits VIc and VIII (liver-type) in a rat genomic library in λ phage EMBL 3, the following different types of genes have been identified.

1. The functional gene for rat liver subunit VIc has been characterized. It contains three introns, one in the 5′-noncoding, one in the coding, and one in the 3′-noncoding region [38].
2. Processed retroposons characterized by all features which are required for being expressed have been identified for subunit VIc [39] and VIII (liver-type) [37]. These processed genes are characterized by the absence of introns, the presence of a direct repeat in the flanking regions, and a poly(A) tail [respectively, a poly(G) tail for the subunit VIc gene] after the polyadenylation signal. Both genes contain promoter-like structures in the 5′-region and an open reading frame coding either for subunit VIc, containing 16 amino acid substitutions, or subunit VIII (liver-type) with a single amino acid substitution. We do not know whether and where these retroposons are expressed, and whether they could substitute a defective functional gene.

3. Two processed retropseudogenes derived from the functional gene for liver-type subunit VIII have been identified. They possess also a direct repeat and no introns [37]. But because of several mutations these pseudogenes cannot be expressed into a protein resembling subunit VIII. In order to differentiate these nonfunctional structures from the above described retroposons we denote them retropseudogenes. It was suggested that retropseudogenes may represent an intermediate stage in the evolution of new genes for new regulatory mechanisms of key enzymes of metabolism [37].

5 The Genetic Basis of Mitochondrial Myopathies

Mitochondrial myopathies comprise a large group of different muscle degenerative diseases associated with mitochondrial defects. In several cases the brain and/or other tissues are also affected. In most affected tissues an aggregation of mitochondria is found by the Gomori-modified trichrome stain ("ragged-red fibers"). Often the patients have an elevated serum lactate. Some cases with neurodegenerative symptoms have recently been grouped into three phenotypically distinct syndromes [40]: KSS (Kearns-Sayre syndrome), MERRF (myoclonic epilepsy and ragged red fibers), and MELAS (mitochondrial encephalopathy, myopathy, lactic acidosis, and stroke-like episodes). In addition, juvenile forms with a lethal or benign course have been described, which are often associated with a defect of cytochrome *c* oxidase. The various clinical pictures have been recently grouped by DiMauro et al. [41]: (1) Myopathy is the main if not exclusive manifestation; (2) brain involvement predominates (including Leigh's syndrome, Alpers disease and Menkes disease); (3) less well-defined clinical entities associated with partial cytochrome *c* oxidase deficiency. Mitochondrial myopathies may be due to a defect of any mitochondrial enzyme or transport protein. The various biochemical defects have also been classified by DiMauro et al. [42].

In many cases of mitochondrial myopathies a reduced or no activity of cytochrome *c* oxidase was found in the affected tissue. This could principally be due to a defective mitochondrial or a defective nuclear gene, coding for a cytochrome *c* oxidase subunit or for a protein or RNA which is essential for the biosynthesis or assembly of the enzyme complex. Since the mitochondrial genome is maternally inherited, the pedigree of the disease could help to identify the genetic basis of the disease. An extensive discussion of the genetics and the various possible types of inheritance of mitochondrial myopathies has been given by Wallace [43].

Recently, deletions in mitochondrial DNA (mtDNA) from muscle tissue but not from leukocytes of 9 from 25 investigated patients with mitochondrial myopathy were described [44]. This finding could be corroborated by studies on muscle mtDNA from seven patients with Kearns-Sayre syndrome, which all showed, in addition to the intact mtDNA, a second mtDNA of 2–7 kb smaller size [45]. These deletions were not found, however, in ten other patients with MELAS, MERRF, Leigh's disease or other types of mitochondrial myopathy. Although these deletions differ in size between different patients, they appear to be identical in the affected tissue, because sharp bands were obtained in Southern blots of every case. From several pedigrees the inheritance of the Kearns-Sayre syndrome was documented, but maternal as well as paternal transmissions have been observed [43]. Therefore, it may be assumed that the defective mtDNA is maternally inherited together with intact mtDNA (heteroplasmic mtDNA), but an additional nuclear-encoded and possibly tissue-specific component may be necessary for proliferation of the deleted mtDNA (mitotic segregation) and thus for the expression of the disease.

A different genetic origin appears to underly the chronic progressive external ophthalmoplegia (CPEO), a mitochondrial myopathy mainly observed in adult individuals [46,47]. In this disease fibers with defective cytochrome *c* oxidase are scattered in cross-sections of muscle tissue. Between 6 and 66% of the total fibers was found to be affected. Subunit-specific antisera to subunits of human heart cytochrome *c* oxidase were applied in immunohistochemical studies on serial sections of skeletal muscle from 17 patients with CPEO [48]. A different pattern of absent subunits was found for each cell lacking cytochrome *c* oxidase activity, as shown in Fig. 7. About 80% of the affected cells showed no immunoreaction with

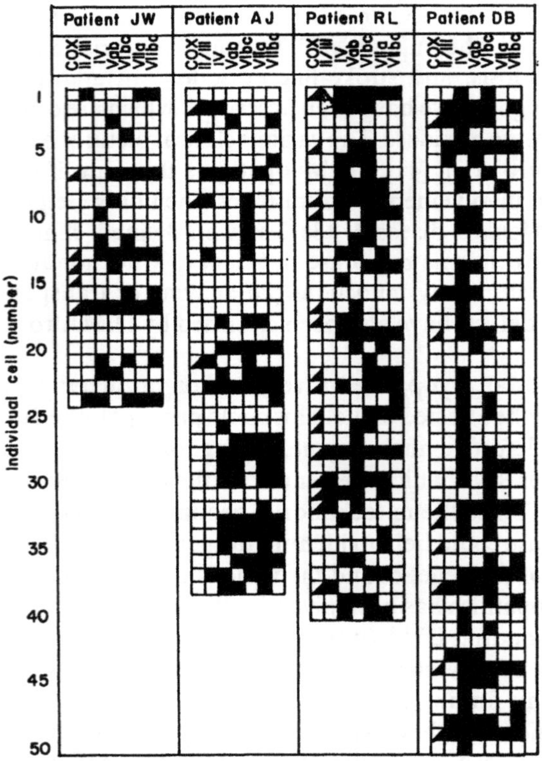

Fig. 7. Immunohistochemical reaction of individual skeletal muscle cells (fibers) lacking cytochrome *c* oxidase activity of four patients with CPEO. Cells with decreased (▲) or no activity (*open squares*) of cytochrome *c* oxidase (*COX*) were analyzed immunohistochemically with antisera to the indicated subunits of human heart cytochrome *c* oxidase [48]. The presence (*closed squares*) or absence (*open squares*) of immunoreaction is indicated. Affected cells concern all types of muscle fibers and are scattered mosaic-like in the cross-sections of the muscle biopsy samples

the antiserum to subunit II/III, but only 20% showed no immunoreaction to subunit IV. The different appearance of the defective subunit pattern in each enzymatic defective cell cannot be explained by a single mutation event. Rather, a statistical and different change in each affected cell must be assumed. We therefore propose a different and statistical mutation of the mitochondrial genome in each affected cell. The more rapid proliferation of the mutated mtDNA as compared to the abundance of intact mtDNA could be explained by a higher replication rate of the mutated mtDNA. However, the mtDNA defect appears to represent only the secondary event. The primary genetic defect could concern a gene normally involved in mtDNA repair, which could be expressed as tissue-specific.

A defective gene for a nuclear-encoded subunit of cytochrome *c* oxidase was assumed in two siblings with fatal infantile mitochondrial myopathy, where enzyme activity was almost absent in muscle tissue [49]. The immunohistochemical study with antisera to subunits of the human heart enzyme revealed the presence of most subunits in skeletal muscle. Only antisera to subunits VIIbc and subunits II/III showed no immunoreactivity. Normal immunoreactivity was found in fibres of the muscle spindle, heart, vessel walls and the gastrointestinal tract. The tissue-specific expression of the defect supports the assumption of a defective nuclear-encoded gene of cytochrome *c* oxidase. The combined defective expression of subunit VIIbc with the mitochondrial-encoded subunits II/III could be explained by assuming degradation of subunits II/III in mitochondria when the holoenzyme complex cannot be assembled due to the lack of subunit VIIbc.

The above data and discussion clearly indicate the enormous number of possible gene defects associated with more or less tissue-specific respiratory failures. Much more knowledge is required of the genetics and inheritance of mtDNA, of the cytoplasmic-nuclear interactions, and of the tissue-specific genes for nuclear-encoded subunits of cytochrome *c* oxidase, for the elucidation of the molecular basis of individual mitochondrial myopathies.

Acknowledgements. Roswitha Roller-Müller is gratefully acknowledged for preparation of the manuscript. This work was supported by the Deutsche Forschungsgemeinschaft (SFB 103, A2 and DFG Ka 192/17-2) and Fonds der Chemischen Industrie.

References

1. Erecinska M, Wilson D (1982) The oxygen dependence of cellular energy metabolism. J Membr Biol 70:1–14
2. Brunori M, Sarti P, Colosimo A, Antonini G, Malatesta F, Jones MG, Wilson MT (1985) Mechanism of control of cytochrome oxidase activity by the electrochemical-potential gradient. EMBO J 4:2365–2368
3. Murphy MP, Brand MD (1987) The control of electron flux through cytochrome oxidase. Biochem J 243:499–505
4. Kadenbach B (1986) Mini Review. Regulation of respiration and ATP synthesis in higher organisms: hypothesis. J Bioenerg Biomembr 18:39–54
5. Van Dam K, Westerhoff HW, Krab K, van der Meer R, Arents JC (1980) Relationship between chemiosmotic flows and thermodynamic forces in oxidative phosphorylation. Biochim Biophys Acta 591:240–250

6. Pietrobon D, Zoratti M, Azzone GF (1983) Molecular shipping in redox and ATPase H^+ pumps. Biochim Biophys Acta 723:317–321
7. Soboll S, Stucki J (1985) Regulation of the degree of coupling of oxidative phosphorylation in intact rat liver. Biochim Biophys Acta 807:245–254
8. Rottenberg H, Hashimoto K (1986) Fatty acid uncoupling of oxidative phosphorylation in rat liver mitochondria. Biochemistry 25:1747–1755
9. Rottenberg H, Steiner-Mordoch S (1986) Free fatty acids decouple oxidative phosphorylation by dissipating intramembranal protons without inhibiting ATP synthesis driven by the proton electrochemical gradient. FEBS Lett 202:314–318
10. Luvisetto S, Pietrobon D, Azzone GF (1987) Uncoupling of oxidative phosphorylation. 1. Protonophoric effects account only partially for uncoupling. Biochemistry 26:7332–7338
11. Pietrobon D, Luvisetto S, Azzone GF (1987) Uncoupling of oxidative phosphorylation. 2. Alternative mechanisms: intrinsic uncoupling or decoupling? Biochemistry 26:7339–7347
12. Lehninger AL, Reynafarje B, Costa L (1985) Action of DCCD on the H^+/O stoichiometry of mitoplast cytochrome *c* oxidase. J Inorg Biochem 23:335–340
13. Papa S, Capitanio N, DeNitto E (1987) Characteristics of the redox-linked proton ejection in beef-heart cytochrome *c* oxidase reconstituted in liposomes. Eur J Biochem 164:507–516
14. Kadenbach B, Kuhn-Nentwig L, Büge U (1987) Evolution of a regulatory enzyme: cytochrome *c* oxidase (complex IV). Curr Top Bioenerg 15:113–161
15. Kadenbach B, Merle P (1981) On the function of multiple subunits of cytochrome *c* oxidase from higher eucaryotes. FEBS Lett 135:1–11
16. Jarausch J, Kadenbach B (1985) Structure of the rat liver cytochrome *c* oxidase complex. I. Studies on nearest neighbour relationship of polypeptides with crosslinking reagents. Eur J Biochem 146:211–217
17. Reimann A, Hüther FJ, Berden JA, Kadenbach B (1988) Anions induce conformation changes and influence the activity and photoaffinity-labelling by 8-azido-ATP of isolated cytochrome *c* oxidase. Biochem J 254:723–730
18. Hüther FJ, Kadenbach B (1986) Specific effects of ATP on the kinetics of reconstituted bovine heart cytochrome *c* oxidase. FEBS Lett 207:89–94
19. Hüther FJ, Berden JA, Kadenbach B (1988) The influence of 8-azido-ATP and other anions on the activity of cytochrome *c* oxidase. J Bioenerg Biomembr 20:503–516
20. Hüther FJ, Kadenbach B (1987) ADP increases the affinity for cytochrome *c* by interaction with the matrix side of bovine heart cytochrome *c* oxidase. Biochem Biophys Res Commun 147:1268–1275
21. Hüther FJ, Kadenbach B (1988) Intraliposomal nucleotides change the kinetics of reconstituted cytochrome *c* oxidase from bovine heart but not from *Paracoccus denitrificans*. Biochem Biophys Res Commun 153:525–534
22. Capaldi RA (1988) Structure and function of cytochrome oxidase: an overview. Ann N Y Acad Sci USA 550:3–7
23. Berden JA, Van Dongen MBM, Slothaak JB, Hartog AF (1985) Differentiation of three types of nucleotide-binding sites on mitochondrial S1 using 8-azido- and 2-azido-adenine nucleotide. In: Quagliariello E, Slater EC, Palmieri F, Saccone C, Kroon AM (eds) Achievements and perspective of mitochondrial research, vol 1. Bioenergetics. Elsevier, Amsterdam, pp 257–267
24. Labonia N, Müller M, Azzi A (1988) The effect of non-esterified fatty acids on the proton-pumping cytochrome *c* oxidase reconstituted in liposomes. Biochem J 254:139–145
25. Merle P, Kadenbach B (1980) On the heterogeneity of vertebrate cytochrome *c* oxidase polypeptide chain composition. Hoppe Seyler's Z Physiol Chem 361:1257–1259
26. Kadenbach B, Büge U, Jarausch J, Merle P (1981) Composition and kinetic properties of cytochrome *c* oxidase from higher eucaryotes. In: Palmieri F et al. Quagliariello E, Siliprandi N, Slater EC (eds) Vectorial reactions in electron and ion transport in mitochondria and bacteria. Elsevier/North-Holland Biomedical Press, Amsterdam New York, pp 11–23
27. Kadenbach B, Hartmann R, Glanville R, Buse G (1982) Tissue-specific genes code for polypeptide VIa of beef liver and heart cytochrome *c* oxidase. FEBS Lett 138:236–238
28. Capaldi RA (1988) Mitochondrial myopathies and respiratory chain proteins. Trends Biochem Sci 13:144–148
29. Jarausch J, Kadenbach B (1982) Tissue-specificity overrides species specificity in cytochrome *c* oxidase polypeptides. Hoppe-Seyler's Z physiol Chem 363:1133–1140

30. Kuhn-Nentwig L, Kadenbach B (1985) Isolation and properties of cytochrome *c* oxidase from rat liver and quantification of immunological differences between isozymes from various rat tissues with subunit-specific antisera. Eur J Biochem 149:147–158
31. Schlerf A, Droste M, Winter M, Kadenbach B (1988) Characterization of two different genes (cDNA) for cytochrome *c* oxidase subunit VIa from heart and liver of the rat. EMBO J 7:2387–2391
32. Merle P, Kadenbach B (1982) Kinetic and structural differences between cytochrome *c* oxidases from beef liver and heart. Eur J Biochem 125:239–244
33. Büge U, Kadenbach B (1986) Influence of membrane lipids, buffer composition and proteases on the kinetics of reconstituted cytochrome *c* oxidase from bovine liver and heart. Eur J Biochem 161:383–390
34. Kadenbach B, Stroh A, Becker A, Eckerskorn C, Lottspeich F (1988) Brown fat cytochrome *c* oxidase is different from the isozymes of liver and heart (submitted)
35. Giles RE, Blanc H, Cann HM, Wallace DC (1980) Maternal inheritance of human mitochondrial DNA. Proc Natl Acad Sci USA 77:6715–6719
36. Meinecke L, Buse G (1985) Studies on cytochrome *c* oxidase. XII. Isolation and primary structure of polypeptide VIb from bovine heart. Biol Chem Hoppe-Seyler 366:687–694
37. Cao X, Hengst L, Schlerf A, Droste M, Mengel T, Kadenbach B (1988) Complexity of nucleus-encoded genes of mammalian cytochrome *c* oxidase. Ann N Y Acad Sci 550:337–347
38. Suske G, Mengel T, Cordingley M, Kadenbach B (1987) Molecular cloning and further characterization of cDNAs for rat nuclear-encoded cytochrome *c* oxidase subunits VIc and VIII. Eur J Biochem 168:233–237
39. Suske G, Enders C, Schlerf A, Kadenbach B (1988) Organization and nucleotide sequence of two chromosomal genes for rat cytochrome *c* oxidase subunit VIc: a structural and a processed gene. DNA 7:163–171
40. DiMauro S, Bonilla E, Zeviani M, Nakagawa M, DeVivo DC (1985) Mitochondrial myopathies. Ann Neurol 17:521–538
41. DiMauro S, Zeviani M, Servidei S, Bonilla E, Miranda AF, Prelle A, Schon EA (1986) Cytochrome oxidase deficiency: clinical and biochemical heterogeneity. Ann N Y Acad Sci 488:19–32
42. DiMauro S, Bonilla E, Zeviani M, Servidei S, DeVivo DC, Schon EA (1987) Mitochondrial myopathies. J Inherit Metab Dis 10 Suppl 1:113–128
43. Wallace DC (1987) Maternal genes: mitochondrial diseases. Birth Defects 23/3:137–190
44. Holt JJ, Harding AE, Morgan-Hughes JA (1988) Deletions of muscle mitochondrial DNA in patients with mitochondrial myopathies. Nature (London) 331:717–719
45. Zeviani M, Moreas CT, DiMauro S, Nakase H, Bonilla E, Schon EA, Rowland LP (1988) Deletions of mitochondrial DNA in Kearns-Sayre syndrome. Neurology 38:1339–1346
46. Johnson MA, Turnbull DM, Dick DJ, Sherratt HSA (1983) A partial deficiency of cytochrome *c* oxidase in chronic progressive external ophthalmoplegia. J Neurol Sci 60:31–53
47. Müller-Höcker J, Pongratz D, Hübner G (1983) Focal deficiency of cytochrome-*c*-oxidase in skeletal muscle of patients with progressive external ophthalmoplegia. Cytochemical-fine-structural study. Virchows Arch 402:61–71
48. Johnson MA, Kadenbach B, Droste M, Old SL, Turnbull DM, Sherratt HSA (1988) Immunocytochemical studies of cytochrome *c* oxidase subunits in skeletal muscle of patients with partial cytochrome oxidase deficiencies. J Neurol Sci 87:75–90
49. Müller-Höcker J, Droste M, Kadenbach B, Pongratz D, Hübner G (1988) Fatal mitochondrial myopathy with cytochrome-*c*-oxidase deficiency and subunit-restricted reduction of enzyme protein in two siblings. An autopsy-immunocytochemical study. Human Pathol (in press)

Fibroblasts and Cytochrome *c* Oxidase Deficiency

D.M. GLERUM, B.H. ROBINSON and R. CAPALDI

1 Introduction

Adenosine triphosphate (ATP) is the universal cellular energy currency, and is produced primarily by the enzymes of the respiratory chain. These enzymes are integral, multisubunit components of the inner mitochondrial membrane. They are responsible for the coupled oxidation of metabolites and phosphorylation of ADP (adenosine diphosphate). Any errors in this process produced by mutation have severe implications for survival.

Cytochrome *c* oxidase (Complex IV) is the terminal component of the respiratory chain, responsible for reducing molecular oxygen and vectorially translocating protons to drive ATP synthesis (reviewed by Azzi 1980). The mammalian enzyme consists of 13 polypeptide subunits (Capaldi et al. 1986; Merle and Kadenbach 1981), compared to the seven subunits of the yeast enzyme (Power et al. 1984), and the two to three subunits of the enzyme in *Paracoccus* (Ludwig and Schatz 1980). The mammalian enzyme is the product of both the nuclear and mitochondrial genomes (Hare et al. 1980; Anderson et al. 1981). The three largest subunits are mitochondrial gene products and form the catalytic core of the enzyme (Capaldi et al. 1983). The remaining ten subunits are all encoded in the nucleus, synthesized on cytoplasmic ribosomes and then transported into the mitochondria (Schatz and Butow 1983). All subunits, of both mitochondrial and cytoplasmic origin, are then assembled into an active complex at the inner mitochondrial membrane. The nuclear encoded subunits are thought to play a regulatory role (Kadenbach 1986), although no specific functions have yet been assigned to the individual subunits.

Given the involvement of two different genomes and two separate transcription and translation systems, it is scarcely surprising to encounter errors in this process. Inherited disorders of the respiratory chain in general, and particularly of cytochrome *c* oxidase, have been widely reported in the literature and many excellent and exhaustive treatises have been written on this subject (DiMauro et al. 1985, 1987a, 1987b, 1988; Scholte et al. 1987; Capaldi 1988). In many cases, patients are shown to have mitochondrial myopathies, abnormal morphological features of the mitochondria. Diagnoses for respiratory chain deficiencies have traditionally been based on enzyme activity and cytochrome spectral analysis,

Departments of Biochemistry and Pediatrics, University of Toronto and Research Institute, Hospital for Sick Children, Toronto, Ontario, Canada, Institute of Molecular Biology, University of Oregon, Eugene, OR, USA

A. Azzi, Z. Drahota and S. Papa
Molecular Basis of Membrane-Associated Diseases

using isolated muscle mitochondria. Using these techniques, the first case of cytochrome *c* oxidase deficiency was reported by Willems et al. in 1977. The defect in this case was associated with a presentation of Leigh's disease. Since then, Leigh's disease or subacute necrotizing encephalomyelopathy, has been documented regularly in conjuction with Complex IV defects (Miyabayashi et al. 1983; Robinson et al. 1987a), although Leigh's disease has also been reported as being involved with other enzyme defects (Robinson et al. 1987b). Other cytochrome *c* oxidase associated syndromes include chronic progressive external ophthalmoplegia (CPEO) (Reichmann 1988), Kearns-Sayre syndrome (KSS) (Tanaka et al. 1986), mitochondrial encephalomyelopathy with lactic acidosis and stroke-like episodes (MELAS) (Johnson et al. 1983), and myoclonus epilepsy with ragged red fibres (MERFF) (Rosing et al. 1985). In many of the reported Complex IV deficiencies, lactic acidosis is one of the primary clinical features. It occurs either as a chronic, persistent acidosis or as an exercise-induced symptom. The latter is of course consistent with a defect which is muscle-specific (see later). Our work has focused on patients with lacticacidemia with, in most cases, multitissue involvement. These patients also present with neurological impairment. We have used cultured skin fibroblasts from these affected pediatric patients to study the nature of the observed cytochrome *c* oxidase defects.

2 Clinical Presentations and Diagnosis

Cytochrome *c* oxidase deficiency is the most widely reported of the inherited respiratory chain defects, and the clinical presentations the most varied. Again, many extensive reviews exist on this subject (DiMauro et al. 1987a, 1988). Our work in this area has involved studies of Complex IV deficiency in cultured skin fibroblasts, obtained by skin biopsies from affected pediatric patients (Table 1). There are three basic classifications for which Complex IV defects have been demonstrable in fibroblasts. These are: (1) fatal infantile mitochondrial myopathy with cardiomyopathy and lacticacidemia, (2) subacute necrotizing encephalomyelopathy (Leigh's disease), and (3) Kearns-Sayre syndrome with lacticacidemia (Robinson). Typically, the patients in the first two categories present anywhere from early in the neonatal period up to 5 or 6 years of age, with hypotonia and lacticacidemia. They also suffer from psychomotor retardation, often displaying a gradual loss of acquired motor function. Nystagmus, ataxia, anorexia and apnea are also common features in these patients (Robinson et al. 1986). In cases where autopsy was permitted, the brains of patients in group 2 were found to have the necrotic and cystic lesions in the brainstem and basal ganglia which are characteristic of Leigh's disease. In three of the cases which we have published (Robinson et al. 1987a), the deficiency has also been documented in muscle. Miyabayashi et al. (1987) have also reported two patients with Leigh's disease and cytochrome *c* oxidase deficiency in which the defect was demonstrated in both fibroblasts and muscle.

Table 1. Cytochrome *c* oxidase deficient patients

Line[a]	Clinical presentation	Activity MITO(%)	W.C.[d](%)	Class[b]	Ref.[c]
890(a)	Leigh's/lacticacidemia	24		S	Robinson et al. 1987a
1489	Leigh's/lacticacidemia	27		P	Robinson et al. 1987a
1630	Leigh's/lacticacidemia	56		P	Glerum et al. 1987
1909	Leigh's/lacticacidemia	19		S	Robinson et al. 1987a
1947	KSS[e]/lacticacidemia		50	P	U
2130	Neonatal death from hypertrophic cardio-myopathy/lacticacidemia	41		P	DiMauro et al. 1987b
2161	Ataxia/lacticacidemia	17		S	Robinson et al. 1987a
2233	Leigh's/lacticacidemia	45		P	U
2246(a)	Leigh's/lacticacidemia	25		S	U
2288	Neonatal death from respiratory failure/ lacticacidemia		30	P	U
2417	Leigh's/lacticacidemia		9	S	U
2434	Neonatal death from respiratory failure/ lacticacidemia		12	S	U
2440	Hypotonia/cardiomyopathy/ lacticacidemia		46	P	U
2550(b)	Ataxia/lacticacidemia		17	S	U
2576	Hypotonia/lacticacidemia		48	P	U
2612	Hypotonia/lacticacidemia		10	S	U
2693(b)	5 months, normal at present		3	S	U

[a](a) and (b) = Two sets of sibs.
[b]S, Severe deficiency; P, partial deficiency.
[c]U, unpublished.
[d]W.C., Whole cell homogenate.
[e]KSS = Kearns-Sayre syndrome.

2.1 Fibroblasts as Diagnostic Tools

As previously mentioned, diagnosis of cytochrome *c* oxidase deficiency has traditionally been made using biopsy tissue from muscle. Defects in Complex IV have also been demonstrated in other tissues such as liver, heart and brain, from biopsy and/or autopsy.

There is now a well-recognized tissue-specific pattern of expression of cytochrome *c* oxidase deficiency (DiMauro et al. 1985) in which certain tissues carry the defective enzyme whereas others are unaffected. This pattern is in keeping with the occurrence of tissue-specific subunits of the cytochrome *c* oxidase complex, demonstrated by SDS-polyacrylamide gel electrophoresis (Merle and Kadenbach 1980), immunological cross-reactivity of subunit specific antisera (Jarausch and Kadenbach 1982; Kuhn-Nentwig and Kadenbach 1985), and by partial protein sequencing of individual subunits (Yanamura et al. 1988). Such tissue-specific expression of Complex IV defects has traditionally given cultured skin fibroblasts a low profile as a diagnostic tool. In some cases of tissue-specific defective Complex IV activity, fibroblast activity appears to be normal. However, in cases of multitissue involvement of the defect, skin fibroblasts can be used for diagnosis. In our laboratory we have found different levels of residual Complex IV activity in a variety of tissues (Table 2), especially when the activity in fibroblasts demonstrates a partial defect.

Another level of complexity involved with Complex IV is the possible existence of developmentally-regulated isoenzymes in specific tissues. There have been reports of a benign reversible cytochrome *c* oxidase deficiency (DiMauro et al. 1983; Zeviani et al. 1987b), in which the enzyme has steadily increasing activity levels as the patient grows older. This developmental specificity was corroborated in a study of immunological differences between adult and fetal rat tissues (Kuhn-Nentwig and Kadenbach 1985). However, a recent report, using protein sequencing of all subunits, has shown no differences between the fetal and adult bovine heart enzymes (Takamiya et al. 1987). Thus, the existence of developmentally-regulated subunits remains to be confirmed. The impact of developmental regulation upon fibroblasts therefore remains unknown. There have been other barriers to the use of skin fibroblasts in studying respiratory chain defects. The paucity of mitochondria in these cells and their lower titer of respiratory chain enzymes have made it difficult to institute reproducible assay systems, particularly for Complex I.

There are, however, two persuasive reasons for using skin fibroblasts for diagnosis. The first is that one is not limited to the small and finite amounts of material obtained from biopsy (or autopsy, after death). The other is that any deficiency demonstrated in successive cultured generations of cells proves unequivocally that the defect is genetic, given that fibroblasts are a somatic cell.

2.2 *Diagnosis*

In all cases of Complex IV deficiency which we have presented so far, the cultured skin fibroblasts have shown decreased levels of pyruvate oxidation but normal pyruvate dehydrogenase complex (PDHC) activity. Cell lines with Complex IV deficiency were subsequently shown to exhibit an abnormal NAD/NADH redox state as measured by the lactate to pyruvate ratio (L/P), after incubation with glucose (Robinson et al. 1987a). An abnormal L/P ratio is used in our laboratory as an indicator of a respiratory chain defect, either in Complex IV or Complex I (NADH:CoQ oxidoreductase). The elevated L/P ratio is also useful in directing a more detailed study if the defect in the enzyme is more subtle, either partial or

Table 2. Typical cytochrome *c* oxidase activity in various tissues[a] (nmol/min/mg protein)

Cell Line	Control	2440	
Tissue			
Fibroblast	4.53 ± 0.41(4)	2.09 ± 0.32(4)	46%
Muscle	1.08	0.811	75%
Liver	4.05	0.00	0%

[a]All assays were performed on homogenates of the tissue under investigation.

kinetic in nature. The diagnosis of a Complex IV deficiency is based on activity assays of both sonicated whole-cell preparations (Glerum et al. 1987) and isolated mitochondria.

In all of the cytochrome *c* oxidase deficient cell lines which we have studied to date, 17 of which are presented here, there is a striking variety with respect to clinical presentation. Some of the patients expired in the neonatal period whereas the others presented later in the first year of life and died a few years later, after a slowly regressing clinical course. There was also a large range of L/P ratio values as well as residual Complex IV activity. There was no apparent correlation between the clinical course, with respect to age of onset or death and severity of symptoms, and the degree of enzyme deficiency. When digitonin-treated fibroblasts were assayed for their ability to synthesize ATP, there was again a striking variation in levels of ATP production, which also did not seem to correlate with either the residual activity or L/P ratios. This large heterogeneity of both clinical and biochemical findings suggests that there are a large number of different mutations being presented. Activity measurements, L/P ratios and ATP production alone are thus not sufficient to characterize these defects. More detailed studies at the protein and molecular level have thus been undertaken.

3 Defect Characterization

To elucidate the precise defects involved in cases of Complex IV deficiency, it is desirable to demonstrate the deletion, depletion or alteration of one specific polypeptide in this multisubunit enzyme complex. At this time there have been no reports of such specific findings.

3.1 Immunodetection

Both immunoblotting and immunoprecipitation experiments, using subunit-specific antibodies for Complex IV as well as antibody to the holoenzyme (native or denatured), have been carried out. Fibroblasts and liver have thus far been the tissues used. The Western blotting technique used both by Miyabayashi et al. (1987) and ourselves (Glerum et al. 1988) has shown a generalized decrease in titratable subunits of cytochrome *c* oxidase in fibroblast mitochondria, obtained from patient cell lines. Our work has revealed two basic classifications with respect to the levels of immunotitratable subunits present, a severe class and a partial class (Fig. 1). These classifications bear a relationship to the measurements of Complex IV activity, and as will be discussed later, also correlate very well with the heme spectra obtained. Those cell lines which lacked any titratable subunits when probed with antibody to subunits MtII and cIV (nomenclature of Capaldi et al. 1986) had activity levels which fell in the range of 9% to 25% of those for normal control cell lines. These cell lines are thus in the severe category. The partial class comprises cell lines with 30% to 50% of normal activity, in which the amount of titratable subunits appears to be either somewhat decreased (by approximately an amount which matches the residual activity) or present at relatively normal levels. At least one of the cell lines in this latter category appears to suffer from a kinetic

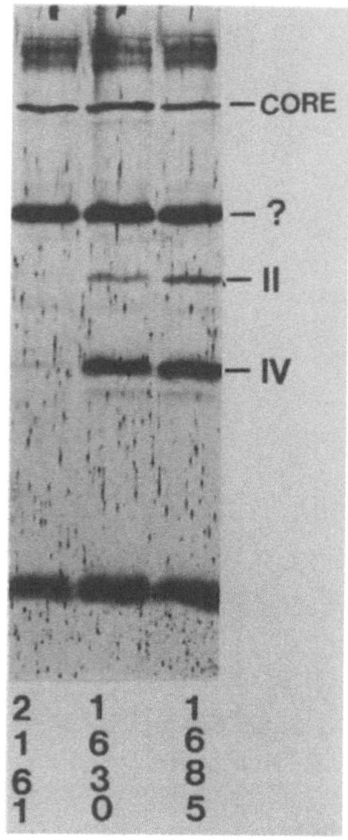

Fig. 1. Immunoblot of mitochondria obtained from cultured skin fibroblasts. *?* is the unidentified, fibroblasts-specific band of M_r = 33 kD; *Core* is the core proteins of Complex III; *II* and *IV* are subunits MtII and cIV of cytochrome *c* oxidase

defect (Glerum et al. 1987). This loss of titratable protein by immunoblotting does not, however, imply that there is a total absence of cytochrome *c* oxidase from affected mitochondria.

It has already been shown that cytochrome *c* oxidase assembles in a sequential manner and that a nuclear encoded subunit is required for this process (Wielburski and Nelson 1983). Studies with the enzyme from *S. cerevisiae* (Wright et al. 1986) have also shown that deletion mutants of COX VIIa fail to assemble properly, resulting in a nonfunctional holoenzyme. This suggests that for those deficient cell lines in the severe category, a mutation in one of the nuclear-encoded subunits prevents either its synthesis or its incorporation into the holoenzyme. This missing or altered subunit then prevents assembly of the enzyme, or reduces the percentage of assembled complex to a level below that detectable by Western blotting.

Immunoprecipitation with antibody to the purified Complex IV enzyme, after labeling of cell protein with ^{35}S-methionine, can be used to better assess whether a specific subunit is being synthesized and/or assembled into the complex. DiMauro et al. (1987a) have shown that in immunoprecipitates of solubilized liver mitochondria from a patient with Leigh's disease, all subunits were present, and at normal levels. In our laboratory, preliminary results suggest that in all our cases of Complex IV deficiency, the subunits are all synthesized and correctly assembled into the holoenzyme (Fig. 2). In some cases there seems to be a decreased synthesis, but again, this does not seem to be specific to any particular subunit; even cell lines from the severe class show synthesis and assembly of all subunits. There were also no detectable abnormalities of subunit size. This lack of specificity is in sharp

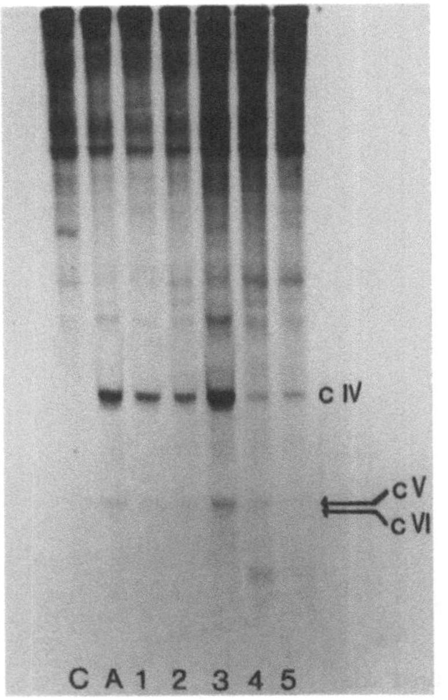

Fig. 2. Immunoprecipitation of ^{35}S-methionine labeled cultured skin fibroblasts. Fibroblasts are labeled for 48 h and immunoprecipitated with 1:1500 dilution of antibody to the denatured holoenzyme. *C* control serum; *A* control cell line 1206; *1* 1947; *2* 2130; *3* 2417; *4* 2434 and *5* 2612

contrast to the results obtained with both Western blotting and immunoprecipitation for Complexes I and III. Schapira and colleagues (1988) have reported a case of a distinct and specific reduction of the 24 kD iron-sulfur (Fe-S) protein in muscle mitochondria. Likewise, Moreadith et al. (1987) have shown selective reduction of the 75 and 13 kD polypeptides. There was, however, a concomitant decrease in the overall amount of titratable protein. A recent report of a specific deletion of the 17 kD polypeptide of Complex III in yeast (Schoppink et al. 1988) shows that its loss does not affect assembly of the complex but reduces the electron transfer activity to 40% of normal levels. This seems to suggest that, in contrast to Complexes I and III, cytochrome *c* oxidase has a more sensitive and demanding assembly process, whereby deletion or alteration of a specific subunit prevents holoenzyme assembly at normal levels.

3.2 Cytochrome Measurement

Cytochrome spectra have traditionally been measured in muscle mitochondria as a means of assessing the presence or absence of certain moieties of both Complexes III and IV. Cytochrome b and c_1 give an indication of the amount of ubiquinone-cytochrome *c* reductase (Complex III) which is present, and cytochrome aa_3 represents the amount of cytochrome *c* oxidase present. Deficiencies to both Complexes III (Darley-Usmar et al. 1983; Kennaway et al. 1984; Morgan-Hughes et al. 1985) and IV (Hoganson et al. 1984) have been assigned on the basis of these heme determinations. We have now shown that these spectra can also be obtained using fibroblast mitochondria (Glerum et al. 1988). As previously mentioned, the spectra added further support to the classification of the cytochrome *c* oxidase

Table 3. Cytochrome *c* oxidase activity and heme content of deficient fibroblast cell lines

Line	Activity (nmol/min/mg protein)		Heme Content (pmol/mg mito protein)			
	Mitochondria	Whole cell	b	c_1	aa_3[a]	$bc_1:aa_3$
Control	40.8 ± 4.9(7)	4.3 ± 0.5(5)	57	54	21	5:1
1630	23.6 ± 4.1(6)	–	77	53	16	8:1
2440	–	2.1 ± 0.3(5)	76	57	11	12:1
2417	–	0.4 ± 0.2(5)	89	91	Ø	–
1909	7.7 ± 1.2(9)	–	25	21	Ø	–

[a]Ø, Undetectable.

deficient lines into the partially or severely deficient categories. The spectral peak for cytochrome aa_3 (at 605 nm) was completely eradicated in cell lines in the severe class (Table 3), while the peaks for cytochromes b and c_1 were present at comparatively normal levels. The α band in the Soret region at 442 nm was also missing. Those cell lines in the partial class presented a measurable aa_3 peak at 605 nm which was reduced from control levels. In contrast to the severely deficient cell lines, the latter cell lines showed some reduced aa_3 absorption at 442 nm which appeared as a shoulder. Again, for these partially deficient lines, the levels of cytochromes b and c_1 were normal. Thus, fibroblasts have also proven adaptable to traditional spectral analysis in assessing cytochrome *c* oxidase deficiency.

3.3 Northern Analysis

Given the rather ambiguous results which have been obtained thus far through biochemical and immunological means, an analysis of Complex IV deficient cells at the molecular level has been undertaken. At present there has been only one published report of such a molecular analysis. This involved in situ hybridization (DiMauro et al. 1988) of muscle sections from a patient with reversible cytochrome *c* oxidase deficiency. The ragged-red fibres were shown to be transcriptionally active for subunit cIV.

We have undertaken an analysis of message levels for various nuclear-encoded subunits of the enzyme, and preliminary results have indicated no detectable alterations at the mRNA level. Northern blotting experiments were carried out on total cellular RNA and probed with full-length cDNA clones for cIV (Zeviani et al. 1987a), cV (Rizzuto et al. 1988), cVI (Zeviani et al. 1988) and cIX (gifts from E.A. Schon). There are no cell lines which are without message for any of cIV, cV, cVI, and cIX, but this is hardly surprising given that the immunoprecipitation showed that all subunits were being synthesized. The elucidation of the precise defects thus await a detailed genetic analysis.

4 Summary

The elucidation of the underlying causes of Leigh's disease and lacticacidemia, resulting from cytochrome *c* oxidase deficiency, has made tremendous strides in the last few years, yet the precise defects involved remain obscure. The use of cultured skin fibroblasts from affected pediatric patients has just begun to emerge as a diagnostic and research tool. The fibroblast cell lines have provided us with what appears to be a variety of different mutations producing variable residual activities and clinical presentations. As the defects appear to be inherited in an autosomal recessive manner, the nuclear-encoded subunits appear to be the site of the different mutations. There has been no evidence for maternal inheritance in any of the cases we have reported, and as such we have concentrated on studying the cytoplasmically synthesized subunits. The first findings of deletions in mtDNA were recently announced (Holt et al. 1988), however, this was again based on studies in muscle. The other possible site for a mutation could be in the gene for an organizer protein required for cytochrome *c* oxidase assembly. It has recently been proposed that mitochondria contain a groEL-type protein which is required in refolding of newly imported proteins (McMullin and Hallberg 1988). This type of protein has now been found to be homologous with an abundant protein in chloroplasts which is implicated in the assembly of the oligomeric ribulose bis-phosphate carboxylase enzyme (Hemmingsen et al. 1988). These homologues have been termed "chaperonins" and one of this class might be involved in cytochrome *c* oxidase assembly, although this has yet to be proven. Given the fact that the low amounts of residual activity in some of the cell lines appear to result from a low amount of assembled and functioning enzyme, a factor involved in assembly which was defective, would explain the lack of assembled enzyme.

Thus, fibroblasts with defects in cytochrome *c* oxidase activity provide a unique opportunity to study the biogenesis and functioning of this important enzyme.

Acknowledgments. We thank the Medical Research Council and National Foundation March of Dimes for financial support.

References

Anderson S, Bankier AT, Barrell BG, DeBruijn MHL, Coulson AR, Drouin J, Eperon IC, Nierlich DP, Roe BA, Sanger F, Schreier PH, Smith AJH, Staden R, Young IG (1981) Sequence and organization of the human mitochondrial genome. Nature (London) 290:457–465

Azzi A (1980) Cytochrome *c* oxidase: towards a clarification of its structure, interactions and mechanism. Biochim Biophys Acta 594:231–252

Capaldi RA (1988) Mitochondrial myopathies and respiratory chain proteins. Trends Biochem Sci 13:144–148

Capaldi RA, Malatesta F, Darley-Usmar VM (1983) Structure of cytochrome *c* oxidase. Biochim Biophys Acta 726:135–148

Capaldi RA, Gonzales-Halphen D, Takamiya S (1986) Sequence homologies and structural similarities between the polypeptides of yeast and beef heart cytochrome *c* oxidase. FEBS Lett 207:11–17

Darley-Usmar VM, Kennaway NG, Buist NRM, Capaldi RA (1983) Deficiency in ubiquinone cytochrome *c* reductase in a patient with mitochondrial myopathy and lactic acidosis. Proc Natl Acad Sci USA 80:5103–5106

DiMauro S, Nicholson JF, Hays AP, Eastwood AB, Papadimitriou A, Koenigsberger R, DeVivo DC (1983) Benign infantile mitochondrial myopathy due to reversible cytochrome *c* oxidase deficiency. Ann Neurol 14:226–234

DiMauro S, Bonilla E, Zeviani M, Nakagawa M, DeVivo DC (1985) Mitochondrial myopathies. Ann Neurol 17:521–538

DiMauro S, Bonilla E, Zeviani M, Servidei S, DeVivo DC, Schon EA (1987a) Mitochondrial myopathies. J Inherit Metab Dis 10 Suppl 1:113–128

DiMauro S, Servidel S, Zeviani M, DiRocco M, DeVivo DC, DiDonato S, Uziel G, Berry K, Hoganson G, Johnsen SD, Johnson PC (1987b) Cytochrome *c* oxidase deficiency in Leigh syndrome. Ann Neurol 22:498–506

DiMauro S, Zeviani M, Servidei S, Prelle A, Miranda AF, Bonilla E, Schon EA (1988) Biochemical and molecular aspects of cytochrome *c* oxidase deficiency. Adv Neurol 48:93–105

Glerum M, Robinson BH, Spratt C, Wilson J, Patrick D (1987) Abnormal kinetic behaviour of cytochrome *c* oxidase in a case of Leigh disease. Am J Human Genet 41:584–593

Glerum DM, Yanamura W, Capaldi RA, Robinson BH (1988) Characterization of cytochrome *c* oxidase mutants in human fibroblasts. FEBS Lett 236:100–104

Hare JF, Ching E, Attardi G (1980) Isolation, subunit composition, and site of synthesis of human cytochrome *c* oxidase. Biochemistry 19:2025–2030

Hemmingsen SM, Woolford C, Vandervies SM, Tilly K, Dennis DT, Georgopoulus CP, Hendrix RW, Ellis RJ (1988) Homologous plant and bacterial proteins chaperone oligomeric protein assembly. Nature (London) 333:330–334

Hoganson GE, Paulson DJ, Chun R, Sufit RL, Shug AL (1984) Deficiency of muscle cytochrome *c* oxidase in Leigh's disease. Pediat Res 18:222A

Holt IJ, Harding AE, Morgan-Hughes JA (1988) Deletions of muscle mitochondrial DNA with patients with mitochondrial myopathies. Nature (London) 331:717–719

Jarausch J, Kadenbach B (1982) Tissue-specificity overrides species-specificity in cytoplasmic cytochrome *c* oxidase polypeptides. Hoppe-Seyler's Z Physiol Chem 363:1133–1140

Jonhson MA, Turnbull DM, Dick DJ, Sherratt HSA (1983) A partial deficiency of cytochrome *c* oxidase in chronic progressive external ophthalmoplegia. J Neurol Sci 60:31–53

Kadenbach B (1986) Mini review. Regulation of respiration and ATP synthesis in higher organisms: hypothesis. J Bioenerg Biomembr 18:39–54

Kadenbach B, Merle P (1981) On the function of multiple subunits of cytochrome *c* oxidase from higher eukaryotes. FEBS Lett 135:1–11

Kennaway NG, Buist NR, Darley-Usmar VM, Papadimitriou A, DiMauro S, Kelly RI, Capaldi RA, Blank NK, D'Agostino A (1984) Lactic acidosis and mitochondrial myopathy associated with deficiency of several components of complex III of the respiratory chain. Pediat Res 18:992–999

Kuhn-Nentwig L, Kadenbach B (1985) Isolation and properties of cytochrome *c* oxidase from rat liver and quantification of immunological differences between isozymes from various rat tissues with subunit-specific antisera. Eur J Biochem 149:147–158

Ludwig B, Schatz G (1980) A two-subunit cytochrome *c* oxidase (cytochrome aa_3) from *Paracoccus denitrificans.* Proc Natl Acad Sci USA 77:196–200

McMullin WT, Hallberg RL (1988) A highly evolutionarily conserved mitochondrial protein is structurally related to the protein encoded by the *Escherichia coli* groEL gene. Mol Cell Biol 8:371–380

Merle P, Kadenbach B (1980) On the heterogeneity of vertebrate cytochrome *c* oxidase polypeptide chain composition. Hoppe-Seyler's Z Physiol Chem 361:1257–1259

Miyabayashi S, Narisawa K, Tada K, Sakai K, Kobayashi E, Kobayashi V (1983) Two siblings with cytochrome *c* oxidase deficiency. J Inherit Metab Dis 6:121–122

Miyabayashi S, Ito T, Abukawa D, Narisawa K, Tada K, Tanaka M, Ozawa T, Droste M, Kadenbach B (1987) Immunochemical study in three patients with cytochrome *c* oxidase deficiency presenting Leigh's encephalomyelopathy. J Inherit Metab Dis 10:289–292

Moreadith RW, Cleeter MWJ, Ragan CI, Batshaw ML, Lehninger AL (1987) Congenital deficiency of two polypeptide subunits of the iron-protein fragment of mitochondrial complex I. J Clin Invest 79:463–467

Morgan-Hughes JA, Hayes DJ, Cooper M, Clark JB (1985) Mitochondrial myopathies: deficiencies localized to complex I and complex III of the mitochondrial respiratory chain. Biochem Soc Trans 13:648–650

Murphy JV, Isohashi F, Weinberg MB, Utter MT (1981) Pyruvate carboxylase deficiency – an alleged biochemical cause of Leigh's disease. Pediatrics 68:401–404

Power SD, Lochrie MA, Patterson TE, Poyton RO (1984) The nuclear-coded subunits of yeast cytochrome *c* oxidase. II. The amino acid sequence of subunit VIII and a model for its disposition in the inner mitochondrial membrane. J Biol Chem 259:6571–6574

Reichmann H (1988) Enzyme activity measured in single muscle fibers in partial cytochrome *c* oxidase deficiency. Neurology 38:244–249

Rizzuto R, Nakase H, Zeviani M, DiMauro S, Schon EA (1988) Subunit Va of cytochrome *c* oxidase is highly conserved between human and bovine. Gene 69:245–254

Robinson BH, De Meirleir L, Glerum M, Sherwood WG, Becker L (1986) Mitochondrial respiratory chain defects identified in cultured skin fibroblasts from patients with lacticacidemia, hypotonia and growth failure. SSIEM, Proc 24th Annu Symp, p 116

Robinson BH, De Meirleir L, Glerum M, Sherwood G, Becker L (1987a) Clinical presentation of mitochondrial respiratory chain defects in NADH-coenzyme Q reductase and cytochrome oxidase: clues to the pathogenesis of Leigh disease. J Pediat 110:216–222

Robinson BH, MacMillan H, Petrova-Benedict R, Sherwood WG (1987b) Variable clinical presentation in patients with deficiency of the pyruvate dehydrogenase complex. A review of 30 cases with a defect in the E_1 component of the complex. J Pediat 111:525–533

Robinson BH (1989) Lacticacidemia: biochemical, clinical and genetic considerations. Adv Human Genet (in press)

Robinson BH Lacticacidemia. In: Scriver CR, Beaudet AL, Sly WS, Valle D (eds) The metabolic basis of inherited disease. McGraw-Hill, New York (in press)

Rosing HS, Hopkins LC, Wallace DC, Epstein CM, Weidenheim K (1985) Maternally inherited mitochondrial myopathy and myoclonic epilepsy. Ann Neurol 17:228–237

Schapira AHV, Morgan-Hughes JA, Cleeter MJW, Cooper JM, Patel SD, Ragan CI, Clark JB (1988) Molecular basis of mitochondrial myopathies: polypeptide analysis in complex I deficiency. Lancet 8584:500–503

Schatz G, Butow RA (1983) How are proteins imported into mitochondria? Cell 32:316–318

Scholte HR, Busch HFM, Luyt-Houwen IEM, Vaandrager-Verduin MHM, Przyrembel H, Arts WFM (1987) Defects in oxidative phosphorylation: biochemical investigations in skeletal muscle and expression of the lesion in other cells. J Inherit Metab Dis 10 Suppl 1:81–97

Schoppink PJ, Hemrika W, Reynen JM, Grivell LA, Berden JA (1988) Yeast ubiquinol: cytochrome *c* oxidoreductase is still active after inactivation of the gene encoding the 17-kDa subunit VI. Eur J Biochem 173:115–122

Takamiya S, Lindorfer M, Capaldi RA (1987) Purification of all thirteen polypeptides of bovine heart cytochrome *c* oxidase from one aliquot of enzyme. Characterization of bovine fetal heart cytochrome *c* oxidase. FEBS Lett 218:277–282

Tanaka M, Nishikimi M, Suzuki H, Ozawa T, Nishizawa M, Tanaka K, Miyatake T (1986) Deficiency of subunits in heart mitochondrial NADH-ubiquinone oxidoreductase of a patient with mitochondrial encephalomyopathy and cardiomyopathy. Biochem Biophys Res Commun 140:88–93

Wielburski A, Nelson BD (1983) Evidence for the sequential assembly of cytochrome *c* oxidase subunits in rat liver mitochondria. Biochem J 212:829–834

Willems JL, Monnens LAW, Trijbels JMF, Veerkamp JH, Meyer AEFH, Van Dam K, Van Haelst U (1977) Leigh's encephalomyopathy in a patient with cytochrome *c* oxidase deficiency of muscle tissue. Pediatrics 60:850–857

Wright RM, Dircks LK, Poyton RO (1986) Characterization of cox 9, the nuclear gene encoding the yeast mitochondrial protein cytochrome *c* oxidase subunit VIIa. Subunit VIIa lacks a leader peptide and is an essential component of the holoenzyme. J Biol Chem 261:17183–17191

Yanamura W, Zhang Y-Z, Takamiya S, Capaldi RA (1988) Tissue-specific differences between heart and liver cytochrome *c* oxidase. Biochemistry 27:4909–4914

Zeviani M, Nakagawa M, Herbert J, Lomax MI, Grossman LI, Sherbany AA, Miranda AF, DiMauro S, Schon EA (1987a) Isolation of a cDNA clone encoding subunit IV of human cytochrome *c* oxidase. Gene 55:205–217

Zeviani M, Peterson P, Servidei S, Bonilla E, DiMauro S (1987b) Benign reversible muscle cytochrome *c* oxidase deficiency. Neurology 37:64–67

Zeviani M, Sakoda S, Sherbany AA, Nakase H, Rizzuto R, Samitt CE, DiMauro S, Schon EA (1988) Sequence of cDNAs encoding subunit Vb of human and bovine cytochrome *c* oxidase. Gene 65:1–11

Human Cytochrome *c* Oxidase Deficiencies; Structural and Functional Aspects

K.M.C. SINJORGO[1], A.O. MUIJSERS[1], H.R. SCHOLTE[2], A.B.P. VAN KUILENBURG[1], J.J. VAN BEEUMEN[3], E.M. BROUWER-KELDER[1], J.A. BERDEN[1], P.A. BOLHUIS[4] and J.M. TAGER[1]

1 Introduction

Mitochondria play a key role in aerobic energy metabolism in eukaryotic cells. The main function of the organelles is the oxidation of fuel molecules by oxygen and the concomitant energy production in the form of ATP.

The first reports of a mitochondrial disease by Luft and co-workers [1,2], describing a patient who suffered from hypermetabolism due to the defective coupling of oxidation and phosphorylation in muscle mitochondria, initiated the investigation of metabolic disorders that result from mitochondrial dysfunctions. Considering the structural and functional complexity of the mitochondrion it is not surprising that this class of diseases, generally referred to as mitochondrial myopathies, is extremely heterogeneous with respect to clinical manifestation, morphology and biochemical origin of the defects. As recently reviewed by Scholte [3], defects in over 60 different mitochondrial enzymes have been reported at this time. Other reviews on various aspects of mitochondrial diseases can be found in references [4–10].

In this chapter we focus on disorders of one particular mitochondrial protein: cytochrome *c* oxidase (Complex IV), the terminal component of the respiratory chain. Investigation of defects in this protein is of interest for several reasons.

Firstly, mitochondrial defects are often restricted to one or more tissues. An exceptional feature of cytochrome *c* oxidase, which might be of importance in this respect, is the occurrence of organ-specific isoforms, which have been demonstrated in various mammals, but not as yet in man [11–13].

Secondly, mitochondrial myopathies can be due to defects of either the nuclear or the mitochondrial genome and can therefore be transmitted by Mendelian or maternal inheritance, respectively. The various subunits of cytochrome *c* oxidase, like those of Complex I, III and V of the respiratory chain, are encoded both in the mitochondrial and in the nuclear genome. Consequently, studies of cytochrome *c* oxidase deficiencies may eventually yield information of general importance regarding (defects in) mitochondrial biogenesis.

[1]Laboratory of Biochemistry, University of Amsterdam, P.O. Box 20151, 1000 HD Amsterdam, The Netherlands
[2]Department of Biochemistry I, Erasmus University Rotterdam, The Netherlands
[3]Laboratory of Microbiology and Microbial Genetics, University of Ghent, Belgium
[4]Department of Neurology, University of Amsterdam, Academic Medical Centre, The Netherlands

A. Azzi, Z. Drahota and S. Papa
Molecular Basis of Membrane-Associated Diseases

After an introduction on structure, function and biogenesis of mammalian cytochrome *c* oxidase isoenzymes, we will discuss our investigations of the molecular structure of human cytochrome *c* oxidase in organs of patients and controls, in the light of other data published in the field.

2 Mammalian Cytochrome *c* Oxidase

2.1 Structure and Biogenesis

Mammalian cytochrome *c* oxidase, an intrinsic protein of the mitochondrial inner membrane, has a complex molecular structure. The protein is composed of 13 different subunits [14]. Furthermore, cytochrome *c* oxidase contains four reducible prosthetic groups; two haem *a* groups and two copper atoms [15,16].

The three largest subunits, I, II and III are coded for by mitochondrial genes and synthesized on mitochondrial ribosomes. At this time the complete mitochondrial genome of a variety of mammals, including man [17], has been sequenced. Mitochondrial transcription and translation have been reviewed in [18,19].

The ten smaller subunits are encoded in nuclear genes, synthesized in the cytoplasm and subsequently imported into the mitochondrion. Most of our knowledge of these processes stems from investigations on yeast (for detailed reviews, see [18,20,21]). Briefly, most proteins destined for the mitochondrial inner membrane are synthesized as precursors with an amino-terminal extension that contains the topogenic signal [21]. These leader sequences, which are generally rich in basic and hydroxylated amino acids and devoid of acidic amino acids, probably form amphiphilic α-helices [22]. Import occurs posttranslationally [23,24]; it is probably receptor-mediated [25,26] and is generally dependent on a membrane potential across the inner membrane [27]. Precursor proteins are processed to the mature form by a matrix protease [28]. Some polypeptides are processed in two steps [29]. Finally, recent evidence shows that import requires not only a membrane potential but also extramitochondrial ATP [30,31] which is probably necessary for unfolding of the precursors during import [32,33]. It has been suggested that this reaction is catalyzed by so-called unfoldases [34,35].

Studies of the biogenesis of the nuclear-encoded subunits of mammalian cytochrome *c* oxidase indicate that these processes are similar, if not identical, to the ones in lower eukaryotes [19]. It has been demonstrated that newly synthesized subunit IV [36,37] and V [38] of rat liver cytochrome *c* oxidase are present as precursors with an N-terminal extension. cDNA's for cytochrome *c* oxidase subunit IV from human liver [39] and bovine liver [40,41], subunit Va from human liver [42], subunit Vb from human fetal muscle and bovine brain [43] and subunits VIc and VIII from rat hepatomic cells have been isolated and (fully or partly) sequenced [44]. Remarkably, the subunit VIc cDNA did not show evidence for an N-terminal extension [44].

Least understood are the processes that coordinate the expression of the two genetic systems, and the assembly of cytochrome *c* oxidase (see [45]). Since the attention of several investigators has been drawn to this field, new information may be expected in the near future.

2.2 *Tissue-Specific Isoenzymes*

A striking characteristic of mammalian cytochrome *c* oxidase is the occurrence of organ-specific isoforms. This phenomenon was first reported by Kadenbach and co-workers [11–13] who observed for a number of mammals differences in migration rates in SDS gel electrophoresis of subunits VIa, VIIa and VIII of cytochrome *c* oxidase isolated from various organs of one species (see Fig. 1). Additional indications for organ-specific differences in nuclear-encoded subunits were found in the reaction of subunits with monospecific antibodies [46] and in cysteine-thiol group reactivity with N-ethyl maleimide [47].

Finally, N-terminal sequencing yielded definite proof of differences in the amino acid sequence in subunits VIa, VIIa and VIII of the bovine heart and liver enzyme [48,49]. This demonstrates that the organ-specific differences in the polypeptides are not the result of posttranslational modifications. Little is known about the molecular processes underlying the biogenesis of organ-specific cytochrome *c* oxidase isoforms.

2.3 *The Function of the Organ-Specific Subunits*

The primary catalytic functions of cytochrome *c* oxidase have all been assigned to the mitochondrially encoded subunits of the enzyme. Subunits I and II are

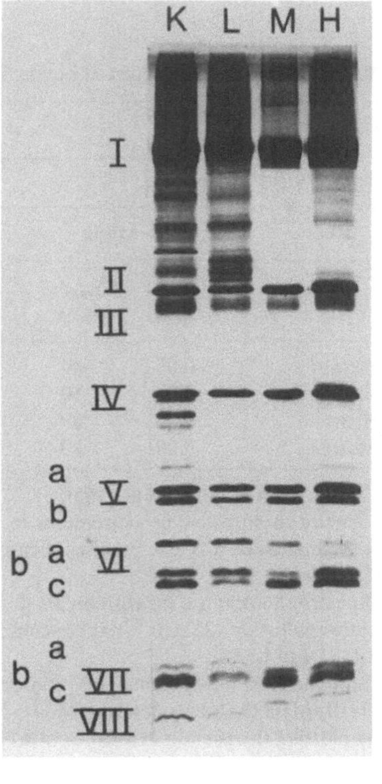

Fig. 1. Comparison of the subunit pattern of cytochrome *c* oxidase purified from bovine kidney (*K*), liver (*L*), skeletal muscle (*M*) and heart (*H*) (Sinjorgo et al. [105])

considered to contain all four prosthetic groups [50,51]. The cytochrome *c* binding domain has been located on subunit II [52–54] and subunit III is thought to be involved in the proton-translocating function of the oxidase [55,56].

In contrast, no specific functions have been ascribed to the nuclear-encoded subunits of the enzyme. Based on the discovery of organ-specific isoforms of these subunits, Kadenbach and co-workers proposed that they might act as organ-specific modulators of enzyme activity [12]. However, comparison by polarographic measurements of the steady-state kinetics of several mammalian isoenzymes, has not yielded consistent results [12,13,57]. Ascorbic acid and TMPD, present in the polarographic system, are known to induce artificial side reactions that may become rate-limiting [58].

We have used a spectrophotometric assay, in which only reduced cytochrome *c* is present as substrate, in order to compare the kinetics of isoenzymes purified from bovine heart, skeletal muscle, liver and kidney (see Fig. 1).

Firstly, the steady-state kinetics of the electron transfer reaction have been studied at low ionic strength (I=27 mM). Under these conditions non-linear Eadie-Hofstee plots are obtained [59], which can be ascribed to independent high and low affinity reactions with cytochrome *c* [60]. The high and low affinity K_m and TN_{max} values determined for the four cytochrome *c* oxidase preparations showed only minor differences (Table 1).

Secondly, similar studies were performed at high ionic strength (I=226 mM). Under these conditions, which give a closer approximation of the situation in vivo, linear Eadie-Hofstee plots are obtained [61]. Table 1 shows that again the kinetic

Table 1. Kinetic parameters of bovine cytochrome *c* oxidase preparations from various organs[a]

Cytochrome *c* oxidase from	Steady state						Presteady state
	I=27 mM[b]				I=226 mM[c]		I=226 mM[d]
	High Affinity		Low Affinity				
	K_m (μM)	TN_{max} (s^{-1})	K_m (μM)	TN_{max} (s^{-1})	K_m (μM)	TN_{max} (s^{-1})	$K_1 \times 10^{-7}$ ($M^{-1}s^{-1}$)
Kidney	0.05	40	12	265	17	350	1.0
Liver	0.07	50	8	281	22	364	0.7
Heart	0.05	39	15	280	23	386	0.7
Muscle	0.05	39	11	258	25	328	0.7

[a] Taken from Sinjorgo et al. [105].

[b] Spectrophotometric measurements in 10 mM potassium phosphate (pH 7.4)/1 mM EDTA/0.05% laurylmaltoside (25°C). Final concentration cyt. *c* oxidase was 1 mM; that of ferrocytochrome *ca.* 0.12–60 μM.

[c] Spectrophotometric measurements in 100 mM potassium-phosphate (pH 7.4)/1 mM EDTA/0.05% laurylmaltoside (25°C). Final concentration cyt. *c* oxidase was 1 mM; that of ferrocytochrome *ca.* 5.1–51.3 μM.

[d] Stopped-flow measurement of initial reduction of cytochrome *c* oxidase (1 μM) by ferrocytochrome *c* (4–10 μM) in the buffer described under Footnote *c*. k_1 was calculated from the slope of the line obtained by plotting the pseudo first-order rate constants versus the concentration of ferrocytochrome *c*.

parameters calculated for the various isoenzymes were comparable. Since both at low and at high ionic strength an equal variation in data was obtained when the kinetics of several bovine heart cytochrome *c* oxidase preparations were studied (not shown), we conclude that no differences exist between the electron transfer activities of the oxidase purified from various tissues.

Finally, we performed presteady-state kinetic studies in which we determined the rate constants of the association (k_1) of cytochrome *c* to the various isoenzymes by stopped-flow measurements (for details of the method, see [62]). No significant differences in k_1 values were found.

Thus, we conclude that in our studies differences in the small subunits of cytochrome *c* oxidase did not cause differences either in the association rate of cytochrome *c* or in the kinetics of the overall electron transfer reaction. Consequently, our results do not support the idea of a role for the organ-specific, nuclear-encoded subunits as modulators of the electron transfer reactions.

Of course, the possibility cannot be excluded that the isoenzymes possess different kinetics when embedded in the membrane. The organ-specific subunits may also be involved in binding of ligands that affect the enzyme activity, thus providing the possibility of an organ-specific response to regulators.

Furthermore, it remains conceivable that organ-specific subunits play a regulatory role in proton translocation, in coupling of electron and proton transfer or in anchoring the enzyme to the membrane (which shows a tissue specific phospholipid composition [13]).

3 Human Cytochrome *c* Oxidase Isoenzymes

Organ-specific isoforms of cytochrome *c* oxidase have not been demonstrated in man. In order to investigate this point we have recently initiated a study aimed at the purification of nuclear encoded cytochrome *c* oxidase subunits from human organs and establishing the amino acid sequence of the polypeptides.

3.1 Amino Acid Sequences of Human Heart Cytochrome c Oxidase Subunits

We have isolated the oxidase from human heart by a modification of the selective solubilization method of submitochondrial particles with deoxycholate. The purified enzyme was treated with organic solvent mixtures of decreasing polarity along the lines developed by the group of Poyton [63] for the purification of yeast cytochrome *c* oxidase subunits. This procedure resulted in a fraction containing subunits Va, Vb, VIb,c and VIII and a second fraction containing the bulk of the subunits IV, VIa and VIIa,b,c.

Each of these fractions was further separated by means of high pressure liquid chromatography on a reversed phase column. At least seven of the small subunits were obtained in a pure form, suitable for protein sequence determination with the pulsed liquid sequencer. Preliminary amino terminal sequence data were obtained for human heart cytochrome *c* oxidase subunits IV and Va, while the full sequence was determined for subunit VIII, the smallest subunit.

The amino terminal sequence of human heart subunit IV as determined by us is probably identical to that deduced from the nucleotide sequence of a human liver bank cDNA [39]. Some ambiguities in the protein sequence data might point to a slight heterogeneity, but as yet there is no other evidence for isoforms of subunits IV in mammals.

We have so far sequenced 40 residues of the amino terminus of human heart subunit Va. We have found only three amino acid substitutions as compared to the sequence of the corresponding bovine heart oxidase subunit [64]. The fact that the sequence is highly conserved suggests an essential role for subunit Va in the enzyme. Subunit Va is probably located close to subunit II at the intermembrane face of the inner membrane, where cytochrome *c* binds [65]. Again, the heart subunit Va protein sequence data seem to match those found by the group of Schon [42] for the liver cDNA nucleotide sequence.

In animals the small subunit VIII is very variable and tissue-dependent; in the first 20 residues the subunits VIII from bovine heart and liver have only 8 residues in common [49]. We determined the sequence for human heart subunit VIII and found that for the amino-terminal 20 residues there were only 6 residues identical to those of bovine heart. Thirteen residues of the human heart subunit VIII were identical to those of bovine liver.

As far as we know our amino acid sequences of the human enzyme are the only ones available at present. Comparison of our data with published cDNA nucleotide sequences did not yield evidence for the existence of tissue-specific isoenzymes of human cytochrome *c* oxidase. Many subunits, however, remain to be sequenced. Combined protein and nucleotide sequence work will be necessary to settle this point. Studies on mitochondrial ATPase [66] has demonstrated that differences may occur in presequences only; these will not be detectable in the mature enzyme. Finally, if two isoforms of a subunit are strongly dissimilar, a DNA probe based on the sequence of one of the isoforms may fail to detect the presence of the other isoform.

4 Human Cytochrome *c* Oxidase Deficiencies

4.1 Clinical Heterogeneity

Like mitochondrial myopathies in general, the ones primarily characterized by a cytochrome *c* oxidase deficiency form a heterogeneous group of disorders. In the past decade the following classification of cytochrome *c* oxidase defects, based on the most prominent clinical symptoms, has evolved (for more details on clinical and morphological signs and symptoms, see [67,68]):

1. Infantile myopathies. These patients show myopathy alone [69,70], myopathy associated with renal dysfunction [71–76] or myopathy associated with cardiopathy [77–79]. Generally, the diseases manifest shortly after birth and the patients usually die during their first year of life. Organ specificity of these types of cytochrome *c* oxidase defects has been reported. In contrast to the decreased cytochrome *c* oxidase activity in muscle (all patients) or kidney [72], normal enzyme functions have been found in heart, brain, liver and fibroblasts [70,71,77].

When tested, the decreased cytochrome *c* oxidase activity in tissue samples could be ascribed to decreased concentrations of immunologically detectable enzyme [70,76], although all cytochrome *c* oxidase subunits seemed to be present [70]. In spectral studies, decreased concentrations of the haem aa_3 prosthetic groups have been found. Remarkably, this was always accompanied by a decrease in cytochrome *b* and normal $(c + c_1)$ concentrations [70–72,74,78]. Two of the families showed autosomal recessive inheritance [71,74]. Three patients with cytochrome *c* oxidase deficiency have been reported who recovered after severe myopathy at infancy [80–82]. In these reversible forms of cytochrome *c* oxidase deficiencies, the concentration of the oxidase was only mildly decreased [82].

2. *Encephalomyopathies.* In these patients, not only the muscle, but also the central nervous system is affected [83–95]. Most common is subacute necrotizing encephalomyelopathy (Leigh's syndrome), a disease that may have other biochemical causes besides cytochrome *c* oxidase deficiency [83–92].

The disease may manifest at birth but infantile, juvenile and late onset forms have also been reported.

In encephalomyopathies, the cytochrome *c* oxidase deficiency appears to be more generalized than in infantile myopathies. Various studies indicate that apart from muscle and brain, also liver, heart, leucocytes, platelets, kidney and cultured fibroblasts can show decreased cytochrome *c* oxidase activity [83–91]. However, non-affected tissues have also been reported [83,88,89].

Immunologically detectable cytochrome *c* oxidase was found to be reduced in muscle tissue of three patients, but, like in infantile myopathies, all subunits of the oxidase seemed to be present [89]. The severe decrease in haem aa_3 content of heart and muscle tissue reported for one patient was not accompained by decreases in cytochromes b or $(c + c_1)$ [83].

Leigh's syndrome shows autosomal recessive inheritance.

3. *Ophthalmoplegia.* These patients have a chronic, progressive, external ophthalmoplegia (CPEO) [7].

Cytochrome *c* oxidase deficiencies have been reported in only some of the muscle fibres of patients suffering from this disease [96–99]. Immunological investigations indicated no decrease in enzyme concentration [99]. However, decreased haem aa_3 levels have been reported [98]. As discussed by Scholte [3] cytochrome *c* oxidase deficiency in CPEO probably is not the primary biochemical cause of the disease, but a result of secondary mitochondrial degeneration developing during the course of the disease.

In most patients no direct evidence has been brought forward, indicating that there is a genetic defect.

A point that might be overlooked in the discussion of the phenotypes of patients is the following. Like many other metabolic pathways, the mitochondrial respiratory chain has a large overcapacity, or in other words, it exerts little control on aerobic energy metabolism. Illustrative in this respect are the studies of Van den Bogert, Kroon and co-workers [100–102], who inhibited mitochondrial protein synthesis in rat by prolonged intravenous infusion of oxytetracycline. The resulting decrease in activities of cytochrome *c* oxidase and other respiratory chain components (up to 80%) had hardly any energetic or metabolic consequences in the tissues investigated.

These observations have several implications. Only severe respiratory chain deficiencies can be expected to cause clinical symptoms. Consequently, hereditary investigation can be expected to be complicated by asymptomatic patients and carriers.

Pathology will develop only if the respiratory chain activity is reduced to such an extent that mitochondrial oxidative phosphorylation cannot cope with the energy demand, which differs in various tissues and at different ages.

Thus, even if there is a generalized deficiency of cytochrome *c* oxidase in all cells, the pathology may be restricted to tissues in which the energy demand is so great that the capacity of the respiratory chain is insufficient to provide an adequate amount of ATP.

As pointed out by Scholte [3], the defect will then probably be amplified by second-order processes leading to mitochondrial doom. Furthermore, we expect that the classification of phenotypes described above will prove to be too restricted. In fact, we recently studied an adult patient who showed myopathy due to cytochrome *c* oxidase deficiency without any signs of encephalopathy (H. Przyrembel and H.R. Scholte, unpublished observations).

4.2 The Protein Structure of Cytochrome c *Oxidase in Patients*

Many different types of genetic defects can conceivably cause primary disorders of cytochrome *c* oxidase. Defects cannot only be located in the mitochondrial or nuclear genes that code for subunits of cytochrome *c* oxidase, but also in one of the many genes that code for proteins involved in biogenesis, transport, or assembly of the enzyme. All other cytochrome *c* oxidase defects, for example in CPEO, should be considered secondary [3,103]. At the protein level the mutations mentioned above will in the first instance lead to alterations in, or absence of one or more subunits. The final protein structure of the enzyme may undergo further changes by impaired assembly or increased proteolysis.

We studied the protein structure of cytochrome *c* oxidase in muscle biopsies of nine patients with severely decreased activity ($<$ 20% of controls) of the enzyme in muscle (Sinjorgo et al. in preparation). The patients were of different ages (2 months to 53 years) and showed various clinical symptoms (e.g. myopathy alone, or Leigh's syndrome).

Cytochrome *c* oxidase was immunopurified from muscle biopsies (15–50 mg) using protein A-Sepharose and a polyclonal antiserum raised against human heart cytochrome *c* oxidase. After SDS-PAGE the subunits were visualized by immunoblotting with the same antiserum. Similar procedures were performed for cytochrome *c* and Complex III of the respiratory chain. The most prominent subunit of Complex III on our blots (13.4 kD) was taken as a reference. Based on our studies two types of cytochrome *c* oxidase defects could be distinguished.

1. Type-A deficiencies. In this class of defects all subunits were present in normal concentrations, indicating decreased catalytic activity of the enzyme, without degradation (see Fig. 2 for an example). In the muscle biopsies of these patients cytochrome *c* and the 13.4 kD Complex III subunit were present in normal or increased connentrations. The latter observation coincided with abnormal high carnitine palmitoyltransferase activity values, indicative of increased mitochondrial content of the cells.

2. *Type-B deficiencies.* In this class of defects the concentrations of all subunits were decreased to levels that corresponded with the decrease in enzyme activity (for an example, see Fig. 3). In these patients, the concentration of the 13.4 kD (and other) Complex III subunits was decreased as well, whereas cytochrome *c* levels were normal. This observation is in line with the fact that in several spectral studies of mitochondria, showing cytochrome *c* oxidase defects, there was a concomitant decrease in the levels of cytochrome aa_3 and b, whereas the concentrations of cytochrome $(c+c_1)$ were normal.

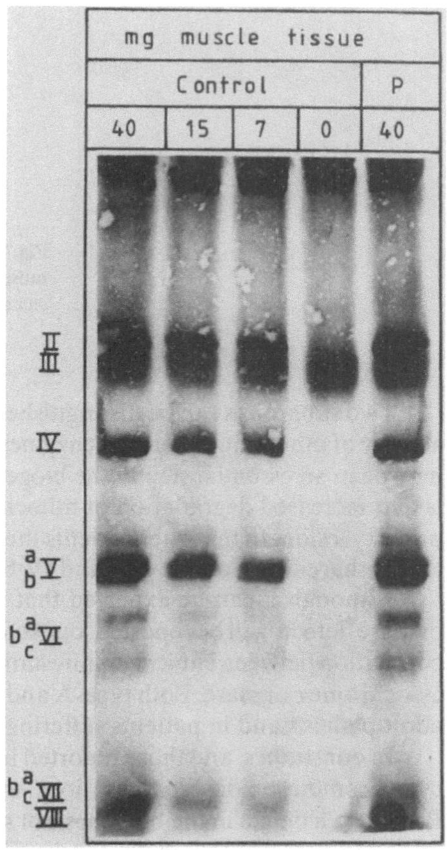

Fig. 2. Subunit pattern of cytochrome *c* oxidase in a 40 mg muscle biopsy of a patient with cytochrome *c* oxidase deficiency (Type A). The subunits of the control and patient enzyme were visualized by immunopurification with protein A Sepharose and a polyclonal antibody against human heart cytochrome *c* oxidase, followed by SDS-PAGE and immunoblotting. The bands observed in the 0 mg control lane represent subunits of the antibodies used

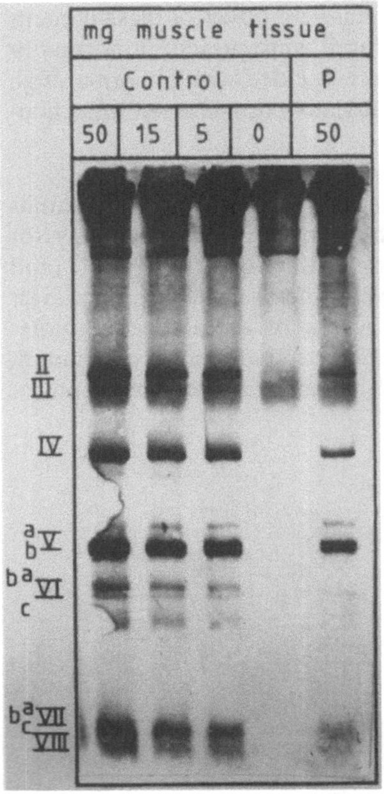

Fig. 3. Subunit pattern of cytochrome *c* oxidase in a 50 mg muscle biopsy of a patient with cytochrome *c* oxidase deficiency (Type B). For further details see legend to Fig. 2

Two subgroups can be distinguished, one with decreased and one with normal activity of other mitochondrial enzymes. In the former patients the primary defect may be in an essential step in the biogenesis of mitochondria. Another possibility is that increased degradation of mitochondria occurs as a secondary effect of the primary lesion. In the latter patients the primary defect may be in a biogenetic step that is shared by Complex III and IV but not by cytochrome *c*.

Although it can be expected that in the future other types of cytochrome *c* oxidase defects will be reported, our studies already clearly indicate that there is no correlation between clinical manifestation and the molecular structure of defective cytochrome *c* oxidase. Both type-A and -B deficiencies were found in infantile and adult patients and in patients suffering from Leigh's syndrome.

In our studies, and those reported in the literature [70,89] none of the defective cytochrome *c* oxidase preparations showed a specific absence of one subunit. Probably, a defect in the biogenesis or structure of individual subunits will lead to impaired assembly or increased proteolysis of all subunits. Since the primary defects are not easily detectable in the mature enzyme, we have initiated an investigation of the biogenesis of the enzyme in cultured cells. The potency of these techniques was recently demonstrated by Bolhuis et al. [104]. By using emitine, an inhibitor of cytoplasmic protein synthesis, they studied mitochondrial protein synthesis in myoblasts of a patient who suffered from a disorder in Complex I, and identified a defect in one of the mitochondrially encoded subunits of this complex.

We believe that metabolic labelling studies, genome cloning and DNA sequence analysis will be the next steps towards elucidation of the molecular basis of cytochrome *c* oxidase deficiencies.

Acknowledgements. Some of the studies described in this work were supported by grants from the Prinses Beatrix Fund, The Netherlands, from The Netherlands Organisation for Pure Scientific Research (N.W.O.) under the auspices of the Netherlands Foundation for Chemical Research (S.O.N.) and from the Belgian National Incentive-Program on Fundamental Research in Life Sciences initiated by the Belgian State – Prime Minister's Office – Science Policy Programming.

We thank Coby van den Bogert and Bert Groen for helpful discussions. We are particularly grateful to the physicians who provided material from patients for our studies.

References

1. Ernster L, Ikkos D, Luft R (1959) Enzymic reactions of human skeletal muscle mitochondria; a tool in clinical metabolic research. Nature (London) 184:1851–1854
2. Luft R, Ikkos D, Palmieri G, Ernster L, Afzelius B (1962) A case of severe hypermetabolism of non-thyroid origin with a defect in the maintenance of respiratory control: a correlated clinical, biochemical and morphological study. J Clin Invest 41:1776–1804
3. Scholte HR (1988) The biochemical basis of mitochondrial diseases. J Bioenerg Biomembr 20:161–191
4. Sengers RCA, Stadhouders AM, Trijbels JMF (1984) Mitochondrial myopathies. Clinical, morphological and biochemical aspects. Eur J Pediatr 141:192–207
5. DiMauro S, Bonilla E, Zeviani M, Nakagawa M, DeVivo DC (1985) Mitochondrial myopathies. Ann Neurol 17:521–538
6. Petty RKH, Harding AE, Morgan-Hughes JA (1986) The clinical features of mitochondrial myopathy. Brain 109:915–938
7. Morgan-Hughes JA (1986) The mitochondrial myopathies. In: Engel AG, Banker BQ, (eds) Myology. McGraw-Hill, New York, pp 1709–1743
8. Stadhouders AM, Sengers RCA (1987) Morphological observations in skeletal muscle from patients with a mitochondrial myopathy. J Inherit Metab Dis 10 (Suppl 1):62–80
9. DiMauro S, Bonilla E, Zeviani M, Servidei S, DeVivo DC, Schon EA (1987) Mitochondrial myopathies J Inherit Metab Dis 10 (Suppl 1):113–128
10. Scholte HR, Busch HFM, Luyt-Houwen IEM, Vaandrager-Verduin MHM, Przyrembel H, Arts WFM (1987) Defects in oxidative phosphorylation. Biochemical investigations in skeletal muscle and expression of the lesion in other cells. J Inherit Metab Dis 10 (Suppl 1):81–97
11. Merle P, Kadenbach B (1980) On the heterogeneity of vertebrate cytochrome *c* oxidase polypeptide chain composition. Hoppe-Seyler's Z Physiol Chem 361, 1257–1259
12. Kadenbach B, Büge U, Jarausch J, Merle P (1981) Composition and kinetic properties of cytochrome *c* oxidase from higher eucaryotes. In: Palmieri F, Quagliariello E, Siliprandi N, Slater EC (eds) Vectorial reactions in electron and ion transport in mitochondria and bacteria. Elsevier/North-Holland Biomedical Press, Amsterdam New York, pp 11–23
13. Merle P, Kadenbach B (1982) Kinetic and structural differences between cytochrome *c* oxidases from beef liver and heart. Eur J Biochem 125:239–244
14. Jarausch J, Kadenbach B (1982) Tissue-specificity overrides species specificity in cytoplasmic cytochrome *c* oxidase polypeptides. Hoppe-Seylers Z Physiol Chem 363:1133–1140
15. Van Gelder BF, Slater EC (1963) Titration of cytochrome *c* oxidase with NADH and phenazine methosulphate. Biochim Biophys Acta 73:663–665
16. Van Gelder BF, Muijsers AO (1966) Cytochrome oxidase. 2. Ratio of cytochrome *a* to a_3. Biochim Biophys Acta 118:47–57

17. Anderson S, Bankier AT, Barrell BG, De Bruijn MHL, Coulson AR, Drouin J, Eperon IC, Nierlich DP, Roe BA, Sanger F, Schreier PH, Smith AJH, Staden R, Young IG (1981) Sequence and organization of the human mitochondrial genome. Nature (London) 290:457–465
18. Tzagoloff A, Meyers AM (1986) Genetics of mitochondrial biogenesis. Annu Rev Biochem 55:249–285
19. Nelson BD (1987) Biogenesis of mammalian mitochondria. In: Lee CP (ed) Current topics in bioenergetics, vol 15. Academic Press, New York London San Diego, pp 221–272
20. Schatz G, Butow RA (1983) How are proteins imported into mitochondria? Cell 32:316–318
21. Schatz G (1987) Signals guiding proteins to their correct locations in mitochondria. Eur J Biochem 165:1–6
22. Von Heijne G (1986) Mitochondrial targeting sequences may form amphiphilic helices. EMBO J 5:1335–1342
23. Maccecchini MC, Rudin Y, Blobel G, Schatz G (1979) Import of proteins into mitochondria: precursor forms of the extramitochondrially made F_1-ATPase subunits in yeast. Proc Natl Acad Sci USA 76:343–347
24. Reid GA, Schatz G (1982) Import of proteins into mitochondria. J Biol Chem 257:13062–13067
25. Riezman H, Hay R, Witte C, Nelson N, Schatz G (1983) Yeast mitochondrial outer membrane specifically binds cytoplasmically-synthesized precursors of mitochondrial proteins. EMBO J 2:1113–1118
26. Ohba M, Schatz G (1987) Protein import into yeast mitochondria is inhibited by antibodies raised against 45-kd proteins of the outer membrane. EMBO J 6:2109–2115
27. Pfanner N, Neupert W (1985) Transport of proteins into mitochondria: a potassium diffusion potential is able to drive the import of ADP/ATP carrier. EMBO J 4:2819–2825
28. Böhni PC, Daum G, Schatz G (1983) Import of proteins into mitochondria. J Biol Chem 258:4937–4943
29. Daum G, Grasser SH, Schatz G (1982) Import of proteins into mitochondria. Cytochrome b_2 and cytochrome *c* peroxidase are located in the intermembrane space of yeast mitochondria. J Biol Chem 257:13075–13080
30. Pfanner N, Neupert W (1986) Transport of Fl-ATPase subunit β into mitochondria depends on both a membrane potential and nucleoside triphosphates. FEBS Lett 209:152–156
31. Chen WJ, Douglas MG (1987) Phosphodiester bond cleavage outside mitochondria is required for the completion of protein import into the mitochondrial matrix. Cell 49:651–658
32. Eilers M, Schatz G (1988) Protein unfolding and the energetics of proteins translocation across biological membranes. Cell 52:481–483
33. Pfanner N, Pfaller R, Kleene R, Ito M, Tropschug M, Neupert W (1988) Role of ATP in mitochondrial protein import. J Biol Chem 263:4049–4051
34. Rothman JE, Kornberg RD (1986) An unfolding story of protein translocation. Nature (London) 322:209–210
35. Deshaies RJ, Koch BD, Werner-Washburne M, Craig EA, Schekman R (1988) A subfamily of stress proteins facilitates translocation of secretory and mitochondrial precursor polypeptides. Nature (London) 332:800–805
36. Schmelzer E, Heinrich PC (1980) Synthesis of a larger precursor for the subunit IV of rat liver cytochrome *c* oxidase in a cell-free wheat germ system. J Biol Chem 255:7503–7506
37. Hatalova I, Kolarov J (1983) Synthesis and intracellular transport of cytochrome oxidase subunit IV and a ADP/ATP translocator protein in intact hepatoma cells. Biochem Biophys Res Commun 110:132–139
38. Schmelzer E, Northemann W, Kadenbach B, Henrich PC (1982) Cell-free synthesis of a large-molecular-weight precursor of cytochrome *c* oxidase subunit V from rat liver and the distribution of its mRNA between free and membrane-bound polysomes. Eur J Biochem 127:177–183
39. Zeviani M, Nakagawa M, Herbert J, Lomax MI, Grossman LI, Sherbany AA, Miranda AF, DiMauro S, Schon EA (1987) Isolation of a cDNA clone encoding subunit IV of human cytochrome *c* oxidase. Gene 55:205–217
40. Lomax MI, Bachman NJ, Nasoff MS, Caruthers MH, Grossman LI (1984) Isolation and characterization of a *c*DNA clone for bovine cytochrome *c* oxidase subunit IV. Proc Natl Acad Sci USA 81:6295–6299
41. Bachman NJ, Lomax MI, Grossman LI (1987) Two bovine genes for cytochrome *c* oxidase subunit IV: a processed pseudogene and an expressed gene. Gene 55:219–229

42. Rizzuto R, Nakase H, Zeviani M, DiMauro S, Schon EA (1988) Subunit Va of human and bovine cytochrome *c* oxidase is highly conserved. Gene (in press)
43. Zeviani M, Sakoda S, Sherbany AA, Nakase H, Rizzuto R, Samitt CE, DiMauro S, Schon EA (1988) Sequence of cDNAs encoding subunit Vb of human and bovine cytochrome *c* oxidase. Gene 65:1–11
44. Suske G, Mengel T, Cordingley M, Kadenbach B (1987) Molecular cloning and further characterization of cDNAs for rat nuclear-encoded cytochrome *c* oxidase subunits VIc and VIII. Eur J Biochem 168:233–237
45. De Vries H, Van't Sant P (1982) Interplay between different genetic systems in eukaryotic cells: nucleo-cytoplasmic-mitochondrial interrelations. In: Quagliariello E, Palmieri F (eds) Horizons in biochemistry and biophysics, vol 7. John Wiley & Sons, New York Chichester, England, pp 249–278
46. Kuhn-Nentwig L, Kadenbach B (1985) Isolation and properties of cytochrome *c* oxidase from rat liver and quantification of immunological differences between isozymes from various rat tissues with subunit-specific antisera. Eur J Biochem 149:147–158
47. Stroh A, Kadenbach B (1986) Tissue-specific and species-specific distribution of -SH groups in cytochrome *c* oxidase subunits. Eur J Biochem 156:199–204
48. Kadenbach B, Ungibauer H, Jarausch J, Büge U, Kuhn-Nentwig L (1983) The complexity of respiratory complexes. Trends Biochem Sci 8:398–400
49. Capaldi RA (1988) Mitochondrial myopathies and respiratory chain proteins. Trends in Biochem Sci 13:144–148
50. Ludwig B, Schatz G (1980) A two-subunit cytochrome *c* oxidase (cytochrome aa_3 from *Paracoccus denitrificans*. Proc Natl Acad Sci USA 77:196–200
51. Buse G (1982) Cytochrome *c* oxidase. In: Lontie R (ed) Copper proteins and copper enzymes. CRC, Boca Raton, pp 119–149
52. Briggs MM, Capaldi RA (1978) Cross-linking studies on cytochrome *c*-cytochrome *c* oxidase complex. Biochem Biophys Res Commun 80:553–559
53. Bisson R, Azzi A, Gutweniger H, Colonna R, Montecucco C, Zanotti A (1978) Interaction of cytochrome *c* with cytochrome *c* oxidase. Photoaffinity of beef heart cytochrome *c* oxidase with arylazido cytochrome *c*. J Biol Chem 253:1874–1880
54. Millett F, De Jong C, Paulson L, Capaldi RA (1983) Identification of specific carboxylase groups on cytochrome *c* oxidase that are involved in binding cytochrome *c*. Biochemistry 22:546–552
55. Saraste M, Pentillä T, Wikström M (1981) Quarternary structure of bovine cytochrome oxidase. Eur J Biochem 115:261–268
56. Prochaska LJ, Bisson R, Capaldi RA, Steffens GCM, Buse G (1981) Inhibition of cytochrome *c* oxidase function by dicyclohexyl-carbodiimide. Biochim Biophys Acta 637:360–373
57. Kadenbach B, Stroh A, Ungibauer M, Kuhn-Nentwig L, Büge U, Jarausch J (1986) Isozymes of cytochrome *c* oxidase: characteristization and isolation from different tissues. Meth Enzymol 126:32–45
58. Ferguson-Miller S, Brautigan DL, Margoliash E (1978) Definition of cytochrome *c* binding domains by chemical modification. III. Kinetics of reaction of carboxydinitrophenyl cytochromes *c* with cytochrome oxidase. J Biol Chem 253:149–159
59. Errede B, Haight GP Jr, Kamen MD (1976) Oxidation of ferrocytochrome *c* by mitochondrial cytochrome *c* oxidase. Proc Natl Acad Sci USA 73:113–117
60. Sinjorgo KMC, Meijling JH, Muijsers AO (1984) The concept of high- and low-affinity reactions in bovine cytochrome *c* oxidase steady-state kinetics. Biochim Biophys Acta 767:48–56
61. Sinjorgo KMC, Steinebach OM, Dekker HL, Muijsers AO (1986) The effects of pH and ionic strength on cytochrome *c* oxidase steady-state kinetics reveal a catalytic and a non-catalytic interaction domain for cytochrome *c*. Biochim Biophys Acta 850:108–115
62. Veerman ECI, Wilms J, Casteleijn G, Van Gelder BF (1980) The pre-steady state reaction of ferrocytochrome *c* with the cytochrome *c* cytochrome aa_3 complex. Biochim Biophys Acta 590:117–127
63. Power SD, Lochrie MA, Poyton RD (1983) Reversed-phase high-performance liquid chromatographic purification of subunits of oligomeric membrane proteins. Nuclear coded subunits of yeast cytochrome *c* oxidase. J Chromatogr 266:585–598
64. Tanaka M, Haniu M, Yasunobu KT, Yu C-A, Yu L, Wei Y-H, King TE (1979) Amino acid sequence of subunit V of bovine heart cytochrome oxidase, the heme *a*-containing subunit. J Biol Chem 254:3879–3885

65. Capaldi R, Malatesta F, Darley-Usmar VM (1988) Structure of cytochrome *c* oxidase. Biochim Biophys Acta 726:135–148
66. Gay MJ, Walker JE (1985) Two genes encoding the bovine mitochondrial ATP synthase proteolipid specify precursors with different import sequences and are expressed in a tissue-specific manner. EMBO J 4:3519–3524
67. DiMauro S, Zeviani M, Bonilla E, Bresolin N, Nakagawa M, Miranda AF, Moggio M (1985) Cytochrome *c* oxidase deficiency. Biochem Soc Trans 13:651–653
68. DiMauro S, Zeviani M, Servidei S, Bonilla E, Miranda AF, Prelle A, Schon EA (1986) Cytochrome oxidase deficiency: clinical and biochemical heterogeneity. Ann NY Acad Sci 488:19–32
69. Trijbels F, Sengers R, Monnens L, Janssen A, Willems J, Ter Laak H, Stadhouders A (1983) A patient with lactic acidaemia and cytochrome oxidase deficiency. J Inherit Metab Dis 6 (Suppl 2):127–128
70. Bresolin N, Zeviani M, Bonilla E, Miller RH, Leech RW, Shanske S, Nakagawa M, DiMauro S (1985) Fatal infantile cytochrome *c* oxidase deficiency: decrease of immunologically detectable enzyme in muscle. Neurology 35:802–812
71. Van Biervliet JPGM, Bruinvis S, Ketting D, De Bree PK, Van der Heiden C, Wadman SK, Willems JL, Bookelman H, Van Haelst U, Monnens LAH (1977) Hereditary mitochondrial myopathy with lactic acidemia, a DeToni-Fanconi-Debré syndrome, and a defective respiratory chain in voluntary striated muscles. Pediatr Res 11:1088–1093
72. DiMauro S, Mendell JR, Sahenk Z, Bachman D, Scarpa A, Scofield RM, Reiner C (1980) Fatal infantile mitochondrial myopathy and renal dysfunction due to cytochrome *c* oxidase deficiency. Neurology 30:795–804
73. Heiman-Patterson TD, Bonilla E, DiMauro S, Foreman J, Schotland DL (1982) Cytochrome *c* oxidase deficiency in a floppy infant. Neurology 32:898–900
74. Minchom PE, Dormer RL, Hughes IA, Stansbie D, Cross AR, Hendry GAF, Jones OTG, Johnson MA, Sherratt HSA, Turnbull DM (1983) Fatal infantile mitochondrial myopathy due to cytochrome *c* oxidase deficiency. J Neurol Sci 60:453–463
75. Müller-Höcker J, Pongratz D, Deufel Th, Trijbels JMF, Endres W, Hübner G (1983) Fatal lipid storage myopathy with deficiency of cytochrome *c* oxidase and carnithine. A contribution to the combined cytochemical fine structural identification of cytochrome *c* oxidase in long term frozen muscle. Virchows Arch 399:11–23
76. Zeviani M, Nonaka I, Bonilla E, Okino E, Moggio M, Jones S, DiMauro S (1985) Fatal infantile mitochondrial myopathy and renal dysfunction caused by cytochrome *c* oxidase deficiency: immunological studies in a new patient. Ann Neurol 17:414–417

Rimoldi M, Bottachi E, Rossi L, Cornelio F, Uziel G, Di Donato S (1982) Cytochrome *c* oxidase deficiency in muscles of a floppy infant without mitochondrial myopathy. J Neurol 227:201–207

78. Sengers RCA, Trijbels JMF, Bakkeren JAJM, Ruitenbeek W, Fischer JC, Janssen AJM, Stadhouders AM, Ter Laak HJ (1984) Deficiency of cytochromes *b* and aa_3 in muscle from a floppy infant with cytochrome *c* oxidase deficiency. Eur J Pediatr 141:178–180
79. Kennaway NG, Wagner ML, Capaldi RA, Takamiya S, Yanamura W, Ruitenbeek W, Sengers RCA, Trijbels JMF (1987) Combined deficiencies of complexes III and IV of the respiratory chain, involving both nuclear and mitochondrial gene products, in skeletal muscle of a patient with lactic acidosis. J Inherit Metab Dis 10 (Suppl 2):247–251
80. DiMauro S, Nicholson JF, Hays AP, Eastwood AB, Papadimitriou A, Koeningsberger R, DeVivo DC (1983) Benign infantile mitochondrial myopathy due to reversible cytochrome *c* oxidase deficiency. Ann Neurol 14:226–234
81. Roodhooft AM, Van Acker KJ, Martin JJ, Ceuterick C, Scholte HR, Luyt-Houwen IEM (1986) Benign mitochondrial myopathy with deficiency of NADH-CoQ reductase and cytochrome *c* oxidase. Neuropediatrics 17:221–226
82. Zeviani M, Peterson P, Servidei S, Bonilla E, DiMauro S (1987) Benign reversible muscle cytochrome *c* oxidase deficiency: a second case. Neurology 37:64–67
83. Willems JL, Monnens LAH, Trijbels JMF, Veerkamp JH, Meijer AEFH, Van Dam K, Van Haelst U (1977) Leigh's encephalomyelopathy in a patient with cytochrome *c* oxidase deficiency in muscle tissues. Pediatrics 60:850–857
84. Miyabayashi S, Narisawa K, Tada K, Sakai K, Kobayashi K, Kobayashi Y (1983) Two siblings with cytochrome *c* oxidase deficiency. J Inherit Metab Dis 6:121–122

85. Miyabayashi S, Narisawa K, Iinuma K, Tada K, Sakai K, Kobayashi K, Kobayashi Y, Morinaga S (1984) Cytochrome *c* oxidase deficiency in two siblings with Leigh encephalomyelopathy. Brain Dev 6:362–372
86. Miyabayashi S, Ito T, Narisawa K, Iinuma K, Tada K (1985) Biochemical study in 28 children with lactic acidosis, in relation to Leigh's encephalomyelopathy. Eur J Pediatr 143:278–283
87. Robinson BH, Ward J, Goodyer P, Baudet A (1986) Respiratory chain defects in the mitochondria of cultured skin fibroblasts from three patients with lactic acidemia. J Clin Invest 77:1422–1427
88. Servidei S, Zeviani M, DiMauro S, DiRocco M, DeVivo DC, DiDonato S, Uziel G, Berry KE, Norman MG, Hoganson G, Johnsen SD, Johnson PC (1986) Cytochrome oxidase deficiency in Leigh's syndrome. Ann Neurol 20:400
89. DiMauro S, Servidei S, Zeviani M, DiRocco M, DeVivo DC, DiDonato S, Uziel G, Berry K, Hoganson G, Johnsen SD, Johnson PC (1987) Cytochrome *c* oxidase deficiency in Leigh syndrome. Ann Neurol 22:498–506
90. Miyabayashi S, Ito T, Abukawa D, Narisawa K, Tada K, Tanaka M, Ozawa T, Droste M, Kadenbach B (1987) Immunochemical study in three patients with cytochrome *c* oxidase deficiency presenting Leigh's encephalomyelopathy. J Inherit Metab Dis 10:289–292
91. Arts WFM, Scholte HR, Loonen MCB, Przyrembel H, Fernandes J, Trijbels JMF, Luyt-Houwen IEM (1987) Cytochrome *c* oxidase deficiency in subacute necrotizing encephalomyelopathy. J Neurol Sci 77:103–115
92. Robinson BH, De Meirleir L, Glerum M, Sherwood G, Becker L (1987) Clinical presentation of mitochondrial respiratory chain defets in NADH-coenzyme Q reductase and cytochrome oxidase: clues to pathogenesis of Leigh disease. J Pediatr 110:216–222
93. Prick MJJ, Gabreëls FJM, Trijbels JMF, Janssen AJM, Le Coultre R, Van Dam K, Jaspar HHJ, Ebels EJ, Op De Coul AAW (1983) Progressive poliodystrophy (Alper's disease) with a defect in cytochrome aa_3 in muscle: a report of two unrelated patients. Clin Neurol Neurosurg 85:57–70
94. Angelini C, Bresolin N, Pegolo G, Bet L, Rinaldo P, Trevisan C, Vergani L (1986) Childhood encephalomyopathy with cytochrome *c* oxidase deficiency, ataxia, muscle washing, and mental impairment. Neurology 36:1048–1052
95. Servidei S, Lazaro RP, Bonilla E, Barron KD, Zeviani M, DiMauro S (1987) Mitochondrial encephalomyopathy and partial cytochrome *c* oxidase deficiency. Neurology 37:58–63
96. Müller-Höcker J, Pongratz D, Hübner G (1983) Focal deficiency of cytochrome *c* oxidase in skeletal muscle of patients with progressive external ophthalmoplegia. Virchows Arch 402:61–71
97. Johnson MA, Turnbull DM, Dick DJ, Sherratt HSA (1983) A partial deficiency of cytochrome *c* oxidase in chronic progressive external ophthalmoplegia. J Neurol Sci 60:31–53
98. Byrne E, Dennett X, Trounce I, Henderson R (1985) Partial cytochrome oxidase (aa_3) deficiency in chronic progressive external ophthalmoplegia. Histochemical and biochemical studies. J Neurol Sci 71:257–271
99. Bresolin N, Moggio M, Bet L, Galanti A, Prelle A, Nobile-Orazio E, Adobbati L, Ferrante C, Pellegrini G, Scarlato G (1987) Progressive cytochrome *c* oxidase deficiency in a case of Kearns-Sayre syndrome: morphological, immunological, and biochemical studies in muscle biopsies and autopsy tissues. Ann Neurol 21:564–572
100. Kroon AM, Van Den Bogert C, Moen AA, Wassenaar JS (1981) The biogenesis of mitochondria and its inhibition by some antibiotics. In: Busch HFM, Jennekens FGI, Scholte HR (eds) Mitochondria and muscular diseases. Mefar Beetsterzwaag, pp 89–95
101. Van den Bogert C, Lont M, Mojet, Kroon AM, (1983) Impairment of liver regeneration during inhibition of mitochondrial protein synthesis by oxytetracycline. Biochim Biophys Acta 722:393–400
102. Kroon AM, Van den Bogert C (1985) The mitochondrial genome as a target for chemotherapy of cancer. In: Quagliariello E, Slater EC, Palmieri F, Saccone E, Kroon AM (eds) Achievements and perspectives of mitochondrial research, vol 2. Elsevier, Amsterdam, pp 21–34
103. Sengers RCA, Stadhouders AM (1987) Secondary mitochondrial pathology. J Inherit Metab Dis 10 (Suppl 1):98–104
104. Bolhuis PA, Barth PG, Wijberg FA, Sinjorgo KMC, Ruitenbeek W (1988) Molecular basis of mitochondrial myopathies. Lancet i:884
105. Sinjorgo KMC, Durak I, Dekker HL, Edel CM, Hakvoort TBM, Van Gelder BF, Muijsers AO (1987) Bovine cytochrome *c* oxidase isoenzymes, which differ in the small subunits, show the same reaction kinetics with cytochrome *c*. Biochim Biophys Acta 893:251–258

4 Mitochondrial Metabolic Diseases

Brown Adipose Tissue, Diet-Induced Thermogenesis and Genetic Obesity

T. Soukup, H. Rauchová, J. Dosoudilová and Z. Drahota

Brown adipose tissue is a thermogenic organ that effectively participates in the maintenance of body temperature in newborn mammals and in adult hibernating animals. The high triglyceride content of brown adipose tissue cells, the high oxidative capacity and specific mitochondrial proton channel which enables energy dissipation in the form of heat are involved in the protection of the animal body against cold (Lindberg 1970; Jánský 1973; Nichols 1979; Cannon and Needergard 1985).

However, recent data showed that the brown adipose tissue may also participate in the protection of the animal organism against overfeeding. The enzyme equipment of the brown adipose tissue cells, together with highly developed vascularization and the neurohumoral regulatory system, can be used as a device for dissipation of excess of energy consumed as food and for avoidance of this deposition in the form of triglycerides (Rothwell and Stock 1979).

We may therefore consider brown adipose tissue as a specialized organ that plays an important role in the maintenance of energy balance in a mammalian organism (Himms-Hagen 1983). Its thermogenic function can be used not only for maintenance of thermal homeostasis of vital organs during cold stress but also for the maintenance of energy homeostasis in situations when the mammalian body is overloaded with high-energy substrates. Disturbances of the functional activity of the brown adipose tissue may result not only in the inability to survive in cold but also in the inability to dissipate the excess energy consumed. Defects of brown adipose tissue function may be thus one of the factors that participates in the development of obesity (Himms-Hagen 1984).

For the evaluation of various factors that may promote adiposity of the mammalian organism, various animal models were developed, classified and characterized (Sclafani 1984). For the evaluation of the role of brown adipose tissue in obese animals, three approaches were used:

1. Surgical or pharmacological disturbances of hypothalamic regulatory centers;
2. Increased, spontaneous food uptake;
3. Genetic manipulations.

Animals with genetic obesity represent a very heterogeneous group of various animal types with defects of the brown adipose tissue function. But they can be a useful tool for the examination of various types of human obesity which are also characterized by a very high heterogeneity.

Institute of Physiology, Czechoslovak Academy of Sciences, Videňská 1083, 142 00 Prague 4, Czechoslovakia

A. Azzi, Z. Drahota and S. Papa
Molecular Basis of Membrane-Associated Diseases

A defective function of brown adipose tissue was identified in three types of animals with genetic obesity, i.e. Zucker obese fa/fa rat, in obese ob/ob mouse and in diabetic obese db/db mouse. In these models defects associated with disturbances of cell membrane systems transmitting hormonal signals to the cell or with disturbances of the mitochondrial thermogenic function were found (Himms-Hagen 1984). Concerning the mitochondrial defects, the reduced quantity of the uncoupling protein, which forms the regulated proton channel (Kopecký et al. 1987; Ježek et al. 1988) through which energy is dissipated, was found (Trayhurn and Mercer 1986).

To obtain more information about the role of brown adipose tissue in the development of obesity we studied another type of obese animal in which the role of brown adipose tissue has not yet been evaluated, viz. Koletski obese rats (Koletski 1973; Hromadová et al. 1980). These animals were selected as biological models of spontaneously hypertensive rats and, after selective inbreeding, an abnormal phenotype characterized by marked obesity appeared among some of the litters (Koletski 1973; NIH Rodents Catalogue 1981).

In our experiments we confirmed previous findings that hyperphagia accompanied by an increased weight of the animal can be detected in 40- to 60-day-old rats (Hromadová et al. 1980). However, the obese animals may be differentiated from their lean litter mates sooner, i.e. 30- to 40-day-old animals. At this developmental period most of the metabolic indicators, such as plasma cholesterol and triglycerides, are still in the range of normal control animals (Koletski 1973). We concentrated our attention especially on this developmental period because we tried to obtain more information indicating that the decreased thermogenesis of the brown adipose tissue occurs during development, preceding the deposition of excess fat (Goodbody and Trayhurn 1981, 1982; Bazin et al. 1984). We found that the weight of brown adipose tissue of obese animals is very much increased in 30- to 40-day-old animals classified as obese according to their body shape (Drahota et al. 1987). The increase, however, was due to a great increase of intracellular triglycerides. The other main tissue components remained unchanged. This indicated very large changes in the structural organization of the brown adipose tissue cells in obese animals.

Electron microscope observations revealed that, contrary to the lean litter mates, brown adipose tissue cells from obese animals lost the character of structural organization typical of brown adipose tissue, i.e. multilocular organization of triglyceride deposits inside the cell (Figs. 1,2). The adipocytes from the brown adipose tissue of obese animals contained only a few large lipid deposits surrounded by a thin layer of cytoplasm. Such an arrangement is typical of white adipose tissue cells. Only a few cells maintaining the multilocular organization of intracellular triglycerides could be detected in brown adipose tissue from obese animals (Fig. 3). Thus, according to the structural organization of intracellular triglycerides, the brown adipose tissue of obese animals resembled white adipose tissue.

Another typical feature of brown adipose tissue cells was the numerous mitochondria with many cristae. When we examined mitochondria in the cytoplasm of lean and obese litter mates we found that in cells from obese animals most of the mitochondria maintained structural characteristics typical of brown adipose

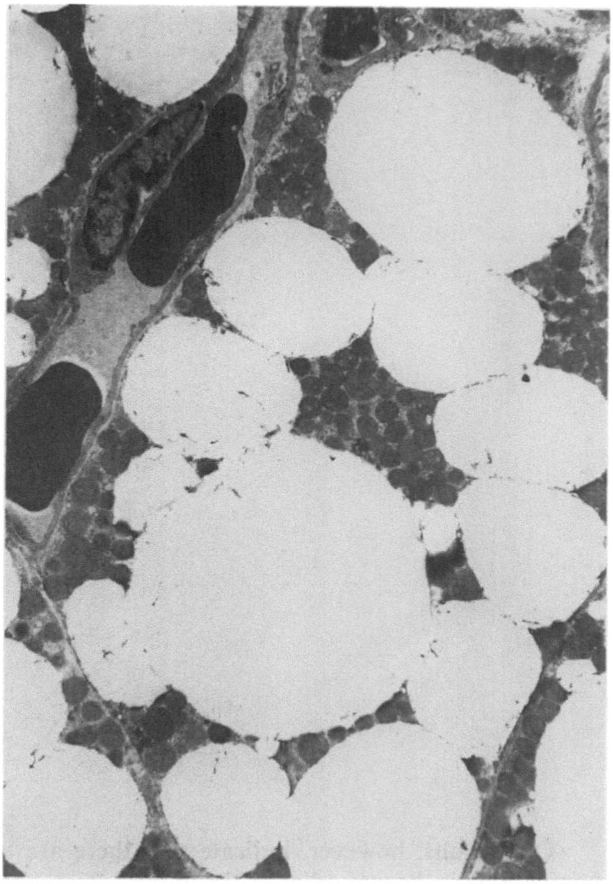

Fig. 1. Brown adipose tissue of lean 45-day-old rats. Multilocular deposition of intracellular triglycerides

tissue but in those cells we found a higher frequency of mitochondria with modified structure.

The mitochondria from brown adipose tissue of obese animals had a lower density of cristae or a completely changed intramitochondrial organization (Figs. 4, 5). These findings, obtained on intact tissue samples, were further confirmed in preparations of isolated mitochondria. In mitochondrial preparations from obese animals there was at least 30% of the mitochondria with a modified arrangement of intramitochondrial cristae, quite similar to that found in intact tissue preparations. Such modified mitochondria were never detected in preparations isolated from brown adipose tissue of the lean litter mates. Instead of densely packed, regularly arranged, cristae an irregular arrangement of intramitochondrial membranes was observed.

On measuring ^{14}C-GDP binding by isolated mitochondria from brown adipose tissue we found a 30% reduction in preparations from obese animals, which indicates a lower quantity of the uncoupling protein in these mitochondria (Drahota et al. 1987).

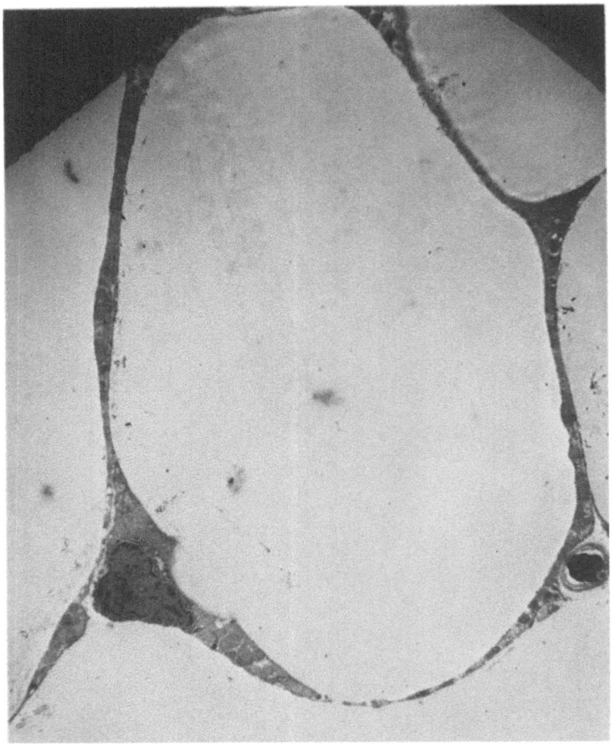

Fig. 2. Brown adipose tissue of obese litter mates. Unilocular deposition of intracellular triglycerides

Our results, however, indicate that there are two different mitochondrial populations in brown adipose tissue cells from obese animals. Therefore, enzymological data should be reevaluated, because, on the basis of morphological observations, a decrease of the uncoupling protein may be connected only with the defective part of the mitochondrial population in the cell.

It may be concluded that Koletski obese rats can be used as a useful model for evaluation of the role of brown adipose tissue in the development of genetic obesity. Changes in the structural organization of the brown adipose tissue cells and modification of cell metabolism can be observed already at the developmental period when typical indicators of obesity are not yet developed. Also, hyperphagia starts at the developmental period when the brown adipose tissue function is already defective. This is a very important fact because it means that brown adipose tissue cannot be used as an organ that helps the organism in its struggle against overfeeding.

Morphological data indicate the inability of brown adipose tissue cells to maintain the intracellular triglyceride deposits in the form of small droplets. This simple physico-chemical mechanism is very important for the metabolic function because it increases the surface of triglyceride deposits and makes them more accessible to lipase action. We found that membrane adenylate cyclase activity is not modified in obese animals (Drahota et al. 1987) which means that the hormonal

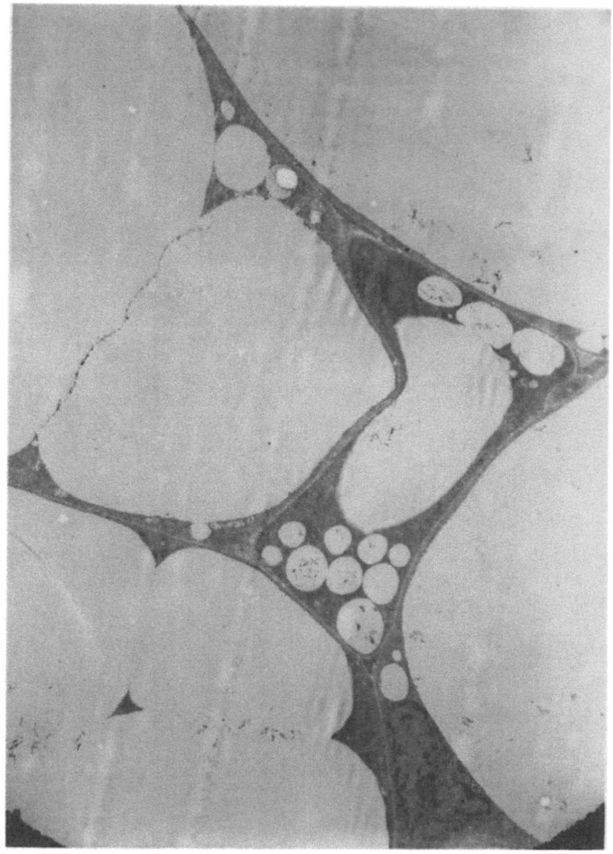

Fig. 3. Brown adipose tissue of obese rat. Residual cells with multilocular deposition of intracellular triglycerides

signal activating lipase action is not disturbed. Rearrangement of intracellular triglycerides can thus deprive the rat of free fatty acid release and their oxidation by mitochondrial enzymes.

Another defect of brown adipose tissue in obese rats is the mitochondrial function. It is quite evident that the thermogenic function is disturbed because the quantity of the uncoupling protein is diminished. However, it is possible that the defect is manifested only in part of the mitochondrial population in the cell. Thus, for the elucidation of the mitochondrial disturbances in obese animals two different mitochondrial populations should be separated and characterized.

On the basis of the data presented, it may be speculated that under certain conditions a partial transformation of the brown adipose tissue into white adipose tissue may occur. Transformation in the opposite direction, i.e. white adipose tissue into brown adipose tissue, was described in young cats exposed to cold treatment (Lončar et al. 1986). Also, in this case the most pronounced changes were connected with intracellular triglyceride reorganization. In white adipose tissue of cold-exposed animals an increase of cells with small lipid droplets was found in comparison with unilocular triglyceride localization in control animals. The other

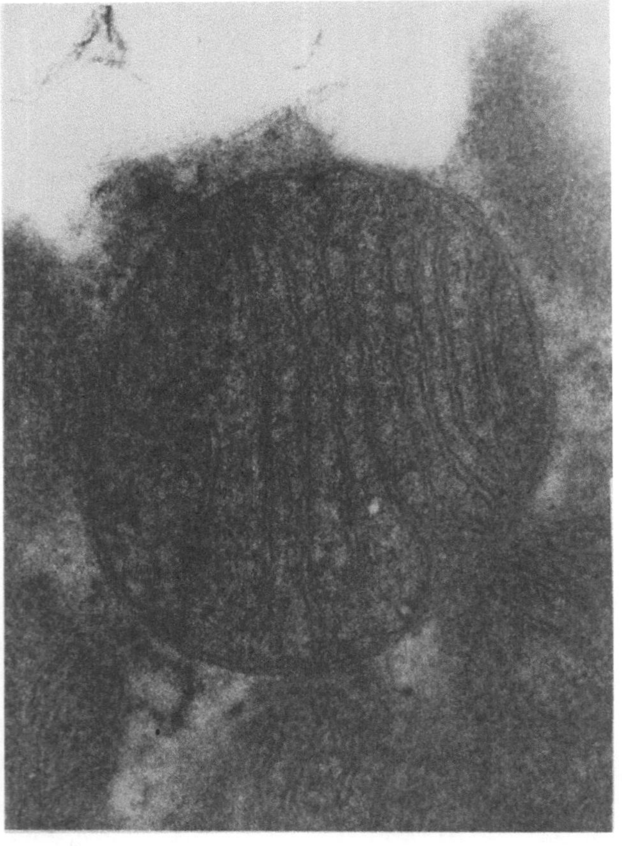

Fig. 4. Brown adipose tissue of lean 45-day-old rat. Typical mitochondrion in cytoplasm

important change found in cold-exposed animals was an increase of mitochondria and rearrangement of their cristae.

In conclusion, it appears that data accumulate indicating that the brown adipose tissue is an important organ which participates in the maintenance of energy balance in the mammalian organism. This is an important problem especially for human beings because increased food intake and decreased physical activity result in an increased formation of triglyceride deposits. It is now quite clear that by its thermogenic function, brown adipose tissue can effectively help to balance the negative effects of increased food intake and prevent obesity. However, more data will be required to understand all the molecular mechanisms which participate in the modification of the thermogenic function of brown adipose tissue in various types of obesity.

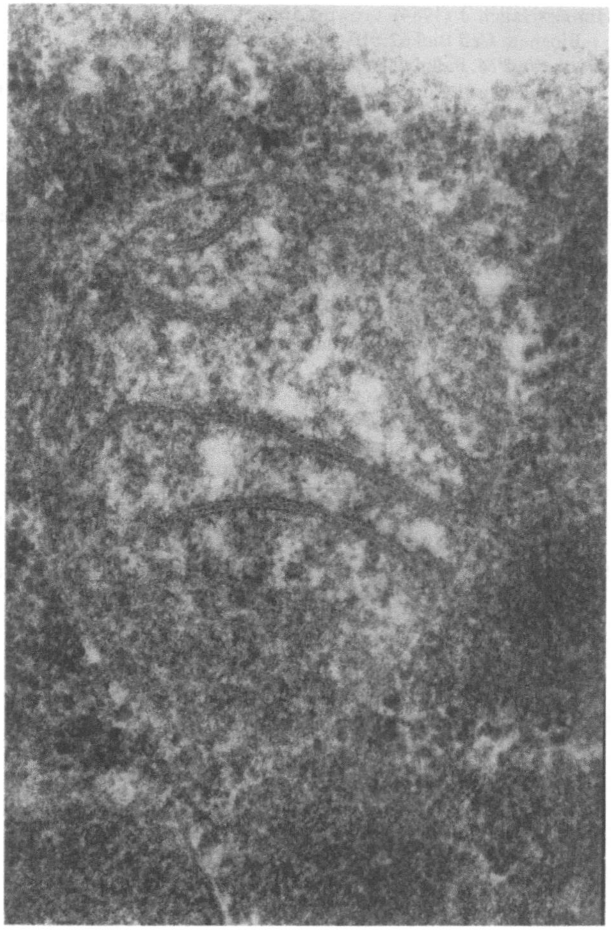

Fig. 5. Brown adipose tissue of obese litter mates. Part of the mitochondrial population has modified the mitochondrial structure

References

Bazin R, Eteve D, Lavau M (1984) Evidence for decreased GDP binding to brown adipose tissue mitochondria of obese Zucker fa/fa rats in the very first days of life. Biochem J 221:241–245

Cannon B, Nedergaard J (1985) Brown adipose tissue: molecular mechanisms controlling activity and thermogenesis. In: Cryer A, Van RLR (eds) New perspectives in adipose tissue: structure, function and development. Butterworths, London, pp 223–270

Drahota Z, Svoboda P, Soukup T, Rauchová H, Dosoudilová J, Vitek V, Kotyk A (1987) Structural and functional alterations of brown adipose tissue in animals with hereditary induced obesity. Proc Symp Membr Diss, Bari

Goodbody AE, Trayhurn P (1981) GDP binding to brown adipose mitochondria of diabetic-obese db/db mice. Decreased binding in both the obese and pre-obese states. Biochem J 194:1019–1022

Goodbody AE, Trayhurn P (1982) Studies on the activity of brown adipose tissue in suckling, pre-obese, ob/ob mice. Biochem Biophys Acta 680:119–126

Himms-Hagen J (1983) Role of thermogenesis in brown adipose tissue. In: Angel A, Hollenberg CH, Roncari DAK (eds) The regulation of energy balance in the adipocyte and obesity: cellular and molecular mechanisms. Raven, New York, pp 259–270

Himms-Hagen J (1984) Brown adipose tissue thermogenesis, energy balance and obesity. Can J Biochem Cell Biol 62:610–617

Hromadová M, Palkovič M, Skottová N (1980) Plasma lipoproteins and activities of several enzymes involved in the lipid metabolism in spontaneously obese rats. In: Palkovič M (ed) Advances in Physiological Science vol 35. Hormones, lipoproteins and atherosclerosis. Akad Kiadó, Budapest, pp 347–360

Jánský L (1973) Non-shivering thermogenesis and its thermoregulatory significance. Biol Rev 43:85–132

Ježek P, Houštěk J, Kotyk A, Drahota Z (1988) Distinct affinity and effector residues in the binding site for a regulatory ligand. Eur Biophys J 16:101–108

Koletski S (1973) Obese spontaneously hypertensive rats – a model for study of atherosclerosis. Exp Mol Pathol 19:53–60

Kopecký J, Ježek P, Drahota Z, Houštěk J (1987) Control of uncoupling protein in brown fat mitochondria by purine nucleotides. Chemical modification by diazobenzenesulfonate. Eur J Biochem 164:687–694

Lindberg O (ed) (1970) Brown adipose tissue. Elsevier, Amsterdam New York

Lončar D, Bedrica L, Mayer J, Cannon B, Nedergaard J, Afzelius BA, Švajger A (1986) The effect of intermittent cold treatment on the adipose tissue of the cat. J Ultrastruct Mol Struct Res 97:119–129

Nichols DG (1979) Brown adipose tissue mitochondria. Biochim Biophys Acta 549:1–29

NIH Rodents Catalogue, NIH Publication (1981) No81-606, Bethesda:253

Rothwell NJ, Stock MJ (1979) A role for brown adipose tissue in diet induced thermogenesis. Nature (London) 281:31–35

Sclafani A (1984) Animal models of obesity: classification and characterization. Int J Obes 8:491–508

Trayhurn P, Mercer SW (1986) Brown adipose tissue in obese animals. Biochem Soc Trans 14:236–239

Acute and Long-Term Regulation of Brown Adipose Tissue Thermogenesis: Physiological and Pathological Implications

J. HOUŠTĚK[1], J. KOPECKÝ[1], D. JANÍKOVÁ[1], J. BEDNÁR[1], M. HOLUB[2], T. SOUKUP[1], M. MIKOVÁ[3] and K. VÍZEK[3]

1 Introduction

Brown adipose tissue has long been recognized as an important site of nonshivering thermogenesis in homeotherms (Smith 1961), which must maintain their body temperature within a very narrow range, regardless of the environmental temperature. Probably the best understood is the function of brown adipose tissue in the *mammalian neonate*, which loses heat at most environmental temperatures far in excess of normal metabolic heat production. The thermal requirements of the newborn exposed for the first time to extrauterine life are mostly covered by heat production in brown adipose tissue, which appears to be the best and preferred way to thermoregulate (Nedergaard et al. 1986). For example, in a newborn rabbit the brown adipose tissue is responsible for more than 60% of increased heat production in the cold (Heim and Hull 1966). In general, after birth brown adipose tissue represents the dominating site of nonshivering thermogenesis, which remains active for several weeks, until other thermoregulatory mechanisms develop (Nedergaard et al. 1986).

Depending on the changing demands of an organism, resulting from its developmental state, thermal environment and dietetic conditions, the brown adipose tissue function can also be reactivated in adult mammals. In particular, chronic cold exposure, i.e. *cold adaptation*, activates brown adipose tissue thermogenesis when other mechanisms of thermoregulation become insufficient (Himms-Hagen 1986). A somewhat special case are the hibernators, where the brown adipose tissue is essential for arousal from hibernation, and which stays active throughout life.

Thermogenesis of brown adipose tissue can be activated in some adult rodents also by dietetic factors, and *diet-induced thermogenesis* plays a significant role in energy expenditure and whole-body energetics (Rothwell and Stock 1986; Trayhurn 1986). Brown adipose tissue then serves as an "energy buffer" and helps to expend excessive caloric intake during overfeeding.

Similarly, to other mammals, brown adipose tissue plays a crucial role in the thermoregulation of *human newborns*, where it represents the major mechanism of

[1]Institute of Physiology, Czechoslovak Academy of Sciences, Videňská 1083, 142 20 Prague, Czechoslovakia
[2]Institute for Clinical and Experimental Medicine, Videňská 800, 140 00 Prague, Czechoslovakia
[3]Research Institute for the Care of Mother and Child, Nábřeži K. Marxe 157, 147 10 Prague, Czechoslovakia

A. Azzi, Z. Drahota and S. Papa
Molecular Basis of Membrane-Associated Diseases

defence against cold. As calculated from in vitro measurements of oxidative and thermogenic activity of isolated brown adipose tissue cells and mitochondria, heat production by 20–30 g of brown adipose tissue present in a full-term newborn infant (Hull 1977) is completely sufficient to account for a two- to three-fold increase of the total metabolic rate induced by exposure to approximately 10°C below the thermoneutral temperature (Heim 1971; Hey 1974; Cannon and Johansson 1980). In recent years more and more attention has been paid to brown adipose tissue thermogenesis in humans and the topic has become rather broad, ranging from thermoregulation of physiological and pathological newborns, hormonally and genetically based alterations of brown adipose function, to a controversial but attractive contribution of brown adipose tissue in adult humans to temperature control, overall energy expenditure and some types of obesity (see Drahota et al., this vol.).

In all the above summarized conditions brown adipose tissue is considered as a unifunctional, specialized thermogenic organ which makes it possible for the chemical energy of substrate oxidation to be quantitatively and instantly converted into heat. Very recently, however, the discovery of intensive 5′-deiodination of tyroxine in brown adipose tissue indicated that the function of brown adipose tissue is more complex and may also influence the metabolism of other tissues.

In this report we discuss the results of our recent studies, as well as other evidence related to (1) the direct and indirect role of brown adipose tissue in heat production, (2) mechanisms by which brown adipose tissue is regulated, (3) perinatal development of brown adipose tissue, (4) hypo- and hyperfunction of brown adipose tissue in pathological states.

2 Direct and Indirect Role of Brown Adipose Tissue in Heat Production

In brown adipose tissue, the chemical energy of the substrates, mainly free fatty acids, oxidized by mitochondrial respiration, is quantitatively converted into heat. The *direct thermogenic function* of the brown adipose tissue thus contrasts with the energetics of other tissues, where the energy of substrates is converted mostly into the energy of ATP (Nicholls 1976). The thermogenic ability of brown adipose tissue mitochondria is based on dissipation of the respiration-generated electrochemical potential gradient of protons via a regulatable H^+ channel (Fig. 1), the so-called uncoupling protein (Lin and Klingenberg 1980) or thermogenin (Cannon and Nedegaard 1984). The resulting physiological uncoupling of oxidative phosphorylation, due to short-circuiting of the proton gradient, permits the high oxidative capacity to be unconstrained by a diminished phosphorylating capacity of mitochondria (Houštěk and Drahota 1977; Cannon and Vogel 1977; Houštěk et al. 1978). The brown adipose tissue is thus endowed with the ability to generate heat at a rate of about 240 mW/g tissue while, for example, liver generates only 4 mW/g (Foster and Frydman 1978; Nicholls et al. 1986).

The *uncoupling protein*, a polypeptide with a monomeric molecular weight of 32 000, is exclusively found in the inner mitochondrial membrane of brown adipose tissue (Cannon et al. 1982; Ricquier et al. 1983a) where it represents up to 14% of the membrane protein (Lin and Klingenberg 1982). The functional unit of

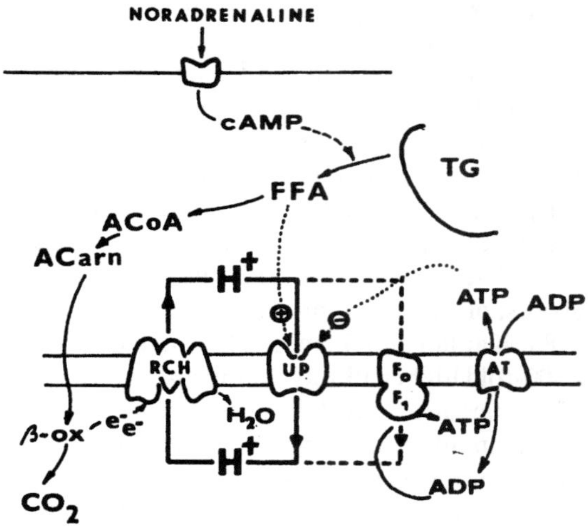

Fig. 1. Integrative scheme of cellular events during brown adipose tissue thermogenesis. Abbreviations: *TG* triglycerides; *FFA* free fatty acids; *ACoA* acyl coenzyme A; *ACarn* acyl carnitine; *β-ox* β-oxidation; *RCH* respiratory chain; *UP* the uncoupling protein; F_0F_1 ATP synthase; *AT* adenine nucleotide translocator

the uncoupling protein is a dimer, with a single nucleotide binding site oriented to cytosol which displays a high affinity for purine nucleoside di- and triphosphates. Their inhibitory effect on uncoupling protein transport activity is involved in the regulation of the uncoupling protein function (see below).

Sequencing of the isolated uncoupling protein and of its nuclear gene, as well as the use of specific antibodies, showed a high degree of homology between different animal species and man (Ricquier and Bouillaud 1986). A significant structural similarity was also found between the uncoupling protein, mitochondrial adenine nucleotide translocator, oxoglutarate/malate carrier and phosphate carrier (Walker et al. 1988) and it was concluded that all these dimeric transport proteins originated from a common ancestral gene (Aquila et al. 1985). The uncoupling protein is also responsible for the unusually high permeability of brown adipose tissue mitochondria to chloride, bromide and some other anions (Nicholls and Lindberg 1973), but the physiological importance of the anion transport remains to be elucidated.

The mechanism of the key reaction in brown adipose tissue thermogenesis, ion translocation via the uncoupling protein, is still not fully understood. Most of the recent experimental evidence (Kopecký et al. 1984, 1987; Rial and Nicholls 1986) supports the view that protons and chloride anions are translocated by two distinct and independent pathways within each single functional unit of the uncoupling protein. In analogy with the adenine nucleotide translocator, a model for the function of the uncoupling protein has recently been proposed in which two independent binding sites for proton and chloride anion reorient between the matrix and cytosolic sides of the membrane during transport, and uncoupling protein exists either in a free form or with a bound proton or anion (Rial and Nicholls 1988).

It has been proposed as early as 1817 (see Afzelius 1970) that brown adipose tissue might be also hormonally active, but convincing evidence that it is involved in control of thyroid hormone action appeared only recently.

The biologically active thyroid hormone, triiodothyronine (T_3) is derived from inactive thyroxine (T_4) mostly by 5′-deiodination in peripheral tissues (Schwartz et al. 1971; Larsen et al. 1981). In 1982 it was shown for the first time that, besides the peripheral deiodinase of thyroxine which is found in liver, kidney and some other tissues (type I deiodinase), there exists another form of the enzyme (type II) (Silva et al. 1982; Visser et al. 1982, 1983), which is found in brain, pineal and pituitary glands, placenta and brown adipose tissue (Leonard et al. 1982; Visser et al. 1982; 1983; Kaplan and Shaw 1984; Tanaka et al. 1986). Type II deiodinase has at least a 400-fold higher affinity for thyroxine (K_m in nM range), exhibits a sequential type of kinetics and is much less sensitive to propylthiouracyl (Goswami and Rosenberg 1986), an inhibitor of type I deiodinase (Silva et al. 1982; Visser et al. 1982, 1983). Based on measurements of the total activity of type I and II deiodinases, it was shown that in an adult euthyroid rat under euthermic conditions the type II deiodinase amounts to at least 40%, and in a 2-week-old rat, to 80% of thyroxine deiodination, respectively (Frumess and Larsen 1975; Larsen and Frumess 1977; Silva et al. 1982; Silva and Matthews 1984). In both adult and neonatal hypothyroid animals it accounts for almost all of the conversion of T_4 to T_3 (Silva et al. 1984; Silva and Matthews 1984). Demonstration of cold-induced activation of type II deiodinase in brown adipose tissue and considerably low activity of enzyme in other tissues suggested that brown adipose tissue deiodinase might represent a major source of plasma T_3 during acute and chronic cold exposure.

As shown in Fig. 2, both the specific and total enzyme activity is low in the warm-adapted hamster. After exposure to cold the activity rapidly increases resulting in 2–3 days in a 25-fold increase of specific activity and an 80-fold increase of the total activity. A similarly dramatic activation of deiodinase was induced by cold in the rat, where specific activity increased 150 times, total activity 210 times and the maximum was reached even faster, in 10 h (Kopecký et al. 1986a).

The cold-induced increase in hamster brown adipose tissue was accompanied by a significantly less pronounced increase of thermogenic activity, determined by the uncoupling protein content (Kopecký et al. 1986a). This indicates that both the T_4 deiodination and direct thermogenic function of brown adipose tissue are activated, although not equally, by cold adaptation. Parallel measurements of serum T_3 and T_4 concentration showed (Fig. 2) that the increase in deiodinase activity was accompanied by a pronounced (three-fold) increase of T_3 level and, in correspondence, by an increase of the T_3/T_4 ratio, suggesting that T_4 produced in brown adipose tissue is exported into the serum. These results strongly support the view (Silva and Larsen 1983, 1985) that 5′-deiodination of T_4 in brown adipose tissue may play an important role in controlling serum T_3. These considerations are further supported by recent direct measurements of the arterio-venous difference in T_3 and T_4 in rat brown adipose tissue (Fernandez et al. 1987).

5′-Deiodination of T_4 in brown adipose tissue is under complex hormonal control. Its stimulation by cold is mediated through α_1-adrenergic receptors (Silva and Larsen 1983, 1986). Adrenergic stimulation thus may be crucial for induction of both the direct thermogenic function and 5′-deiodination of T_4 in brown adipose tissue. Little is known about the mechanisms by which the enzyme is activated. The

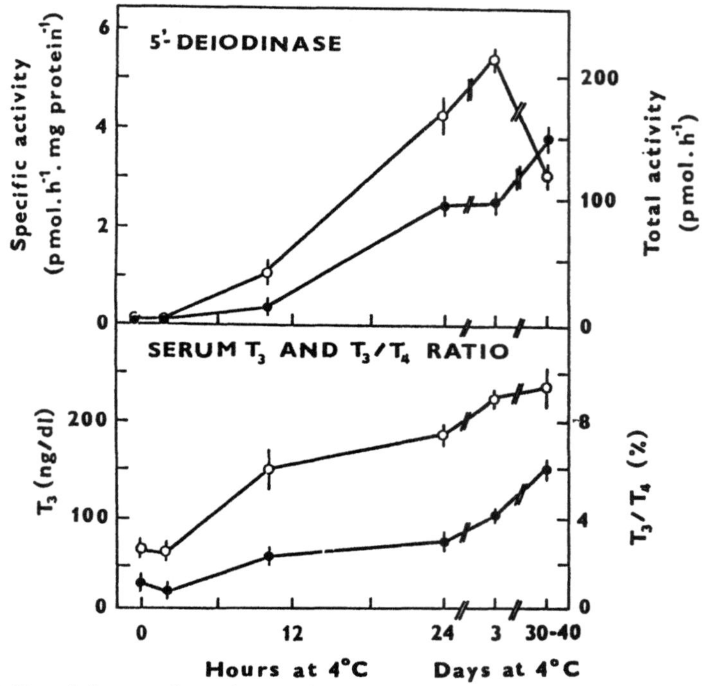

Fig. 2. Thyroxine 5′-deiodinase in brown adipose tissue and concentration of thyroid hormone in serum of hamsters during acute and long-term exposure to cold. Adult golden Syrian hamsters, acclimated to 24°C were exposed to 4°C as indicated. Specific (-o-) and total (-•-) activity of thyroxine 5′-deiodinase was measured in the homogenate of interscapular brown adipose tissue. Ratio T_3/T_4 (-•-) was calculated from T_3 (-o-) and T_4 concentrations determined in serum. Values shown are means ± SE(n = 4–14). For details, see Kopecký et al. (1986a)

observed increase of enzyme activity by cold, amounting to two orders of magnitude, as well as prevention of the activation by cycloheximide and actinomycin D (Jones et al. 1986; Silva and Larsen 1986), indicate that mainly the increased quantity of the enzyme due to activated gene transcription and protein synthesis is responsible for the changes observed.

In several states where the thermogenic function of brown adipose tissue is impaired the 5′-deiodinase activity of the tissue decreases. In myopathic hamsters, which have less brown adipose tissue than normal (Himms-Hagen and Gwilliam 1980), the capacity of peripheral T_3 production is lower (Kopecký et al. 1986b). Also, in pregnant or lactating rats, which have a lower capacity for nonshivering thermogenesis in brown adipose tissue, catecholamine-induced activation of 5′-deiodinase in brown adipose tissue was significantly decreased (Giralt et al. 1986). In genetically obese (ob/ob) mice, which exhibit an altered thermogenic function of the brown adipose tissue, cold adaptation was without effect on the deiodinase activity (Kates and Himms-Hagen 1985).

Maintaining homeothermy in mammals involves a significant portion of the basal metabolic rate, the so-called endothermic thermogenesis, which occurs in most organs and is under slow, but exclusive control of the thyroid hormone

(Himms-Hagen 1983; Girardier and Stock 1983). This is in contrast to a rapidly inducible component of nonshivering thermogenesis which is regulated by noradrenaline from the nervous system and which predominantly occurs in brown adipose tissue (see below). The substantial contribution of brown adipose tissue to peripheral thyroxine deiodination may therefore control the intensity of basal metabolism and thus the accompanying heat production in other tissues. Such a regulatory role, which might be important not only in adult animals but also in newborns where, in parallel with the development of brown adipose tissue, thermogenic function deiodinase activity increases (Iglesias et al. 1987; Houštěk et al. 1988a), has to be considered as an *indirect thermogenic function* of the brown adipose tissue.

Thyroxine deiodination in the brown adipose tissue is essential for differentiation of the brown adipose tissue itself (see below) and it may also be important for maturation and development of other organs and tissues during the perinatal period, since intensive growth and increased proteosynthetic activity in most tissues requires the presence of T_3 (Oppenheimer and Samuels 1983). Interestingly, a very high deiodinase activity is found in the brown adipose tissue of fetal rat shortly before birth (Iglesias et al. 1987). Even more interesting is the question as to whether human brown adipose tissue has an analogous activity and, if present, to what extent it participates in the regulation of serum T_3. No data on human brown adipose tissue are available and, due to the lability of the enzyme and problems with obtaining fresh samples of tissue, precise measurements may require some time.

3 Mechanisms of Regulation of the Brown Adipose Tissue

Under all conditions when the brown adipose tissue is active, a precise regulation is absolutely essential because intensive, uncontrolled heat production would immediately disbalance the thermal homeostasis of an organism. Since proton transport through uncoupling protein represents the exothermic and rate-limiting step in brown adipose tissue thermogenesis (Nicholls and Locke 1984), it is quite obvious that regulation concerns primarily the uncoupling protein. Depending on the developmental state, thermal environment, diet and other conditions, three types of regulatory events can be recognized: acute and immediate activation and inhibition of heat production, long-term regulation and an intermediate form, called unmasking or initiation.

When the thermogenic potential of brown adipose tissue is developed to the required extent, the dominant form or regulation is the *acute regulation*. In this case the initial signal at the level of brown adipose tissue cells is the release of noradrenaline from numerous sympathetic nerve endings which are in contact with each brown adipocyte (Girardier and Seydoux 1986). Upon activation of receptors (Fig. 1) the increased level of cAMP activates, through a cascade of reactions, hormone-sensitive lipase and free fatty acids released from tissue triglycerides serve as a substrate for respiration (Nicholls and Locke 1984). At the same time, the H^+-translocating activity of the uncoupling protein must be activated at a rate sufficient to allow reentry of protons extruded from mitochondria by intensive respiration (Fig. 1).

Two ligands, purine nucleotide and free fatty acid, interact specifically with the uncoupling protein and appear to be primarily responsible for the physiological modulation of its activity (Nicholls and Locke 1984; Nicholls et al. 1986). The originally proposed regulatory function of fatty acid CoA esters (Cannon et al. 1977) seems to be unlikely (Locke et al. 1982a,b; Ježek and Houštěk 1986). Activation of the uncoupling protein by free fatty acids has been extensively studied and documented (see Nicholls et al. 1986). In contrast to acyl CoA esters, the concentrations of free fatty acids needed for activation of H^+ transport in in vitro systems are in good agreement with free fatty acid levels in brown adipose tissue after noradrenaline stimulation (Locke et al. 1982a,b; Norman and Flatmark 1984), although a specific free fatty acid-binding site on the uncoupling protein still remains to be characterized. On the other hand, the interaction of inhibitory adenosine or guanosine di- and triphosphates with the nucleotide binding site of uncoupling protein (Heaton et al. 1978; Lin and Klingenberg 1982; Nicholls and Locke 1984; Klingenberg 1988) is well established but their in vivo action is still unclear. This is mainly because the millimolar concentrations of cytosolic ADP and ATP are well above the micromolar K_d values of nucleotide binding to the uncoupling protein (Rial et al. 1983; Lin and Klingenberg 1982), while the concentrations required for half-maximal inhibition (K_i) of the proton conductance are under all conditions studied always higher (Cannon et al. 1973; LaNoue et al. 1982; Locke et al. 1982b; Kopecký et al. 1984, 1987) than the K_d values of the binding.

Using isolated intact brown adipose tissue mitochondria and mitochondria in which the nucleotide binding site was chemically modified (Kopecký et al. 1984, 1987; Houštěk et al. 1987), we have demonstrated that the relationship (Fig. 3) between the occupation of the uncoupling protein nucleotide binding sites in the membrane and resulting inhibitory effect on H^+ transport is clearly nonlinear, in contrast to Cl^- transport. In up to some 50% fractional saturation of binding sites no inhibitory effect of bound nucleotides is observed, but at near-maximal saturation the inhibiton becomes very steep. This indicates that there is no direct link between the nucleotide binding and induced inhibition. As different amino acid

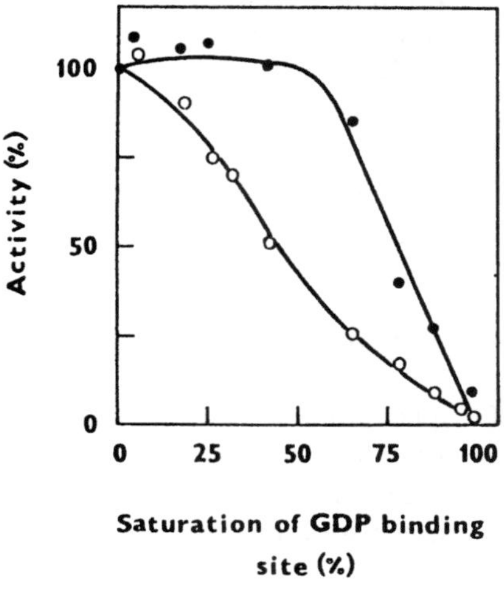

Fig. 3. Relationship between occupancy of the uncoupling protein nucleotide-binding site by GDP and the rate of uncoupling protein-mediated H^+ – (-●-) and Cl^- – (-○-) transport in isolated hamster brown adipose tissue mitochondria. H^+ and Cl^- transport activities and binding of ^{3}H-GDP were measured as described in Kopecký et al. (1987)

residues were shown (Kopecký et al. 1987) to ensure the affinity of nucleotide binding and the resulting gating effect, it is probable that different conformational states of the uncoupling protein are involved in purine nucleotide control, similarly as for example in the adenine nucleotide translocator (Vignais et al. 1985). The sharp inhibition of transport occurring close to saturation of the binding would be advantageous for in vivo regulation as only a small fraction of the H^+ translocation capacity of the uncoupling protein molecules in the membrane is sufficient for continuous dissipation of the proton gradient generated at the maximum rate of respiration (Nicholls 1976; Lin and Klingenberg 1982). This indicates that in cooperation with changing levels of free fatty acids which act at a site different from nucleotide binding (Kopecký et al. 1987) and possibly also the changes of cytosolic pH (Klingenberg 1988), only relatively small differences in cytosolic nucleotides and/or their ratio around the basal nucleotide level (LaNoue et al. 1986) would open and close the uncoupling protein proton channel. This may be sufficient to switch thermogenesis on and off.

Similarly to acute regulation, the process of "masking-unmasking" is the form of regulation which activates already existing molecules of the uncoupling protein. This phenomenon was first noted by Desautels and co-workers (1978) and later confirmed by others (Gribskov et al. 1986; Nedergaard and Cannon 1987; Peachey et al. 1988), who showed that within few minutes after cold exposure, the number of uncoupling protein nucleotide binding sites significantly increases. As the unmasking is not connected with the increased content of uncoupling protein in the membrane (Gribskov et al. 1986) and does not depend on its de novo synthesis, the mechanism has to be based on an activation of "latent" uncoupling proteins. Unmasking might operate via conformational changes of the protein and/or transverse movements of the protein in the phospholipid bilayer. Unmasking is accompanied by a significant increase in H^+ transport activity (Cannon and Nedergaard 1988) and can be induced by volume changes of mitochondria (Nedergaard and Cannon 1987), possibly involving anion transport activity of uncoupling protein.

On a long-term basis, the regulation of brown adipose tissue and its thermogenic potential is achieved by biosynthetic and degradative processes which modulate the amount of tissue, the number of mitochondria and their specific enzymic equipment, including the uncoupling protein (Ricquier and Bouillaud 1986). Due to the *long-term regulation*, the thermogenic potential increases by cold adaptation or overfeeding of adult animals and decreases by chronic exposure to a warm environment. This type of regulation is responsible for quantitative and qualitative changes of brown adipose tissue during the ontogenetic development.

4 Perinatal Development of Brown Adipose Tissue

Of special interest are the changes in brown adipose tissue during the perinatal period which represents an original development of brown adipose tissue thermogenesis per se, following the "programmed" ontogenetic recruitment in accordance with the degree of maturity of the newborn (Nedergaard et al. 1986).

The main prerequisite of the thermogenically active brown adipose tissue is represented by fully competent, functional *thermogenic mitochondria* with a high oxidative capacity matched by sufficient capacity of the uncoupling protein-mediated proton conductivity, while the phosphorylating activity is low (Nicholls 1976; Houštêk et al. 1978; Nicholls and Locke 1984; Lin and Klingenberg 1982). Therefore, the content of cytochrome oxidase, the uncoupling protein and F_1-ATPase should be a sensitive criterion of the existing changes and functional state of the tissue. To characterize the profile and changes in the energetics of brown adipose tissue occurring during the perinatal period we studied the embryonic mouse and rat, and newborn hamster. While mouse and rat are born with a well-developed thermogenic function of brown adipose tissue and represent an altricial type of neonate, the function of brown fat in hamster, an immature type of neonate develops after birth (Nedergaard et al. 1986). As a methodological approach we used specific antibodies against cytochrome oxidase, uncoupling protein and F_1-ATPase which not only allow the quantification of these proteins in milligram amounts of brown adipose tissue by immunoblotting, but also permit the investigation of their cellular location by immuno-electronmicroscopy.

In embryonic mouse and rat (Houštěk et al. 1987, 1988b) brown adipose tissue could be macroscopically identified and dissected after 14–15 days of fetal development. Already on the 15–16th day the tissue contains some, not very well-differentiated mitochondria. As shown in Fig. 4, cytochrome oxidase is also present and its quantity continuously increases towards birth following an increasing number of mitochondria. F_1-ATPase is also present on the 15–16th day but it reaches its maximum already on the 19th day, when the uncoupling protein appears for the first time. Between the 19th day and birth the content of the uncoupling protein rapidly increases and parallels the increase of cytochrome oxidase. Consequently, the original "immature" mitochondria with only a few and

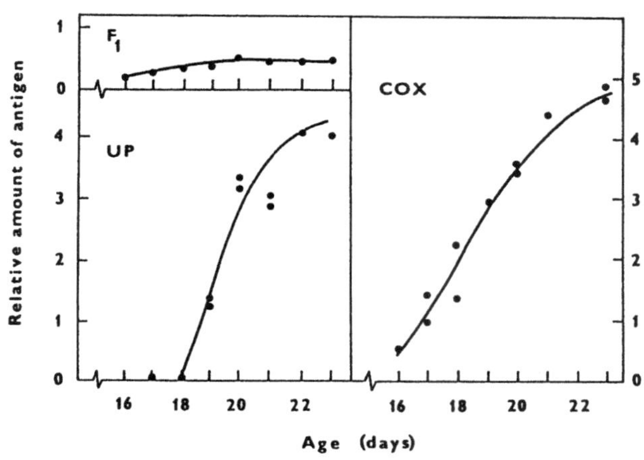

Fig. 4. Quantitative evaluation of F_1-ATPase (*F_1*), the uncoupling protein (*UP*) and cytochrome oxidase (*COX*) in mouse embryonic brown adipose tissue. Immunoblotting analysis using specific rabbit antisera was performed in homogenates of interscapular brown adipose tissue of mouse fetuses of indicated age (days after mating), delivered by ceasarian section. For details, see Houštěk et al. (1988b)

irregular cristae are continuously replaced by differentiated, highly branched mitochondria. It is evident that the mitochondria present before the 19th day of pregnancy are apparently nonthermogenic as they lack the uncoupling protein but contain comparable amounts of cytochrome oxidase and ATPase. They resemble the mitochondria from other, phosphorylating tissues.

In hamster brown adipose tissue (Fig. 5) fully competent thermogenic mitochondria are formed postnatally, in agreement with the inability of newborn hamster to maintain body temperature when cold-stressed (Rink 1969), and the absence of ^{3}H-GDP binding sites at the time of birth (Sundin et al. 1981). As indicated by the content of cytochrome oxidase, the uncoupling protein and F_1-ATPase, and their respective ratios in homogenates and isolated mitochondria, before the ninth postnatal day when the uncoupling protein appears for the first time, ATPase is again in quantitative relation with cytochrome oxidase. Importantly, the immunochemical quantification is in perfect agreement with the measurements of cytochrome oxidase and ATPase activities (Houštěk et al. 1988a).

Therefore, both in altricial mouse and rat and in immature hamster the ontogenic development of thermogenic function in brown adipose tissue results

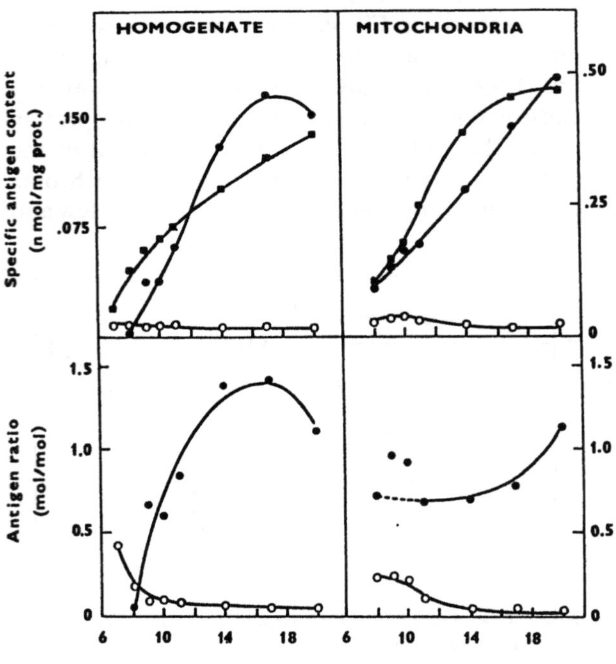

Fig. 5. The oxidative, phosphorylating and thermogenic potential of brown adipose tissue of the developing golden Syrian hamster. The specific content of cytochrome oxidase (COX, -■-), F_1-ATPase (F_1, -○-) and the uncoupling protein (UP, -●-) was determined by immunoblotting analysis using specific rabbit antisera in homogenates and isolated mitochondria of interscapular brown adipose tissue. The data were used to calculate the molar ratios (UP/COX, -●-; F_1/COX, -○-) between these proteins. For details, see Houštěk et al. (1988a)

from a *transformation of preexisting, nonthermogenic mitochondria into thermogenic mitochondria*. It is obvious that this can only result from abruptly induced asynchronous gene expression and protein synthesis of enzymes of mitochondrial energy conversion.

Biogenesis of brown adipose tissue mitochondria depends on stimulation of both α_1- and β-adrenergic receptors (Jacobsson et al. 1986) and requires also thyroid hormones (Himms-Hagen 1983). Intracellular conversion of T_4 to T_3 was essential for optimal thermogenic function and induction of synthesis of the uncoupling protein and other enzymes during cold adaptation (Bianco and Silva 1987a,b), and the altered response of hypothyroid rats was abolished by administration of T_4 but not T_3. Intracellular 5′-deiodination of T_4 in brown adipose tissue seems to be required also for the development of brown adipose tissue and the formation of thermogenic mitochondria in the newborn since fetal rat brown adipose tissue exhibits a ten times higher activity of 5′-deiodinase of T_4 with respect to the adult animal (Iglesias et al. 1987). Also, the formation of thermogenic mitochondria in postnatal hamster was paralleled by a rapidly increasing deiodinase activity (Houštěk et al. 1988a).

As documented by specific mRNA levels (Freeman and Patel 1984; Ricquier and Bouillaud 1986), the synthesis of nucleus-encoded uncoupling protein is regulated at the transcription level, although translational regulation is not excluded. Two species of mRNA (around 18 S and 15 S) have been found using uncoupling protein cDNA probes (Jacobsson et al. 1985; Ricquier and Bouillaud 1986; Ridley et al. 1986) in small animals but it remains to be elucidated whether they correspond to different transcription units or to different splicing. Newly synthesized uncoupling protein is able to incorporate also into mitochondria isolated from other than brown adipose tissues (Freeman et al. 1983; Ricquier and Bouillaud 1986) which indicates that no specific receptor site is required for its insertion.

The highly pronounced differences in the synthesis of cytochrome oxidase and ATPase require even more complex regulation, as they are based on the well-balanced cooperation of the nuclear and mitochondrial genomes and proteosynthetic machineries, since they contain subunits of both nuclear and mitochondrial origin.

At variance with immunoblotting, immuno-electronmicroscopy (Fig. 6) is capable of detecting individual protein molecules in the cell. With the aid of immuno-electronmicroscopy it was possible to confirm the changes in the uncoupling protein during perinatal development of brown adipose tissue, as well as the higher content of F_1-ATPase in nonthermogenic brown adipose tissue mitochondria (Houštěk et al. 1988a). It was also noted that the beginning of intensive incorporation of the uncoupling protein into the membrane is shortly preceded by a significant presence of its antigen in the cytosol. In case of F_1-ATPase the cytosolic location was found only at the early stage of development when the mitochondria are nonthermogenic. Extramitochondrially located antigens of mitochondrial proteins may of course be detected more sensitively than the membrane localized ones, since more numerous epitopes may be accessible for the polyclonal antibodies used. Nevertheless, immuno-electronmicroscopy proves to be a suitable technique for following the fate of newly synthesized proteins in the cell.

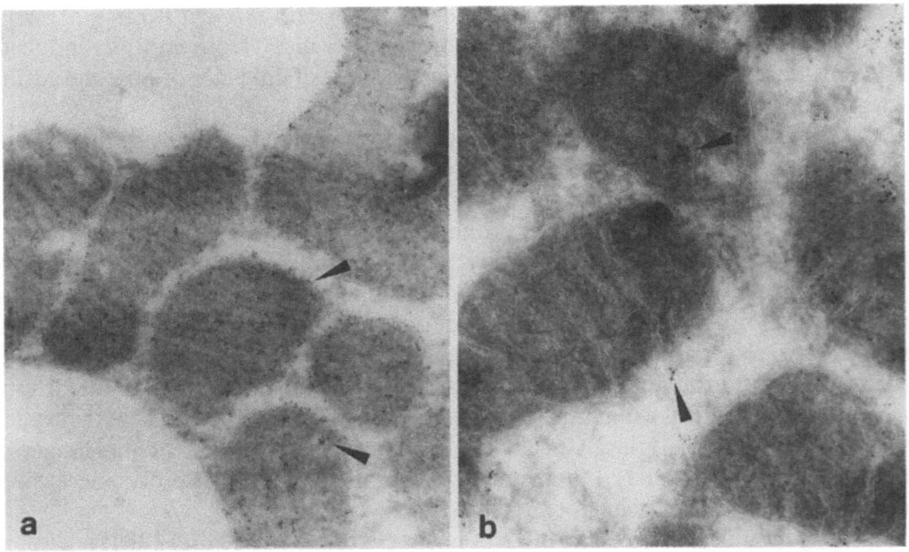

Fig. 6. Detection of the uncoupling protein by immuno-electronmicroscopy. The uncoupling protein was visualized in ultrathin cryosections of interscapular brown adipose tissue of adult (**A**, magnification 12 000 times) and 8-day-old (**B**, magnification 50 000 times) hamster using specific rabbit anti-uncoupling protein antibody and secondary goat anti-rabbit, antibody-colloidal gold probe. For details, see Houštěk et al. (1988a)

5 Hypo- and Hyperfunction of Brown Adipose Tissue in Pathological States

There is an increasing number of well-documented cases where the function of brown adipose tissue differs from that in normal controls. It can be either elevated or depressed and an abnormal function of brown adipose tissue occurs both in the neonate and in adults. The aetiology of pathological states is rather heterogeneous and includes genetic factors, hormonal regulation, hypoxia and other variables that are only conjectured at present (Lean and James 1986; Lean et al. 1986a; Rothwell and Stock 1986; Trayhurn 1986). In most cases the changes in the thermogenic function of brown adipose tissue are manifested by changes in the amount of tissue, the triglyceride content, the content of mitochondria and the enzymic equipment. They result mainly from a long-term type of regulatory event. At the time of writing, the main parameter being followed in many studies is the specific and total content of uncoupling protein which is assessed either directly using specific antibodies or indirectly by means of ^{3}H-GDP binding to isolated mitochondria.

The *hypofunction* of brown adipose tissue is conclusively documented, for example, in young and adult animals with genetic obesity (ob/ob mouse) (Himms-Hagen and Desautels 1978; Goodbody and Trayhurn 1982), where a decreased amount of the uncoupling protein results from the decreased turnover of noradrenaline in the tissue (Knehans and Romsos 1982). Uncoupling protein is also decreased in diabetic, obese (db/db) mice (Goodbody and Trayhurn 1981).

Malfunction is recognized in several pathological states of newborn animals and humans. It is caused by insufficient trophic stimuli due to thyroidectomy or decreased T_4 deiodination in brown adipose tissue (Polk et al. 1987; Bianco and Silva 1987a,b), insulin-resistant diabetes (Mercer and Trayhurn 1983), or Cushing's syndrome (Lean et al. 1986a). Insufficient function of brown adipose tissue is expected in stillborns and in pre-term infants as well as in small-for-day newborns (Lean and James 1986; Lean et al. 1986a), which have serious difficulties in maintaining homeothermy. It might be related also to infants dying of "cold syndrome" (Aherne and Hull 1966). Available data on the change of the uncoupling protein content in these states are convincing but have to be considered with caution because the number of documented cases is low. Up to now no example of hypofunction has been found due to the synthesis of defective uncoupling protein, but in accordance with a large number of well-known, inherited enzyme defects, it can be expected to exist.

The unwanted *hyperfunction* of brown adipose tissue has been well described in patients with chronically elevated circulating catecholamines from pheochromocytoma (Aherne and Hull 1966; Ricquier et al. 1983b; Ricquier and Mory 1984; Lean et al. 1986b). It is probable that also other hormonal disorders may induce a hyperfunction of brown adipose tissue. Hyperthermia of brown adipose tissue origin seems to be involved in some cases of cot death (Naeye 1974a, b; Stanton 1984) and chronic hypoxia of different causes (Teplitz and Lim 1974), in connection with elevated noradrenaline levels (Perrin et al. 1984). Brown adipose tissue can also become neoplasmic and benign types of brown adipose tissue tumors, hibernomas, were found in both animals and man, while the malignant forms were found only in animals (Gaffney et al. 1983; Stefanski et al. 1987).

A special case of elevated metabolism and increased heat production is the febrile response of animals and man to bacterial pyrogens. Mild fever has possibly a beneficial role in the defence against infection (Eiger and Kluger 1983) and physiological activation of thermogenesis in brown adipose tissue has to be expected. Indeed, brown fat was recently shown to be involved in heat production during development of fever in young rabbits but not in the maintenance of elevated body temperature after the fever had peaked (Harris et al. 1984). The mechanism of induction of brown adipose tissue thermogenesis involves a centrally-acting tumor necrosis factor, a polypeptide secreted by macrophages stimulated by the bacterial pyrogens (Coombes et al. 1987).

A better understanding of different pathological states necessitates the precise determination of tissue changes occurring, particularly those which affect the enzymic equipment of mitochondria. For this purpose sensitive immunochemical methods seem to be particularly suited. We have used the immunoblotting technique (Houštěk et al. 1988a, b) to characterize two animal models in which the thermogenic function of brown adipose tissue differs substantially from normal controls.

In the first case we studied hypotrophic fetal rats (delivered on the 21st day of gestation by caesarian section) where hypotrophy was induced by experimental ischaemia due to unilateral constriction of the uterine artery (performed on the 17th day). Fetuses from the constricted side were significantly smaller than the controls from the unconstricted side and from sham-operated animals. Even more than the body weight, the amount of brown adipose tissue was reduced. As also the specific content of the uncoupling protein decreased, the resulting thermogenic capacity of brown adipose tissue with respect to body weight differed between hypotrophic and control animals almost three-fold (Fig. 7). The hypotrophy is quite frequent in human newborns. Although it has a multifactorial aetiology, hypoxia and hyponutrition belong to the factors involved. It may be expected that a similar malfunction of brown adipose tissue also occurs in hypotrophic newborn infants.

In the second case we used genetically nude mice (nu/nu) which are completely hairless and also immunologically deficient (Holub et al. 1988). As shown in Table 1, already at near-thermoneutral temperature (28°C), brown adipose tissue of the nude mouse exhibited a significantly increased content (2.6 times) of uncoupling protein with respect to the controls. Exposure to very moderate cold stress (gradient of 6°C) further increased the uncoupling protein content (6.6-fold difference). The change is almost completely due to the increased content in the mitochondrial membrane which exceeds nearly three times a parallel increase of oxidative capacity. As apparent from the content of F_1-ATPase and the ratios

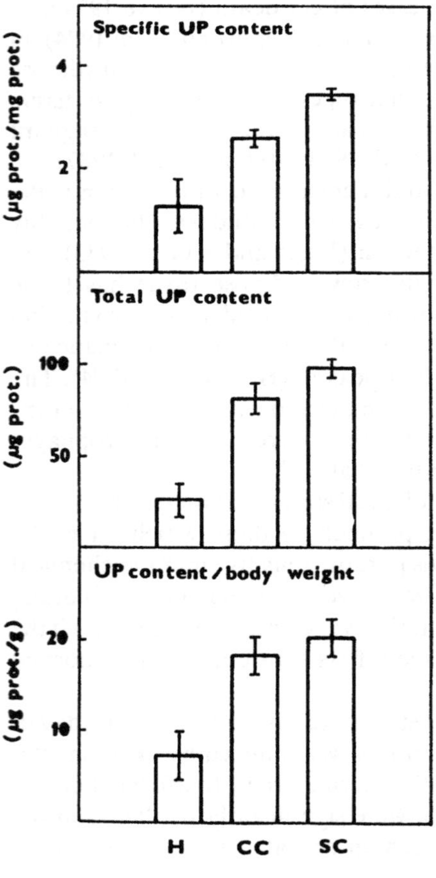

Fig. 7. Thermogenic activity of brown adipose tissue in hypotrophic rats. Unilateral constriction of the uterine artery was performed on the 17th day of gestation and the ceasarian section on the 21st day, respectively. The uncoupling protein was measured by immunoblotting in homogenates of interscapular brown adipose tissue (as in Fig. 5) from hypotrophic fetuses (*H*), fetuses from the unconstricted side of the same animal (contralateral control, *CC*) and fetuses of sham-operated rats (sham controls, *SC*). Values shown are means ± SE (n = 4–8)

Table 1. Oxidative, phosphorylating and thermogenic potential of brown adipose tissue of the genetically nude (nu/nu) mouse[a]

	Homogenate				Isolated mitochondria		
	UP	UP	COX	F_1	UP/COX	F_1/COX	UP/F_1
	(Specific content in percent control)				(Relative to control)		
+/+ − 28°C	100	100	100	100	1	1	1
nu/nu − 28°C	260	190	200	110	0.9	0.5	1.7
nu/nu − 22°C	660	605	230	80	2.6	0.3	7.6

[a] Specific pathogen – free adult BALB/C nu/nu mice and haired (+/+) control mice were exposed for 4 weeks to an ambient temperature of 28°C or 22°C (cage individually). Homogenates and isolated mitochondria were prepared from interscapular brown adipose pooled from 3–6 animals in each group. Specific content of the uncoupling protein (UP), cytochrome oxidase (COX) and F_1-ATPase (F_1), determined by immunoblotting as in Fig. 6, was used to calculate UP/COX, F_1/COX and UP/F_1 ratios.

between the three enzymes, the capacity to phosphorylate ATP markedly decreased. Brown adipose tissue of nude mice therefore undergoes changes similar to those observed in neonate animals (see above); mitochondria become better equipped for intensive heat production required to compensate increased loss of heat from the body. Of the parameters calculated, the uncoupling protein/F_1-ATPase ratio appears to be the most sensitive measure of the state of the tissue.

One of the main obstacles in quantifying the uncoupling protein content in human brown adipose tissue is the enormous heterogeneity of tissue from multiple locations. We have investigated more than 20 samples from autopsies of pathological newborns using immunoblotting and found that the content of the uncoupling protein may differ more than ten times (1.1–13.1 μg/mg protein) in various samples from the same individual. This is clearly more critical than the problem of obtaining samples soon after death. A parallel determination of F_1-ATPase and cytochrome oxidase showed that the ratios between these enzymes might be a much better parameter for assessing the thermogenic potential. These ratios are also advantageous as they are not influenced by the possible inaccuracy in the determination of sample protein concentration.

References

Afzelius BA (1970) Brown adipose tissue: its gross anatomy, histology, and cytology. In: Lindberg O (ed) Brown adipose tissue. Elsevier, New York, pp 1–31

Aherne W, Hull D (1966) Brown adipose tissue and heat production in the newborn infant. J Pathol Bacteriol 91:223–233

Aquila H, Link TA, Klingenberg M (1985) The uncoupling protein from brown fat mitochondria is related to mitochondria ADP/ATP carrier. Analysis of sequence homologies and of folding of the protein in the membrane. EMBO J 4:2369–2367

Bianco AC, Silva JE (1987a) Intracellular conversion of thyroxine to triiodothyronine is required for the optimal thermogenic function of brown adipose tissue. J Clin Invest 79:295–300

Bianco AC, Silva JE (1987b) Optimal response of key enzymes and uncoupling protein to cold in BAT depends on local T_3 generation. Am J Physiol 253:E225–E263

Cannon B, Johansson BW (1980) Nonshivering thermogenesis in the newborn. In: Baum H, Gergely J (eds) Molecular aspects of medicine, vol 3. Pergamon, Oxford, pp 119-223
Cannon B, Nedergaard J (1984) The biochemistry of an inefficient tissue: brown adipose tissue. Essays Biochem 20:110-164
Cannon B, Nedergaard J (1988) Adaptive, "initiating" and acute regulation of the amount and activity of the uncoupling protein thermogenin in brown adipose tissue. In: Azzi A, Nałęcz KA, Nałęcz MJ, Wojtczak L (eds) Anion carriers of mitochondrial membranes. Springer, Berlin Heidelberg New York Tokyo (in press)
Cannon B, Vogel G (1977) The mitochondrial ATPase of brown adipose tissue: purification and comparison with the mitochondrial ATPase of beef heart. FEBS Lett 76:284-289
Cannon B, Nicholls DG, Lindberg O (1973) Purine nucleotides and fatty acids in energy coupling in mitochondria from brown adipose tissue. In: Azzone GF, Ernster L, Papa S, Quagliariello E, Siliprandi N (eds) Mechanisms in bioenergetics. Academic Press, New York London, pp 357-363
Cannon B, Sundin U, Romert L (1977) Palmitoyl CoA: a possible physiological regulator of nucleotide binding to brown adipose tissue mitochondria. FEBS Lett 74:43-46
Cannon B, Hedin A, Nedergaard J (1982) Exclusive occurrence of thermogenin antigen in brown adipose tissue. FEBS Lett 150:129-132
Coombes RC, Rothwell NJ, Shah P, Stock MJ (1987) Changes in thermogenesis and brown fat activity in response to tumour necrosis factor in the rat. Biosci Rep 7:791-799
Desautels M, Zaror-Behrens G, Himms-Hagen J (1978) Increased purine nucleotide binding, altered polypeptide composition, and cold thermogenesis in brown adipose tissue mitochondria of cold-acclimated rats. Can J Biochem 56:378-383
Eiger SM, Kluger MJ (1983) Thermogenesis and fever. In: Girardier L, Stock MJ (eds) Mammalian thermogenesis. Chapman & Hall, London New York, pp 305-320
Fernandez JA, Mampel T, Villaroya F, Inglesias R (1987) Direct assessment of brown adipose tissue as a site of systemic tri-iodothyronine production in the rat. Biochem J 243:281-284
Foster DO, Frydman M (1978) Nonshivering thermogenesis in the rat: II measurements of blood flow with microspheres point to brown adipose tissue as the dominant site of calorigenesis induced by noradrenaline. Physiol Pharmacol 56:110-122
Freeman KB, Patel HV (1984) Biosynthesis of the 32 Kdalton uncoupling protein in brown adipose tissue of developing rabbits. Can J Biochem 62:479-485
Freeman KB, Chien SM, Litchfield D, Patel HV (1983) Synthesis in vitro of rat brown adipose tissue 32000 M_r protein. FEBS Lett 158:325-330
Frumess RD, Larsen PR (1975) Correlation of serum triiodothyronine (T_3) and thyroxine (T_4) with biological effects of thyroid hormone replacement in propylthiouracil-treated rats. Metab Clin Exp 24:547-554
Gaffney EF, Hargreaves HK, Semple E (1983) Hibernoma: distinctive light and electron microscopic features and relationship to brown adipose tissue. Human Pathol 14:677-687
Giralt M, Villarroya F, Mampel T, Iglesias R (1986) Impaired basal and noradrenaline-induced iodothyronine 5′-deiodinase activity in brown adipose tissue from pregnant and lactating rats. Biochem Biophys Res Commun 138:1315-1321
Girardier L, Seydoux J (1986) Neural control of brown adipose tissue. In: Trayhurn P, Nicholls DG (eds) Brown adipose tissue. Arnold, London, pp 122-151
Girardier L, Stock MJ (1983) Mammalian thermogenesis. An introduction. In: Girardier L, Stock MJ (eds) Mammalian thermogenesis. Chapman & Hall, London New York, pp 1-7
Goodbody AE, Trayhurn P (1981) GDP-binding to brown adipose tissue mitochondria of diabetic-obese (db/db) mice. Biochem J 194:1019-1022
Goodbody AE, Trayhurn P (1982) Studies on the activity of brown adipose tissue in suckling pre-obese ob/ob mice. Biochim Biophys Acta 680:119-126
Goswami A, Rosenberg IN (1986) Iodothyronine 5′-deiodinase in brown adipose tissue: thiol activation and propylthiouracil inhibition. Endocrinology 119:916-923
Gribskov CL, Henningfield MF, Swick AG, Swick RW (1986) Evidence for unmasking of rat brown-adipose tissue mitochondrial GDP-binding sites in response to acute cold exposure. Biochem J 233:743-747
Harris WH, Foster DO, Nadean BE (1984) Evidence for a contribution by brown adipose tissue to the development of fever in the young rabbit. Can J Physiol 63:595-598

Heaton GM, Wagenvoord RJ, Kemp A, Nicholls DG (1978) Brown adipose tissue mitochondria: photoaffinity labelling of the regulatory site for energy dissipation. Eur J Biochem 82:515–521
Heim T (1971) Thermogenesis in newborn infant. Clinical obstetrics and gynecology. Perinat Biol 14:790–820
Heim T, Hull D (1966) The blood flow and oxygen consumption of brown adipose tissue in the newborn rabbit. J Physiol 186:42–55
Hey E (1974) Heat loss from animals and man. Butterworth, London
Himms-Hagen J (1983) Thyroid hormone and thermogenesis. In: Girardier L, Stock MJ (eds) Mammalian thermogenesis. Chapman & Hall, London, pp 141–177
Himms-Hagen J (1986) Brown adipose tissue and cold acclimation. In: Trayhurn P, Nicholls DG (eds) Brown adipose tissue. Arnold, London, pp 214–268
Himms-Hagen J, Desautels M (1978) A mitochondrial defect in brown adipose tissue of the obese (ob/ob) mouse: reduced binding of purine nucleotides and a failure to respond to cold by an increased binding. Biochim Biophys Res Commun 83:628–633
Himms-Hagen J, Gwilliam C (1980) Abnormal brown adipose tissue in hamsters with muscular dystrophy. Am J Physiol 239:G18–G22
Holub M, Houštěk J, Janiková D, Rychter Z, Větvička V, Vrána A, Kazdová L (1988) Influence of ambient temperature on nude mouse metabolic and immune status. In: Bing-quan WU, Zhang Jie (eds) Immune-deficient animals in biomedical research. 6th Int Worksh Immune-deprivated animals, Beijing 1988. Karger, Basel (in press)
Houštěk J, Drahota Z (1977) Purification and properties of mitochondrial ATPase of hamster brown adipose tissue. Biochim Biophys Acta 484:127–139
Houštěk J, Kopecký J, Drahota Z (1978) Specific properties of brown adipose tissue mitochondrial membrane. Comp Biochem Physiol 608:209–214
Houštěk J, Ježek P, Kopecký J (1987) Methods for studying the structure and function of the mitochondrial uncoupling protein. In: Bertoli E, Chapman D, Cambria A, Scapagnini U (eds) Biomembrane and receptor mechanisms. Fidia research series, vol 7. Liviana, Padua, pp 176–191
Houštěk J, Janiková D, Bednár J, Kopecký J, Soukup T (1988a) Postnatal formation of thermogenic mitochondria in hamster brown fat. (in preparation)
Houštěk J, Kopecký J, Rychter Z, Soukup T (1988b) Uncoupling protein in embryonic brown adipose tissue – existence of nonthermogenic and thermogenic mitochondria. Biochim Biophys Acta (in press)
Hull D (1977) Brown adipose tissue and the newborn infant's response to cold. In: Phillip EE, Barnes J, Newton M (eds) Scientific foundations of obstetrics and gynecology, 2nd edn. Heinemann, London, pp 540–550
Iglesias R, Fernandez JA, Mampel T, Obregon MJ, Villaroya F (1987) Iodothyronine 5′-deiodinase activity in rat brown adipose tissue during development. Biochim Biophys Acta 923:233–240
Jacobsson A, Stadler U, Glotzer MA, Kozak LP (1985) Mitochondrial uncoupling protein from mouse brown fat. Molecular cloning, genetic mapping, and mRNA expression. J Biol Chem 260:16250–16254
Jacobsson A, Nedergaard J, Cannon B (1986) α- and β-adrenergic control of thermogenin mRNA expression in brown adipose tissue. Biosci Rep 6:621–631
Ježek P, Houštěk J (1986) Palmitoyl-CoA is not a physiological regulator of the uncoupling protein of brown adipose tissue mitochondria. EBEC Rep 4:388
Jones R, Henschen L, Mohell N, Nedergaard J (1986) Requirement of gene transcription and protein synthesis for cold- and norepinephrine-induced stimulation of thyroxine deiodinase in rat brown adipose tissue. Biochim Biophys Acta 889:366–373
Kaplan MM, Shaw EA (1984) Type II iodothyronine 5′-deiodination by human and rat placenta in vitro. J Clin Endocrinol Metab 59:253–257
Kates AL, Himms-Hagen J (1985) Defective cold-induced stimulation of thyroxine 5′-deiodinase in brown adipose tissue of the genetically obese (ob/ob) mouse. Biochem Biophys Res Commun 130:188–193
Klingenberg M (1988) Nucleotide binding to uncoupling protein. Mechanism of control by protonation. Biochemistry 27:781–791
Knehans AW, Romsos DR (1982) Reduced norepinephrine turnover in brown adipose tissue of ob/ob mice. Am J Physiol 242:253–261

Kopecký J, Guerrieri F, Ježek P, Drahota Z, Houštěk J (1984) Molecular mechanism of uncoupling in brown adipose tissue: the noneidentity of proton and chloride conducting pathways. FEBS Lett 170:186–190
Kopecký J, Sigurdson L, Park IRA, Himms-Hagen J (1986a) Thyroxine 5′-deiodinase in hamster and rat brown adipose tissue: effect of cold and diet. Am J Physiol 251:E1–E7
Kopecký J, Sigurdson L, Park IRA, Himms-Hagen J (1986b) Thyroxine 5′-deiodinase in brown adipose tissue of myopatic hamsters. Am J Physiol 251:E8–E13
Kopecký J, Ježek P, Drahota Z, Houštěk J (1987) Control of uncoupling protein in brown-fat mitochondria by purine nucleotides. Eur J Biochem 164:687–694
LaNoue KF, Strzelecki D, Koch C (1986) Regulation of the uncoupling protein in brown adipose tissue. J Biol Chem 261:298–305
LaNoue KF, Koch CD, Meditz RB (1982) Mechanism of action of norepinephrine in hamster brown adipocytes. J Biol Chem 257:13740–13748
Larsen PR, Frumess RD (1977) Comparison of the biological effects of thyronine and triiodothyronine in the rat. Endocrinology 100:980–988
Larsen PR, Silva JE, Kaplan MM (1981) Relationships between circulating and intracellular thyroid hormones: physiological and clinical implications. Endocrinol Rev 2:87–102
Lean MEJ, James WPT (1986) Brown adipose tissue in man. In: Trayhurn P, Nicholls DG (eds) Brown adipose tissue. Arnold, London, pp 339–365
Lean MEJ, Trayhurn P, Murgatroyd PR, Dixon AK (1986a) The case for brown adipose tissue function in humans: biochemistry, physiology and computed tomography. In: Berry EM, Bloudheim SH, Eilahon ME, Shafrir E (eds) Recent advances in obesity research, vol 5. Libbey, London Paris, pp 109–116
Lean MEJ, James WPT, Jennings G, Trayhurn P (1986b) Brown adipose tissue in patients with phaeochromocytoma. Int J Obesity 10:219–227
Leonard JL, Mellen SA, Larsen PR (1982) Thyroxine 5′-deiodinase activity in brown adipose tissue. Endocrinology 112:1153–1155
Lin CS, Klingenberg M (1980) Isolation of the uncoupling protein from brown adipose tissue mitochondria. FEBS Lett 113:299–303
Lin CS, Klingenberg M (1982) Characteristics of the isolated purine nucleotide binding protein from brown rat mitochondria. Biochemistry 21:2950–2956
Locke RM, Rial E, Scott ID, Nicholls DG (1982a) Fatty acids as acute regulators of the proton conductance of hamster brown fat mitochondria. Eur J Biochem 129:373–380
Locke RM, Rial E, Nicholls DG (1982b) The acute regulation of proton conductance in cell and mitochondria from the brown fat of cold- and warm-adapted guinea-pigs. Eur J Biochem 129:381–387
Mercer SW, Trayhurn P (1983) Developmental changes in fatty acid synthesis in interscapular adipose tissue of lean and genetically obese (ob/ob) mouse. Biochem J 212:393–398
Naeye RL (1974a) Hypoxaemia and the sudden infant death syndrome. Science 186:837–839
Naeye RL (1974b) Evidence of antecedent hypoxia and hypoxaemia. In: Robinson RR (ed) FE Camps Symp SIDS, Toronto, pp 1–6
Nedergaard J, Cannon B (1987) Apparent unmasking of [H^3]GDP binding in rat brown-fat mitochondria is due to mitochondrial swelling. Am J Biochem 164:681–686
Nedergaard J, Connolly E, Cannon B (1986) Brown adipose tissue in the mammalian neonate. In: Trayhurn P, Nicholls DG (eds) Brown adipose tissue. Arnold, London, pp 152–213
Nicholls DG (1976) The bioenergetics of brown adipose tissue mitochondria. FEBS Lett 61:103–110
Nicholls DG, Lindberg O (1973) Brown adipose tissue mitochondria: the influence of albumin and nucleotides on passive ion permeabilities. Eur J Biochem 37:523–530
Nicholls DG, Locke RM (1984) Thermogenic mechanism in brown fat. Physiol Rev 64:1–64
Nicholls DG, Cunningham SA, Rial E (1986) The physiological mechanisms of brown adipose tissue thermogenesis. In: Trayhurn P, Nicholls DG (eds) Brown adipose tissue. Arnold, London, pp 52–85
Norman PT, Flatmark T (1984) Increase in mitochondrial content of long chain acyl CoA in brown adipose tissue during cold acclimation. Biochim Biophys Acta 794:225–233
Oppenheimer JH, Samuels HM (1983) Molecular basis of thyroid hormone action. Academic Press, New York London
Peachey T, French RR, York DA (1988) Regulation of GDP binding and uncoupling protein concentration in brown fat mitochondria. Biochem J 249:451–457

Perrin DG, Cutz E, Becker LE, Bryan AC, Madapallimatum A, Sole MJ (1984) Sudden infant death syndrome: increased carotic-body dopamine and adrenaline content. Lancet ii:535–537
Polk DH, Callegari CC, Newnham J, Padbury JF, Reviczky A, Fischer DA, Klein AH (1987) Effect of fetal thyroidectomy on newborn thermogenesis in lamb. Pediat Res 21:453–457
Rial E, Nicholls DG (1986) Chemical modification of the brown-fat-mitochondria uncoupling protein with tetranitromethane and N-ethylmaleimide: a cysteine residue is implicated in the nucleotide regulation of anion permeability. Eur J Biochem 161:689–694
Rial E, Nicholls DG (1988) On the mechanism of transport by the uncoupling protein from brown adipose tissue mitochondria. In: Azzi A, Nałęcz KA, Nałęcz MJ, Wojtczak L (eds) Anion Carriers of Mitochondrial Membranes Mitochondrial transport proteins. Springer (in press)
Rial E, Poustie A, Nicholls DG (1983) Brown adipose tissue mitochondria: the regulation of the 32 000 Mr uncoupling protein by fatty acids and purine nucleotides. Eur J Biochem 137:197–203
Ridley RG, Patel HV, Parfett CLJ, Olynyk KA, Reichling S, Freeman KB (1986) Immunological detection of cDNA clones encoding the uncoupling protein of brown adipose tissue: evidence for an antigenic determinant within c-terminal eleven amino acids. Biosci Rep 6:87–94
Ricquier D, Bouillaud F (1986) The brown adipose tissue mitochondrial uncoupling protein. In: Trayhurn P, Nicholls DG (eds) Brown adipose tissue. Arnold, London, pp 86–104
Ricquier D, Mory G (1984) Factors affecting brown adipose tissue activity in animals and man. Endocrinol Metab 13:501–520
Ricquier D, Barlet JP, Garel JM, Combes-George M, Dubois MP (1983a) An immunological study of the uncoupling protein of brown adipose tissue mitochondria. Biochem J 210:859–866
Ricquier D, Mory G, Nechad M, Combes-George M, Thibauld J (1983b) Development and activation of brown fat in rats with pheochromocytoma PC 12 tumours. Am J Physiol 245:172–177
Rink RD (1969) Oxygen consumption, body temperature and brown adipose tissue in the postnatal golden hamster (*Mesocricetus auratus*). J Exp Zool 170:117–123
Rothwell NJ, Stock MJ (1986) Brown adipose tissue and diet-induced thermogenesis. In: Trayhurn P, Nicholls DG (eds) Brown adipose tissue. Arnold, London, pp 269–298
Schwartz HL, Surks MI, Oppenheimer JH (1971) Quantitation of extrathyroidal conversion of L-thyroxine to 3,5,3-triiodo-L-thyronine in the rat. J Clin Invest 50:1124–1130
Silva JE, Larsen PR (1983) Adrenergic activation of triiodothyronine production in brown adipose tissue. Nature (London) 305:712–713
Silva JE, Larsen PR (1985) Potential of brown adipose tissue type II thyroxine 5′-deiodinase as a local and systemic source of triiodothyronine in rats. J Clin Invest 76:2296–2305
Silva JE, Larsen PR (1986) Hormonal regulation of iodothyronine 5′-deiodinase in rat brown adipose tissue. Am J Physiol 251:639–643
Silva JE, Matthews PG (1984) Thyroid hormone metabolism and source of plasma triiodothyronine in 2-week-old rats: effects of thyroid status. Thyroid hormone metabolism in neonatal period. Endocrinology 115:2394–2405
Silva JE, Leonard JL, Crantz FR, Larsen PR (1982) Evidence for two tissue specific pathways for in vivo thyroxine 5′-deiodination in the rat. J Clin Invest 69:1176–1184
Silva JE, Gordon MB, Crantz FR, Leonard JL, Larsen PR (1984) Qualitative and quantitative differences in the pathways of extrathyroidal T_3 generation between euthyroid and hypothyroid rats. J Clin Invest 73:898–907
Smith RE (1961) Thermogenic activity of hibernating gland in the cold acclimated rat. Physiologist 4:113–120
Stanton AN (1984) Overheating and cot death. Lancet ii:1199–1201
Stefanski SA, Elwell MR, Yoshitomi K (1987) Malignant hibernoma in a Fischer 344 rat. Lab Anim Sci 37:347–350
Sundin U, Herron D, Cannon B (1981) Brown fat thermoregulation in developing hamsters (*Mesocricetus auratus*): a GDP-binding study. Biol Neonate 39:141–149
Tanaka K, Murakami M, Greer MA (1986) Type II thyroxine 5′-deiodinase is present in the rat pineal gland. Biochem Biophys Res Commun 137:863–868
Teplitz C, Lim YC (1974) The diagnostic significance of diffuse brown adipose tissue (BAT) transformation on adult humans periadrenal fat: a morphological indicator of severe chronic hypoxaemia. Lab Invest 30:390
Trayhurn P (1986) Brown adipose tissue and energy balance. In: Trayhurn P, Nicholls DG (eds) Brown adipose tissue. Arnold, London, pp 299–338

Vignais PV, Block MR, Boulay F, Brandolin F, Lauquin GJ-M (1985) Molecular aspects of structure-function relationship in mitochondrial adenine nucleotide carrier. In: Benga G (ed) Structure and properties of cell membranes, vol 2. CRC, Boca Raton, pp 139–179

Visser TJ, Leonard JL, Kaplan MM, Larsen PR (1982) Kinetic evidence suggesting two mechanisms for iodothyronine 5′-deiodination in rat cerebral cortex. Proc Natl Acad Sci USA 79:5080–5084

Visser TJ, Kaplan MM, Leonard JL, Larsen PR (1983) Evidence for two pathways of iodothyronine 5′-deiodination in rat pituitary that differ in kinetics, propylthiouracil sensitivity and response to hypothyroidism. J Clin Invest 71:992–1002

Walker JE, Dyer MR, Fearnley IM, Lutter R, Medd SM, Pilkington SJ, Runswick MJ, Todd RJ (1988) Molecular genetics of the ATP synthase and metabolite transport from mammalian mitochondria. EBEC Rep 5:27

Molecular Composition, Fluidity of Membranes and Functional Properties of Human Liver Mitochondria and Microsomes

GH. BENGA

1 Introduction

Liver is a particularly useful tissue for the characterization of subcellular organelles since the purified fractions prepared from liver are derived mainly from a single cell type, the hepatocyte. This is not only because hepatocytes constitute 78% of the volume of liver (Blouin et al. 1977) but also because they are the major cell type broken by homogenization of the tissue (Dallner et al. 1966).

A great deal is known about the composition and functional properties of subcellular fractions of rat liver (see Rouser et al. 1968, for review). However, the structure of membranes, in particular the association between lipids and proteins, is a matter of considerable discussion (Benga and Holmes 1984; Benga 1985). The significance of the fatty acid composition of lipids is poorly understood. The analysis of the composition of mitochondria from different sources, combined with the knowledge of their properties, should lead to a better understanding of the interaction between lipids and proteins in membranes, particularly in regard to their functional significance.

On the other hand, although it was considered that in humans the liver is the best enzymologically studied organ, investigations have been performed mainly on homogenates (Schmidt and Schmidt 1970) and have given conflicting results. Studies on subcellular fractions of human liver therefore represent an advancement in this field. The study of human liver mitochondria, for example, is important both for better knowledge of the metabolic properties of human liver and to illustrate differences in structure and function between liver mitochondria under normal and pathological conditions (cholecystitis, cirrhosis, jaundice, or diseases in which the general metabolism is deranged such as diabetes and hyperlipidemia).

To begin the study of human liver mitochondria we first had to determine the optimal conditions for the isolation and study of enzymic properties (Benga et al. 1971, 1972). In subsequent studies many structural and functional peculiarities of mitochondria and microsomes were found (Benga 1974, 1975; Benga and Muresan 1974; Benga and Borza 1975; Benga et al. 1978, 1979, 1983). This review is aimed to describe the most interesting features of human liver subcellular membranes uncovered by our studies, to present some previously unpublished results and to discuss the significance of our findings for some human diseases.

Department of Cell Biology, Faculty of Medicine, Medical and Pharmaceutical Institute Cluj-Napoca, and Institute of Hygiene and Public Health, 6 Pasteur St., 3400 Cluj-Napoca, Rumania

A. Azzi, Z. Drahota and S. Papa
Molecular Basis of Membrane-Associated Diseases

2 Lipid Composition of Human Liver Mitochondria and Microsomes

The main characteristic of human liver mitochondria proved to be their lipid composition (Benga et al. 1978). This is illustrated in Table 1, which shows the total lipid content as well as the proportion of different lipid classes present in human and rat liver mitochondria. It can be seen that human liver mitochondria contain twice as much lipid as those from the rat (the difference was statistically highly significant, $p > 0.001$, Student's unpaired t-test). This increased lipid content of human liver mitochondria is not associated with an increase in content of a particular lipid class. On the contrary, the proportions between different lipid and phospholipid classes are similar in both human and rat liver mitochondria. Likewise the human liver microsomes contain almost two-fold more lipid compared to those from the rat ($p < 0.001$, Student's unpaired t-test), with almost identical proportions of different lipid and phospholipid classes.

The fatty acid composition of the total lipids as well as that of the phospholipids appear to be different in human compared to rat liver mitochondria and microsomes (Table 2). The differences were related to the essential fatty acids; it was thus interesting to note that the total lipids of human liver mitochondria and microsomes contain more linoleic acid and less arachidonic acid than those of the rat. These differences appeared in the phospholipid fraction of mitochondrial lipids and not in triglyceride and free fatty acid fractions. As the phospholipids are found predominantly in cell membranes, it is clear that the lipid composition of mitochondrial membranes in human and rat livers are different. The same pattern of fatty acid composition was observed in the major phospholipid classes, that is, more linoleic acid and less arachidonic acid in human compared to rat liver mitochondria.

The fatty acid composition of mitochondrial phospholipids is a challenging problem. Marked variations in the fatty acid composition of mitochondria from different species have been reported (Rouser et al. 1968). A comparison of the fatty acid composition of liver mitochondria obtained from different species gives an interesting picture. Chicken liver mitochondria contain more oleic acid and less stearic and arachidonic acids than rat liver mitochondria. Fish liver mitochondria contain very little linoleic, linolenic acid, and arachidonic acids, however, they have a high amount of unsaturation because they contain substantial amounts of very long-chain polyenoic acyl residues (20:5, 22:5, 22:6). Although dietary differences may be responsible for some of the species variation in fatty acid composition, it is clear that mitochondria may have widely different fatty acid compositions (Rouser et al. 1968). Richardson et al. (1961, 1962) showed that fish-eating birds retain linoleic and arachidonic acids in mitochondria, although acids of the linolenate family were also found, demonstrating that the diet only influences to some extent the fatty acid composition of mitochondria.

Human liver mitochondria, with a high linoleate and a low arachidonate content, appear to be another interesting case of species variation in fatty acid composition. It is noteworthy that such a pattern of mitochondrial fatty acids has not been reported for any other species. Moreover, when dietary alterations of the mitochondrial fatty acid composition are induced in the rat, the linoleate and arachidonate increase or decrease in parallel (Hayashida and Portman 1960; Stancliff et al. 1969). The pattern of fatty acid composition of human liver mitochondria found by us does not seem to

Table 1. Lipid composition of liver mitochondria and microsomes[a]

Lipid	Mitochondria				Microsomes			
	(μg/mg protein)		(% of total lipid)		(μg/mg protein)		(% of total lipid)	
	Human	Rat	Human	Rat	Human	Rat	Human	Rat
Total lipids	401.0	205.3	100	100	1032	610		
Phospholipids	344.0	180.0	86.0	88.0			77.2	81.9
Phosphatidylcholine			40.5	41.5			47.8	51.3
Phosphatidyl-ethanolamine			31.8	32.7			24.0	24.1
Cardiolipin			14.0	14.2			—	—
Phosphatidylinositol			6.2	5.6			11.6	10.8
Phosphatidylserine			4.0	3.9			5.7	4.4
Sphingomyelin			2.4	2.1			4.8	3.5
Phosphatidic acid			1.0	0.9			0.7	1.2
Triglycerides	20.7	12.4	5.2	6.0			8.6	8.2
Free fatty acids	24.0	16.0	6.0	7.9			3.9	2.9
Cholesterol	11.2	5.4	2.9	2.6			2.5	2.0

Source: Data for mitochondria from Benga et al. (1978) and for microsomes from Benga et al. (1983).
[a]The results are the mean ± SEM for 8–30 determinations.

Table 2. Fatty acid composition of rat and human liver mitochondrial and microsomal membranes[a]

Fatty	Mitochondria				Microsomes			
	Total	Lipids	Phospholipids		Total	Lipids	Phospholipids	
acid[b]	Human	Rat	Human	Rat	Human	Rat	Human	Rat
14:0	2.2	1.8	1.4	1.1	1.6	0.7	1.4	0.5
16:0	19.2	18.0	21.2	17.7	23.3	22.9	29.1	19.1
16:1	3.8	3.0	2.9	1.9	2.4	3.6	2.0	3.8
18:0	15.0	17.0	14.2	14.0	12.4	17.9	27.7	20.0
18:1	17.2	18.1	16.0	16.6	14.7	13.7	15.0	13.2
18:2	35.6	26.2	38.1	25.7	36.4	26.5	22.1	28.6
20:4	3.6	12.0	4.0	14.5	6.8	13.8	3.4	12.9

Source: Data for mitochondria from Benga et al. (1978) and for microsomes from Benga et al. (1983).

[a] The results are expressed in relative percentages of the sum of fatty acids and represent the mean of 9–12 determinations for total lipids and 4–8 determinations for phospholipids.

[b] Fatty acids are designated as follows: 14:0 myristate, 16:0 palmitate, 16:1 palmitoleate, 18:0 stearate, 18:1 oleate, 18:2 linoleate, 20:4 arachidonate.

be related to the diet of the patients investigated; the determinations were performed over a period of several years on subjects with no special diets or alimentary habits. The variations among different subjects were rather small, that is, of the same order of magnitude as the variations for the determinations on rats fed a standard laboratory diet.

Taking into account the known roles of endoplasmic reticulum in lipid metabolism and membrane biogenesis, it was interesting to study the fatty acid composition of microsomes. This appeared also to be different in human compared to rat liver microsomes (Table 2). The total lipids of human liver microsomes contain less stearic acid, arachidonic acid and 4-, 7-, 10-, 13-, 16-, 19-docosahexaenoic acid and more linoleic acid than those of the rat. In phospholipids the differences between rat and human liver microsomes were statistically significant only for arachidonic acid and 4-, 7-, 10-, 13-, 16-, 19-docosahexaenoic acid. In addition, a higher amount of palmitic acid was noted in phospholipids of human liver compared to rat liver microsomes.

The degree of saturation of the lipids was quantitated by calculating the unsaturation index and the ratio of unsaturated fatty acids to saturated fatty acids in the particular lipid fraction. The unsaturation index of total lipids and of various phospholipid classes was greater in the rat than in the human liver mitochondrial and microsomal membranes (Table 3). This means that human liver membranes contain more saturated lipids than those of the rat.

The pattern of fatty acid composition and particularly the higher linoleate and lower arachidonate in human liver subcellular membranes suggest peculiarities in the phospholipid metabolism of human liver. It is known that there are two pathways of phospholipid biosynthesis (Van den Bosch 1974) using linoleic acid. It can be incorporated in phospholipid by de novo synthesis. On the other hand, it can be transformed into arachidonic acid (by fatty acid elongation), which could be then utilized for the acylation of lysophospholipids. It is possible that in human liver the fatty acid elongation and acylation of lysophospholipids are less active relative to de novo synthesis.

Table 3. Unsaturation of fatty acids in liver mitochondria and microsomal membranes

Lipid	Source of membranes	Unsaturation index[a]	
		Mitochondria	Microsomes
Total lipids	human	106.6	117.1
	rat	121.5	132.1
Phospholipids	human	111.1	74.8
	rat	152.7	132.2
Phosphatidyl-	human	114.1	
choline	rat	125.3	
Phosphatidyl	human	105.3	
ethanolamine	rat	151.0	
Cardiolipin	human	154.3	
	rat	181.9	

Source: Data for mitochondria from Benga et al. (1978) and for microsomes from Benga et al. (1983).

[a]Unsaturation index = $\sum_{a=1}^{k}$ [(number of double bonds in a) × (wt% occurence of a)] for each fatty acid in a group of k fatty acids.

3 Amino Acid Composition of Mitochondrial Membranes

Comparative studies of the amino acid composition of membrane proteins vs non-membrane proteins have been made from the composition data available on purified proteins (Vanderkooi and Capaldi 1972; Von Heijne 1981). The purpose of such estimations is to contribute to the understanding of the forces involved in protein-lipid interactions in biological membranes.

A comparison of the average amino acid composition of the membrane proteins from human and rat liver mitochondria has been made by calculating the number of residues of each amino acid (n_i) for a hypothetical molecular weight of 100 000 in Table 4. It is immediately obvious that for threonine, valine, methionine, isoleucine, tyrosine and histidine there is no significant difference between mitochondrial membranes of the two species. Whereas the human membrane contains more arginine, the rat membrane contains more lysine and these two changes may cancel each other more or less as far as the net charge of the protein is concerned. The negative values in column 6 indicate that the human membranes contain less (aspartic acid plus asparagine), serine (glutamic acid plus glutamine) and phenylalanine, all of which except phenylalanine are hydrophilic amino acid residues.

Table 4. Comparison of amino acid compositions of human and rat liver mitochondrial protein[a]

Amino acid	Human		Rat		Human minus rat
	n_i	SE n_i	n_i	SE n_i	
Aspartic acid + asparagine	77.4	0.4	79.5	0.2	−2.1
Threonine	50.5	0.2	51.2	0.9	N.S.
Serine	58.2	0.5	60.0	1.1	−1.8
Glutamic acid + glutamine	88.8	0.4	90.7	0.4	−1.9
Proline	45.2	0.4	43.5	0.7	+1.7
Glycine	78.4	0.2	74.3	0.2	+4.1
Alanine	79.4	0.3	76.6	0.2	+2.8
Valine	67.3	0.1	67.0	0.5	N.S.
Methionine	23.4	0.5	23.8	0.2	N.S.
Isoleucine	51.7	0.6	51.6	0.6	N.S.
Leucine	89.6	0.4	88.0	0.5	+1.6
Tyrosine	28.9	0.3	28.9	0.8	N.S.
Phenylalanine	39.2	0.3	40.6	0.2	−1.4
Lysine	56.6	0.3	58.5	0.3	−1.9
Histidine	21.0	0.1	20.8	0.4	N.S.
Arginine	45.5	0.2	44.2	0.5	+1.3
1/2 Cystine	13.0	0.1	10.3	0.1	+2.7

Source: Data from Benga and Ferdinand (1977) *Int. J. Biochem.*

[a] Values are expressed as numbers of residues (n_i) for a hypoythetical mol. wt. of 100 000 excluding ethanolamine. The standard error of n_i (SE n_i) is scaled from the SEM values reported by Benga and Ferdinand (1977). In the extreme right column in difference between human and rat is given where statistically significant (N.S. = not significantly different).

The consequence of changes in the dicarboxylic acid plus amide content is hard to assess because the change may occur entirely in the amide content, in which case the net charge of the membranes would not differ; or entirely in the dicarboxylic acid content in which case the rat membrane protein would be more negatively charged; or the change may comprise alterations to both. In all the other cases the value given in column 6 of Table 4 is positive, that is, human membrane contains more of the amino acids of a hydrophobic type.

The nature of protein-lipid interactions in biological membranes is of great importance to the understanding of the membrane structure and function. According to the fluid mosaic model (Singer and Nicolson 1972), the proteins can be either loosely associated, probably through electrostatic bonds, with the polar head groups of the bilayer, or more strongly associated, probably by interacting with the hydrocarbon interior of the bilayer (the "peripheral" and "integral" membrane proteins, respectively). One might expect the "integral" membrane proteins to possess a higher proportion of hydrophobic amino acids than non-membrane proteins, thus allowing their hydrophobic interaction with the lipid bilayer.

Following sonication and osmotic shock, used in our procedure for preparation of mitochondrial membranes (Benga and Ferdinand 1977), it is probable that most of the loosely associated proteins are removed. As mentioned above our data clearly indicated a higher lipid and phospholipid content of human, in comparison with rat, liver mitochondrial membranes. In relation to their lipid composition the comparison of the amino acid compositions of rat and human mitochondrial membranes is interesting. The two sets of results bear a striking resemblance when the mole fraction of each amino acid is compared. On the whole it is clear that the membrane protein in the human liver mitochondria is more "hydrophobic" than its counterpart in the rat, although the differences (statistically significant) are really quite small, as might be expected for such specialized structures as mitochondrial membrane proteins. We have also to consider that the hydrophobic amino acid residues in membrane proteins reside in transbilayer helices (see Engelman et al. 1986 for review) rather than being distributed over the complete sequence. However, it appears that even working with whole mitochondrial membrane preparations, where there is a high lipid content (as in human liver mitochondria), there is an increased content of hydrophobic amino acid residues in membrane proteins. This finding supports the importance of hydrophobic interactions between proteins and lipids in human mitochondrial membranes.

4 Fluidity of Membranes as Probed by Spin Labeling

In order to evaluate the consequences of the particular composition of human subcellular membranes on their fluidity, the spin labeling method (see Benga 1983, for review) has been used. In the case of mitochondrial membranes the spin labels were the N-oxyl-4,4′-dimethyloxazoline derivatives of 5- and 16-ketostearic acid (they will be referred to as 5 NS and 16 NS). Since the spectra of 5 NS in mitochondrial suspensions were indicative of strong impedance of spin label motion, the splitting between the low and high field extrema ($2T_{\parallel}$) was used as an index for comparing degrees of mobility (Benga et al. 1978). A greater value of $2T_{//}$ reflects more restricted rotational motion, therefore indicating a more viscous environment around the spin probe (McConnell and McFarland 1970).

The rotational correlation time (τ_c) is a defined motion parameter which is reasonably accurate for isotropic motion in the fast tumbling region (Benga 1983) and can be expressed as:

$$\tau_c = 6.5 \times 10^{-10} W_0(\sqrt{h_0/h_1} - 1) \text{ s},$$

where W_0 (in Gauss) is the width of the mid-field line and h_0 and h_1 are the heights of the mid-field and high-field lines of the first derivative spectrum (Benga et al. 1978). A large value of τ_c indicates a rigid microenvironment around the spin label. The rotational correlation time has been used as a motion parameter for the spectra of 16 NS in mitochondrial suspensions.

It is obvious from Table 5 that at all temperatures, the mobility of 5 NS (evaluated by $2T_{//}$) as well as that of 16 NS (evaluated by τ_c) is higher in liver mitochondria from the human than in those of the rat. This indicates a higher fluidity of lipid in human mitochondrial membranes.

Table 5. Comparison of the mobility of spin labeled fatty acid derivatives in human and rat liver mitochondrial membranes

Spin label	ESR parameter	Source of mitochondria	Temperature (°C)							
			5	10	15	20	25	30	35	40
5 NS	$2T_{//}$	human	60.9	59.8	58.8	57.8	56.9	53.7	51.2	48.0
		rat	61.6	60.9	60.0	59.1	58.7	56.2	54.3	51.8
16 NS	τ_c	human	27.5	26.2	18.7	16.9	15.6	14.0	11.6	00.0
		rat	36.2	34.3	28.7	24.3	20.0	17.4	15.6	14.0

Note: Preparation and labeling of membranes and ESR measurements were performed as described in Benga et al. (1978).

In the case of microsomes the spin label TEMPO (2,2,6,6, tetramethylpiperidine-l-oxyl) was used to evaluate the fluidity of membranes. A typical ESR spectrum of TEMPO in a suspension of human liver microsomes (Fig. 1) is a superposition of two spectra: one is due to TEMPO dissolved in the fluid, hydrophobic region of membrane bilayer, the other is due to TEMPO in the aqueous phase. To a first approximation, H is proportional to the amount of spin label dissolved in the membrane bilayer and P is proportional to the amount dissolved in the aqueous region. The TEMPO spectral parameter, f, equal to H/(H + P) (see notations in Fig. 1), is thus equal to the fraction of TEMPO dissolved in the fluid lipid phase of the sample at any one temperature (Shimshick and McConnell 1973a, b).

The plots of f versus temperature showed discontinuities at the two characteristic temperatures, t_h and t_l, described by Linden et al. (1973). Figure 2 shows examples of curves obtained with rat and human liver microsomes, respectively. The lower temperature, t_l, at which a discontinuity in the partitioning plot of TEMPO occurred, was not always observed, probably due to the insensitivity of the method (Kleemann

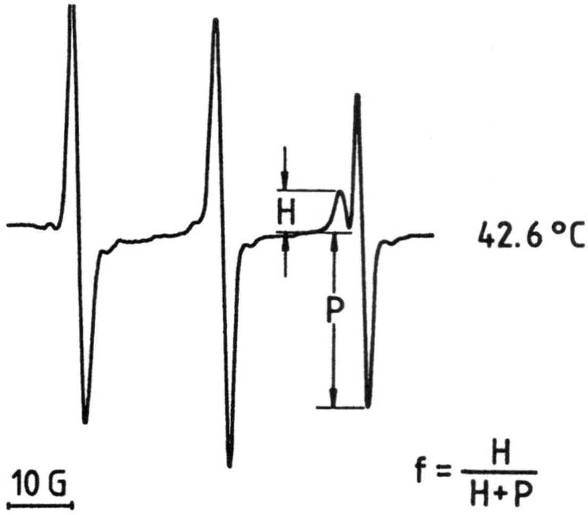

Fig. 1. A typical ESR spectrum of TEMPO in human liver microsomes

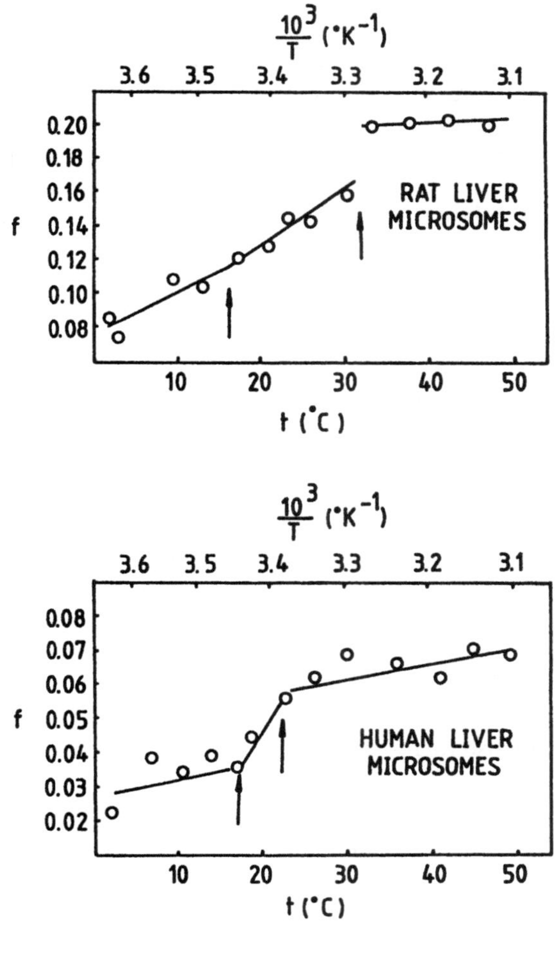

Fig. 2. Plots of the TEMPO spectral parameter *f* as a function of temperature in rat and human liver microsomes. The protein concentration of the microsomal suspension was 20 mg/ml for human liver and 80 mg/ml for rat liver

and McConnell 1974). However, the higher temperature, t_h, was always noticeable. It was obvious that in the case of human liver microsomes the discontinuity in the partitioning plot occurs at lower temperatures than in the case of rat liver microsomes. There is an almost 10° C difference in t_h for human compared to rat liver microsomes (21.6 ± 2.1 and 31.2 ± 3°C, respectively; $p < 0.001$). This temperature was considered to be that at which the formation of solid patches of membrane lipids is first detected (Kleemann and McConnell 1974) and it was decreased in the membranes of *Escherichia coli* grown in a medium supplemented with linoleic acid or oleic acid, compared to cells grown in a medium supplemented with elaidic acid. These studies indicate that the more unsaturated the fatty acid, the lower is the t_h. The membrane with more unsaturated fatty acids is more fluid. On a similar line of thinking, the lower t_h in human liver microsomes would indicate a higher fluidity of these membranes compared to the rat liver microsomes.

It appears that both mitochondrial and microsomal membranes in human liver have a higher fluidity than their counterparts in the rat. There are three main factors that control the membrane fluidity: the chain length and unsaturation of fatty acids, the cholesterol and the proteins. We have seen that the unsaturation of fatty acids in human liver membranes is not higher, but even smaller than in rat liver membranes and the cholesterol content is not different between the two. On the other hand, the ratio of lipid to protein is double in human compared to rat subcellular membranes. Therefore, we interpret the fluidity of human membranes as being due to a lesser immobilization of lipids by proteins in comparison with rat liver membranes. It might be possible that the higher lipid to protein ratio in human membranes compared to rat membranes is reflective of a higher proportion of the surface area occupied by the lipid bilayer relative to the surface area occupied by the integral membrane proteins.

5 Relation Between Membrane Molecular Composition and Some Enzymic Activities of Mitochondria

The correct fluidity of cell membranes is of great importance for the required rate of diffusional and metabolic processes of the tissue. It has been suggested that a possible function of the fatty-acid chain distribution in membrane lipids is to match membrane fluidity, environmental temperatures and the rate of membrane processes (see Benga 1985, for reviews). However, the same fluidity may be obtained by different combinations of chain length and unsaturation of fatty acids in phospholipids. Therefore, it is possible that a certain fatty acid composition of membrane lipids (species specificity) is required for a specific interaction between the paraffinic chains of lipids and the hydrophobic regions of membrane proteins (enzymes).

Taking into account the molecular composition of human liver mitochondrial membranes, it seems reasonable that they might exhibit certain functional properties and also might share some functional features with other lipid-rich mitochondria. This appeared to be the case indeed. The use of defatted serum albumin in the homogenization and washing medium proved to be essential in order to obtain maximum respiratory rates and high values of the acceptor control index (Benga et al. 1972). This is similar to the effects of albumin on hamster brown adipose tissue mitochondria (Nicholls 1974). The concentration of EDTA required to yield a maximal P/O ratio was found to depend on the Mg^{2+} concentration and as for rat brown adipose tissue mitochondria (Grav et al. 1970), it must be nearly equal to that of Mg^{2+} (Fig. 3). It was also interesting to find that albumin and also α-oxoglutarate stimulated adenine nucleotide translocation in human liver mitochondria (Table 6) as described by Christiansen et al. (1973) for mitochondria from brown adipose tissue of guinea pigs and rats. The effect of serum albumin was explained by these authors to be due to the binding of the bulk of fatty acids in the mitochondria, while the effect of oxoglutarate was related to the increase of exchangeable ATP and ADP at the expense of AMP. Defatted albumin also markedly affects the 3-hydroxy-butyrate dehydrogenase activity of human liver mitochondria, the activity being increased by a factor of 2 (Table 7).

The adenosine triphosphatase (ATPase) activity of mitochondria in fact proved to be a useful index of mitochondrial damage (Benga and Mureşan 1974). It is well known from studies of mitochondria from different sources that intact mitochondria exhibit little or no hydrolytic activity towards adenosine triphosphate (ATP), referred to as the "latency" of the ATPase. The classical activator of the "latent" ATPase is

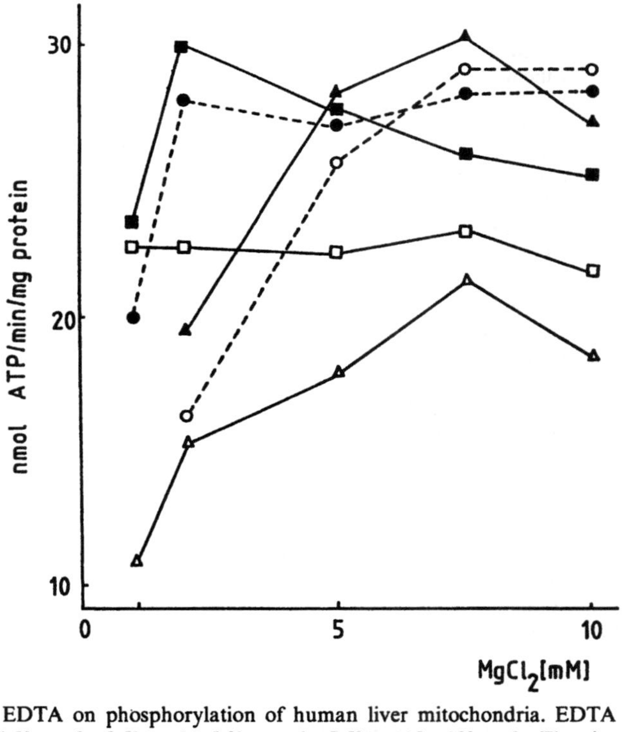

Fig. 3. Influence of Mg^{2+} and EDTA on phosphorylation of human liver mitochondria. EDTA concentration [mM]: 0[△—△]; 1.0[□—□]; 2.5[■—■]; 5.0[○—○]; 7.5[▲—▲]; 10[●—●]. The incubation medium (1 ml final volume) contained: 180 mM sucrose, 50 mM KCl, 16 mM Tris-HCl (pH 7.4), 10 mM glutamate, 10 mM NaF, 20 mM glucose, 2 mg yeast hexokinase (3.7 EU)/mg, 10 mg BSA, 0.1 mM ADP. The reaction was initiated by addition of 0.3 mg mitochondrial protein. After 10 min of incubation at 37°C, 0.5 ml of 1.5 N perchloric acid was added. The glucose-6-phosphate was measured enzymatically in the neutralized extract by examining the reduction of NADP with glucose-6-phosphate dehydrogenase at 340 nm

Table 6. Effect of serum albumin and oxoglutarate on ADP translocation in human liver mitochondria

Additions during preincubation	ADP translocated (nmol/mg protein)
None	10.4
Oxoglutarate (5mM)	25.9
Albumin (1%)	33.5

Note: The translocation was measured at 0°C in 240 mM KCl, 40 mM Tris-HCl, pH 7.4, 2 mM EDTA. The reaction was initiated by addition of 60 μM [^{14}C] ADP and stopped after 5 min with 60 μM atractyloside and dilution with 5 ml ice-cold 250 mM sucrose. The mitochondria were isolated by Millipore filtration, washed with 120 mM KCl, 20 mM Tris-HCl, pH 7.4, 1 mM EDTA, then dissolved and counted for radioactivity. The results are the means of 4 experiments.

Table 7. Enzymic activities of rat and human liver mitochondria[a]

Enzyme	Human liver	Rat liver
ATPase[b]		
Basal Activity		
Without BSA	12.6 ± 3.1	
With 1% BSA	7.7 ± 3.0	
5 mM Mg^{2+}		
Without BSA	57.2 ± 10.2	
With 1% BSA	26.6 ± 8.9	
0.1 mM DNP		
Without BSA	53.2 ± 15.1	
With 1% BSA	69.2 ± 16.5	
3-Hydroxybutyrate-dehydrogenase[b]		
Without BSA	33.7 ± 4.1	110 ± 25
With 1% BSA	70.5 ± 2.2	115 ± 23
Cytochrome oxidase[c]		
With horse cytochrome *c*	9.1 ± 2.3	106 ± 21
With human cytochrome *c*	47.1 ± 7.2	92 ± 18
Succinate-cytochrome *c* reductace[c]		
With horse cytochrome *c*	3.8	29.5
With human cytochrome *c*	20.6	30.5

Source: Data from Benga and Mureşan (1974); Benga and Borza (1975) and Benga et al. (1980).
[a]Results are mean and SEM for 5–14 experiments.
[b]The specific activity is expressed as nanomoles of substrate transformed per mg protein per minute.
[c]The enzymic activity is expressed in terms of the first-order kinetic constant (milliliters per mg protein per minute).

2,4-dinitrophenol (DNP). When mitochondria are damaged, for example, by aging, osmotic shock, or addition of surface active agents, activation by DNP is lost and a simultaneous activation by Mg^{2+} appears.

A marked effect of the presence of albumin on ATPase activity of human liver mitochondria was observed. The addition of albumin to the incubating media led to a decrease of the basal and magnesium-simulated activity, whereas DNP simulation was considerably greater compared to that in the absence of albumin (Table 7). The effect of albumin was more pronounced with lipid-rich mitochondria isolated from fatty livers. This action of albumin could not be replaced by EDTA. The addition of albumin to the incubating media enables the distinction of a "reversible" and an "irreversible" type of damage to the mitochondria. The pattern of the ATPase activity of mitochondria with preserved integrity is characterized by a low activity, a small simulation by Mg^{2+} (about two-fold), and a marked simulation by DNP (up to 20 times). When the basal and Mg-ATPase decrease upon the addition of albumin and the DNP-ATPase rises so that the ratio DNP-ATPase to Mg-ATPase becomes greater than 1, a reversible damage of mitochondria is assumed. This type of damage may occur during the homogenization of the tissue and the isolation of mitochondria possibly due to the deleterious action of agents such as free fatty acids, bilirubin, or bile acids or due to the activation of phospholipases and lipases found in mitochondria, lysosomes and microsomes. In case of irreversible damage of the mitochondria, the normal pattern of the ATPase activity is not restored by adding albumin. This type of damage is probably due to an in vivo alteration of the mitochondria.

One of the most interesting observations was that human liver mitochondria exhibited a much greater cytochrome oxidase activity with human cytochrome *c* than with horse cytochrome *c*, whereas rat liver mitochondria oxidized the horse and human cytochrome *c* at the same rate (Benga and Borza 1975). In other words, human liver mitochondria show a specificity for human cytochrome *c* (Table 7). The concentrations of cytochromes aa_3 were practically the same in human liver mitochondria (0.27 nmol/mg protein) and rat liver mitochondria (0.24 nmol/mg protein).

Our findings can be discussed taking into account the small differences in the amino acid sequence between horse and human cytochrome *c* (Dickerson 1972). There are only 12 replacements of amino acids of 104 residues, 6 of those being with amino acids with similar chemical properties (Dickerson 1972). Only six replacements involve a change in the functional groups of the side chain at positions 12,50,58,60,83,89 and 92. If we look at the three-dimensional structure of horse cytochrome *c* (Dickerson 1972), we can see that residue 12 is in the upper part of the heme crevice next to lysine 13. Okunuki (in Dickerson 1972) has shown that blocking the lysine 13 reduced the reactivity of cytochrome *c* by a factor of 2. The other five residues are all on the left-hand side of the cytochrome *c* molecule (the left and right sides of the molecule are as described by Dickerson 1972). Our data suggest the importance of the top and left side of the cytochrome *c* for the oxidase reaction. It was considered for some time that the cytochrome *c* is oxidized on one side and is reduced on the other side. However, we noted that the same specificity towards the human cytochrome *c* (compared with horse cytochrome *c*) was exhibited by human liver mitochondria also in the reduction reaction, namely by the succinate-cytochrome *c* reductase (Table 7). This suggested that the same part of the cytochrome *c* molecule is involved in both oxidation and reduction reactions. Our suggestion proved to be true, as Rieder and Bosshard (1978a) reported that a binding site for cytochrome bc_1 on cytochrome *c* is very similar or even identical to the binding site for the oxidase. The same authors (Rieder and Bosshard 1978b) proposed a binding site which spans over the top and part of the left-hand site of the cytochrome *c* molecule. This corresponds exactly to the location of the six positions where the significant amino acid differences between the human and horse cytochrome *c* occur. The finding of a single binding site is important for the mechanism of the cytochrome *c* catalyzed electron transfer from the cytochrome bc_1 complex to cytochrome oxidase. Two contrasting mechanisms can be investigated. In the so-called static mechanism, cytochrome *c* is bound simultaneously to cytochrome bc_1 and to cytochrome oxidase accepting electrons from cytochrome c_1 and donating electrons to the oxidase. Two spatially separated binding sites for bc_1 and oxidase on the cytochrome *c* molecule are predicted by the "static" mechanism. In the alternative, "dynamic" mechanism, cytochrome *c* oscillates between bc_1 and oxidase carrying electrons to and from. Identical or overlapping binding sites would conform to the dynamic mechanism. Our data with human liver mitochondria are in support of the dynamic mechanism.

On the other hand, we can take into account the hydrophobicity of all 12 replacements in the amino acid sequence. The horse cytochrome *c* has a charged residue (Glu), whereas the human cytochrome *c* has an Ala at position 92. There is another charged residue in horse cytochrome *c*, Lys 60, replaced by Gly in human cytochrome *c*; but the human cytochrome *c* has a charged residue (Glu 89), whereas horse cytochrome *c* has Thr at position 89. If we consider the hydrophobicity of the other replacements according to the hydrophobicity scales (Segrest and Feldmann 1974; Engelman et al. 1986), the human cytochrome *c* has a greater hydrophobicity than the horse cytochrome *c*.

To function correctly, cytochrome *c* must interact with the mitochondrial membrane at the right point. As we have shown, human liver mitochondria contain two to three times more lipids than rat liver mitochondria. All 12 replacements in the amino acid sequence are found on the outside of the cytochrome *c* molecule and are possibly involved in the penetration into the membrane of human liver mitochondria, which is more "hydrophobic" than its counterpart in the rat.

6 Medical Applications

We found some differences in oxygen uptake and the acceptor control index (ACI) values among subjects with ulcers, biliary diseases and those with no digestive disease (Toader et al. 1976). There were significant differences between the ulcer and biliary tract disease groups when succinate was the substrate; the oxygen uptake without ADP was higher, and consequently the ACI was lower, for patients with biliary diseases. Because it was considered that the liver was damaged in any digestive disease, we tried to study mitochondria from patients with no digestive diseases, such as epigastric or umbilical hernia and uterine fibroma. In comparison with patients with no digestive diseases, for subjects with ulcers or biliary diseases, smaller values of oxygen uptake with ADP and of ACI were noted with glutamate and α-ketoglutarate plus malate as substrate.

The differences in oxygen uptake and ACI values among subjects with ulcers, biliary diseases, and those with no digestive diseases impose a restriction on the selection of subjects with digestive diseases as controls. The liver may be affected by many agents even in the case of subjects with no digestive diseases. We have noted steatosis and fibrosis in some cases with umbilical hernia and obesity. Therefore, we consider that patients undergoing abdominal surgery, who are selected as controls in studies on isolated fractions of human liver, should fulfill the following conditions: absence of any hepatic antecedents, no clinical evidence of liver involvement, no abnormality in routine liver function tests, a histologic aspect free of pathological conditions, and a normal aspect of the tissue during the homogenization and fractionation procedure. A higher content of lipids is indicated by a more whitish color of the tissue and the homogenate and by the accumulation of lipid at the top of the tube during the centrifugation. A macroscopic fibrosis can be seen during the mincing of the tissue; such areas should be removed.

The liver of patients with biliary diseases has a greater lipid content as compared with that of subjects with ulcers. The lipid content of liver mitochondria was about 0.25 mg lipids/mg protein in patients with ulcers and 0.40 mg lipids/mg protein in patients with biliary diseases (Toader et al. 1976). In such patients the pattern of ATPase activity typical of "reversible damage" was found. Moreover, the respiratory activities are considerably increased if 1% albumin is added to the incubation media. This can be obtained even in cases of liver mitochondria isolated from patients with obstructive jaundice of short duration (up to 4 weeks) (Table 8).

In patients with prolonged jaundice "irreversible" damage of mitochondria was noted. In such cases the oxygen uptake with the nicotinamide adenine dinucleotide (NAD)-linked substrates and the ACI were lowered (Table 9). The damage of

Table 8. Respiratory activities of liver mitochondria isolated from patients with obstructive jaundice of short duration (less than 4 weeks)[a]

Substrate	Case No.	Albumin in the incubation medium: Absent	Albumin in the incubation medium: Present
Succinate plus rotenone	1	48 (1.7)	86 (3.2)
	2	–	91 (1.8)
Pyruvate plus malate	1	7.5 (1.1)	17.5 (2.3)
	2	–	19.7 (1.9)
Alpha-ketoglutarate	1	10.8 (1)	21.3 (1.5)
		–	27.5 (2.1)

[a]Results are the values of oxygen uptake in nmol/min/mg protein and of the acceptor control index (in parentheses).

Table 9. Respiratory activity of liver mitochondria isolated from patients with hepatitis and jaundice lasting over 6 weeks[a]

Substrate	No. of determinations	Oxygen uptake (nmol/min/mg) protein)	Acceptor control index
Succinate plus rotenone	4	66.0 ± 20.0	1.9 ± 0.5
Pyruvate plus malate	4	14.7 ± 4.3	1.7 ± 0.2
Glutamate	4	14.4 ± 6.8	1.7 ± 0.1
Isocitrate plus malate	2	19.5	1.8

[a]Results are mean ± SD.

mitochondria in such patients is also documented by electron microscopy. In the tissue mitochondria are swollen or condensed with bizzare forms. Spin-label studies have indicated a higher fluidity of the mitochondrial membranes in patients with jaundice.

If we look at biological membranes not only in terms of permeability barriers, but also as catalytic surfaces, the membranes constitute a uniquely ordered system for metabolic reactions. Human liver subcellular membranes with their peculiar composition (higher lipid content, a particular fatty acid composition, more hydrophobic amino acid residues in membrane proteins in human compared to rat liver mitochondria), particularities of enzymic activities, and higher fluidity as inferred by spin-label motion provide an interesting system of protein-lipid interactions in liver membranes of great physiological and pathological significance.

It is generally recognized that the availability of direct precursor acids (free fatty acids) is one of the limiting factors in the biosynthesis of endoperoxide metabolites in animal tissues. The amount of the precursor acids (i.e. free arachidonic acid) released from tissue phospholipids by the action of phospholipase would depend on the fatty acid composition of tissue lipids. Hwang et al. (1982) showed that reduction in the tissue arachidonic acid level resulted in concomitant decrease in serum levels of endoperoxide metabolites. These free oxygen radicals are injurious to tissues and may cause lysis of red cells, peroxidation of lysosomal and mitochondrial membranes, and destruction of endothelial cells (see Halliwell and Gutteridge 1984, for review). Consequently, free oxygen radicals are involved in a variety of diseases, including neoplasia where alterations of membrane lipids have been reported (Mountford and Wright 1988) and atherosclerosis. The low content of arachidonic acid in human membrane phospholipids might be considered as an adaptive mechanism in humans in order to maintain a rather low level of endoperoxide metabolites.

Acknowledgements. The results reviewed above have been obtained in collaboration with Victoria Borza, Adriana Hodârnău, Letiţia Mureşan, Rozalia Tilinca (Department of Biochemistry), Victor I. Pop (Department of Cell Biology), Faculty of Medicine, Medical and Pharmaceutical Institute Cluj-Napoca, Rumania; Dorin Poruţiu (Laboratory of Electron Microscopy, Medical and Pharmaceutical Institute Cluj-Napoca, Rumania), Ion Petrescu (Department of Biochemistry, Faculty of Biology, University of Cluj-Napoca, Rumania), William Ferdinand (Department of Biochemistry, University of Sheffield, UK), and John Wrigglesworth (Department of Biochemistry, Chelsea College University of London, UK).

References

Benga Gh (1974) Fracţionarea subcelulară a ţesutului hepatic uman. I. Izolarea fracţiunii mitocondriale. St Cerc Biochim 17:123–135

Benga Gh (1975) Fracţionarea subcelulară a ficatului uman. II. Caracterizarea mitocondriilor in condiţii patologice. St Cerc Biochim 18:83–91

Benga Gh, Borza V (1975) Differences in reactivity of cytochrome oxidase from human liver mitochondria with horse and human cytochrome *c*. Arch Biochem Biophys 169:354–357

Benga Gh (1983) Spin labelling. In: Wrigglesworth J (ed) Biochemical research techniques. John Willey & Sons, New York London, pp 79–117

Benga Gh (ed) (1985) Structure and properties of cell membranes, vol 1-3. CRC, Boca Raton

Benga Gh, Ferdinand W (1977) Increased content of hydrophobic amino acid residues in lipid-rich mitochondrial membranes: a comparison of rat and human liver mitochondria. Int J Biochem 8:17–20

Benga Gh, Holmes RP (1984) Interactions between components in biological membranes and their implications for membrane function. Prog Biophys Mol Biol 43:195–257

Benga Gh, Mureşan L (1974) Human liver mitochondria. III. ATPase activity as an index of mitochondrial damage. Biochem Med 10:131–145

Benga Gh, Tărmure C, Hodârnău A (1971) The determination of proteins in lipid-rich mitochondrial suspensions. Enzyme 12:574–577

Benga Gh, Mureşan L, Hodârnău A, Dancea S (1972) Conditions for isolation and study of enzymic properties of human liver mitochondria. Biochem Med 6:508–521

Benga Gh, Hodârnău A, Böhm B, Borza V, Tilinca R, Dancea S, Petrescu I (1978) Human liver mitochondria: relation of a particular lipid composition to the mobility of spin-labelled lipids. Eur J Biochem 84:625–633

Benga Gh, Hodârnău A, Tilinca R, Porutiu D, Dancea S, Pop V, Wrigglesworth J (1979). Fractionation of human liver mitochondria: enzymic and morphological characterization of the inner and outer membranes as compared to rat liver mitochondria. J Cell Sci 35:417–429

Benga Gh, Petrescu I, Tărmure C, Pop V (1980) Species related functional properties of mitochondria; comparison between rat and human liver mitochondria. Rev Roum Biol 25:147–154

Benga Gh, Pop VI, Ionescu M, Hodârnău A, Tilinca R, Frangopol PT (1983) Comparison of human and rat liver microsomes by spin label and biochemical analyses. Biochim Biophys Acta 750:194–199

Blouin A, Bolender RP, Weibel ER (1977) Distribution of organelles and membranes between hepatocytes and non-hepatocytes in the rat liver parenchyma. J Cell Biol 72:441–445

Christiansen EN, Drahota Z, Duszynski J, Wojtczak L (1973) Transport of adenine nucleotides in mitochondria from the brown adipose tissue. Eur J Biochem 34:506–512

Dallner G, Siekevitz P, Palade GE (1966) Biogenesis of endoplasmic reticulum membranes. J Cell Biol 30:97–117

Dickerson RE (1972) The structure and history of an ancient protein. Sci Am 226:58–72

Engelman DM, Steitz TA, Goldman A (1986) Identifying nonpolar transbilayer helices in amino acid sequences of membrane proteins. Annu Rev Biophys Chem 15:321–353

Grav HJ, Pedersen JI, Christiansen EN (1970) Conditions in vitro which affect respiratory control and capacity for respiration-linked phosphorylation in brown adipose tissue. Eur J Biochem 12:11–23

Halliwell B, Gutteridge JMC (1984) Oxygen toxicity, oxygen radicals, transition metals and disease. Biochem J 219:1–14

Hayashida T, Portman OW (1960) Swelling of liver mitochondria from rats fed diets deficient in essential fatty acids. Proc Soc Exp Biol Med 103:656–659

Hwang DH, Chanmugam P, Anding R (1982) Effects of dietary 9-trans, 12-trans linoleate on arachidonic acid metabolism in rat platelets. Lipids 17:307–313

Kleemann W, McConnell MM (1974) Lateral phase separations in *Escherichia coli* membranes. Biochim Biophys Acta 345:220–230

Linden CD, Wright KL, McConnell HM, Fox CF (1973) Lateral phase separations in membrane lipids and the mechanism of sugar transport in *E. coli*. Proc Nat Acad Sci USA 70:2271–2275

McConnell HM, McFarland BG (1970) Physics and chemistry of spin labels. Q Rev Biophys 3:91–136

Mountford CE, Wright LC (1988) Organization of lipids in the plasma membranes of malignant and stimulated cells: a new model. Trends Biochem Sci 13:172–177

Nicholls DG (1974) Hamster brown-adipose-tissue mitochondria. The control of respiration and the proton electrochemical potential gradient by possible physiological effectors of the proton conductance of the inner membrane. Eur J Biochem 49:573

Richardson T, Tappel AL, Gruger EH (1961) Essential fatty acids in mitochondria. Arch Biochem Biophys 94:1–6

Richardson T, Tappel AL, Smith LM, Houle CR (1962) Polyunsaturated fatty acids in mitochondria. J Lipid Res 3:344–350

Rieder R, Bosshard HR (1978a) Cytochrome bc_1 and cytochrome oxidase can bind to the same surface domain of the cytochrome *c* molecule. FEBS Lett 92:223–226

Rieder R, Bosshard HR (1978b) The cytochrome *c* oxidase binding site on cytochrome *c*. Differential chemical modification of lysine residues in free and oxidase-bound cytochrome *c*. J Biol Chem 253:6045–6053

Rouser E, Nelson GJ, Fleischer S, Simon G (1968) Lipid composition of animal cell membranes, organelles and organs. In: Chapman D (ed) Biological membranes. Physical fact and function, vol 1. Academic Press, New York London, pp 5–69

Schmidt E, Schmidt FW (1970) Enzyme activities in human liver. Enzymol Biol Clin 11:67–129
Segrest JP, Feldmann RJ (1974) Membrane proteins: amino acid sequence and membrane penetration. J Mol Biol 87:853–858
Shimshick EJ, McConnell HM (1973a) Lateral phase separations in phospholipid membranes. Biochemistry 12:2351–2360
Shimshick EJ, McConnell HM (1973b) Lateral phase separations in binary mixtures of cholesterol and phospholipids. Biochem Biophys Res Commun 53:446
Singer SJ, Nicolson GL (1972) The fluid mosaic model of the structure of cell membranes. Science 175:720–731
Smith L, Davies HC, Nava ME (1976) Evidence for binding sites on cytochrome *c* for oxidases and reductases from studies of different cytochromes *c* of known structure. Biochemistry 15:5827–5831
Stancliff RC, Williams MA, Utsumi K, Packer L (1969) Essential fatty acid deficiency and mitochondrial function. Arch Biochem Biophys 131:629–642
Toader C, Acalovschi I, Toader I, Manta I, Hodârnău A, Benga Gh (1976) Factors influencing the establishment of the normal values of the respiratory activities of human liver mitochondria. Enzyme 21:232–242
Van Den Bosch H (1974) Phosphoglyceride metabolism. Annu Rev Biochem 43:243–277
Vanderkooi G, Capaldi RA (1972) A comparative study of the amino acid composition of membrane proteins and other proteins. Ann NY Acad Sci 195:135–138
Von Heijne G (1981) Membrane proteins. The amino acid composition of membrane penetrating segments. Eur J Biochem 120:275–278

Changes in Erythrocyte and Platelet Membrane Fluidity in Childhood Obesity

G. FERRETTI, A. TANGORRA, P.L. GIORGI, M. DOTTI, G. CURATOLA and E. BERTOLI

1 Introduction

Obesity has been associated with various functional membrane abnormalities in animals (York 1985) and in humans (Table 1). Alterations in insulin receptor binding have been described in adipocytes (Amatruda et al. 1975) as well as in erythrocytes (Kappy and Plotnik 1980) and in monocytes (Misbin et al. 1983). Moreover, the studies of Sundaresan et al. (1983) and Rey Stolle et al. (1986) have shown that in adult obese patients, platelet sensitivity to aggregating agents, such as ADP and adrenalin, is altered. Changes in Na^+-K^+-ATPase activity have been also described in liver (Bray et al. 1981), in the skeletal muscle (Charalambous et al. 1984a) and in the erythrocyte membranes (De Luise et al. 1980; Charalambous et al. 1984b). All the forementioned abnormalities could be related to abnormalities in the composition and structural properties of the cell membranes, in fact, it is well known that membrane fluidity and composition remarkably affect the binding of a variety of hormones to their receptors and the activity of membrane-bound enzymes (Shinitzky 1984). Obesity is characterized by abnormalities of lipid metabolism, an increase of synthesis and turnover of cholesterol, and alterations of lipoprotein metabolism have been described (Nestel et al. 1973; Freedman and Burke 1985). Lipoprotein abnormalities include very low density lipoprotein production (VLDL), increased low density lipoprotein (LDL) turnover and reduced plasma cholesterol associated with high density lipoproteins (Freedman and Burke 1985).

Table 1. Alterations in membrane functions in human obesity

Receptor binding	*Cell membrane*	*Reference*
Insulin	Erythrocyte	Kappy and Plotnik (1980)
	Monocyte	Misbin et al. (1983)
	Adipocyte	Amatruda et al. (1975)
Thrombin	Platelet	Tangorra et al. (1988)
Adrenergic	Platelet	Sundaresan et al. (1983)
Enzymes		
Na^+-K^+-ATPase	Erythrocyte	Charalambous et al. (1984b)
	Skeletal muscle	Charalambous et al. (1984a)
	Liver	Bray et al. (1981)
Acetylcholinesterase	Erythrocyte	Dotti et al. (1988 and pers. commun.)

Institute of Biochemistry and Pediatrics, School of Medicine, University of Ancona, Italy

A. Azzi, Z. Drahota and S. Papa
Molecular Basis of Membrane-Associated Diseases

Changes of erythrocyte and platelet membrane lipid composition could be related to alterations of plasma lipids and/or to modifications of exchange processes between plasmatic and cell-membrane compartments (Owen et al. 1984).

A causal linkage between modifications of lipoprotein metabolism and erythrocyte and platelet membrane abnormalities in obesity has not yet been clearly established. However, such a relationship is supported by experimental evidence in other inborn or acquired pathologies (Cooper 1977; Owen et al. 1984).

In a previous study (Curatola et al. 1987), we have shown that erythrocyte membranes from obese children are characterized by a decrease of membrane fluidity; alterations in erythrocyte membrane properties have also been found in obese adults (Beguinot et al. 1985). Membrane fluidity is strictly related to membrane composition and structure. In the present study, we investigated erythrocyte and platelet membrane fluidity by means of fluorescence spectroscopy of 1,6-diphenyl-1,3,5-hexatriene (DPH), and its cationic derivate, the 1-(4-trimethylamino phenyl)-6-phenyl-1,3,5-hexatriene (TMA-DPH). The results have been discussed in relation to membrane compositional changes and to modifications of clinical parameters.

2 Methods

The subjects of this study were 30 obese children and 15 normal weight children matched to give a similar distribution of age and sex. The body-mass index (BMI) (weight in kg divided by the square of height in meters) taken as a measure of obesity (Keys et al. 1972) ranged from 22.5–44.8 kg m^{-2} in the obese. In the controls the BMI ranged from 10.48–21.2 kg m^{-2}.

2.1 Plasma Lipids

Total cholesterol, triglycerides and phospholipids were measured respectively by the methods of Allain et al. (1974), Bucolo and David (1973) and Takayama et al. (1977).

The determination of high density lipoprotein cholesterol (HDL-C) was performed on supernatant after precipitation of the apolipoprotein B containing lipoproteins using a phosphotungstic acid/$MgCl_2$ reagent followed by cholesterol measurement in the supernatant (Assmann et al. 1981). Low density lipoprotein cholesterol (LDL-C) concentration was calculated according to the formula of Friedewald et al. (1972). Lipoprotein analysis was performed using cellulose acetate electrophoresis.

2.2 Membrane Preparation

Ten ml blood was drawn from each subject after a 12-h overnight fast. Five ml was introduced into heparinized tubes and used for the preparation of erythrocyte membranes by hemolyzing the washed erythrocytes in 5 mM phosphate buffer, pH

8, according to the method of Steck and Kant (1974). Membrane protein content was estimated by the method of Lowry et al. (1951).

Five ml blood was mixed with EDTA 0.25 M (0.15% final concentration. Platelets were isolated by centrifugation of platelet-rich plasma (PRP) following the method of Kubina et al. (1987).

2.3 Membrane Lipid Extraction and Analysis

Lipids were extracted from erythrocyte membranes and platelets as described by Rose and Oklander (1965). Cholesterol content was assayed enzymatically using cholesterol oxidase (Allain et al. 1974). Phospholipid content was determined as inorganic phosphate by the method of Fiske and Subbarow (1965).

2.4 Fluorescence Polarization Measurements

In this study the hydrophobic fluorescent probe 1,6-diphenyl-1,3,5-hexatriene (DPH) (Aldrich Chemical Company, FRG), and a cationic derivative of DPH, the 1-(4-trimethylamino phenyl)-6-phenyl-1,3,5-hexatriene (TMA-DPH) (Molecular Probes Inc., Junction City, OR, USA) were used to investigate membrane fluidity (Kuhry et al. 1983). In comparison with DPH, TMA-DPH is rapidly incorporated into the plasma membrane of intact cells and it does not penetrate for 20 min into the cell, therefore, its use has been recommended for studies in living cells (Kuhry et al. 1983).

The fluorescent probe DPH from a 2×10^{-3} M stock solution in tetrahydrofuran, was added to erythrocyte membranes to give a DPH final concentration of 10^{-6} M. Platelet incubation with the fluorescent probe TMA-DPH was performed as described by Kubina et al. (1987). Steady-state fluorescence polarization measurements of DPH or TMA-DPH were performed using a Perkin-Elmer MPF-66 equipped with two glass prism polarizers. Excitation and emission wavelengths were 365 and 430 nm, respectively, for both probes. The steady-state fluorescence polarization (P), corrected for light scattering, was obtained following Shinitzky and Barenholz (1978).

3 Results

The physical characteristics and the lipid and lipoprotein data of controls and obese subjects are shown in Table 2. The obese children had higher triglyceride concentrations and no significant difference in plasma total cholesterol with respect to the controls. Moreover, in obese subjects an increase in LDL and LDL-C and a decrease in HDL and HDL-C levels with a consequent increase of the LDL/HDL were observed, however, the differences were not significant.

The mean value of fluorescence polarization (P) of DPH in erythrocyte membranes of control subjects was 0.230 ± 0.010. The corresponding value in obese subjects was significantly increased with respect to the controls (0.252 ± 0.013, $p < 0.001$, Table 3) and indicates an overall decrease of membrane fluidity

Table 2. Physical characteristics, plasma lipids and lipoproteins in obese and control children

	Controls	Obese[a]
BMI (kg/m²)	18 ± 3	28 ± 3**
Plasma lipids		
Cholesterol (mg/dl)		
Total	157 ± 17	153 ± 26
LDL	88 ± 13	93 ± 21
HDL	48 ± 9	45 ± 14
Triglycerides (mg/dl)	54 ± 18	84 ± 34*
Lipoproteins LDL (%)	56 ± 8	61 ± 9
HDL (%)	43 ± 8	37 ± 9
LDL/HDL	1.4 ± 0.3	1.8 ± 0.8

[a] $^{**}p < 0.001$; $^{*}p < 0.05$.

Table 3. Fluorescence polarization (P) of DPH and cholesterol to phospholipid (C/PL) molar ratio in erythrocyte membranes of obese and control subjects

	P[a]	C/PL[a]
Controls	0.230 ± 0.01	0.81 ± 0.12
Obese	0.252 ± 0.013**	1.01 ± 0.23*

[a] Values are expressed as mean ± SD.
$^{*}p < 0.05$; $^{**}p < 0.001$.

in agreement with our previous results (Curatola et al. 1987). The erythrocyte membrane lipid analysis has evidenced an increase of the cholesterol to phospholipid (C/PL) molar ratio (0.81 ± 0.12 vs 1.01 ± 0.23) (Table 3), and indicates that changes in the erythrocyte membrane lipid composition are present in obese children as observed in obese adults (Beguinot et al. 1985).

Using TMA-DPH as fluorescent probe, we observed alterations in the fluidity of platelets of obese subjects with an increase in the P value (0.435 ± 0.01) in obese compared to the value of 0.406 ± 0.02 of the controls (Table 4). These changes in platelet membrane fluidity were not associated with any relevant modifications of the C/PL molar ratio (0.58 ± 0.1 vs 0.59 ± 0.06; Table 4).

4 Discussion

In agreement with our previous studies (Curatola et al. 1987), the present results have confirmed changes in erythrocyte membrane properties in childhood obesity. Moreover, we have shown a decreased membrane fluidity in platelets from obese subjects. The cholesterol to phospholipid (C/PL) molar ratio and the degree of fatty acid unsaturation influence membrane fluidity; in general, a decrease in the unsaturation and an increase in the C/PL cause a decrease in membrane fluidity.

Table 4. Fluorescence polarization (P) of TMA-DPH and cholesterol to phospholipid (C/PL) molar ratio in platelets of obese and control subjects

	P[a]	C/PL
Controls	0.406 ± 0.02	0.58 ± 0.1
Obese	0.435 ± 0.01**	0.59 ± 0.06

[a] Values are expressed as mean ± SD; ** $p < 0.001$.

In erythrocyte membranes, the increase in fluorescence polarization may be likely a consequence of compositional changes; in fact, the C/PL in erythrocyte membranes of the obese was higher with respect to the controls. Mature red blood cells lack the ability to synthesize lipids de novo so that the cholesterol content and phospholipid composition depend on the exchange with the plasma lipoprotein surface. Platelets, unlike red cells, can synthesize membrane lipids, however, it is known that lipoproteins appear important in regulating also the lipid composition of platelet membranes (Owen et al. 1984).

It should be stressed that the changes in platelet and erythrocyte membrane fluidity were present in obese patients even in the absence of remarkable changes of the plasma lipid pattern.

The results suggest that erythrocytes as well as platelets respond to small variations in plasma lipids with changes in composition and/or physical state and support the concept that membranes of obese subjects present structural abnormalities. Although we studied only erythrocyte and platelet membranes, defects in membrane fluidity have been described also in other cells. In particular, Neufeld et al. (1986) described alterations in membrane composition and physico-chemical state in mononuclear leukocytes of obese patients. Alterations in the composition and structural properties of biological membranes have striking effects on membrane functional activities (Shinitzky 1984), therefore, changes of membrane fluidity could be of physiological significance in explaining the several alterations in membrane functions that have been described in obesity.

References

Allain CC, Poon LS, Chan CS (1974) Enzymatic determination of total serum cholesterol. Clin Chem 20:470–475

Amatruda JM, Livingston JH, Lockwood DA (1975) Insulin receptor: role in the resistance of human obesity in insulin. Science 188:264–268

Assmann G, Schriewer H, Funke H (1981) Zur Richtigkeit der HDL-Cholesterin- und HDL-Apolipoprotein-A-1-Bestimmung nach Phosphorwolframsäure/$MgCL_2$-präzipitation Apolipoprotein-B-haltiger Lipoproteine. J Clin Chem Clin Biochem 19:273–278

Beguinot F, Tramontano D, Duilio C, Formisano S, Beguinot L, Mattioli P, Mancini M, Aloj SM (1985) Alteration of erythrocyte membrane lipid fluidity in human obesity. J Clin Endocrinol Metab 60:1226–1230

Bray GA, Kral JG, Bjorntorp P (1981) Hepatic sodium-potassium-dependent ATPase in obesity. N Engl J Med 304:1580–1582

Bucolo G, David H (1973) Quantitative determinations of triglycerides by the use of enzymes. Clin Chem 19:476–482

Charalambous BM, Webster DJT, Mir MA (1984a) Elevated skeletal muscle sodium-potassium ATPase in human obesity. Clin Chim Acta 141:189–195

Charalambous BM, Morgan K, Spurlock G, Mir MA (1984b) Erythrocyte sodium pump activity in human obesity. Clin Chim Acta 141:179–187

Cooper RA (1977) Abnormalities of cell-membrane fluidity in the pathogenesis of disease. Semin Med Beth Isr Hosp 297:371–377

Curatola G, Ferretti G, Bertoli E, Dotti M, Bartolotta E, Giorgi PL (1987) Changès in membrane fluidity in erythrocytes of obese children: a spin label study. Pediatr Res 22:141–144

De Luise M, Blackburn GL, Flier JS (1980) Reduced activity of red cell sodium-potassium pump in human obesity. New Engl J Med 303:1017–22

Dotti M, Ferretti G, Curatola G, Bari S, Giorgi PL (1988) Changes of the acetylcholinesterase activity in erythrocyte membranes of obese children. Proc Congr Pediatria oggi, Ancona 10–13 Maggio 1988

Fiske C, Subbarow Y (1965) The colorimetric determination of phosphorus. J Biol Chem 66:375–380

Freedman SD, Burke GL (1985) Relationship of changes in obesity to serum lipid and lipoprotein changes in childhood and adolescence. JAMA 254:515–520

Friedewald WT, Levy RI, Fredrickson DS (1972) Estimation of the concentration of low-density lipoprotein cholesterol in plasma, without use of preparative ultracentrifuge. Clin Chem 18:499–502

Kappy MS, Plotnick L (1980) Erythrocyte insulin binding in obese children and adolescents. J Clin Endocrinol Metab 51:1440–1446

Keys A, Fidanza F, Karvonen MJ, Kimura N, Taylor HL (1972) Indices of relative weight and obesity. J Chronic Dis 25:329–333

Kubina M, Lanza F, Cazenave JP, Laustriat G, Kuhry JG (1987) Parallel investigation of exocytosis kinetics and membrane fluidity changes in human platelets with the fluorescent probe, trimethyl-ammonio-diphenylhexatriene. Biochim Biophys Acta 901:138–144

Kuhry JG, Fontenau P, Duportail G, Maechling C, Laustriat G (1983) TMA-DPH: A suitable fluorescence polarization probe for specific plasma membrane fluidity studies in intact living cells. Cell Biophys 5:129–140

Lowry OH, Rosebrough NJ, Farr AL, Randall RJ (1951) Protein measurement with the folin phenol reagent. J Biol Chem 193:265–275

Misbin RI, Moffa AM, Kappy MS (1983) Insulin binding to monocytes in obese patients treated with carbohydrate restriction and changès in physical activity. J Clin Endocrinol Metab 56:273–278

Nestel PJ, Schreibman PH, Ahrens EH (1973) Cholesterol metabolism in human obesity. J Clin Invest 52:2389–2395

Neufeld ND, Ezrin C, Corbo L, Long D, Bush MA (1986) Effects of caloric restriction and exercise on insulin receptors in obesity: association with changes in membrane lipids. Metabolism 35:580–587

Owen JS, McIntyre N, Gillet MPT (1984) Lipoproteins cell membrane and cellular functions. Trends Biochem Sci 5:238–242

Rey Stolle AM, Rodriguez PM, Diez Marques ML, Navarro Ruiz A (1986) Platelet aggregation in obese women on a hypocaloric diet. Ir Med Sci 14:272–273

Rose HG, Oklander M (1965) Improved procedure for the extraction of lipids from human erythrocytes. J Lipid Res 6:428–431

Shinitzky M (ed) (1984) Membrane fluidity and cellular functions. Physiology of membrane fluidity. CRC, Boca Raton, pp 1–51

Shinitzky M, Barenholz Y (1978) Fluidity parameters of lipid regions determined by fluorescence polarization. Biochim Biophys Acta 515:367–394

Steck TL, Kant JA (1974) Preparation of impermeable ghosts and inside out vesicles from human erythrocyte membranes. In: Colowick SP, Kaplan NO (eds) Methods in enzymology. Academic Press, New York London, pp 172–173

Sundaresan PR, Weintraub M, Hershey LA, Kroening BH, Hasday JD, Banerjee SP (1983) Platelet α-adrenergic receptors in obesity: alteration with weight loss. Clin Pharmacol Ther 33:776–785

Takayama M, Itoh S, Nagasaki T (1977) A new enzymatic method for determination of serum choline-containing phospholipids. Clin Chim Acta 79:93–98

Tangorra A, Ferretti G, Curatola G, Guerrieri A, Catassi C, Giorgi PL, Bertoli E (1988) Altered properties of platelet membrane in childhood obesity. Biochem Int 17:837–846

York DA (1985) Alterations in membrane function, organization and composition in the obese ob/ob mouse. Proc Nutrit Soc 44:189–200

Skeletal Muscle Mitochondria and Phospholipase Activity in Malignant Hyperthermia

K. S. CHEAH[1]

1 Introduction

Malignant hyperthermia (MH), a genetically inherited potentially fatal disorder, affects primarily the skeletal muscle of humans and pigs. In apparently healthy humans [1–5] and in stress-susceptible pigs [6–12], the MH syndrome can be induced with halothane [1–12], suxamethonium [3,13–18] or halothane plus suxamethonium [4]. In stress-susceptible pigs, the syndrome can also be triggered by physiological or environmental stress such as excitement and changes in temperature [7,18–20]. Once initiated, the classical symptoms of the syndrome are gross muscular rigidity, a rapid rise in body temperature, hyperventilation, severe metabolic acidosis and elevated levels of serum metabolites [21–23]. The body temperature can increase at a rate of 1 °C per 5 min [4], and the metabolic rate can exceed 17 times normal [20]; if uncontrolled with dantrolene [4], death occurs [4,8,19,24]. The occurrence of anaesthetic-induced MH in apparently healthy patients is about 1 in 15 000 [3]. In healthy pigs, the incidence of MH is dependent on the breed and can be as high as 88% in highly stress-susceptible breeds [25]. The MH syndrome shows striking similarities in humans and pigs [26] and genetically-selected pigs are thus frequently used as experimental models for investigating the human MH syndrome.

The development of muscular rigidity in human and porcine MH is generally believed to be due to an increased level of myoplasmic Ca^{2+} [14,27–29]. A significantly higher than normal myoplasmic Ca^{2+} was indeed observed in MH-susceptible humans [30] and pigs [31] and in MH-susceptible pigs early post-mortem [32].

The purpose of this chapter is to discuss the probable role of skeletal muscle mitochondria in the molecular mechanism of MH, reviewing pertinent findings on respiration, Ca^{2+} transport, membrane permeability and phospholipase A_2 activity. Evidence suggests that fatty acids, liberated by an enhanced phospholipase A_2 activity, play a key role in the development of MH by directly causing Ca^{2+} release from the Ca^{2+}-regulating organelles to account for the increased level of myoplasmic Ca^{2+}, which in turn is responsible for the muscle hyper-rigidity and enhanced rate of glycolysis.

[1] Agricultural and Food Research Council, Food Research Institute, Bristol, UK and Department of Anesthesiology, Hahnemann University, Philadelphia, NJ, USA

A. Azzi, Z. Drahota and S. Papa
Molecular Basis of Membrane-Associated Diseases

2 Mitochondria

Various investigators have suggested that skeletal muscle mitochondria are associated in the development of MH [8,28–29,32–36]. Mitochondria are mainly responsible for ATP synthesis by oxidative phosphorylation [37]; the ATP generated is essential for many metabolic processes such as muscle contraction and Ca^{2+} transport [38]. Mitochondria can accumulate Ca^{2+} at the expense of ATP formation [39], and these organelles can also act as efficient Ca^{2+} buffers under various physiological and pathological conditions [40].

2.1 Respiration

Reports on the respiratory functions of mitochondria isolated from post-mortem and biopsy skeletal muscles of MH-susceptible humans and pigs (Table 1) were conflicting; present evidence suggests that there is no abnormality in isolated mitochondria from MH skeletal muscle when ADP is used to stimulate substrate oxidation. The discrepancies reported could be attributed mainly to differences in the quality of the isolated mitochondria; for example, the isolated organelles were mostly swollen and/or damaged [41] thereby showing an abnormal respiratory control index with ADP [42]. The hypothesis of uncoupling of oxidative phosphorylation [8,34] and the proposed existence of an aberrant mitochondrial respiratory chain system [42] in MH-skeletal muscle mitochondria were, however, not generally accepted. The principal objections against these hypotheses were: firstly, intact and tightly-coupled mitochondria could be isolated either immediately post-mortem [28,45–47] or from biopsied porcine [44,46–49] and human [50–51] MH-skeletal muscle; secondly, carefully isolated MH-skeletal muscle mitochondria showed normal oxidative phosphorylation and respiratory control (Table 1); and thirdly, skeletal muscle of MH-susceptible pigs contained normal amounts of mitochondria [49] and cytochrome components [45,48] in the respiratory chain system.

The effect of halothane on mitochondrial respiration of MH-susceptible pigs and humans was also investigated. Halothane inhibited the State 3 respiration of NAD^+-linked substrate oxidation of pigs [48–49,52] and humans [50,53], and this was accompanied by uncoupling of oxidative phosphorylation [21,42,44,49,53]. The main effect of halothane was inhibition of electron transport of NAD^+-linked substrate oxidation [43,50], and succinate oxidation was either slightly inhibited [43] or unaffected [50]. The effect of halothane on MH-skeletal muscle mitochondrial respiration was thus essentially similar to that reported for mitochondria from normal beef-cardiac muscle [54–55], rat-skeletal muscle [56], rat liver [52,57] and human skeletal muscle [53,58] following treatment with halothane.

2.2 Ca^{2+} Transport

Mitochondria can accumulate large amounts of Ca^{2+} by an active transport system through energy derived either by hydrolysis of ATP or by electron transport [59–63], and can also act as efficient Ca^{2+} storage systems in various pathological conditions linked with a large influx of Ca^{2+} into cells [40]. This remarkable ability

Table 1. Respiration, oxidative phosphorylation and respiratory control of skeletal muscle mitochondria isolated from biopsy (B) and post-mortem (PM) samples of MH-susceptible pigs (*) and humans (**)

Mitochondria	Substrate	Respiration (State 3)	Oxidative phosphorylation	Respiratory control	Ref.
Skeletal muscle (B)*	Glutamate + malate		Uncouple	Decrease	8
Biceps femoris (B)*	Glutamate + malate	Decrease	Decrease	Decrease	42
Biceps femoris (B)*	Malate + pyruvate	Normal	Normal	Normal	48
M. longissimus dorsi (B)*	Malate + pyruvate	Decrease	Normal	Normal	46
M. longissimus dorsi (PM)*	Malate + pyruvate Succinate Ascorbate + TMPD	Normal	Normal	Normal	28,45
Skeletal muscle (B)*	Glutamate	Normal	Normal	Normal	49
Semimembranosus (B, PM)*	Glutamate Succinate	Increase	Increase	Normal	44
Semitendinosus (B)*	Glutamate + malate	Decrease	Normal	Decrease	43
	Succinate	Decrease	Normal	Normal	43
Vastus lateralis (B)**	Glutamate + malate Succinate	Normal	Normal	Normal	50
Vastus lateralis (B)**	Succinate	Normal	Normal	Normal	51

of mitochondria to sequester Ca^{2+} has influenced research on the Ca^{2+} uptake by MH-skeletal muscle mitochondria (Table 2). The main aim of all these investigations has been to determine whether there is a malfunction in the mitochondrial Ca^{2+} uptake which could explain the increase in the level of myoplasmic Ca^{2+} in MH.

Studies on the Ca^{2+} transport system of MH-skeletal muscle mitochondria have been concentrated mainly on uptake rather than release using either Ca^{2+}-binding dyes such as murexide or radioactive $^{45}Ca^{2+}$.

Table 2. Ca^{2+} uptake in MH-skeletal muscle mitochondria of pigs (*) and humans (**) (PM, post-mortem; B, biopsy; +, prior to development of MH; + +, during MH syndrome)

Mitochondria	Substrate	Ca^{2+} uptake	References
M. longissimus dorsi (PM)*	Succinate	Normal (20°C)	28
	Succinate	Decrease (40°C)	28
Vastus lateralis (B)*	Succinate	Decrease (25°C)	47
M. longissimus dorsi (PM)*	ATP	Normal (24°C)	41
+ Semitendinosus (B)*	Glutamate + malate	Decrease (37°C)	43
+ + Semitendinosus (B)*	ATP	Decrease (25°C)	64
+ Biceps femoris (B)*	Endogenous (in situ)	Normal	36
+ + Biceps femoris (B)*	Endogenous (in situ)	Increase	36
Skeletal muscle (B)**	Glutamate + malate	Normal (25°C)	65
Vastus lateralis (B)**	Succinate	Normal (25°C)	51
	Succinate	Decrease (40°C)	51

Porcine MH-skeletal muscle mitochondria isolated from post-mortem samples contain twice the normal amount of endogenous Ca^{2+} [28]. Significantly greater than normal amounts of endogenous Ca^{2+} have also been observed by X-ray microprobe analysis in MH-skeletal muscle mitochondria in biopsied porcine muscle obtained after 10 min in vivo halothane plus succinylcholine-induced MH [36]. At 20°C, the rate and the amount of exogenous Ca^{2+} accumulated by tightly-coupled MH-skeletal muscle mitochondria during succinate oxidation is normal [28]. Ca^{2+} accumulation is also normal in mitochondria isolated at 5 min post-mortem [41], in mitochondria from biopsied samples of skeletal muscle of MH-susceptible pigs treated in vivo with neuroleptic drugs (azaperone or spiperone) to delay the onset of halothane-induced MH [64], and from biopsied human MH-skeletal muscle [51,65]. In MH-susceptible humans, Ca^{2+} uptake by tightly-coupled mitochondria is significantly ($P < 0.001$) lower than normal at 40°C [51] with succinate as substrate. In MH-susceptible pigs, Ca^{2+} uptake also decreases significantly ($P < 0.001$) at 40°C [66], and at 25°C with organelles isolated either following the development of MH [64] or at 30 min post-mortem [67]; the altered properties of mitochondria from 30 min post-mortem are the result and not the cause of MH. The rate of Ca^{2+} uptake by MH-skeletal muscle mitochondria has also been reported to be less than 40% of normal with either succinate [43,44] or glutamate plus malate [43] as substrate at 25°C [44] and at 37°C [43]. Halothane inhibits mitochondrial Ca^{2+} uptake with glutamate plus malate [43] and either stimulates [44] or has no effect [43] with succinate at 37°C.

Skeletal muscle mitochondria of MH-susceptible humans [51] and pigs [28] are easily uncoupled by exogenous Ca^{2+} at 40°C but not at 25°C. In these mitochondria Ca^{2+}-induced uncoupling occurs at the same concentration of exogenous Ca^{2+}, whereas normal mitochondria are still tightly coupled at about twice the amount of added exogenous Ca^{2+} during succinate oxidation. The Ca^{2+}-induced uncoupling is prevented by fatty acid-free bovine serum albumin and by phospholipase A_2 inhibitors (spermine and tetracaine). In MH-susceptible humans [51], skeletal muscle mitochondria behave normally at 40°C up to a level of 701 ± 145 (n = 5) nmol Ca^{2+} per mg protein, beyond which an increase in State 4 respiration occurs, resulting in a progressive decline in the respiratory control. Fatty acid-free bovine serum albumin and oligomycin could restore the State 4 respiration to normal, suggesting that the increase in the rate is due to fatty acids stimulating the ATPase activity. A correlation was observed between the respiratory control ($r = -0.73$; $P < 0.001$) and State 4 respiration ($r = 0.73$; $P < 0.001$) with the maximal tension induced in muscle biopsy strips by halothane (3%), a major triggering agent for MH syndrome [1–12].

Skeletal muscle mitochondria of MH-susceptible pigs, under aerobic or anaerobic conditions, release their endogenous Ca^{2+} significantly ($P < 0.001$) faster than normal in the presence of the uncoupling agent, FCCP [26,68], and faster than normal exogenous Ca^{2+} release is also observed at the onset of anaerobiosis [28,32]. Halothane also enhances the exogenous Ca^{2+} release of porcine MH-skeletal muscle mitochondria under both aerobic and anaerobic conditions [69].

2.3 *ATPase Activity*

Very limited studies have been reported on the mitochondrial ATPase activity of skeletal muscle mitochondria of MH-susceptible humans and pigs. In pigs, the mitochondrial ATPase activity at 25°C shows no difference between normal and MH-susceptible pigs when these organelles are isolated at 5 min post-mortem [67]. However, the ATPase activity of MH-susceptible pigs shows a 37% reduction when isolated at 30 min post-mortem but remains stable in normal mitochondria even when isolated at 1 h post-mortem. Mitochondria of MH-susceptible pigs contain significantly ($P < 0.001$) higher oligomycin-insensitive ATPase activity than normal [33]. This was postulated to be due to a higher mitochondrial endogenous Ca^{2+}-stimulated phospholipase A_2 activity [28,29], which could destroy the oligomycin sensitivity of the mitochondrial ATPase activity [70]. In MH-susceptible humans, an increase in ATPase activity due to fatty acids is shown to be responsible for an increase in the State 4 rate in the Ca^{2+}-induced State 3–State 4 transition of succinate oxidation at 40°C [51].

2.4 Transition Temperature

A change in mitochondrial membrane permeability could be inferred from changes in the transition temperature observed in the Arrhenius plots of either ADP or Ca^{2+}-stimulated respiration of substrate oxidation [71]. The Arrhenius plots of the Ca^{2+}-stimulated respiration of succinate oxidation show that there is a significant ($P < 0.001$) difference in the transition temperature between genetically selected, MH-susceptible and normal pigs, but with only small changes in the energy of activation above and below the transition temperature [28]. Mitochondria from MH-susceptible pigs show a transition temperature [26.3 ± 0.8°C ($n = 5$)] 9°C higher than normal [17.3 ± 0.8°C ($n = 4$)] and this could be restored to normal with either spermine or Mg^{2+} [28]. No difference in the transition temperature was detected between the two types of pigs when ADP was used instead of Ca^{2+} to stimulate succinate oxidation [28]. The transition temperature of skeletal muscle mitochondria of MH-susceptible and normal humans has yet to be investigated.

2.5 Phospholipase A_2 Activity

At 40°C but not at 25°C, intact and tightly-coupled skeletal muscle mitochondria of MH-susceptible humans [51] and pigs [28] accumulate only half the normal amount of exogenous Ca^{2+} during succinate oxidation in the presence of rotenone. A further increase in the amount of exogenous Ca^{2+} in the medium results in uncoupling [28,51] and large amplitude swelling [28,66]. In the presence of either fatty acid-free bovine serum albumin or phospholipase A_2 inhibitors at 40°C, however, MH-skeletal muscle mitochondria accumulate the same amount of exogenous Ca^{2+} as normal mitochondria and do not show Ca^{2+}-induced uncoupling and large amplitude swelling [28,51,72]. The 9°C higher transition temperature for porcine MH-skeletal muscle mitochondria observed in the Ca^{2+}-stimulated respiration is restored to normal with either spermine, a phospholipase A_2 inhibitor [73] or Mg^{2+}, a cation specific for inhibiting Ca^{2+} binding [74–76]. Spermine can also affect the kinetics of the Ca^{2+} transport by interacting with the Ca^{2+}-binding sites of the mitochondrial membranes [77], and Mg^{2+} can protect and stabilize the structural integrity of the mitochondrial membranes [78–79]. On stimulation by Ca^{2+} [80–83], the phospholipase A_2 causes the hydrolysis of the phospholipids in the mitochondrial membranes and this results in the liberation of unsaturated, long-chain fatty acids and the formation of lyso-derivatives [82,84–85]. The fatty acids induce uncoupling of the mitochondria [86–87], and the fatty acids and lyso-derivatives of the mitochondrial membrane phospholipids cause destabilization, resulting in swelling of the mitochondria [88–91]. The skeletal muscle mitochondria of MH-susceptible humans [92–98] and pigs [22,36,67,99–100] tend to exhibit large amplitude swelling with disrupted cristae, but when isolated intact these organelles show no apparent impairment in their capacity for oxidative phosphorylation [28,43,45–50]. The Ca^{2+}-induced large amplitude swelling in porcine MH-skeletal muscle mitochondria can be prevented by phospholipase A_2 inhibitors [28] and trifluoperazine (Cheah and Cheah, unpublished), implying that the phospholipase A_2 activity could be calmodulin-dependent. Phospholipase A_2

activity is inhibited 70% with 21 nmol spermine per mg protein [33] and 45% with 22 nmol trifluoperazine per mg protein. Under the latter conditions (Cheah and Cheah, unpublished), the amount of fatty acids formed is very similar to that produced by normal mitochondria in the absence of the inhibitor, indicating that the enzyme activity of porcine MH-skeletal muscle mitochondria is reduced to a normal level. This concentration of trifluoperazine is also sufficient to prevent the Ca^{2+}-induced uncoupling and large amplitude swelling of porcine MH-skeletal muscle mitochondria.

Mitochondria from M. longissimus dorsi of genetically selected MH-susceptible pigs contain a normal amount of phospholipase A_2 protein but its activity is significantly ($P < 0.001$) higher than normal [33]. This enzyme activity increases significantly at 12 weeks of age onwards, whereas that of normal pigs remains constant during growth [101]. Similar to phospholipase A_2 activity, the endogenous calmodulin content only increases significantly with growth in MH-susceptible pigs [101], and the level of mitochondrial calmodulin is significantly ($P < 0.01$) higher in adult MH-susceptible pigs [29]. Mitochondrial calmodulin was determined on the basis of its ability to stimulate calmodulin-deficient phosphodiesterase activity by estimating the amount of P_i released by 5′-nucleotidase [102] in boiled mitochondrial extracts. The activation profile of the Ca^{2+}-dependent phosphodiesterase activity of the isolated calmodulin of MH-skeletal muscle mitochondria is similar to that of normal pigs [72], and both are identical to those of purified rat-liver mitochondrial matrix calmodulin [103] and bovine cardiac mitochondrial matrix calmodulin [104].

3 Phospholipase Activity and Skeletal Muscle Mitochondria in Malignant Hyperthermia

It is generally agreed that in porcine and human MH there is an apparent loss of Ca^{2+} control in the skeletal muscles [14,27–29], resulting in a rise in myoplasmic Ca^{2+} [30–32]. This could occur because of either a malfunction in the Ca^{2+}-regulating organelles or a defect in the cell membrane. The increase in the level of myoplasmic Ca^{2+} [30–32] could be due either to Ca^{2+} efflux from the sacroplasmic reticulum and/or mitochondria or to an increased influx from the extracellular fluid into the muscle cells via the sarcolemma. In porcine MH, the data suggest that the increase in myoplasmic Ca^{2+} is due to Ca^{2+}-regulating organelles rather than to a Ca^{2+} influx from the extracellular fluid [36] since the total muscle Ca^{2+} starts to decrease within 5 min of the induction of MH with halothane, and is accompanied by an increase in plasma Ca^{2+} [21]. An increase in total serum Ca^{2+} [9,23] and whole blood ionized Ca^{2+} [4] also occurs during halothane-induced porcine MH.

Mitochondria have been suggested to play a role in the development of MH mainly by their ability to sequester Ca^{2+}. Uncoupling of mitochondria was initially proposed as a cause of MH since symptoms similar to MH could be experimentally induced by the uncoupling agent, 2,4-dinitrophenol, in halothane-anaesthetized dogs [34]. This hypothesis is, however, not widely accepted. Halothane, at low concentration, uncouples oxidative phosphorylation, but inhibits respiration at high concentration (see Sect. 2.1). Uncoupling of oxidative phosphorylation alone could only partially account for all the observed heat produced during MH syndrome [4,27,105]. Inhibition of respiration would result in the reduction of O_2 consumption and CO_2 production, all of which are known to increase in MH [4,9,24].

The hypothesis of an aberrant respiratory chain system [42] could not be substantiated since skeletal muscle mitochondria of MH-susceptible pigs contain a normal amount of cytochrome components [45,48] and a normal electron transport chain system. Furthermore, the majority of investigators agree that isolated mitochondria from MH-susceptible muscles of humans and pigs show normal oxidative phosphorylation and respiratory control (see Sect. 2.1).

An enhanced phospholipase A_2 activity has been postulated [28,29,33,72] to be responsible for the increase in the myoplasmic Ca^{2+} in the MH syndrome. This enzyme [28,66,72] is responsible for the Ca^{2+}-induced uncoupling, large amplitude swelling and enhanced Ca^{2+} release in MH-skeletal muscle mitochondria (see Sect. 2.5). Phospholipase A_2 and its by-products of long-chain unsaturated fatty acids from the mitochondrial phospholipids are capable of inducing Ca^{2+} release from sarcoplasmic reticulum [33], and MH-skeletal muscle mitochondria or the end products of phospholipase A_2 activity could inhibit the Ca^{2+} uptake of both the normal and MH sarcoplasmic reticulum [33]. No difference in the level of endogenous Ca^{2+} has been observed in the isolated sarcoplasmic reticulum of normal and MH-susceptible pigs [106,107], but significantly higher than normal amounts of endogenous Ca^{2+} have been detected in the isolated MH-skeletal muscle mitochondria [28,36]. These observations suggest that skeletal muscle mitochondria are implicated in the MH syndrome. Skeletal muscle mitochondria of MH-susceptible humans [51] and pigs [28] have a lower than normal capacity to retain Ca^{2+} because of their significantly higher than normal endogenous phospholipase A_2 activity. This results in large amplitude mitochondrial swelling accompanied simultaneously by an enhanced Ca^{2+} release [28]. The excess Ca^{2+} and the long-chain unsaturated fatty acids released would then induce additional Ca^{2+} release from the sarcoplasmic reticulum [33] probably by the mechanism of calcium-induced calcium release [108,109] and/or fatty acid-induced Ca^{2+} release [33], ultimately leading to an increase in the level of myoplasmic Ca^{2+} [30,32]. The increase in myoplasmic Ca^{2+} is then responsible for the increase in muscle hyper-rigidity and the enhanced rate of glycolysis in the MH syndrome since muscle contraction is dependent on the concentration of free Ca^{2+} in the sarcoplasm [110,111], and the rate of glycolysis is enhanced by Ca^{2+} stimulating the myofibrillar ATPase [112] and phosphorylase kinase [113–115]. The enormous rise in the temperature to 42°C or higher [4,9] is due to a combination of several processes. Heat production in the initial 10 min in porcine MH syndrome could be explained by an increase in aerobic metabolism [17] associated with shortening and contraction of muscles [116], and in subsequent periods mainly by an increase in lactate formation [7,17,21].

Skeletal muscle mitochondria have also been suggested to play an important role in halothane-induced porcine MH syndrome without the involvement of Ca^{2+} [35,117]. Halothane is implicated to affect mitochondrial function by completely inhibiting the transport of NADH and the dehydrogenation of pyruvate. The inhibition of mitochondrial respiration by halothane would result in an enhanced production of lactate, a reduction in ATP synthesis and subsequently a decrease in muscle creatine phosphate and pH. The decrease in muscle pH due to lactate has been suggested to stimulate the release of catecholamines which would then further accelerate glycogenolysis [117]. However, this hypothesis is difficult to reconcile. Firstly, O_2 consumption increases rather than decreases during the MH syndrome [4,9,24]. Secondly, drug-induced porcine MH results in an increase in myoplasmic Ca^{2+} in vivo [36]. Thirdly, halothane inhibits the release of catecholamines [118]. It has been suggested that halothane causes gross changes in the defective mitochondrial membranes and thereby inhibits the dehydrogenation of pyruvate [35,117], but no explanation has been offered for the cause of the defective mitochondrial membranes present only in MH-susceptible pigs [117]. The lesion responsible for the occurrence of defective mitochondrial membranes could be explained by an enhanced phospholipase A_2 activity. It is conceivable that in drug-induced porcine MH, the inhibition of mitochondrial dehydrogenation of pyruvate could result in a further increase in the rate of lactate formation [117]. However, this could not be the primary lesion since porcine MH [7,18–20] and to a certain extent human MH [4,119] could be initiated, for example, by high environmental temperature without the use of any volatile anaesthetic. Under these conditions, an increase in membrane permeability through activation of phospholipase A_2 would occur, ultimately resulting in an increase in myoplasmic Ca^{2+} as previously discussed.

Human and porcine MH can be effectively prevented by dantrolene [4], a specific skeletal muscle relaxant [120] which acts by reducing the amount of Ca^{2+} released into the myoplasm [121–123]. Dantrolene (20 μM) has no effect on the Ca^{2+} transport system in purified sarcoplasmic reticulum from both MH-susceptible and normal pigs [124] but inhibits mitochondrial Ca^{2+} release and phospholipase A_2 activity. In MH-susceptible pigs, dantrolene (40 μM) inhibits mitochondrial Ca^{2+} release by 44% and phospholipase A_2 activity by 29% (Cheah and Cheah, unpublished data). A possible mitochondrial defect in MH-susceptible patients was also suggested by a recent non-invasive ^{31}P-NMR study [125]. This report shows that MH-susceptible patients have a significantly higher value for P_i/PCr ratio than normal, and they also show a significantly slower post-exercise recovery rate for the value of PCr/P_i, which is a useful indicator of mitochondrial function [126].

Phospholipase A_2 is widely distributed in various tissues [29]. Free fatty acids generated by this enzyme appear to play a key role in the development of MH syndrome. Significantly higher than normal levels of fatty acids are produced in MH-susceptible humans [127,128] and in pigs ([129], Cheah and Cheah, unpublished data). These fatty acids may directly cause Ca^{2+} release from the Ca^{2+}-regulating organelles [28,29,33] in addition to potentiating the effect of halothane in skeletal muscle [127] and in red blood cells [130].

References

1. Denborough MA, Forster JFA, Lovell RRH, Maplestone PA, Villiers JD (1962) Anaesthetic deaths in a family. Brit J Anaesth 34:395–396
2. Britt BA, Kalow W (1968) Hyper-rigidity and hyperthermia associated with anesthesia. Ann NY Acad Sci 151:947–958
3. Britt BA, Kalow W (1970) Malignant hyperthermia: a statistical review. Can Anaesth Soc J 17:293–315
4. Gronert GA (1980) Malignant hyperthermia. Anesthesiology 53:395–423
5. Denborough MA (1980) The pathopharmacology of malignant hyperpyrexia. Pharmacol Ther 9:357–365
6. Harrison GG, Biebuyck JF, Terblanche J, Dent DM, Hickman R, Saunders SJ (1968) Hyperpyrexia during anaesthesia. Brit Med J 3:594–595
7. Topel DG, Bicknell EJ, Preston KS, Christian LL, Matsushima CY (1968) Porcine stress syndrome. Mod Vet Pract 49:40–41
8. Sybesma W, Eikelenboom G (1969) Malignant hyperthermia syndrome in pigs. Neth J Vet Sci 2:155–160
9. Berman MC, Harrison GG, Bull AB, Kench JE (1970) Changes underlying halothane-induced malignant hyperthermia in Landrace pigs. Nature 225:653–655
10. Hall LW, Trim CM, Woolf N (1972) Further studies of porcine malignant hyperthermia. Brit Med J 2:145–148
11. Nelson TE, Jones EW, Venable JH, Kerr DD (1972) Malignant hyperthermia of Poland China swine. Anesthesiology 36:52–56
12. Clark MG, Williams CH, Pfeifer WF, Blosham DP, Holland PC, Taylor CA, Lardy HA (1973) Accelerated substate cycling of fructose-6-phosphate in the muscle of malignant hyperthermia pigs. Nature 245:99–101
13. Ryan JF, Papper EM (1970) Malignant fever during and following anesthesia. Anesthesiology 32:196–201
14. Moulds RFW, Denborough MA (1974) Biochemical basis of malignant hyperpyrexia. Brit Med J 2:241–244
15. Hall LW, Woolf N, Bradley JWP, Jolly DW (1966) Unusual reaction to suxamethonium chloride. Brit Med J 4:1305
16. Gronert GA, Theye RA (1976) Suxamethonium-induced porcine malignant hyperthermia. Brit J Anaesth 48:513–517
17. Hall GM, Bendall JR, Lucke JN, Lister D (1976) Porcine malignant hyperthermia. II Heat production. Brit J Anaesth 48:305–308
18. Allen WM, Berrett S, Harding JDJ, Patterson DSP (1970) Experimentally-induced acute stress syndrome in Pietrain pigs. Vet Rec 87:64–69
19. Allen WM, Hebert CN, Smith LP (1974) Deaths during and after transportation of pigs in Great Britain. Vet Rec 94:212–214
20. Williams CH, Houchins C, Shanklin MD (1975) Pigs susceptible to energy metabolism in the fulminant hyperthermia stress syndrome. Brit Med J 3:411–413
21. Berman MC, Kench JE (1973) Biochemical features of malignant hyperthermia in Landrace pigs. In: Gordon RA, Britt BA, Kalow W (eds) Int Symp Malignant Hyperthermia. Thomas, Springfield, USA, pp 287–297
22. Brucker RF, Williams, CH, Popinigis J, Galvez TL, Vail WJ, Taylor CA (1973) In vitro studies on liver mitochondria and skeletal muscle sarcoplasmic reticulum fragments isolated from hyperpyrexic swine. In: Gordon RA, Britt BA, Kalow W (eds) Int Symp Malignant Hyperthermia. Thomas, Springfield, USA, pp 238–270
23. Van den Hende C, Lister D, Muuylle E, Ooms L, Oyaert W (1976) Malignant hyperthermia in Belgian Landrace pigs rested or exercised before exposure to halothane. Brit J Anaesth 48:821–829
24. Williams CH, Shanklin MD, Houchins C (1977) Energy metabolism in fulminant hyperthermia stress syndrome in swine. In: Henschel EO (ed) Malignant hyperthermia: currents concepts. Appleton-Century-Crofts, New York, pp 149–156
25. Webb AJ (1980) The incidence of halothane sensitivity in British pigs. Anim Prod 31:101–105
26. Wingaard DW, Gatz EE (1978) Some observations on stress-susceptible patients. In: Aldrete JA, Britt BA (eds) 2nd Int Symp Malignant Hyperthermia. Grune & Stratton, New York, pp 363–372
27. Britt BA (1987) Aetiology and pathophysiology of malignant hyperthermia. In: Britt BA (ed) Malignant Hyperthermia. Nijhoff, Boston, pp 11–42

28. Cheah KS, Cheah AM (1981) Mitochondrial calcium transport and calcium-activated phospholipase in porcine malignant hyperthermia. Biochim Biophys Acta 634:70–84
29. Cheah KS, Cheah AM (1985) Malignant hyperthermia: molecular defects in membrane permeability. Experientia 41:656–661
30. Lopez JR, Alamo L, Caputo C, Wikinski L, Ledezma D (1985) Intracellular ionized calcium concentration in muscles from humans with malignant hyperthermia. Muscle Nerve 8:355–358
31. Lopez JR, Alamo LA, Jones DE, Papp L, Allen PD, Gergely J, Sreter FA (1986) $[Ca^{2+}]_i$ in muscles of malignant hyperthermia susceptible pigs determined in vivo with Ca^{2+} selective microelectrodes. Muscle Nerve 9:85–86
32. Cheah KS, Cheah AM, Crosland AR, Casey JC, Webb AJ (1984) Relationship between calcium release, sarcoplasmic calcium, glycolysis and meat quality in halothane-sensitive and halothane-insensitive pigs. Meat Sci 10:117–130
33. Cheah KS, Cheah AM (1981) Skeletal muscle mitochondrial phospholipase A_2 and the interaction of mitochondria and sarcoplasmic reticulum in porcine malignant hyperthermia. Biochim Biophys Acta 638:40–49
34. Wilson RD, Nichols Jr. RJ, Dent TE, Allen CR (1966) Disturbances of oxidative phosphorylation mechanism as a possible etiological factor in sudden unexplained hyperthermia occurring during anesthesia. Anesthesiology 27:231–232
35. Mitchell G, Heffron JJA (1980) Porcine stress syndromes: a mitochondrial defect. S Afr J Sci 76:546–551
36. Stadhouders AM, Viering WAL, Verburg MP, Ruitenbeek W, Sengers RCA (1984) In vivo induced malignant hyperthermia in pigs. III Localization of calcium in skeletal muscle mitochondria by means of electron microscopy and microprobe analysis. Acta Anaesthesiol Scand 28:14–26
37. Chance B, Williams GR (1956) The respiratory chain and oxidative phosphorylation. Adv Enzymol, Interscience, New York, pp 65–134
38. Mela L (1977) Mechanism and physiological significance of calcium transport across mammalian mitochondrial membranes. In: Bronner F, Kleinzeller A (eds) Current topics in membrane transport. Academic Press, New York, pp 321–366
39. Lehninger AL (1970) Mitochondria and calcium ion transport. Biochem J 119:129–138
40. Martonosi AN (1984) Mechanisms of Ca^{2+} release from sarcoplasmic reticulum of skeletal muscle. Physiol Rev 64:1240–1320
41. Greaser ML, Cassens RG, Briskey EJ, Hoekstra WG (1969) Post-mortem changes in subcellular fractions from normal and pale, soft, exudative porcine muscle. 2. Electron microscopy. J Food Sci 34:125–132
42. Eikelenboom G, Van den Bergh SG (1973) Mitochondrial metabolism in stress-susceptible pigs. J Anim Sci 37:692–696
43. Gronert GA, Heffron JJA (1979) Skeletal muscle mitochondria in porcine malignant hyperthermia: respiratory activity, calcium functions and depression by halothane. Anesth Anal 58:76–81
44. Britt BA, Endrenyi L, Cadman DL, Fan HM, Fung HYK (1975) Porcine malignant hyperthermia: effects of halothane on mitochondrial respiration and calcium accumulation. Anesthesiology 42:292–300
45. Cheah KS (1973) Comparative studies of the mitochondrial properties of longissimus dorsi muscles of Piertrain and large white pigs. J Sci Food Agric 24:51–61
46. Brooks GA, Cassens RG (1973) Respiratory functions of mitochondria isolated from stress-susceptible and stress-resistant pigs. J Anim Sci 37:688–691
47. Campion DR, Olson JC, Topel DG, Christian LL, Kuhlers DL (1975) Mitochondrial traits of muscle from stress-susceptible pigs. J Anim Sci 41:1314–1317
48. Ruitenbeek W, Verburg MP, Janssen AJM, Stadhouders AM, Sengers RCA (1984) In vivo induced malignant hyperthermia in pigs. II. Metabolism of skeletal muscle mitochondria. Acta Anaesthesiol Scand 28:9–13
49. Denborough MA, Hird FJR, King JO, Marginson MA, Mitchelson KR, Nayler WG, Rex MA, Zapf P, Condron RJ (1973) Mitochondria and other studies in Australian Landrace pigs affected with malignant hyperthermia. In: Gordon RA, Britt BA, Kalow W (eds). Int Symp Malignant Hyperthermia. Thomas, Springfield, USA, pp 229–237
50. Britt BA, Kalow W, Endrenyi L (1973) Malignant hyperthermia and the mitochondria in human patients. In: Gordon RA, Britt BA, Kalow W (eds) Int Symp Malignant hyperthermia. Thomas, Springfield, USA. pp 387–398

51. Cheah KS, Cheah AM, Fletcher JE, Rosenberg H (1989) Skeletal muscle mitochondrial respiration of malignant hyperthermia-susceptible patients. Ca^{2+}-induced uncoupling and free fatty acids. Int J Biochem (in press)
52. Mitchelson KR, Hird FJR (1973) Effect of pH and halothane on muscle and liver mitochondria. Amer J Physiol 225:1393–1398
53. Britt BA, Kalow W, Gordon A, Humphrey JG, Rewcastle NB (1973) Malignant hyperthermia: an investigation of five patients. Can Anaesth Soc J 20:431–467
54. Harris RA, Munroe J, Farmer B, Kim KC, Jenkins P (1971) Action of halothane upon mitochondrial respiration. Arch Biochem Biophys 142:435–444
55. Grist EM, Baum H (1974) A possible mechanism for the halothane-induced inhibition of mitochondrial respiration: binding of endogenous calcium to NADH dehydrogenase. FEBS Lett 45:41–44
56. Britt BA, Kalow W, Endrenyi L (1972) The effects of halothane and methoxyflurane on rat skeletal muscle mitochondria. Biochem Pharmacol 21:1159–1169
57. Miller RN, Hunter FE (1970) The effect of halothane on electron transport, oxidative phosphorylation, and swelling in rat liver mitochondria. Mol Pharmacol 6:67–77
58. Hall GM, Kirtland SJ, Baum H (1973) The inhibition of mitochondrial respiration by inhalational anaesthetic agents. Brit J Anaesth 45:1005–1009
59. Bygrave FL (1978) Mitochondria and the control of intracellular calcium. Biol Rev 53:43–79
60. Carafoli E, Crompton M (1978) Regulation of intracellular calcium. Curr Top Membr Transp 10:151–261
61. Fiskum G, Lehninger A (1980) The mechanisms and regulation of mitochondrial Ca^{2+} transport. Fed Proc 39:2432–2436
62. Carafoli E (1982) The transport of Ca^{2+} by mitochondria. In: Martonosi A (ed) Membranes and transport. Plenum Press, New York, Vol 1, pp 611–616
63. Akerman KEO, Nicholls DG (1983) Physiological and bioenergetic aspects of mitochondrial calcium transport. Rev Physiol Biochem Pharmacol 95:149–201
64. Somers CJ, McLoughlin JV (1982) Malignant hyperthermia in pigs: calcium ion uptake by mitochondria from skeletal muscle of susceptible animals given neuroleptic drugs and halothane. J Comp Path 92:191–198
65. Heffron JJA (1984) Mitochondrial and plasma membrane changes in skeletal muscle in malignant hyperthermia syndrome. Biochem Soc Trans 12:360–362
66. Cheah KS, Cheah AM (1978) Calcium movements in skeletal muscle mitochondria of malignant hyperthermic pigs. FEBS Lett 95:307–310
67. Greaser ML, Cassens RG, Briskey EJ, Hoekstra WG (1969). Post-mortem changes in subcellular fractions from normal and pale, soft and exudative porcine muscle. 1. Calcium accumulation and adenosine triphosphatase activities. J Food Sci 34:120–124
68. Cheah KS, Cheah AM (1979) Mitochondrial calcium, erythrocyte fragility and porcine malignant hyperthermia. FEBS Lett 106:265–268
69. Cheah KS, Cheah AM (1976) The trigger for PSE condition in stress-susceptible pigs. J Sci Food Agric 27:1137–1144
70. Kagawa J, Racker E (1966) Partial resolution of the enzymes catalyzing oxidative phosphorylation. VIII. Properties of a factor conferring oligomycin sensitivity of the mitochondrial adenosine triphosphatase. J Biol Chem 241:2461–2466
71. Raison JK (1973) The influence of temperature-induced phase changes on the kinetics of respiratory and other membrane-associated enzyme systems. Bioenergetics 4:285–309
72. Cheah KS (1987) Mitochondria and malignant hyperthermia. In: Britt BA (ed) Malignant Hyperthermia. Nijhoff, Boston, pp 79–102
73. Sechi AM, Cabrini L, Landi L, Pasquali P, Lenaz G (1978) Inhibition of phospholipase A_2 and phospholipase C by polyamines. Arch Biochem Biophys 186:248–254
74. Akerman KEO (1978) Effect of pH and Ca^{2+} on the retention of Ca^{2+} by rat liver mitochondria. Arch Biochem Biophys 189:256–262
75. Jacobus WE, Brierley GP (1969) Ion transport by heart mitochondria. XVI. Cation binding by submitochondrial particles. J Biol Chem 244:4995–5004
76. Sordahl LA (1975) Effects of magnesium, ruthenium red and the antibiotic ionophore A-23187 on initial rates of calcium uptake and release by heart mitochondria. Arch Biochem Biophys 167:104–115

77. Akerman KEO (1977) Effect of Mg^{2+} and spermine on the kinetics of Ca^{2+} transport in rat liver mitochondria. J Bioenerg Biomembr 9:65–72
78. Packer L. Utsumi K, Mustafa G (1966) Oscillatory states of mitochondria. 1. Electron and energy transfer pathways. Arch Biochem Biophys 117:381–393
79. Leblanc P, Clauser H (1974) ADO and Mg^{2+} requirement for Ca^{2+} accumulation by hog heart mitochondria: correlation with energy coupling. Biochim Biophys Acta 347:87–101
80. Bjornstad P (1966) Phospholipase activity in rat liver mitochondria studied by the use of endogenous substrates. J Lipid Res 7:612–620
81. Waite M, Scherphof GL, Boshouwers FMG, Van Deenen LLM (1969) Differentiation of phospholipases A in mitochondria and lysosomes of rat liver. J Lipid Res 10:411–420
82. Nachbaur J, Colbeau A, Vignais PM (1972) Distribution of membrane-confined phospholipase A in the rat hepatocyte. Biochim Biophys Acta 274:426–446
83. Zurini M, Hugentobler G, Gazzotti P (1981) Activity of phospholipase A_2 in the inner membrane of rat-liver mitochondria. Eur J Biochem 119:517–521
84. Scherphof GL, Waite M, Van Deenen LLM (1966) Formation of lysophosphatidyl ethanolamines in cell fractions of rat liver. Biochim Biophys Acta 125:406–409
85. Honjo I, Ozawa K (1968) Lysolecithin inhibition of mitochondrial metabolism. Biochim Biophys Acta 162:624–627
86. Vazquez-Colon L, Ziegler FD, Elliot WB (1966) On the mechanism of fatty acid inhibition of mitochondrial metabolism. Biochemistry 5:1134–1139
87. Weinbach EC, Garbus J, Glaggett CE (1966) The rapid restoration of respiratory control to uncoupled mitochondria. J Biol Chem 241:3708–3717
88. Lehninger AL, Remmert LF (1959) An endogenous uncoupling and swelling agent in liver mitochondria and its enzymic formation. J Biol Chem 234:2459–2464
89. Avi-Dor Y (1960) A spectrophotometric study of the spontaneous and microsome-induced swelling of rat-liver mitochondria. Biochim Biophys Acta 39:53–61
90. Zborowski J, Wojtczak L (1963) Induction of swelling of liver mitochondria by fatty acids of various chain length. Biochim Biophys Acta 70:596–598
91. Aleksandrowicz Z, Swierczynski J, Wrzolkowa T (1973) Protective effect of nupercaine on mitochondrial structure. Biochim Biophys Acta 305:59–66
92. Carpenter GC (1966) Rhabdomyolysis after routine administration of succinylcholine in children. Soc Paediat Res 175:29–30
93. Isaacs H, Frere G, Mitchell J (1973) Histological, histochemical and ultramicroscopic findings in muscle biopsies from carriers of the trait for malignant hyperpyrexia. Brit J Anaesth 45:860–868
94. Schiller HH, Mair WGP (1974) Ultrastructural changes of muscle in malignant hyperthermia. J Neurol Sci 21:93–100
95. Reske-Nielsen E, Haase J, Kelstrup J (1975) Malignant hyperthermia in a family. Acta Path Microbiol Scand Sect A 83:651–660
96. Isaacs M (1987) Myopathy and malignant hyperthermia. In: Aldrete JA, Britt BA (eds) 2nd Int Symp Malignant Hyperthermia. Grune & Stratton, New York, pp 89–102
97. Hull MT, Muller J, Albrecht WH (1978) Morphologic abnormalities in a case of malignant hyperthermia. Anesthesiology 48:223–228
98. Reske-Nielsen E (1978) Malignant hyperthermia in Denmark: survey of a family study and investigations into muscular morphology in ten cases. In: Aldrete JA, Britt BA (eds) 2nd Int Symp Malignant Hyperthermia. Grune & Stratton, New York, pp 287–327
99. Bergmann V (1979) Changes of cardiac and skeletal muscle in pigs following transport stress. An electron microscopic study. Exp Path 17:243–248
100. Cloke JD, Davis EA, Gordon J, Hsieh S-I, Grider J, Addis PB, McGrath CJ (1981) Scanning and transmission electron microscopy of normal and PSE porcine muscle. Scanning electron microscopy, 3 SEM, AMF O'Hare, Chicago, pp 435–446
101. Cheah KS, Cheah AM, Waring JC (1986) Phospholipase A_2 activity, calmodulin, calcium and meat quality in young and adult halothane-sensitive and halothane-insensitive British Landrace pigs. Meat Sci 17:37–53
102. Teo TS, Wang TH, Wang JH (1973) Purification and properties of the protein activator of bovine heart cyclic adenosine 3′,5′-monophosphate phosphodiesterase. J Biol Chem 248:588–595
103. Hatase O, Tokuda M, Itano T, Matsui H, Doi A (1982) Purification and characterization of calmodulin from rat liver mitochondria. Biochem Biophys Res Commun 104:673–679

104. Hatase O, Tokunda M, Sharma RK, Wang JH, Green DE (1983) Purification and characterization of the heat-stable calmodulin-binding protein from the matrix of bovine heart mitochondria. Biochem Biophys Res Commun 113:633–637
105. Wang JK, Moffitt EA, Rosevear JW (1969) Oxidative phosphorylation in acute hyperthermia. Anesthesiology 30:439–442
106. McIntosh DB, Berman MC, Kench JE (1977) Characteristics of sarcoplasmic reticulum from slowly glycolysing and from rapidly glycolysing pigs skeletal muscle post-mortem. Biochem J 166:387–398
107. Ohnishi ST, Taylor S, Gronert GA (1983) Calcium-induced calcium release from sarcoplasmic reticulum of pigs susceptible to malignant hyperthermia. FEBS Lett 161:103–107
108. Endo M (1977) Calcium release from sarcoplasmic reticulum. Physiol Rev 57:71–108
109. Miyamoto H, Racker E (1981) Calcium-induced calcium release at terminal cisternae of skeletal sarcoplasmic reticulum. FEBS Lett 133:235–238
110. Ebashi S, Endo M, Ohtsuki I (1969) Control of muscle contraction. Quart Rev Biophys 2:351–384
111. Weber A, Murray JM (1973) Molecular control mechanisms in muscle contraction. Physiol Rev 53:612–673
112. Scopes RK (1974) Studies with a reconstituted muscle glycolytic system: the rate and extent of glycolysis in simulated post-mortem conditions. Biochem J 142:79–86
113. Ozawa E, Hosoi K, Ebashi S (1967) Reversible stimulation of muscle phosphorylase b kinase by low concentrations of calcium ions. J Biochem (Tokyo) 61:531–533
114. Heilmeyer LMG Jr., Meyer F, Haschke RH, Fischer EH (1970) Control of phosphorylase activity in a muscle glycogen particle. II. Activation by calcium. J Biol Chem 245:6649–6656
115. Brostrom CO, Hunkeler FL, Krebs EC (1971) The regulation of skeletal muscle phosphorylase kinase by calcium ions. J Biol Chem 246:1961–1967
116. McMahon TA (1984) Muscle heat and fuel. In: McMahon TA (ed) Muscles, reflexes and locomotion. Princeton Univ Press, Princeton, New Jersey, pp 27–52
117. Mitchell G, Heffron JJA, Van Rensburg AJJ (1980) A halothane-induced biochemical defect in muscle of normal and malignant hyperthermia-susceptible Landrace pigs. Anesth Analg 59:250–256
118. Rorie KD, Tyce GM, Mackenzie RA (1984) Evidence that halothane inhibits norepinephrine release from sympathetic nerve endings in dog saphenous vein by stimulation of presynaptic inhibitory muscarinic receptors. Anesth Analg 63:1059–1064
119. Britt BA (1974) Malignant hyperthermia: a pharmacogenetic disease of skeletal and cardiac muscle. N Engl J Med 290:1140–1142
120. Snyder HR, Davis CS, Bickerton RK, Halliday RP (1967) 1(5-Arylfurfurylidene)amino hydantoins. A new class of muscle relaxants. J Med Chem 10:807–810
121. Hainaut K, Desmedt JE (1974) Effect of dantrolene sodium on calcium movements in single muscle fibres. Nature 252:728–730
122. Morgan KG, Bryant SH (1977) The mechanism of action of dantrolene sodium. J Pharmacol Exp Ther 201:138–147
123. Desmedt JE, Haunaut K (1977) Inhibition of the intracellular release of calcium by dantrolene in barnacle giant muscle fibres. J Physiol 265:565–585
124. White MD, Collins JG, Denborough MA (1983) The effect of dantrolene on skeletal-muscle sarcoplasmic reticulum function in malignant hyperpyrexia in pigs. Biochem J 212:399–405
125. Olgin J, Argove Z, Rosenberg H, Tuchler M, Chance B (1988) Non-invasive evaluation of malignant hyperthermia susceptibility with phosphorus nuclear magnetic resonance spectroscopy. Anesthesiology 68:507–513
126. Argove Z, Bank WJ, Maris J, Peterson P, Chance B (1987) Bioenergetic heterogeneity of human mitochondrial myopathies as demonstrated by in vivo phosphorus magnetic resonance spectroscopy. Neurology 37:257–262
127. Fletcher JE, Rosenberg H (1986) In vitro muscle contractures induced by halothane and suxamethonium. Brit J Anaesth 58:1433–1439
128. Campbell IT, Ellis FR, Evans RT, Mortimer MG (1983) Studies of body temperatures, blood lactate, cortisol and free fatty acid levels during exercise in human subjects susceptible to malignant hyperthermia. Acta Anaesthesiol Scand 27:349–355
129. Spencer GSG (1980) Relationship between plasma somatomedin activity and levels of cortisol and free fatty acids following stress in pigs. J Endocrinol 84:109–114
130. Fletcher JE, Kistler P, Rosenberg H, Michaux K (1987) Dantrolene and mepacrine antagonize the hemolysis of human red blood cells by halothane and bee venom phospholipase A_2. Toxicol Appl Pharmacol 90:410–419

The Antimitochondrial Antibodies (AMA) of Primary Biliary Cirrhosis (PBC)

H. BAUM[1] and M. FUSCONI[2]

1 Introduction

Primary biliary cirrhosis (PBC) is a chronic liver disease whose etiology is not yet established and which affects predominately middle-aged women, with no obvious geographical clustering or genetic predisposition. It is characterized by progressive inflammatory obliteration of the intrahepatic bile ducts. Nevertheless, the clinical presentation and disease progression are quite variable, so that the disease may be virtually asymptomatic or may lead to inexorable liver failure and death. Clinical features of the disease and evidence of its underlying autoimmune basis have been extensively reviewed elsewhere [1]. Two points, however, are worth stressing. Firstly, notwithstanding the extensive immunological disturbances in the disease, and in particular the striking occurrence of high and persistent titres of antimitochondrial antibodies (AMA) (see below), a cellular and molecular explanation for the damage to the intrahepatic bile ducts remains to be established. Secondly, there are reports that the disease may re-occur in the donor organ after liver transplantation [2].

2 The Antimitochondrial Antibodies in PBC

One of the most characteristic features of PBC, and one that has attracted much attention, is the occurrence of non-organ, non-species-specific AMA of both the IgG and IgM classes. These were first detected as granular cytoplasmic staining in immunofluorescence microscopy (IFL), some years after the original discovery of complement-fixing autoantibodies to tissue extracts [3]. AMA were later shown to be detectable by a variety of techniques including complement fixation, ELISA, radioimmunoassay and immunodiffusion [4,5,6]. However, IFL remained the standard test for AMA and these were shown to have high diagnostic value for PBC, although titres did not correlate to severity or histological stage of the disease [1].

[1]Department of Biochemistry, King's College London, Campden Hill Road, London W8 7AH, UK
[2]Cattedra di Semeiotica Medica, Istituto di Clinica Medica II, Policlinico S Orsola, Via Massarenti 9, 40138 Bologna, Italy

A. Azzi, Z. Drahota and S. Papa
Molecular Basis of Membrane-Associated Diseases

3 The "M" Classification of Mitochondrial Autoantigens

Although IFL is of value in the diagnosis of PBC, it became clear that it is not entirely specific and that, moreover, using a standard block of tissue sections, the appearance and tissue distribution of the cytoplasmic staining is variable. This is partly because some IFL patterns arise from reaction of specific mitochondrial autoantigens with AMA diagnostic of conditions other than PBC. For example, the pattern given by the antigen designated as M1 is due to the anti-cardiolipin antibodies of syphilis, cardiolipin of course being a characteristic constituent of the inner mitochondrial membrane.

In course of time nine "M" antigens have been defined, partly on the basis of IFL patterns but also on the basis of properties and localization of antigens as detected by other techniques such as complement fixation [7]. These, and their clinical correlations are listed in Table 1.

4 M2 and M4 as Markers for PBC

Although the M8 and M9 antigens may prove to be of importance in predicting the prognosis of PBC (with M9 an indicator of an early, benign form of the disease [8]), they are as yet very poorly defined. One important point about M9, however, is that the AMA directed against it are also detectable in close relatives of PBC patients and in laboratory workers studying PBC. A recent and important finding about

Table 1. The 'M' classification of specificities of antimitochondrial antibodies

	Occurrence	Organ specificity	Mitochondrial location[a]
M1	Secondary syphilis	None	Inner membrane (cardiolipin)
M2	Primary biliary cirrhosis	None	Inner membrane (loose attachment)
M3	Venocuran-induced pseudolupus	None	Outer membrane
M4	"Mixed form" of chronic cholestatic liver disease, primary biliary cirrhosis	None	Outer membrane (trypsin insensitive)
M5	Particular subgroup of systemic lupus erythematosus	None	Outer membrane
M6	Iproniazid-induced heptatitis	Liver	Outer membrane
M7	Cardiomyopathies	Myocardium	Inner membrane
M8	Primary biliary cirrhosis (indicator of poor prognosis?)	None	Outer membrane
M9	Primary biliary cirrhosis (indicator of benign form?) healthy subjects (particularly relatives and laboratory workers studying PBC)	Liver, kidney	Outer membrane (trypsin sensitive)

[a] Some "outer membrane antigens" may be associated with other cell membranes.

these AMA will be mentioned at the end of this review. Otherwise we shall now only be concerned with the antigens designated as M2 and M4.

The M2, inner membrane associated antigen is very strongly clinically correlated with PBC. One of its defining characteristics was its disappearance on treatment of tissue sections with trypsin [4]. M4 is trypsin insensitive and probably located on the outer mitochondrial membrane [4]. Some PBC patients exhibit both M2 and M4 reactivity, whilst some sera may show only the M4 pattern. It may be that M4 reactivity reflects an "overlap" of PBC with chronic active hepatitis [4].

The best definition of M2 reactivity is not the interpretation of IFL patterns but the spectrum of bands visualized by PBC sera on immunoblots of mitochondria. These will be discussed below, but a key feature is a prominent band of MW 70–74 kD, the precise MW being species-specific [9]. Virtually all high titre PBC sera react strongly with this autoantigen, so that it has become looked upon as almost the definitive marker for PBC. However, sera from a minority of patients with clinically-defined PBC, and showing strongly positive IFL for AMA, seem to react exclusively with a mitochondrial autoantigen with a non-species-specific MW of around 52 kD [10]. Competitive ELISA experiments suggest that this antigen shares no common epitopes with the M2 antigens, it is trypsin insensitive and the AMA directed against it react poorly with submitochondrial particles or preparations of chloroform released ATPase (see below). Consequently, it has been concluded [11] that the 52 kD protein may represent the outer membrane associated M4 antigen, despite earlier reports that the latter could not be visualized by immunoblotting.

An interesting related observation was that in a comparison of the sera from patients in histologically-defined early and late stages of the disease, the percentage of sera reacting only with the 52 kD band remained constant. However, there was a significant increase, with disease progression, in the number of sera exhibiting "mixed form" reactivity (M2 plus M4), with a corresponding decrease in the number reacting only with the M2 antigens [11]. Despite this indication of the clinical significance of M4, very little further work has yet been reported on its chemical properties or possible identity, with the exception of an intriguing observation in relation to the ageing of human diploid fibroblasts [12]. Little further consideration will therefore be given to this antigen in the present review.

The M2 pattern of autoantigenic bands on immunoblots is of variable complexity (see below), but usually includes prominent bands at 74, 55 and 43 kD, in bovine mitochondria, probably corresponding to major bands at 70, 45 and 39 kD in human mitochondria. On the basis of the frequency distribution of such patterns it has been possible to compare the AMA of PBC with those of subjects with high AMA titres by IFL and ELISA, who are nonetheless asymptomatic and have no histological evidence of PBC. A group of such subjects, including patients with a variety of other autoimmune diseases as well as apparently healthy subjects, detected at routine screening, were shown to be indistinguishable from PBC patients in the pattern of reactivity of their sera with mitochondria on immunoblots [11] (Table 2). There is, as yet, no evidence that in the longer term such subjects may

Table 2. Comparison of immunoblotting and ELISA results of PBC and non-PBC M2-M4 IFL AMA positive sera[a]

			74 kD	58 kD	55 kD	52 kD	51 kD	46 kD	43 kD	IB-
PBC	ELISA++	41(73%)	41(100%)	4(10%)	38(93%)	21(51%)	20(49%)	5(12%)	30(73%)	–
AMA	ELISA+	11(20%)	6(55%)	–	4(36%)	6(55%)	–	–	1(9%)	–
(56)	ELISA–	4(7%)	–	–	1(25%)	4(100%)	–	–	–	–
Non-										
PBC	ELISA++	9(42%)	9(100%)	1(11%)	8(89%)	5(56%)	4(44%)	1(11%)	3(33%)	–
AMA	ELISA+	6(29%)	2(33%)	–	–	2(33%)	–	–	2(33%)	3(50%)
(21)	ELISA	6(29%)	1(17%)	–	2(33%)	2(33%)	–	–	–	3(50%)

[a] All sera were IFL positive for AMA. It should be noted that PBC and non-PBC sera reacting strongly against submitochondrial particles (predominantly inner membrane) by ELISA (++), show very similar frequencies of reaction on immunoblots against mitochondrial peptides separated by SDS-PAGE.

develop PBC. What is clear from these findings is that M2-reactive AMA do not uniquely arise as a secondary consequence of a primary ductular lesion of independent origin.

5 Early Attempts at the Characterization of the M2 Autoantigen

Electron microscopy [13] and submitochondrial fractionation, with monitoring by CFT [14], localized the M2 antigen to the inner mitochondrial membrane [15]. The amount of antigen tended to be related to the respiratory capacity of the mitochondria tested, so that in quantitative CFT, mitochondria from brown adipose tissue were more reactive than those from bovine heart, which were in turn much more antigenic than mitochondria from rat liver. Antigenicity was heat insensitive and insensitive to periodate, glycosidases and nucleases, but significantly sensitive to mercurials [16]. Attempts were made to identify the antigen by inhibition of mitochondrial function with AMA, but PBC sera were without effect on any of the individual or coupled reactions of submitochondrial particles. An apparent advance was the demonstration that the antigen co-purified with the F_1-ATPase, both in its release by chloroform or chaotropic agents from the inner membrane and its migration in native polyacrylamide gel electrophoresis [17]. However, it subsequently proved possible to separate the antigen from the functional ATPase and to demonstrate that it was not one of the associated protein factors [18].

6 Definition of M2 by Immunoblotting

The application of immunoblotting greatly accelerated progress in the definition of the M2 autoantigen. Immunoblots of subcellular fractions of rat liver, obtained by differential or density gradient centrifugation, confirmed the mitochondrial location of the antigen but revealed that it could be resolved into a number of co-purifying peptide bands [19]. Immunoblotting of bovine heart mitochondria also revealed a family of bands, which all co-purified with submitochondrial particles and chloroform-released F_1. However, the banding pattern was different than that for rat liver mitochondria. This led to a detailed examination of the M2 antigens, as detected by individual PBC sera, in different organs of the same species and between species. It was established that for a given species the antigenic pattern on immunoblots was virtually identical for all organs. For example, in the rat, the M2 pattern was the same for mitochondria from liver, kidney, heart, spleen, brain, white adipose tissue, brown adipose tissue and hepatoma cells. However, between species there were marked differences, in particular in the precise MW of the prominent band with a MW of around 70–74 kD for mammalian mitochondria, probably corresponding to an antigen of a much lower MW in yeast [9,20]. It is of some interest that, in this study, mitochondria from normal human liver and from the liver of patients with PBC gave identical patterns on immunoblots.

Notwithstanding the differences in the precise molecular identity of the antigens revealed by immunoblotting, each of the mitochondrial preparations apparently expressed complete sets of corresponding epitopes, since reactivity against mitochondria of each species tested could be absorbed out of PBC sera by mitochondria of each of the other species. It did nevertheless seem to be the case that more than one epitope was involved, rather than that the antigens were merely partial proteolytic products of a single, common precursor. Firstly, isolation of mitochondria in the presence of a "cocktail" of protease inhibitors, whilst eliminating occasional artefactual bands, did not alter the major features of the immunoblots [9]. Secondly, there was a variability in immunoblot pattern as revealed for the same mitochondrial preparation by different sera (as mentioned previously above). Clearly, if only a single epitope were involved, all positively reacting sera would react with the same family of polypeptides. It was of interest in this connection that if sera were classified into groups in terms of complexity of banding pattern against bovine mitochondria, they fell into the same groups as if they had been similarly classified with rat mitochondria [20].

Further proof that the antigenic species in a given mitochondrial preparation were structurally unrelated came from studies of fragmentation patterns on partial proteolysis. Bovine heart mitochondria were subjected to SDS-PAGE, and bands were cut from the gel at the previously located positions for the three major antigens. If the antigens from these gel slices were then re-electrophoresed, blotted and developed with PBC sera, each ran as a single band. However, if the gel extracts were first treated with the SDS-insensitive V_8 protease and then subjected to SDS-PAGE and immunoblotting, each antigen gave rise to a completely different pattern of immunologically-reactive peptide fragments [21].

The immunoblotting techniques also permitted general confirmation of the chemical properties of the antigens as previously established by CFT (see above), although differences were detected between the sensitivities of the individual bands, particularly to heat and to mercurials [20]. Possible involvement of thiol groups as indicated by sensitivity of some of the antigenic bands to mercurials was further indicated by the enhanced intensity of staining as well as resolution of bands, when gels were run in the presence of dithiothreitol. One slightly puzzling feature, in view of the apparent insensitivity of antigenicity to agents that might have been expected to modify carbohydrate epitopes is that treatment with

Concanavalin A of nitrocellulose sheets containing blotted antigens completely abolished reactivity against PBC sera [22]. This may indicate that the antigens are glycoproteins, even if the antigenic epitopes themselves do not involve any carbohydrate moieties.

Further indication that the antigens might be glycoproteins is the resolution of each of them into multiple spots of differing isoelectric points by two-dimensional electrophoresis prior to immunoblotting. Work is currently in progress to determine the relationship between these species of differing patterns of ionization (which might, for example, also reflect varying degrees of phosphorylation, or isoforms of specific proteins).

Immunoblotting confirmed that AMA include immunoglobulins of both the IgG and IgM classes. Overall, IgG AMA were found to bind in greater amounts and to a larger number of antigenic bands than do IgM AMA, with the latter only binding a particular antigen if IgG AMA do so also [20].

7 Cross-Reactivity of M2 AMA with Bacteria

Early work using CFT and ELISA had indicated a cross-reactivity of some PBC sera with *P. denitrificans* [23]. Later studies, with immunoblotting, extended the range of micro-organisms exhibiting such reactivity to include *E. coli, K. pneumoniae* and membranes of the thermophilic bacterium PS3 [20]. With some organisms the pattern of bands reacting with PBC sera was simpler than in the case of mitochondria. However, some sera exhibited very complex reactivities with antigens from *E. coli*. This was not surprising in view of the fact that many PBC patients have a clinical history of chronic, upper urinary tract infection [24] and this raised the question of whether the microbial reaction with PBC sera actually involved AMA or was merely a reflection of prior histories of infection. However, specific cross-reactivity was unambiguously demonstrated for *E. coli, K. pneumoniae, P. mirabilis* and *S. aureus* by eluting bound antibodies from immunoblots of mitochondria and of membranes of the micro-organisms and re-testing these "immunopurified" antibodies against the other preparations [25]. Further reference will be made later to the significance of these experiments.

8 The M2 Antigen in Yeast

The existence of the M2 antigen in yeast mitochondria permitted a number of important studies on the genetics and control of expression of the antigenic peptides. Studies with petite mutants clearly demonstrated that the antigens were nuclearly coded and there was some evidence of a cytoplasmic precursor of higher molecular weight [26]. Furthermore, expression of the antigen could be reversibly repressed by growth in high glucose media [27]. However, impairing membrane assembly with erythromycin, or abolishing the membrane potential with uncouplers, whilst greatly diminishing the amount of antigen found in the mitochondria did not give rise to an accumulation of precursors in the cytoplasm, suggesting that these have a very short half-life and/or that their synthesis is under very tight control.

The identification of yeast strains which poorly expressed the antigen even under aerobic conditions permitted more detailed genetic studies through crossing with 'high producers'. The conclusion was that expression is under complex control by regulatory genes and by the structural/physiological state of the inner mitochondrial membrane [27].

9 Possible Non-Mitochondrial Location of M2 Antigen

Subcellular fractionation studies with tissues other than liver had suggested a more complex distribution of antigen than a simple, unique mitochondrial location [20]. In retrospect, those studies used succinate dehydrogenase as marker enzyme for mitochondria with no confirmation of mitochondrial integrity. The subsequent identification of the 70 kD antigen as not being an integral inner membrane constituent (see below) therefore casts doubt on the significance of that work. Nevertheless, studies with yeast also suggested a variable cellular location, including an apparently clear association with a marker for the plasma membrane [28]. Direct immunofluorescence studies were therefore made with intact hepatocytes, hepatoma cells and human lymphocytes from normal subjects and from patients with PBC. From these studies it appeared that antigen was expressed at the cell surface which reacted with antibodies specifically present in PBC sera that could be absorbed out by mitochondria [28]. It was therefore concluded that these surface antigens were recognized by PBC-specific AMA. Evidence was also presented that in the case of lymphocytes of patients with PBC, these were already coated with the patient's own AMA. These studies have been criticized on technical grounds [29]; and because of the considerable clinical significance of any surface reactivity, it would certainly be desirable for further work to be done on this question.

Another report of a non-mitochondrial location of M2 AMA is also of potential importance. In studies on biopsies of human breast cancer using indirect immunofluorescence microscopy it was found that, in a certain number of cases of advanced infiltrating duct carcinoma, PBC sera stained sections which, on counter-staining for succinate dehydrogenase, showed very little mitochondrial activity [30]. The paucity of mitochondria in these sections was confirmed by electron microscopy. In other biopsies, in the same series of cancers and in all normal breast tissue, PBC immunofluorescence correlated precisely with staining for succinate dehydrogenase and with electron microscopy. It was therefore concluded that in a de-differentiated and rapidly proliferating tissue, the M2 antigen might be inappropriately expressed.

10 Identity of the 70 kD Antigen

Since the demonstration that the M2 antigens were not constituents of the F_1-ATPase, there has been considerable speculation as to their identity, particularly that of the prominent 70 kD antigen. One possibility for that antigen, on the grounds of MW, was one of the non-haem iron proteins of Complex I of the respiratory chain, particularly since antibodies raised against that protein reacted strongly with the chloroform released F_1-ATPase preparation, known to be enriched with the M2 antigens [20]. However, unlike the 70 kD antigen, it was shown that the MW of the Complex I protein did not vary markedly with species. Furthermore, an antiserum against that latter protein did not react against a preparation of immunopurified 70 kD antigen. It was therefore concluded that the proteins are not identical [20]. However, a recent report has again raised the possibility of the non-haem iron protein being the antigen [31]. In this case the variation in MWs is explained in terms of proteolysis and differing epitopes; and an antiserum against the Complex I protein is shown to react with the immunopurified M2 antigen. This last observation is not definitive since this antiserum would have been raised against material purified by SDS-PAGE which would presumably have contained not only the non-haem iron protein but also any other peptide present that migrated similarly. Without further experimental details it is not possible to exclude that antibodies against the 70 kD protein might have been 'accidentally' raised in this manner.

Such speculation, however, now appears redundant, in view of the virtually unambiguous identification of the 70 kD antigen by Yeaman et al., which has recently been reported [32]. This important development was derived from a major programme of work by Gershwin and colleagues, which led firstly to the production of monoclonal antibodies against this peptide and then to the cloning from a rat liver cDNA library of a 1370 base-pair insert coding for a major part of the antigenic molecule [33]. The nucleotide sequence, and the deduced amino acid sequence were not thought to correspond to (or have significant homology with) any known protein. However, detailed inspection of the predicted protein sequence by Yeaman et al. strongly suggested its identity as being the E2 component of the mammalian pyruvate dehydrogenase complex (PDC). The primary sequence around the lipoate attachment site on the E2 component of bovine heart PDC is known, and this ten residue sequence corresponds exactly to a region in the deduced sequence of the cloned antigen as reported by Gershwin et al. Furthermore, there is extensive homology between other regions of the predicted sequence and the previously determined sequence of E2 from *E. coli*. Confirmation of this identity was the specific reactivity on immunoblots (38/40) of PBC sera with purified E2 (although the two non-reactive sera were from patients otherwise indistinguishable from those whose sera reacted positively). E2 is the acetyl transferase component of PDC, and contains covalently-bound lipoic acid. It represents the core of a large (8.5×10^6 kD) complex and itself has a MW of greater than 3×10^6 kD. Within this core are 60 copies of the E2 polypeptide arranged with icosahedral symmetry. The identity of the 70 kD antigen with this molecule is totally convincing. It is compatible with inner membrane association and easy detachment, high MW under non-denaturing conditions, thiol involvement, biological distribution, and location and control of biosynthesis. Interestingly, in a recent study of the reactivity of another set of PBC sera with a PDC preparation, there were again a few sera which were unreactive. Each of these was reactive in immunoblots of whole mitochondria, against the 52 kD M4 antigen [34].

11 The Identity of the 50 kD Antigen

The second major antigen of the M2 family has (in mammalian mitochondria) a MW of 45–55 kD depending on species. Preparations of E2 contain another, closely-associated peptide, "component X". This has a MW of around 50 kD and there is now good evidence that it is not a proteolytic fragment of E2 but a structurally different peptide. It is therefore of considerable interest that every PBC serum that reacted with E2 also reacted with the component X present in the same preparation [32]. It thus seems that the 50 kD antigen has also now been identified. Since X is part of the PDC and since, as previously mentioned, partial proteolysis of the 50 kD antigen revealed it to be a structurally different peptide from the 70 kD antigen, this assignment of identity is compatible with previous studies.

The precise function of X in PDC remains to be established, but it may have a role in the complex physiological control of this key enzyme. PBC sera may therefore turn out to be a useful tool in monitoring possible changes in levels of X in various physiological states.

12 Some Unresolved Problems Relating to M2-AMA

The identity of the other individual members of the M2 family of antigens has not been published at the time of writing of this review, but it seems that two of them are E2s from the other complex dehydrogenases with mechanisms analogous to that of PDC, oxoglutarate dehydrogenase and the branched chain oxo acid dehydrogenase (Yeaman, personal communication). Individual sera reacted with one, two or all three of the E2s, in agreement with the variable patterns of reactivity of PBC sera against whole mitochondria. However, there is no evidence that E2 is either phosphorylated or glycosylated. It therefore remains important to explain the resolution of the 70 kD antigen into a set of spots with differing iso-electric points on two-dimensional immunoblots.

There is a more important problem that will remain to be resolved even after the individual antigens have all been identified. As mentioned previously, they all appear to be peptides of differing structure. Moreover, at least in the case of E2, the epitope(s) recognized by the M2-AMA would seem to be exclusively peptidyl. This is because the cloned gene product (which would have no post-translational additions) has been reported to absorb out from PBC sera reactivity against the 70 kD antigen only, and these immunopurified AMA react only with that band on subsequent immunoblotting [33]. The implication of these observations is that M2-AMA consist of a family of peptide-specific antibodies.

However, we have consistently shown [35,36] that for some PBC sera at least, antibodies eluted from individual bands on immunoblots of mitochondria or bacteria react with all bands on subsequent immunoblots of either preparations. These observations strongly imply the existence of AMA recognizing some common epitope(s). This apparent paradox might be resolved by careful studies of numbers of PBC sera, since it might be that there are both peptide-specific and cross-reactive AMA and that different sera have different proportions of antibodies of these two kinds. (Presumably sera of the class reacting with only a few bands on immunoblots would possess predominately peptide-specific AMA). The existence of two such kinds of AMA, if established, could be of relevance to the pathogenesis and prognosis of the disease, and this could well be studied by comparing sera of asymptomatic subjects with those of patients at various stages of the disease. It would be of interest, if this hypothesis proved valid, to test whether peptide-specific and cross-reacting antibodies were of different immunoglobulin isotypes, and also whether it was only the cross-reacting antibodies that reacted at the cell surface (assuming, of course, that that observation were to be confirmed). It would certainly seem to be more likely that any cell-surface epitope would be a common one rather than being peptide-specific for E2 or X.

If there were indeed a common epitope amongst the M2 family of antigens, it might be, or include, a non-peptidyl moiety. One candidate for this might be lipoic acid, in view of its attachment to all E2s, and to component X. This would also be compatible with the sensitivity of antigenicity to mercurials and its enhancement be dithiothreitol. However, free reduced lipoic acid does not seem to be an antigen, at least in ELISA [36]. Another possibility would be carbohydrate, if the multiple forms identified for each band reflected a hitherto undetected glycolysation of E2 and X, rather than simply iso-forms of the enzymes.

13 Speculation on the Pathogenesis of PBC

A synthesis of what is now known about the AMA of PBC allows for some informed speculation as to the stimulus for AMA production and their possible involvement in the disease process. As has previously been mentioned, there is a strong correlation between PBC and a history of chronic upper urinary tract infection, particularly with *E. coli* [24]. Such infection is almost exclusive to women, as is PBC. Furthermore, it is likely that infection of the kidney will result in a release into the circulation of breakdown products of bacterial cells. This could include fragments of their PDC (and perhaps other complex dehydrogenases), the core E2 aggregates of which are to some extent virus-like and might be expected to be excellent immunogens, notwithstanding that in the case of *E. coli* the E2 core of PDC has only 24 copies of the enzyme and so is smaller than the core of the mitochondrial complex. Depending on the identity of the infective organism and the precise course of the infection, a spectrum of immune responses might be expected.

Some resulting antibodies (or stimulated T-cells) might be pathogen-specific, but in view of strong homologies between microbial and mitochondrial enzymes, some could be autoreactive. These would be expected to be innocuous unless any cross-reacting epitopes also occurred at any cell surface. In such a case, the scene would be set for an autoimmune disease whose progress might be affected by the pattern of any subsequent kidney infections, as well as possibly by subject-specific factors. Other routes of infection might also be relevant.

This is all very speculative, but clarification of the questions of cell-surface reactivity and of peptide-specific versus cross-reactive AMA should indicate whether such ideas are worth pursuing. Other indications will come from current work on the identity of pathogens and on routes of microbial infection in relation to the occurrence of AMA and the epidemiology of PBC. Two supporting items of evidence have already come from this latter work. These are confirmation of our earlier finding [20] that anti-M2 AMA are detectable in sera in certain cases of chronic *E. coli* infection of the upper urinary tract, and the intriguing finding that, in all cases so far tested, recurrent urinary tract infection is associated with antibodies reacting with a mitochondrial peptide with a MW corresponding to that reported for the major M9 antigen. A further related and important finding has recently been reported [37], and that is the occurrence in PBC sera of antibodies reacting with subunits of the acetyl cholinesterase receptor. It has already been established that this receptor shares immunologically cross-reactive epitopes with membrane constituents of *E. coli* [38]. Circumstantial evidence is thus accumulating for a "molecular mimicry" etiology for PBC.

Acknowledgements. Much of the work referred to in this review was supported by generous grants from the Wellcome Trust. We are also grateful for support by the Cancer Research Campaign and Consiglio Nazionale delle Ricerche (CNR) grant 86.00501.04.

References

1. Kaplan MM (1987) Primary biliary cirrhosis. N Engl J Med 316:521–528
2. Neuberger J, Portmann B, MacDougall BRD, Calne RJ, Williams R (1982) Recurrence of primary biliary cirrhosis after liver transplantation. N Engl J Med 306:1–4
3. Walker JG, Doniach D, Roitt IM, Sherlock S (1965) Serological tests in diagnosis of primary biliary cirrhosis. Lancet i:827–831
4. Berg PA, Wiedmann KH, Sayers T, Kloppel G (1980) Serological classification of chronic cholestatic liver disease by the use of two different types of antimitochondrial antibodies. Lancet ii:1329–1332
5. Gupta RC (1985) Specificities and characteristics of mitochondrial protein antigens assessed by enzyme linked immunosorbent assay for antimitochondrial antibodies. Relationship to immunofluorescent and precipitating antibodies. Clin Exp Immunol 59:604–612
6. Manns M, Mayer zum Buschenfelde KH (1982) A mitochondrial antigen-antibody system in cholestatic liver disease detected by radioimmunoassay. Hepatology 2:1–7
7. Berg PA, Klein R, Linderborn-Fotinos J (1986) Antimitochondrial antibodies in primary biliary cirrhosis. J Hepatol 2:123–131
8. Berg PA, Klein R (1985) Clinical and prognostic relevance of different mitochondrial antibody profiles in primary biliary cirrhosis (PBC). Mol Asp Med 8:235–247
9. Ghadiminejad I, Baum H (1987) Reaction pattern of mitochondrial antibodies of primary biliary cirrhosis (PBC) is species specific but not organ specific. J Bioeng Biomenbr 19:245–259
10. Ghadiminejad I, Baum H (1986) Discrimination between M2 and M4 antimitochondrial antigens in primary biliary cirrhosis. Biosci Rep 6:1049–1056
11. Fusconi M, Ghadiminejad I, Bianchi FB, Baum H, Bottazzo GF, Pisi E (1988) Heterogeneity of antimitochondrial antibodies with the M2-M4 pattern by immunofluorescence as assessed by Western immunoblotting and enzyme linked immunosorbent assay. Gut 29:440–447
12. Ghadiminejad I, Harper N, Bittles AH, Baum H (1987) Age-dependent loss of a mitochondrial antigen in cultured human diploid fibroblasts. Biochem Soc Trans 15:1177–1178
13. Bianchi FB, Penfold PL, Roitt IM (1973) Mitochondrial antibodies in primary biliary cirrhosis. V. Ultrastructural localization of the antigen to the inner mitochondrial membrane using a direct peroxidase conjugate. Br J Exp Path 54:652–657
14. Berg PA, Roitt IM, Doniach D, Horne RW (1969) Mitochondrial antibodies in primary biliary cirrhosis. III. Characterization of the inner membrane complement fixing antigen. Clin Exp Immunol 4:511–525
15. Berg PA, Doniach D, Roitt IM (1967) Mitochondrial antibodies in primary biliary cirrhosis. I. Localization of the antigen to inner mitochondrial membranes. J Exp Med 126:277–293

16. Meek F, Khoury EL, Doniach D, Baum H (1980) Mitochondrial antibodies in chronic liver diseases and connective tissue disorders: further characterization of the autoantigens. Clin Exp Immunol 41:43–54
17. Baum H, Davey JM, Elsden J, Leoutsakos A, Meek F, Sayers T (1979) Evidence that adenosine triphosphatase is one of the mitochondrial antigens of autoimmune liver diseases. Biochem Soc Trans 7:213–215
18. Sayers T, Leoutsakos A, Berg P, Baum H (1981) Antimitochondrial antibodies (AMA) in primary biliary cirrhosis. 1 Separation of the PBC antigen activity from the mitochondrial ATPase activity. J Bioeng Biomembr 13:255–268
19. Baum H, Palmer C (1984) Further characterization of the antimitochondrial antibodies of primary biliary cirrhosis. EBEC Rep 3:691–692
20. Baum H, Palmer C (1985) The PBC-specific antigen. Mol Asp Med 8:201–234
21. Chelliah J, Baum H (1988) Immunoblot peptide mapping – a novel application to explore the relationship between mitochondrial antigens. Biochem Soc Trans 16:151–153
22. Ghadiminejad I, Baum H (1985) The possible glycoprotein nature of the mitochondrial antigens of primary biliary cirrhosis. Biochem Soc Trans 13:713
23. Sayers T, Baum H (1976) Possible cross-reactivity of human antimitochondrial antibodies with membrane vesicles of *Paracoccus denitrificans.* Biochem Soc Trans 4:138–139
24. Burroughs AK, Rosenstein IJ, Epstein O, Hamilton-Miller JM, Brumfitt W, Sherlock S (1984) Bacteriuria and primary biliary cirrhosis. Gut 25:133–137
25. Butler P, Chelliah J, Biagini M, Brumfitt W, Burroughs A, Baum H (1988) Specific reactivity of PBC specific mitochondrial antibodies against peptides of *E. coli, K. pneumoniae* and *P. mirabilis,* demonstrated by immunopurification. Proc Eur Soc Stud Liver Hepatol 7 (Suppl 1):580
26. Ghadiminejad I, Baum H (1985) The mitochondrial membrane antigens of primary biliary cirrhosis are proteins transported from the cytoplasm. Biochem Soc Trans (1988) 13:713–716
27. Ghadiminejad I, Wilkie D, Baum H (1988) Expression of the primary biliary cirrhosis antigens in yeast: aspects of mitochondrial control. J Bioeng Biomembr 20:243–259
28. Ghadiminejad I, Baum H (1987) Evidence for the cell-surface localization of antigens cross-reacting with the "mitochondrial antibodies" of primary biliary cirrhosis. Hepatology 7:743–750
29. Lenzi M, Preda P, Bianchi FB, Biagini G, Cassani F, Volta U, Pisi E (1985) Mechanically isolated hepatocytes are unsuitable to detect antibodies directed against plasma membrane determinants. Liver 5:212–220
30. Baum H, Trivedi D, Baum M (1986) Inappropriate expression of mitochondrial antigens in breast cancer. Biochem Soc Trans 14:755
31. Frostel A, Mendel-Hartvig I, Nelson BD, Totterman TH, Bjorkland A, Ragan I, Primary biliary cirrhosis: immunosorbent purification, characterization and identification of the major mitochondrial antigen. (submitted)
32. Yeaman SJ, Fussey SPM, Danner DJ, James OWF, Mutimer DJ, Bassendine MF (1988) Primary biliary cirrhosis; identification of the two major M2 mitochondrial autoantigens. Lancet i:1067–1070
33. Gershwin ME, Sturgess A, Mackay IR, Coppel R (1987) Identification and specificity of a cDNA encoding the 70 kd mitochondrial antigen recognized in primary biliary cirrhosis. J Immunol 138:3525–3531
34. Fusconi M, Baum H, Cassani F, Lenzi M, Ballardini G, Bianchi FB (1988) Antimitochondrial positive but anti M2 negative primary biliary cirrhosis. Lancet i:1464
35. Baum H, Butler P, McCormick PA, Flannery GR, Brumfitt W, Burroughs AK (1988) Antimitochondrial antibodies (AMA) in primary biliary cirrhosis recognise both specific peptides and shared epitopes of the M2 antigen. Proc Int Assoc Stud Liver. Hepataloy 8:1448
36. Flannery GR, Burroughs AK, Butler P, Chelliah J, Hamilton-Miller J, Brumfitt W, Baum H. Antimitochondrial antibodies in primary biliary cirrhosis recognise both specific peptides and shared epitopes of the M2 family of antigens. Hepatoloy (in press)
37. Kyriatsoulis A, Manns M, Gerken G, Lohse AW, Maelicke A, Wessler I, Reske K, Meyer zum Buschenfelde KH (1988) Immunochemical characterization of anti-acetylcholine receptor antibodies in primary biliary cirrhosis. J Hepatol 6:283–290
38. Stefansson K, Dieperink ME, Richman DP, Gomez CM, Marton LS (1985) Sharing of antigenic determinants between the nicotinic acetylcholine receptor and proteins in *Escherichia coli, Proteus vulgaris* and *Klebsiella pneumoniae.* N Engl J Med 312:221–225

Toxicity, Antioxidants, and Metabolism

L. PACKER

1 Introduction

We all realize that oxygen is a dangerous friend. Oxygen is something that we cannot live without, but it is also a substance that we cannot live with indefinitely. This chapter will review two aspects of this subject. First, an overview will be given of some of the points relevant to the idea that oxygen is toxic by pointing out the underlying chemical and biological mechanisms that are important in the toxic effects of oxygen in biological systems. Secondly, some experimental approaches that have been undertaken in our laboratory, using the concept of studying the increased metabolism associated with exercise in order to follow the generation of oxygen-free radicals (oxidants) and their consequences, in terms of changes in antioxidants, will be described. Antioxidants are major cellular defense mechanisms that are important for preventing molecular, cellular and tissue damage due to free radicals and, hence, their understanding is pertinent to relating "cell function and disease."

2 Oxygen is Toxic

This idea gained considerable credibility from work of physiologists more than 4 decades ago, who noted an inverse relationship between species life span and metabolic rate, and a direct relationship between body or brain weight and the mammalian species life span. This concept of the inverse relationship of metabolic rate to life span has been tested with cold-blooded animals such as fish or insects. For example, with insects, by modifying the environmental temperature, their metabolic rate can be set to operate at different levels. As the temperature is raised from 15° to about 30°C, the metabolic rate (oxygen consumption) progressively increases but the life span of the insects decreases proportionally. What is extremely interesting about such experiments is that the total amount of oxygen which the insects consume is unaffected regardless of the life span of the insects (or the metabolic rate). It is as if a certain amount of oxygen is allocated to individual insects during their life span, and the individual can live its life either fast (by flying) or slow (by walking). The slowly metabolizing individuals can benefit from a longer

Membrane Bioenergetics Group, Lawrence Berkeley Laboratory, University of California, Berkeley, CA 94720, USA

A. Azzi, Z. Drahota and S. Papa
Molecular Basis of Membrane-Associated Diseases

life span. Whether such experiments are important for animals or human life span is still unknown.

How then could oxygen be toxic? We know that the consumption of oxygen at the cellular level is due to electron transport reactions, such as the respiratory electron transport system in the mitochondria, where oxygen is reduced to water at the expense of the oxidation of carbohydrate substrates (respiration) (cf. Fig. 1). One problem, however, is that during the substrate oxidation, electrons can be donated by the mitochondrial electron transport chain ubiquinone components (ubiquinone semiquinone) to oxygen, generating a superioxide radical anion. Since all aerobic cells contain the enzyme superoxide dismutase or other substances such as ascorbic acid (Vitamin C) which reduce superoxide to hydrogen peroxide, an accumulation of hydroperoxides occurs. Although hydroperoxides (like hydrogen peroxide) themselves are stable molecules, if transition metals are present in their free reactive form in the cell (e.g., Fe or Cu), hydroperoxides can be decomposed to generate (by the Fenton reaction) either hydroxyl radicals from hydrogen peroxide or alkoxyl radicals from organic hydroperoxides. These radicals are readily reactive and can damage biological macromolecules, including DNA, enzymes, and structural components of cells. Such reactions can be causal to accelerating aging and cancer.

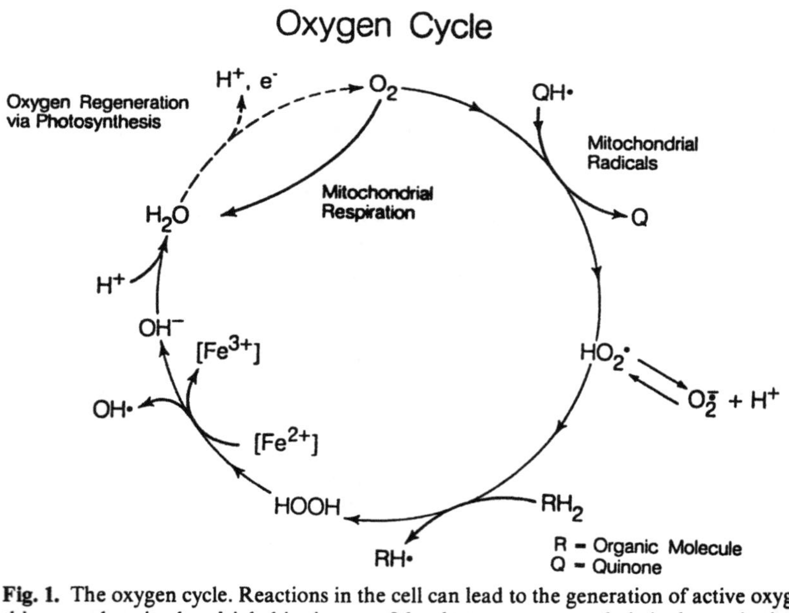

Fig. 1. The oxygen cycle. Reactions in the cell can lead to the generation of active oxygen species. In this example, mitochondrial ubiquinones of the electron transport chain in the semiquinone state react with oxygen to form superoxide. Superoxide radicals may exist either as the anion or as a weak acid. Superoxide dismutase or water-soluble reductants convert superoxide to hydrogen peroxide, and the latter can be removed by hydroperoxide-removing enzymes such as glutathione peroxidase. Alternatively, Fe-catalyzed Fenton reactions can decompose hydroperoxides to form reactive oxygen species such as the hydroxyl radicals. Completion of the reduction cycle and formation of water by the respiratory chain involves a four-electron transfer reaction by cytochrome oxidase

Indeed, let us consider the effects of ionizing radiation. Ionizing radiation is known to accelerate cancer and aging. The main effects of ionizing radiation in biological systems are reactions with water. This is because water is the major component in intracellular and extracellular areas of the body to which ionizing radiation is exposed. Water can be decomposed by ionizing radiation to produce two main free radicals: a protonated electron which, in the presence of oxygen, will immediately form the superoxide radical anion, and a hydroxyl radical. So ionizing radiation works in water to produce the same two major free radical species that are also produced by the oxygen cycle during metabolism. One of my colleagues, Prof. B. Ames, has often been quoted to say that "metabolism is like being irradiated all the time."

It is interesting to note that oxygen itself is also a free radical. In fact, it is a biradical. However, because two oxygen atoms are joined together in an O_2 molecule, and since the free electrons in O_2 spin in the same direction, oxygen is not a very reactive radical. However, the superoxide radical is somewhat more reactive and can propagate lipid peroxidation first forming the lipid peroxy radical. Oxygen can also initiate lipid peroxidation if a lipid first becomes a radical, after which it will more readily react with oxygen. Lipids may become radicals, since the unsaturated fatty acids which they contain are readily susceptible to electron addition or hydrogen extraction by electron transport reactions in the membranes in which they are located. Once lipid peroxy radicals have been formed, they can undergo chain reactions in membranes to form other radicals, transferring their radicals to proteins and nucleic acids to cause oxidative damage to these molecules and cross-linking reactions, reducing their biological activity. The removal of superoxide by superoxide dismutase would reduce the chances of this radical reacting with lipids to form lipid peroxy radicals. Indeed, Cutler's laboratory has demonstrated that species life span is proportional to liver superoxide dismutase activity. This suggests that antioxidant enzymes are important in protecting against oxygen toxicity and aging.

It is well recognized that there is a correlation between inflammation and cancer. Here, the scenario is that normal tissues are damaged by the system in which leukocytes are specifically triggered to cause a respiratory burst of superoxide radicals through a system present in their plasma membranes. This system forms hydrogen peroxide which then can form other toxic products, either by metal-catalyzed Fenton reactions or by the enzyme myeloperoxidase, which forms hypochlorus acid and subsequently more potent oxidizing species such as mono- and di-substituted chloramines. This is a cellular mechanism normally used to remove foreign infectious agents such as bacteria or to target tumor cells, but in the process, normal cells may also be damaged. The high concentration of leukocytes in the lungs of people with asbestos fibers, due to continuous recruitment of neutrophils, is an example where continuing inflammation exists, eventually leading to cancer. In the multistep model system for cancer studied extensively in mouse skin, it is recognized that chemical carcinogens often involve the generation of reactive oxygen species which in turn may damage DNA and contribute to the irritation and inflammation that eventually lead to the progression of precancerous cells into rapidly growing cancer cells.

These considerations immediately suggest that antioxidants, which protect against the propagation of free radical reactions, should have a beneficial effect on health and aging (cf. Fig. 2). Indeed, recent epidemiological studies have shown a correlation between lung cancer cases and number of cigarettes smoked, but people who consume a higher amount of yellow or green vegetables, or have elevated levels of blood vitamin E and beta-carotene, have a lower incidence of lung cancer.

So in the lipid- and water-soluble portions of cells, including both vitamin E and vitamin C, respectively, antioxidants may have important anticancer and anti-aging effects. Vitamin E, for example, which reacts with the lipid peroxy radical to form an organic peroxide, becomes in the process a radical itself. However, the vitamin E (alpha-tocopheroxyl) radical is not very dangerous because the free electron is delocalized around the aromatic ring structure of vitamin E. Vitamin E may decompose as a result of becoming a radical or it may be regenerated in biological systems, such as by vitamin C. When vitamin C reacts to quench free radicals, vitamin C also becomes a slowly reacting free radical (the ascorbyl radical) which readily reacts with itself to regenerate reduced and oxidized ascorbic acid molecules.

There are many antioxidants in the cell besides superoxide dismutase, vitamin E, and vitamin C. Table 1 lists some of the common antioxidants in cells and extracellular fluids. Many of these substances or enzymes react to remove substances which may initiate free radical reactions (like hydroperoxides) rather than directly removing free radicals themselves as in the case of superoxide dismutase.

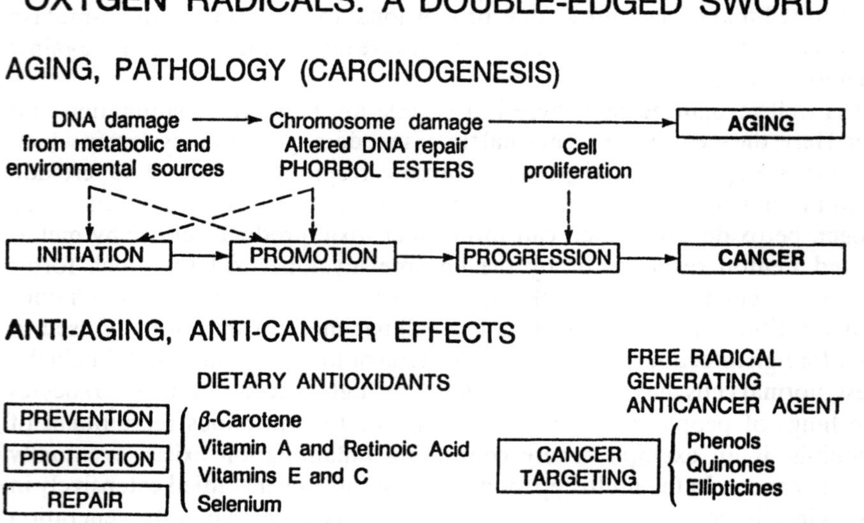

Fig. 2. Oxygen radicals, a double-edged sword. Active oxygen species have been implicated in the multistep process of cancer and in the aging process, although active oxygen species may not be exclusive in this regard. Dietary antioxidants exhibit anticancer effects and effective anticancer agents are quinones which generate first superoxide, and then initiate a cascade of active oxygen species which serve to damage and target cancer

Table 1. Antioxidants

In cellular system membranes:	*In extracellular fluids:*
Vitamin E	Caeruloplasmin
Ubiquinones	Transferrin
Aqueous compartments:	Lactoferrin
Ferritin	Albumin
Vitamin C	Haptoglobin
Glutathione	Superoxide dismutase (trace)
Superoxide dismutase	Uric acid
Catalase	Vitamin C
Glutathione peroxidase	
Glutathione-s-transferase	

Other substances are effective as antioxidants because they can sequester (bind) and render metal ions unreactive for Fenton reactions that can lead to the formation of free radicals. Examples of course are iron-binding substances such as transferrin, ferritin, lactoferrin. So we have many antioxidants-water-soluble substances, and enzymes together with lipid-soluble substances that should reduce free radical damage in aging and pathology. Thus, there is evidence that anticancer effects can be exhibited by dietary antioxidants. An important unanswered question is, what is the right combination of dietary antioxidants to optimize anticancer or anti-aging effects? Another strategy is to use the ideas that we learn about from these reactions of oxidants and antioxidants to design drugs, for example, to specifically target the destruction of parasites or cancer cells. Thus, we know that quinones are among the most effective reagents that are antiparasitic or anticancer.

3 Oxidants and Antioxidants in Exercise

Most people believe that a regular program of exercise is good for you and that regular exercise with proper training programs will improve the performance of the cardiovascular system and the reduction of body fat levels. These changes have been widely documented as being significant in the reduction of cardiovascular disease and in the reduction of age-related diseases. However, a concept which has only recently been introduced is that the increased metabolism associated with exercise may lead to the production of oxidants having potentially harmful effects (Davies et al. 1982; Packer 1986a,b, 1987). Some of our experiments have shown that even mild exercise leads to dramatic changes in tissue oxidative stress. One could paraphrase mild exercise as "the amount of exercise that it takes to stir a margarita or a martini cocktail." Together with my colleagues in the laboratory of G.A. Brooks, we have performed experiments involving mild exercise in humans and studied the response of the glutathione antioxidant system. The glutathione system is centrally important in metabolism because much of the cells' reducing power is utilized to maintain large amounts of glutathione present in aerobic cells (between 1–10 mM) in its reduced form. Metabolism channels its reducing power to keep glutathione more than 95% in its reduced form during normal metabolism. Reduced glutathione is important because it can react chemically or with the enzyme glutathione peroxidase to decompose organic or inorganic hydroperoxides, avoiding the possibility that they may become involved in generating free radicals that can cause tissue damage.

To study the effects of mild exercise, we recruited eight human volunteers who were already well-trained individuals (Gohil et al. 1988). These subjects were set to perform at 65% of their $VO_{2,max}$ on a bicycle ergometer. During the initial 15 min of exercise, their blood glutathione level dramatically changed, up to 60% of the reduced glutathione was oxidized. This more oxidized steady state level of glutathione was maintained for up to 90 min of exercise after which exercise was ceased. During the next few hours, the blood glutathione levels returned to normal. However, an interesting finding was that over a 3-day period, the total amount of glutathione in the blood underwent a net increase. This suggests that a single bout of exercise stimulates an increase in blood glutathione levels which may afford protection against successive bouts of exercise. The oxidation of glutathione indicates that oxidants must have been produced in exercise, because the reaction of glutathione with hydroperoxides causes its oxidation. Since hydroperoxides arise from superoxide, it follows that electron transport reactions have generated superoxide, hydrogen peroxide, and oxidized glutathione, in that sequence. The steps involve the formation of oxidants (superoxide) and a change in the redox status of the major, primary, preventative cellular antioxidant. The source of the superoxide and hydrogen peroxide have not yet, however, been definitely identified (cf. Fig. 3). The glutatione is certainly in the blood cell fraction and not

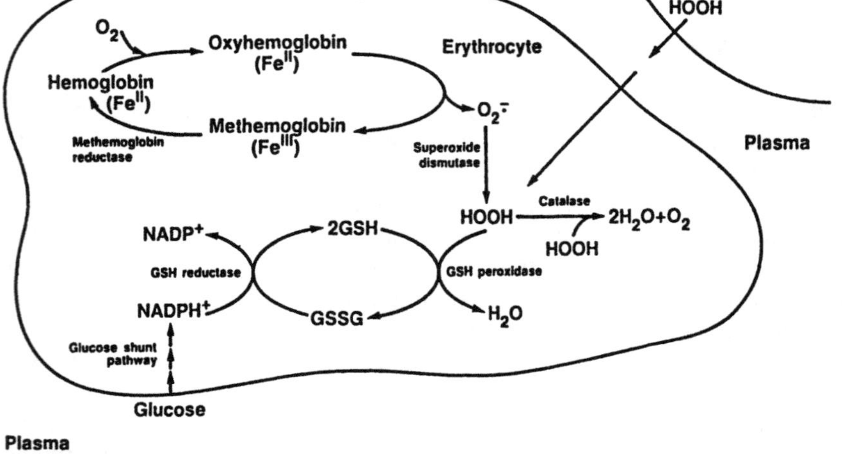

Fig. 3. The oxidative stress of exercise leads to large oxidation changes in the glutathione system of human blood. The diagram illustrates that the sources of active oxygen species, resulting in glutathione oxidation, could arise endogenously in red blood by the formation of methemoglobin or alternatively, hydroperoxides produced in surrounding tissues may leak and enter red blood cells to also serve as the substrate for glutathione oxidation

in the plasma, but whether the reactions that form superoxide in the red blood cells (such as methemoglobin formation) are the main source of superoxide or whether hydrogen peroxide comes from surrounding tissues and enters into the circulating red blood cells is not yet known. Nevertheless, it is clear that a significant "oxidative stress" is generated even during mild exercise.

4 Vitamin E Effects

We, have also been interested in the responses of vitamin E, the cells major chain-breaking antioxidant, during exercise and exercise training (Quintanilha and Packer 1983; Aikawa et al. 1984; Gohil et al. 1986; Packer 1986a,b). In a series of experiments, we have observed that if animals are endurance-trained, a large increase in the biogenesis of mitochondria (Davies et al. 1981; Kirkwook et al. 1987) in red skeletal muscles and a decrease in the tissue vitamin E content of animals are observed. These animals were given a dietary amount of vitamin E that is considered to be normal for rats (40 IU/kg diet). In these animals, the increased amounts of mitochondria could be readily identified in the red muscles by specific increases in enzymes associated with the inner mitochondrial membrane, such as cytochrome-c reductase and by increased levels of ubiquinone and ubiquinols 9 and 10, specific quinones located in the inner mitochondrial membrane.

By comparing the changes in tissue vitamin E content to the ubiquinone content, very spectacular decreases in the vitamin E to ubiquinone ratio were calculated, following exercise endurance training (Gohil et al. 1987; Lang et al. 1987a). This suggests that mitochondrial vitamin E is decreased by exercise training. This can be a sensitive test to use in other situations where oxidative stress may occur. The vitamin E status may change with animal performance (metabolism) or exposure to environmental toxic substances or dietary substances that result in the production of oxidants which deplete the vitamin E pool. The results suggest that perhaps higher dietary vitamin E may be required during regular exercise. Although it cannot be proven from these experiments, the circumstantial evidence is strong that oxidants and free radicals are produced and that free radical reactions are going on during the exercise period. In other studies with animals, it has also been found that glutatione is oxidized in red skeletal muscles accompanying exercise, particularly exercise to exhaustion, which causes large increases in oxidation of reduced glutathione, not only in muscle tissues but in other tissues such as liver.

Since antioxidants are important in preventing certain diseases and aging, it is possible that increased physical activity could accelerate aging, at least of certain tissues that do not experience a positive adaptive increase as a result of exercise. Also, animals or humans might be susceptible to toxic effects of substances that generate free radicals such as certain drugs that are metabolized by the liver. If glutathione in the liver becomes largely oxidized by exercise, the liver may not be able to remove active oxygen species that arise during drug detoxification, an idea that has been advanced by A. Quintanilha and M. Smith (personal communication).

5 Antioxidant Interactions

Water-soluble and lipid-soluble antioxidants are effective mainly by virtue of the fact that they are reducing agents which form relatively nonreactive radical species. They should in principle be able to interact with each other either directly or indirectly through metabolic processes involving the generation of reducing power. These interrelationships involve a central role for superoxide dismutase, glutathione peroxidase (Gohil et al. 1988; Lang et al. 1987b), and substances such as glutathione, vitamin C (ascorbic acid), and vitamin E in these processes.

Vitamin E may be regenerated by some of these metabolic processes such as interaction with vitamin C or enzymes which have the characteristic of being able to reduce free radicals (cf. Packer et al. 1986; Packer 1987). This concept, which has been designated "free radical reductase," could be important to investigate in the future to determine its potential significance for maintaining fully active antioxidant status. We have begun some investigations to establish whether free radical reductase activity exists. We seek to determine the importance of free radical reductase activity in the scavenging of persistent free radicals in cells and tissues with the aim of eventually elucidating their importance in prevention of disease and the slowing of aging.

6 Concluding Remarks

Prevention is the most challenging field of medicine with which we are confronted today. In preventive medicine, we assess a certain degree of risk in adopting a certain life-style or selecting a particular dietary nutrient or antioxidant regime with an anticipated benefit against aging or age-related diseases. So it is a risk/benefit situation, for which the answers will not be known for many years. Therefore, one of the important horizons for molecular medicine is to determine mechanisms of oxidant formation and then to design antioxidant therapy to improve health and slow aging. In the future, these studies will be a central focus of the activities of many research laboratories seeking to advance our understanding of cell function and disease by developing molecular and cellular model systems for studying oxidants and antioxidants aiming to lend more precision to the rapidly developing field of prevention in medicine.

Acknowledgments. Research reported in this work carried out in our laboratory received support from the National Foundation for Cancer Research, Hoffmann-LaRoche Inc., and the National Institute of Health (CA47597).

References

Aikawa KM, Quintanilha AT, de Lumen B, Brooks GA, Packer L (1984) Exercise endurance training alters vitamin E tissue levels and red blood cell hemolysis in rodents. Biosci Rep 4:253–257

Davies KJA, Packer L, Brooks GA (1981) Biochemical adaptation of mitochondria, muscle, and whole-animal respiration to endurance training. Arch Biochem Biophys 209:538–553

Davies KJA, Quintanilha AT, Brooks GA, Packer L (1982) Free radicals and tissue damage produced by exercise. Biochem Biophys Res Commun 107:1198–1205

Gohil K, Packer L, de Lumen B, Brooks GA, Terblanche SE (1986) Vitamin E deficiency and vitamin C supplements: exercise and mitochondrial oxidation. J Appl Physiol 60:1986–1991

Gohil K, Rothfuss L, Lang J, Packer L (1987) Effect of exercise training on tissue vitamin E and ubiquinone content. J Appl Physiol 63:1638–1641

Gohil K, Viguie C, Stanley W, Brooks GA, Packer L (1988) Blood glutathione and oxidation during human exercise. J Appl Physiol 64:115–119

Kirkwood SP, Packer L, Brooks GA (1987) Effects of endurance training on a mitochondrial reticulum in limb skeletal muscle. Arch Biochem Biophys 255:80–88

Lang J, Gohil K, Burk RF, Packer L (1987a) Selenium deficiency, endurance, exercise capacity and antioxidant status in rats. J Appl Physiol 63:2532–5

Lang J, Gohil K, Rothfuss L, Packer L (1987b) Exercise training effects on mitochondrial enzyme activity, ubiquinones, and vitamin E, anticarcinogenesis and radiation protection. In: Nygaard OF, Simic M, Cerutti P (eds) Anticarcinogenesis and radiation protection Plenum, New York, pp 253–257

Packer L (1986a) Vitamin E, physical exercise and tissue damage in animals. Med Biol 62:105–109

Packer L (1986b) Oxygen radicals and antioxidants in endurance exercise. In: Benzi G, Packer L, Siliprandi N (eds) Biochemical aspects of physical exercise. Elsevier, Amsterdam, pp 101–123

Packer L (1987) Exercise, aging, and antioxidants. In: Benzi G (ed) Advances in myochemistry. Proc 2nd ISM Congr, Rome, Italy, 8–10 October, 1987. Libbey, London, pp 37–50

Packer L, Gohil K, de Lumen B, Terblanche SE (1986) A comparative study of the effects of ascorbic acid deficiency and supplementation on endurance and mitochondrial oxidative capacities in various tissues of the guinea pig. Comp Biochem Physiol 83B:235–240

Quintanilha AT, Packer L (1983) Vitamin E, physical exercise and tissue oxidative damage. Biology of vitamin E. In: Porter R, Whelan J (eds) Proc Ciba Found Symp, 7–10 March. Pitman, pp 56–69

5 Cancer and Energy Metabolism

Mitochondrial H^+-ATP Synthase Under Normal and Pathological Conditions

S. PAPA, F. GUERRIERI, F. ZANOTTI and F. CAPUANO

1 Introduction

The H^+-ATP synthase complex is a multisubunit enzyme, associated with the inner mitochondrial membrane, which plays a central role in energy transduction in the cell [1]. It could be classified as an F_oF_1 type ATPase [2]; in aerobic organisms such ATPases participate predominantly in ATP synthesis.

The ATP synthase is structurally and functionally organized in three parts: the catalytic sector or F_1, which catalyzes ATP synthesis or ATP hydrolysis, the membrane sector or F_o, which functions as a transmembrane proton translocator during the catalytic cycle of the enzyme, and the "stalk" connecting F_o and F_1, which is involved in the coupling between chemical catalysis and proton translocation [3–5].

2 Polypeptide Composition

The F_1 catalytic sector consists invariably of five subunits: 3α, 3β, 1γ, 1∂, 1ε [1–3]. The genes of these subunits have been identified in prokaryote genomes and in the nuclei of eukaryotes, and their primary structure and interspecies homologies have been described [6]. The α and β subunits are responsible for chemical catalysis, the role of subunits γ, ∂ and ε remains to be established.

The F_o membrane sector, responsible for proton translocation, consists in *E. coli* of three subunits a, b, c. Their genes are located in the unc or ATP operon [6] and they have been characterized as essential components for the proper assembly and function of F_o in the ATP synthase complex [7–9]. The subunit composition of F_o of eukaryotic ATP synthases appears to be more complex [1–4]; the following proteins have been identified: (1) a nuclear-encoded protein, calculated Mr 24 668 (apparent Mr from gels 24 000–27 000 [10–14]; (2) the oligomycin-sensitivity-conferral protein (OSCP) [15], encoded by nuclear pseudogenes [16], calculated Mr 20 698 (apparent Mr from gels 21 000–25 000) [10,13,17]; (3) a mitochondrial-encoded protein (ATPase 6) [18], calculated Mr 24 816 (apparent Mr from gels 23 000–18 000) [18]; (4) a nuclear-encoded protein, calculated Mr 18 600 (apparent Mr from gels 20 000–19 000) [11]; (5) subunit F_6 [19], nuclear-encoded [16], calculated Mr 9070 (apparent Mr from gels 9000); (6) subunit A6L, mitochon-

Institute of Medical Biochemistry and Chemistry and Centre for the Study of Mitochondria and Energy Metabolism C.N.R., University of Bari, Italy

A. Azzi, Z. Drahota and S. Papa
Molecular Basis of Membrane-Associated Diseases

drial-encoded [18], calculated Mr 7965 (apparent Mr from gels 8100); (7) subunit c (DCCD-binding protein) [20], encoded separately by two nuclear genes [6], calculated Mr 8000 (apparent Mr from gels 9000–8000) [21,22].

Besides these seven subunits certain preparations of the bovine-heart ATP synthase contain a protein with an apparent Mr of 31 000 (uncoupler-binding protein) [23] and one with an apparent Mr, in its monomeric form, of 13 000 (Factor B) [24].

From comparison of the amino acid sequences and hydrophobic profiles it appears that subunit c, ATPase 6 and OSCP of the bovine enzyme are equivalent to subunits, c, a, and F_1 -∂ of the *E. coli* enzyme respectively. On the other hand, subunits A6L, F_6 and the nuclear-encoded subunit Mr 18 600 (subunit d in Walker's terminology) [11] of the bovine enzyme do not have counterparts in *E. coli*.

SDS-polyacrylamide gel electrophoresis (linear gradient 14–20%) of purified F_oF_1 [25,26] and F_o [26,27] from beef heart resolves, in the region between the Mr of γ and ∂ F_1 subunits, four protein bands with apparent Mr's of 31 000, 27 000, 25 000 (which consists, in fact, of a closely spaced doublet) and 23 000 (Fig. 1.). After preparative SDS-polyacrylamide gel electrophoresis, the pure F_o-proteins were obtained by electroelution in glycerol-H_2O [13,27] (Fig. 1) and used for production of polyclonal antibody [13,14].

Our efforts have been directed to the protein with an apparent Mr of 27 000. This was subjected to automated Edman degradation and revealed an amino acid sequence at the N-terminus of PVPPLPPEHGGKVRF [14] which is exactly that determined on the protein with an apparent Mr of 24 000 isolated by Walker et al. [11] and which corresponds to the nucleotide sequence of the cognate cDNA [11]. It may be noted that this cDNA has one codon for cysteine and the protein isolated and sequenced here reacts with thiol reagents [27]. Thus, the protein of apparent Mr 27 000, isolated from F_o, is the same as that described by Walker et al. and denominated subunit b [11]. The difference in the apparent Mr obtained from electrophoretic mobility is clearly due to different conditions in the SDS-PAGE. The protein can be denominated PVP from its first three residues at the N-terminus [14].

Studies on the accessibility to exogenous proteases of ATP-synthase subunits in submitochondrial particles from bovine heart [13–14] showed that the PVP subunit of F_o could be cleaved by trypsin to a fragment of 18 000 (reacting with polyclonal antibodies raised against the intact protein) only when F_1 subunits were removed from the particles by urea. The 18 000 fragment remained associated with the membrane. Thus, it appears that the PVP subunit is an F_o subunit with hydrophilic extension(s) covered by F_1 and possibly involved in the F_1–F_o junction.

Figure 2 shows that progressive digestion of the PVP subunit, effected by incubation of USMP with increasing concentrations of trypsin, resulted in a progressive inhibition of proton conduction and loss of sensitivity of this process to the inhibitory action of oligomycin. It appears, however, that there remained significant proton-conducting activity (about 50% of the control) in USMP with 70% of the PVP subunit digested by trypsin [13,14] but this residual activity was oligomycin-insensitive.

Liposomes reconstituted with F_o, isolated from trypsinized USMP, in which about 75% of the PVP protein had been cleaved to the 18 000 fragment, exhibited

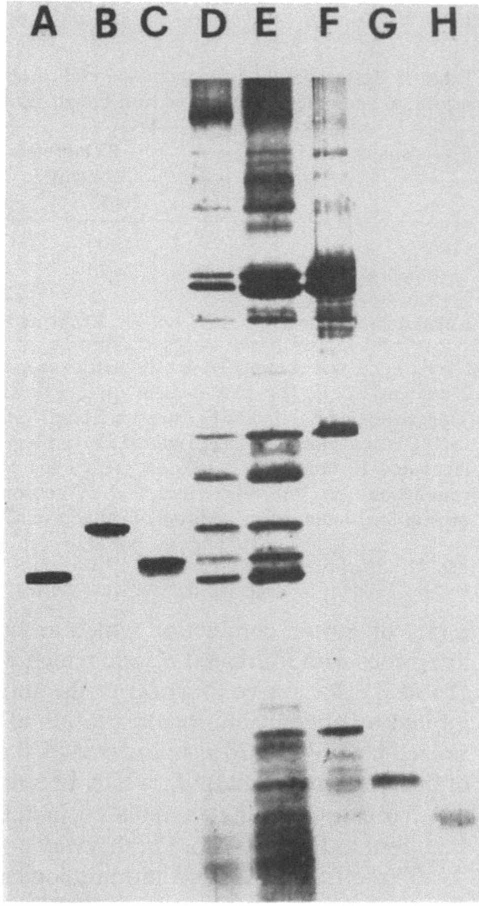

Fig. 1. SDS-polyacrylamide gel electrophoresis of purified F_0F_1, F_0, F_1 and isolated F_0 proteins. F_0F_1 and F_0 were purified as described in [26], F_1 and IF_1 as reported in [40]; F_6 as reported in [19]; all the other F_0 proteins were purified by preparative gel electrophoresis as reported in [13,14]. On SDS-slab gel the following samples were analyzed: **A** 1 μg F_0II protein; **B** 1 μg PVP-F_0I protein; **C** 1.2 μg OSCP; **D** 8 μg protein F_0; **E** 15μg protein F_0F_1; **F** 5 μg protein F_1; **G** 2.5 μg IF_1; **H** 2.5 μg F_6

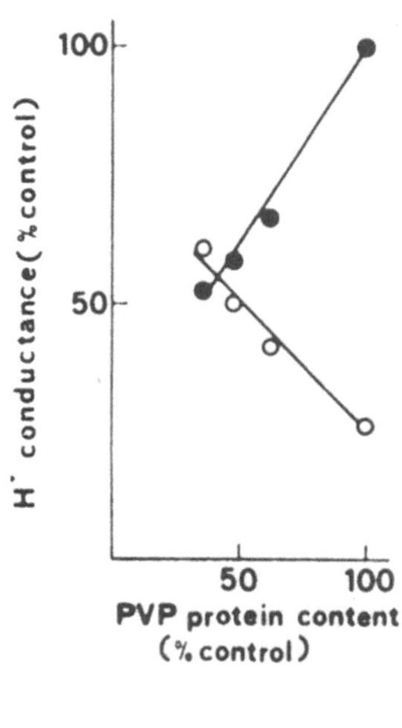

Fig. 2. Relationship between observed passive H^+ conduction in USMP and the content of PVP (F_0I) protein. For USMP preparation, trypsin treatment, and determination of H^+ conduction, see [13,14]; PVP (F_0I) content was determined by densitometric analysis of immunoblot, as described in [13,14]. The content of PVP and the H^+ conduction (as reciprocal value of t½ of anaerobic release of the respiratory proton gradient) of untreated USMP were taken as 100%. Additions: •—•, none; ○—○, + oligomycin 0.5 μg/mg particle proteins after trypsin treatment [14]

Table 1. Restoration of H^+ conduction and of oligomycin sensitivity by addition of isolated PVP protein to liposomes reconstituted with F_0 extracted from trypsin-treated USMP[a]

F_0-liposomes	PVP protein content (%)	Valinomycin induced H^+ release $1/t½$ (s^{-1}) —	+ Oligomycin
Control	100	1.18	0.40
Trypsin-treated F_0	25	0.51	0.59
Trypsin-treated F_0 + Isolated PVP protein	25 + 20 μg/mg F_0	1.40	0.35

[a] PVP content was obtained by densitometric analysis of immunoblots as described in the legend to Fig. 2 and in [13,14]. The PVP content of F_0 extracted from untreated USMP was taken as 100%. Trypsin-treated F_0 refers to F_0 extracted from USMP treated with 50 μg trypsin/mg particle protein as in [14]. Where indicated, 20 μg isolated PVP protein [14] were added to F_0 liposomes (1 mg F_0/ml) and incubated 10 min in KCl 150 mM before the addition of valinomycin for measurement of H^+ conduction (see [14]). Where indicated, F_0 liposomes were incubated with 1 μg oligomycin/mg F_0 protein for 10 min before addition of valinomycin.

a rate of proton conduction which amounted to only 43% of that exhibited by liposomes with untreated F_0 and which was completely insensitive to oligomycin (Table 1). Proton conduction in the untreated F_0-liposomes was markedly inhibited by oligomycin. Reconstitution of the liposomes containing F_0 from trypsinized USMP with 20 μg of isolated PVP protein/mg of F_0 resulted in full recovery of proton conduction by F_0 which became concomitantly oligomycin-sensitive.

It thus appears that the isolated, intact PVP protein is able to replace its inactive fragment in F_0 and restore in this way normal proton conductivity.

These functional and immunological studies clearly indicate that the PVP protein is a functional component of F_0, essential for normal proton conductivity in the ATP synthase and for the inhibition of oligomycin of proton conduction in F_0. Thus, the PVP protein is the second subunit of mitochondrial F_0, in addition to the 8000 DCCD-binding protein [22], which is shown to contribute to the proton-conducting activity of the mitochondrial ATP synthase. It remains to clarify: (1) how these two subunits interact to form the H^+ channel in F_0; (2) the role of the other F_0 polypeptides and (3) how the assembly of the various subunits is organized to give a functional complex.

3 Alteration of the Mitochondrial ATP Synthase Complex in Tissue Regeneration and Neoplasia

As mentioned above, the subunits of the mammalian ATP synthase complex are encoded partly by nuclear and partly by mitochondrial genes, thus normal biosynthesis of ATP synthase in eukaryotes requires a concerted expression of genes located in two different genomes.

Liver regeneration is a useful system for the study of growth regulation in mammalian tissues. During regeneration or in neoplasia, the hepatocytes of the liver of adult animals enter the replicative cycle and it can be demonstrated that most

Table 2. ATPase activity of submitochondrial particles (ESMP) from rat liver, Morris hepatoma 3924A and regenerating rat liver[a]

ESMP	V_{max} (μmol min^{-1}mg prot^{-1}) SE	K_m SE	Respiratory control
Rat liver	1.13 ±0.12	0.17 ±0.01	5.00
Morris hepatoma 3924A	1.08 ±0.10	0.49 ±0.02	5.50
Regenerating rat liver	0.30 ±0.06	0.17 ±0.02	1.00

[a]ATPase activity and respiratory rate were measured as described in [30,31]. V_{max} and K_m were calculated from Lineweaver-Burke plots. Rats weighing 250–300 g were sacrificed 24 h after hepatectomy. Means ± SE from 6 experiments.

cells undergo DNA synthesis and subsequent cell division [28]. In particular, rat-liver regeneration involves a marked reorganization of cell and tissue mass, characterized by a phase of retro-differentiation followed by re-differentiation [29], thus regeneration of rat liver represents a dynamic, in vivo system in which the normal order of assembly of the various subunits of protein complexes may be altered, leading to observable, functional and structural anomalies.

We have studied the alteration of the structure and function of mitochondrial F_oF_1-ATP synthase during liver regeneration and in Morris hepatoma 3924A, a rapidly growing tumor, poorly differentiated and characterized by a high rate of aerobic glycolysis.

Mitochondria from both regenerating rat liver and Morris hepatoma show alterations of activity of the ATPase complex (see Table 2). However, some differences can be noted: (1) regenerating rat liver mitochondria showed no alteration of K_m for ATP but a marked inhibition of the ATPase activity accompanied by a decrease of the respiratory control ratio due to an increase of the state IV respiratory rate (see also [31,32]). (2) the Morris hepatoma mitochondria showed no decrease of respiratory control, very little inhibition of ATPase activity (4%), but a three fold increase of the K_m for ATP.

Kinetic analysis by Eadie-Hofstee plot (Fig. 3) shows the reported cooperative behaviour of the ATP hydrolase activity [33]. This cooperative pattern is explained on the basis of two (or three) catalytic alternating sites [34]. It can be noted that the degree of cooperativity was diminished for the hepatoma ATP hydrolase activity compared to the control enzyme, as indicated by the enhancement of the apparent high affinity constant(s), whilst the low affinity constant remained practically unchanged (Fig. 3).

Oligomycin depresses the catalytic activity of F_1 as a consequence of direct inhibition of proton conduction through the F_o sector of the H^+-ATPase [35]. Titration curves showed that ten times more oligomycin was required to cause half-maximal inhibition of the ATPase activity in hepatoma submitochondrial particles compared to control liver particles (Fig. 4A, B). No difference in the rate of proton conduction between control and hepatoma particles and in the oligomycin titer for half-maximal inhibition of proton translocation was, however, observed (Fig. 4C).

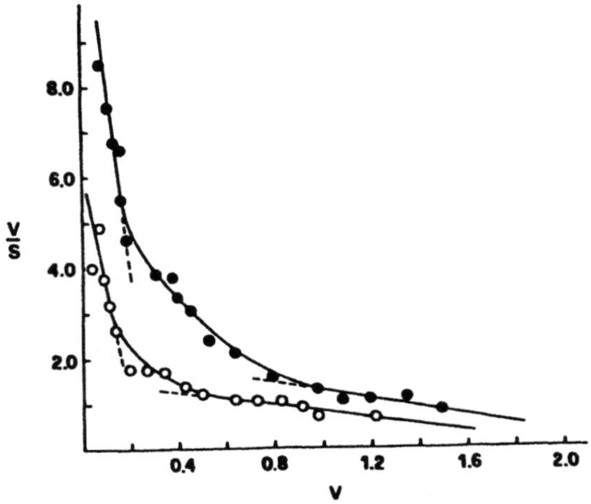

Fig. 3. Eadie-Hofstee plots of the kinetics of ATP hydrolysis by ESMP from control rat liver and Morris hepatoma 3924A. ATPase activity was measured as described in [30,31]. The ATP concentration ranged from 0.01 to 1.8 mM. ●) Control rat liver; (○) Morris hepatoma 3924A

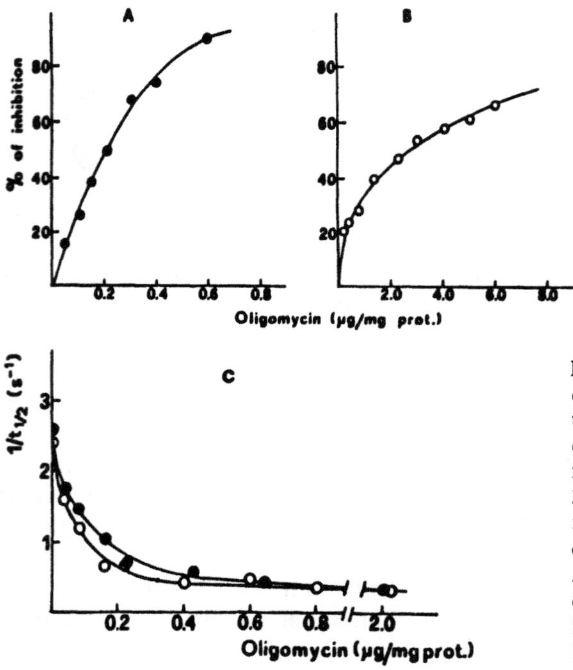

Fig. 4. Titration of the inhibitory effect of oligomycin on ATPase activities (**A,B**) and on the overall rate of proton conduction (**C**) in ESMP from rat liver and Morris hepatoma 3924A. ESMP were preincubated 5 min with oligomycin at the concentration given on the abscissa. ATPase activity and proton conduction were measured as described in (31). (●) Rat liver; (○) Morris hepatoma 3924A

Oligomycin is a specific inhibitor of the proton-conducting pathway in the F_o sector of ATPase, and the extension of the inhibitory effect to ATP hydrolysis (or synthesis) appears to involve one (OSCP) [15] or more proteins at the F_1-F_o junction [13,14,36].

These observations would suggest that in the H^+-ATPase complex from hepatoma, where subunits directly involved in the proton-conducting pathway are not affected, there is an alteration of one or more of the subunits (OSCP, F_6, γ, ∂

and ε), which are thought to be involved in the functional interaction between the two sectors of the complex.

Using an antibody against bovine F_1, which cross-reacted with rat liver α and β subunit but not with subunit γ [31,32], a decrease of the content of the β subunit during rat-liver regeneration 24 h after partial hepatectomy could be shown [21,32]. Gel electrophoresis of inside-out submitochondrial particles showed a clear decrease of the bands in the MW region of the α and β subunits of F_1 in particles prepared from mitochondria isolated from regenerating rat liver 24 h after partial hepatectomy (Fig. 5A). However, F_1 isolated from these particles showed the same polypeptide composition and the same activity as the control (not shown) but the recovery amounted only to 15–20% of the control [31].

These observations suggest that the decrease of ATPase activity observed 24 h after partial hepatectomy is due to a decrease of the F_1 content in mitochondria. This was accompanied by an oligomycin-sensitive increase of the state IV respiratory rate and a decrease of the respiratory control ratio (RCR), suggesting that the uncoupled state was due to enhanced proton conduction by F_0.

Proton conductivity by F_0 is measured in submitochondrial particles by kinetic analysis of anaerobic release of the respiratory proton gradient [37,38]. ESMP prepared from mitochondria isolated from regenerating rat liver 24 h after partial hepatectomy showed an increase of oligomycin-sensitive proton conductivity as shown by the decrease of the t½ of anaerobic proton release (Fig. 5B). In addition, as observed in F_1-deprived particles prepared from beef heart mitochondria [37], the kinetic pattern changed from biphasic to monophasic (Fig. 5B). In the presence of F_0 inhibitors, like oligomycin or DCCD, the proton conductivity was reduced to the same low level in submitochondrial particles from control and regenerating rat liver (Fig. 6).

The data presented suggest that in regenerating rat liver there exist F_0 sectors lacking their F_1 counterparts. The titration for inhibition by DCCD and oligomycin of passive proton conductivity indicates that there is no significant difference in the function of the F_0 sector between regenerating and control liver.

The transient decrease of the F_1 content, occurring in the initial phase of regeneration, suggests that in this phase there can be changes either in the synthesis of F_1 or in the uptake into mitochondria of F_1 subunits. This results in ATPase complexes lacking F_1 moieties, producing a proton pore through the F_0 sector which dissipates the electrochemical proton gradient set up by the respiration across the mitochondrial membrane. This situation is transitory and after 48 to 72 h of regeneration the full complement of F_1 [30–32] is restored.

4 Role of the ATPase Inhibitor Protein (IF_1)

A peculiar role in the regulation of the catalytic activity and coupling between H^+ transport by F_0 and catalytic activity in F_1 seems to be played by the IF_1 protein. This protein, Mr 10 kD [39,40], has an amphiphatic nature and binds specifically by its hydrophobic moiety to the β subunit, thus inhibiting the catalytic activity [41]. The binding of IF_1 to the ATPase complex is suppressed by respiratory $\Delta\mu H^+$, thus allowing ATP synthesis to proceed in the respiring steady state (42–43). In the absence of $\Delta\mu H^+$, IF_1 binds to the complex, thus preventing futile ATP hydrolysis.

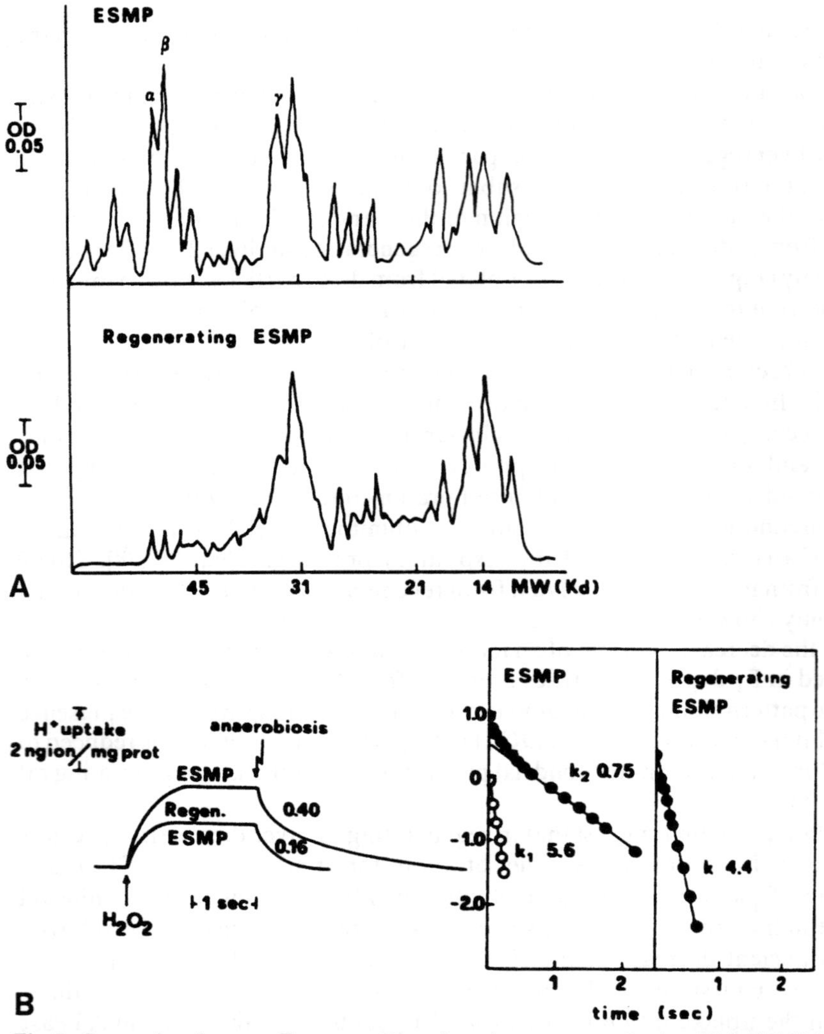

Fig. 5. Comparison between F_1 content (**A**) and passive proton conduction (**B**) in ESMP from control and regenerating rat liver. **A** Densitometric analysis of SDS gel electrophoresis of ESMP prepared from control and regenerating rat liver. For experimental conditions, see [30,31]. **B** Proton translocation induced by oxygen pulses of succinate supplemented ESMP. Mathematical analysis was carried out as in Pansini et al. (37). (Redrawn from 30 and 31) For details, see [30,31]

It has been found in our laboratory that IF_1 also exerts an inhibitory action on passive proton conduction through the F_o moiety [38,40] and that this action has the same titer (Fig. 7) and pH dependence (Fig. 8) of the effect exerted on the catalytic process [38,40]. The inhibitory activity decreases with pH, showing half-maximal activity around 7. This suggests a critical role of histidine residues and, in fact, chemical modification of one of the five histidine residues present in the inhibitor protein [41] prevents the activity of the isolated protein when reconstituted with the enzyme (Fig. 7). The dual inhibitory effect exerted by IF_1 on both the catalytic

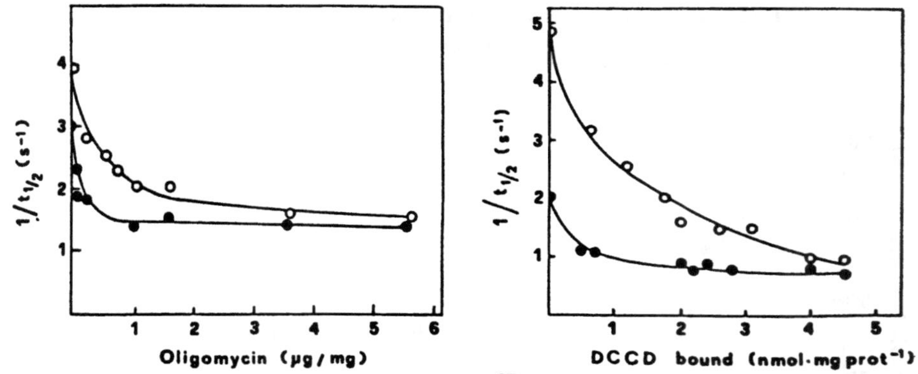

Fig. 6. Effect of oligomycin and DCCD on the overall rate of anaerobic relaxation of respiratory $\Delta\mu H^+$ in ESMP from control (●—●) and regenerating rat liver (○—○). For experimental conditions, see [31] (From [31])

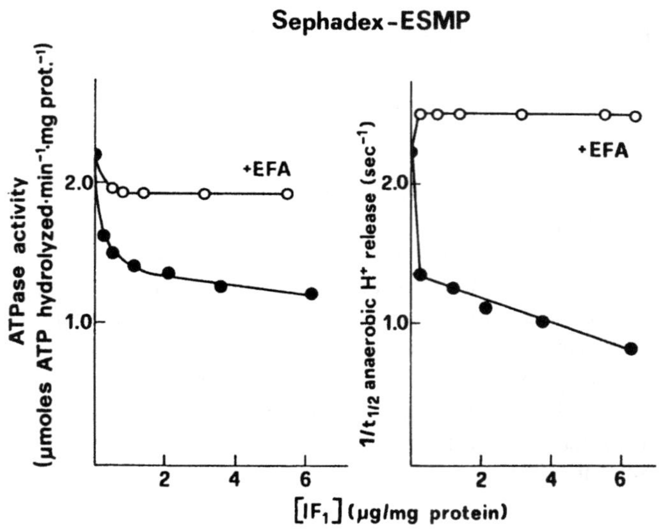

Fig. 7. Effect of diethylpyrocarbonate (*EFA*) treatment of purified IF_1 and its inhibition of ATPase activity and anaerobic release of respiratory proton gradient in submitochondrial particles depleted of IF_1 (Sephadex particles). Submitochondrial particles and IF_1 preparations, treatment of IF_1 with EFA and measurements of ATPase activity and passive H^+ conductivity were carried out as described in [38] (From [38])

activity and proton conduction qualifies this protein as a constituent of the gate. It is possible that the protein, in addition to its binding site, on the β-subunit of F_1, also binds to component(s) of the membrane sector (see also [44,45]). The pH dependence of the inhibitor suggests that IF_1 will be without a significant effect on the enzyme under normal aerobic conditions where the cytosolic pH is around neutrality and that of the matrix space is around 8. However, under conditions leading to cellular ischemia, like intensive work performance by the skeletal

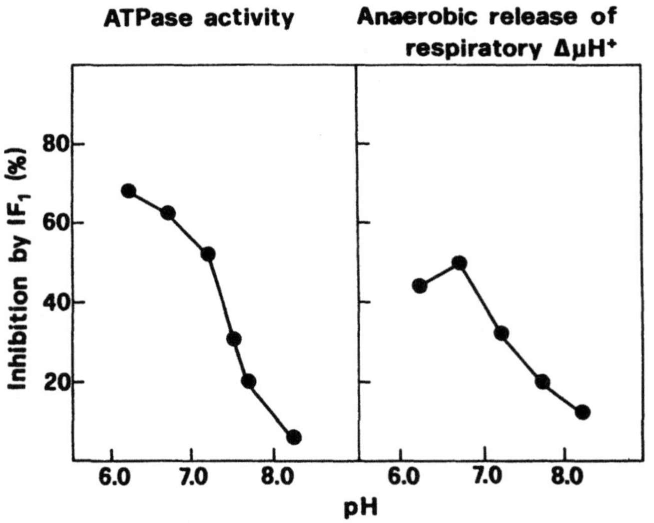

Fig. 8. pH dependence of the inhibition by IF_1 on ATPase activity and anaerobic release of respiratory proton gradient in Sephadex particles. The experiments were carried out as described in [38] (From [38])

muscle, circulatory insufficiency or artery occlusion, respiratory activity may stop with a drop of the transmembrane $\Delta\mu H^+$ and interruption of oxidative phosphorylation. Glycolysis will be activated as a consequence of the decrease in cytosolic phosphate potential with substantial lowering of pH well below neutrality. This acidification will promote inhibition of the H^+-ATPase by IF_1 which otherwise, in the absence of the pressure exerted by $\Delta\mu H^+$, will rapidly casue consumption of cellular ATP. Thus, the inhibitory protein may represent an extremely important specific defence mechanism for animals and in particular humans who unfortunately are more extensively exposed to the dangers of cellular ischemia (see also [46,47]).

Acknowledgement. This work was supported by C.N.R. Grant No. 87.1374.44.

References

1. Papa S, Altendorf K, Ernster L, Packer L (eds) (1984) H^+-ATPase (ATPase synthase): structure, function, biogenesis. The F_o-F_1 complex of coupling membranes. ICSU, Miami/Adriatica Editrice, Bari
2. Pedersen PL, Carafoli E (1987) Ion motive ATPases. I. Ubiquity, properties and significance to cell function. TIBS 12:146–150
3. Amzel LM, Pedersen PL (1983) Proton ATPases, structure and mechanism. Annu Rev Biochem 52:801–824
4. Papa S, Guerrieri F, Zanotti F, Scarfò R (1984) Proton translocation in the F_0 sector of the H^+-ATP synthase of mitochondria. Role of polypeptides and amino acid residues. In: Papa S, Altendorf K, Ernster L, Packer L (eds) H^+-ATPase (ATP synthase): structure, function, biogenesis. The F_0F_1 complex of coupling membranes. ICSU, Miami/Adriatica Editrice, Bari, pp 233–247

5. Gogol IP, Lücken V, Capaldi A (1987) The stalk connecting the F_1 and F_0 domains of ATP synthase visualized by electron microscopy of unstained specimens. FEBS Lett 219:274–278
6. Walker IE, Cozens AL, Dyer MR, Fearnley IM, Powell SJ, Runswick MJ (1987) Genes for ATP synthases from bacteria, chloroplasts and mitochondria. In: Ozawa T, Papa S (eds) Bioenergetics: structure and function of energy transducing systems. Japan Sci Soc Press, Tokio/Springer, Berlin Heidelberg New York, pp 167–178
7. Schneider E, Altendorf K (1984) Subunit b of the membrane moiety (F_0) of ATP synthase (F_1F_0) from *Escherichia coli* is indispensable for H^+ translocation and binding of the water-soluble F_1 moiety. Proc Nat Acad Sci USA 81:7279–7283
8. Schneider E, Altendorf K (1985) All three subunits are required for the reconstitution of an active proton channel (F_0) of *Escherichia coli* ATP synthase (F_1F_0). EMBO J 4:515–518
9. Friedl P, Hoppe J, Gunsalus RP, Michelsen O, von Meyenburg K, Schairer HU (1983) Membrane integration and function of the three F_0 subunits of the ATP synthase of *Escherichia coli* K_{12}. EMBO J 2:99–103
10. Montecucco C, Dabbeni-Sala F, Friedl P, Galante YM (1983) Membrane topology of ATP synthase from bovine-heart mitochondria and *Escherichia coli*. Eur J Biochem 132:189–194
11. Walker JE, Runswick MJ, Poulter L (1987) ATP synthase from bovine mitochondria. The characterization and sequence analysis of two membrane-associated subunits and the corresponding cDNAs. J Mol Biol 197:89–100
12. Torok K, Joshi S (1985) Cross-linking of bovine mitochondrial H^+-ATPase by copper-o-phenantroline. Interaction of the oligomycin-sensitivity-conferring protein with a 24 kDa protein. Eur J Biochem 153:155–159
13. Houstek J, Kopecky J, Zanotti F, Guerrieri F, Jirillo E, Capozza G, Papa S (1988) Topological and functional characterization of the F_0I subunit of the membrane moiety of the mitochondrial H^+-ATP synthase. Eur J Biochem 173:1–8
14. Zanotti F, Guerrieri F, Capozza G, Houstek J, Ronchi S, Papa S (1988) Identification of the nuclear-encoded Mr 24,668 protein of H^+-ATP synthase from bovine heart (F_0I) as a functional component of the F_0 moiety. FEBS Lett (in press)
15. McLennan DH, Tzagoloff A (1968) Studies on the mitochondrial adenosine triphosphatase system IV. Purification and characterization of the oligomycin-sensitivity conferring protein. Biochemistry 7:1603–1610
16. Walker JE, Gay NJ, Powell SJ, Kostine M, Dyer MR (1987) ATP synthase from bovine mitochondria: sequences of imported precursors of oligomycin sensitivity conferral protein, factor 6 and adenosintriphosphatase inhibitor protein. Biochemistry 26:8613–8619
17. Depuis A, Zaccai G, Satre M (1983) Optical properties and small-angle neutron-scattering of bovine heart mitochondrial oligomycin-sensitivity conferring protein. Biochemistry 22:5951–5956
18. Fearnely IM, Walker JE (1986) Two overlapping genes in bovine mitochondrial DNA encode membrane components of ATP synthase. EMBO J 5:2003–2008
19. Kanner BI, Serrano M, Kandrach MA, Racker E (1976) Preparation and characterization of homogenous coupling factor 6 from bovine heart mitochondria. Biochem Biophys Res Commun 69:1050–1056
20. Sebald W, Hoppe J (1981) On the structure and genetics of the proteolipid subunit of the ATP synthase complex. Curr Top Bioenerg 12:1–64
21. Sigrist H, Sigrist-Nelson K, Gitler C (1977) Single-phase butanol extraction: a new tool for proteolipid isolation. Biochim Biophys Res Commun 74:178–184
22. Kopecky J, Guerrieri F, Papa S (1983) Interaction of dicyclohexylcarbodiimide with the proton-conducting pathway of mitochondrial H^+-ATPase. Eur J Biochem 131:17–24
23. Hanstein WG (1976) Uncoupling of oxidative phosphorylation. Biochim Biophys Acta 456:126–148
24. Sanadi DR (1982) Mitochondrial coupling factor B. Properties and role in ATP synthesis. Biochim Biophys Acta 683:39–56
25. Muneyuki E, Ohno K, Kagawa Y, Hirata H (1987) Reconstitution of the proton translocation ATPase from bovine-heart mitochondria into planar phospholipid bilayer membranes. J Biochem (Tokyo) 102:1433–1440
26. Guerrieri F, Capozza G, Houstek J, Zanotti F, Jirillo E, Papa S. Isolation of the membrane sector of the bovine-heart mitochondrial F_0F_1 ATP synthase using the zwitterionic detergent 3-(3-cholamidopropyl)dimethylammonio-1-propane sulfonate. Functional and immunological characterization. (in preparation)

27. Zanotti F, Guerrieri F, Che YW, Scarfò R, Papa S (1987) Proton translocation by the H^+-ATPase of mitochondria. Effect of modification by monofunctional reagents of thiol residues in F_o polypeptides. Eur J Biochem 164:517–523
28. Enrich C, Gahmberg CG (1985) Pre-replicative changes of rat-sinusoidal plasma membrane glycoproteins during hepatic regeneration. FEBS Lett 181:12–16
29. Uriel J (1979) Retro-differentiation and the fetal patterns of gene expression in cancer. Adv Cancer Res 29:127–174
30. Buckle M, Guerrieri F, Papa S (1985) Changes in activity and F_1 content of mitochondrial H^+-ATPase in regenerating rat liver. FEBS Lett 188:345–351
31. Buckle M, Guerrieri F, Pazienza A, Papa S (1986) Studies on polypeptide composition, hydrolytic activity and proton conduction of mitochondrial F_oF_1 H^+-ATPase in regenerating rat liver. Eur J Biochem 155:439–445
32. Guerrieri F, Capuano F, Buckle M, Papa S (1988) Alteration of mitochondrial H^+-ATPase complex in tissue regeneration and neoplasia. In: Albano O, Gorrod J, Papa S (eds) Perspectives in molecular approaches to human diseases. Hahwood, Chichester (in press)
33. Kaschnits RM, Hatefi Y, Morris HP (1976) Oxidative phosphorylation properties of mitochondria isolated from transplanted hepatoma. Biochim Biophys Acta 449:224–235
34. Cross RL, Grubmeyer C, Penefsky HS (1982) Mechanism of ATP hydrolysis by beef-heart mitochondrial ATPase. Rate enhancement resulting from co-operative interactions between multiple catalytic sites. J Biol Chem 257:12101–12105
35. Senior AE (1988) ATP synthesis by oxidative phosphorylation. Physiol Rev 68:177–231
36. Penefsky HS (1985) Reaction mechanism of the membrane-bound ATPase of submitochondrial particles from beef heart. J Biol Chem 260:13728–13734
37. Pansini A, Guerrieri F, Papa S (1978) Control of proton conduction by the H^+-ATPase in the inner mitochondrial membrane. Eur J Biochem 92:545–551
38. Guerrieri F, Zanotti F, Che YW, Scarfò R, Papa S (1987) Inactivation of the mitochondrial ATPase inhibitor protein by chemical modification with diethylpyrocarbonate. Biochim Biophys Acta 892:284–293
39. Pullman ME, Monroy GC (1963) A naturally occurring inhibitor of mitochondrial adenosine triphosphatase. J Biol Chem 238:3762–3769
40. Guerrieri F, Scarfò R, Zanotti F, Che YW, Papa S (1987) Regulatory role of the ATPase inhibitor protein on proton conduction by mitochondrial H^+-ATPase complex. FEBS Lett 213:67–72
41. Harris DA (1984) The mitochondrial ATPase inhibitor protein-structure and function. In: Papa S, Altendorf K, Ernster L, Packer L (eds) H^+-ATPase (ATP synthase): structure, function, biogenesis. The F_oF_1 complex of coupling membranes. ICSU, Miami/Adriatica Editrice, Bari, pp 387–394
42. Lippe G, Sorgato MC, Harris DA (1988) Kinetics of the release of the mitochondrial inhibitor protein. Correlation with synthesis and hydrolysis of ATP. Biochim Biophys Acta 933:1–11
43. Lippe G, Sorgato MC, Harris DA (1988) The binding and release of the inhibitor protein are governed independently by ATP and membrane potential in ox-heart submitochondrial vesicles. Biochim Biophys Acta 933:12–21
44. Hashimoto T, Yoshida Y, Tagawa K (1984) Purification and properties of factors in yeast mitochondria stabilizing the F_1F_o-ATPase-inhibitor complex. J Biochem 95:131–136
45. Hashimoto T, Yoshida Y, Tagawa K (1983) Binding properties of an intrinsic ATPase inhibitor and occurrence in yeast mitochondria of a protein factor which stabilizes and facilitates the binding of inhibitor to F_1F_o-ATPase. J Biochem 94:715–20
46. Rouslin W, Pullman MM (1987) Protonic inhibition of the mitochondrial adenosine 5′-triphosphatase in ischemic cardiac muscle. Reversible binding of the ATPase inhibitor protein to the mitochondrial ATPase during ischemia. J Med Cell Biol 19:661–668
47. Rouslin W (1987) Factors affecting the reactivation of the oligomycin-sensitive adenosine 5′-triphosphatase and the release of ATPase inhibitor protein during the re-energization of intact mitochondria from ischemic cardiac muscle. J Biol Chem 262:3472–3476

Antitumor Effect of Drugs Interfering with Mitochondrial Biogenesis

Š. Kuželа[1], K. Luciaková[1], J. Kolarov[1] and B.D. Nelson[2]

1 Introduction

Most of the rapidly growing tumors have a high capacity of aerobic glycolysis. Warburg's hypothesis explaining this phenomenon by an impaired respiratory capacity of tumor cells attracted much attention in the past. The attempts to define the supposed defect(s) in oxidative metabolism of cancer cells were unsuccessful. Tumor mitochondria differ often very significantly from the organelles of the tissue of origin [review in 1]. These differences probably do not account for the altered, relative contributions of oxidative phosphorylation and glycolysis to the total cellular energy production in rapidly growing tumors. More relevant to the altered pattern of the tumor energy metabolism seems to be the diminished content of mitochondria in tumor cells, especially in the cells of rapidly growing tumors [review in 1]. In spite of their low content the organelles are indispensable to tumor energetics; even in the highly glycolyzing tumors oxidative phosphorylation covers no less than 50% of the energy demands of the cells [2]. Lower capacity of oxidative phosphorylation accompanying the reduced content of mitochondria in tumor cells has been employed in developing a new approach to cancer chemotherapy by using drugs interfering with the formation of functional mitochondria [3–5]. This communication deals with the mechanism by which inhibitors of mitochondrial biogenesis exert their antitumor effect.

2 Results and Discussion

2.1 Inhibition of Mitochondrial Protein Synthesis and Its Consequences

In the formation of functional mitochondria two different genetic (nuclear and mitochondrial DNA) and protein synthesizing (cytoplasmic and mitochondrial) systems are involved [review in 6,7]. Mammalian mitochondrial DNA codes for 13 different peptides [8] all of which are synthesized by a specific mitochondrial protein-synthesizing system [9,10]. The intramitochondrially formed proteins are constituents of several crucial inner mitochondrial membrane enzyme complexes: NADH dehydrogenase, bc_1 complex, cytochrome oxidase and ATP synthase

[1]Cancer Research Institute, Slovak Academy of Sciences, 812 32 Bratislava, Czechoslovakia
[2]University of Stockholm, Arrhenius Laboratory, Department of Biochemistry, 106 91 Stockholm, Sweden

A. Azzi, Z. Drahota and S. Papa
Molecular Basis of Membrane-Associated Diseases

[review in 7]. Consequently, the mitochondrial translation products, though few in number, are essential for the biogenesis of functional mitochondria.

Fate of Extramitochondrially Synthesized Mitochondrial Partner Proteins Under in Vivo Inhibited Mitochondrial Protein Synthesis. The formation of intramitochondrially synthesized proteins can be selectively blocked by several antibacterial antibiotics (e.g. tetracyclines or chloramphenicol and its derivatives) [11]. In vivo administration of these drugs results in a diminished content of the enzymes containing the intramitochondrially synthesized subunits. This can be exemplified by changes in the contents of ATP synthase, cytochrome oxidase and bc_1 complex subunits in mitochondria of rat liver upon in vivo treatment with oxytetracycline (Fig. 1).

From the rat cytochrome oxidase only the subunits I, II and III, and from the bc_1 complex only cytochrome b are synthesized intramitochondrially [12]. Figure 1 thus also shows that the contents of both intra- and extramitochondrially synthesized proteins participating in the formation of the same enzyme complex (partner proteins) decrease upon selective in vivo inhibition of mitochondrial protein synthesis. The mitochondrial content of the inner membrane enzymes lacking intramitochondrially synthesized partner proteins, such as adenine nucleotide translocase, is unaffected by the in vivo inhibition of mitochondrial protein synthesis (Fig. 1). Quantitative evaluation of the amounts of intra- and extramitochondrially synthesized components of the bc_1 complex and cytochrome oxidase in mitochondria and whole cells of rat liver and Zajdela hepatoma revealed that the contents of the partner proteins of the same enzyme complex decrease equally upon the in vivo inhibition of mitochondrial protein synthesis. On the other hand, the contents of respective enzyme complexes consisting of the intra- and extramitochondrially synthesized peptides are not lowered to the same degree. From the enzyme complexes studied the content of cytochrome oxidase displayed the highest, that of the bc_1 complex an intermediate and that of ATP synthase the lowest sensitivity towards the inhibition of mitochondrial protein synthesis (Table

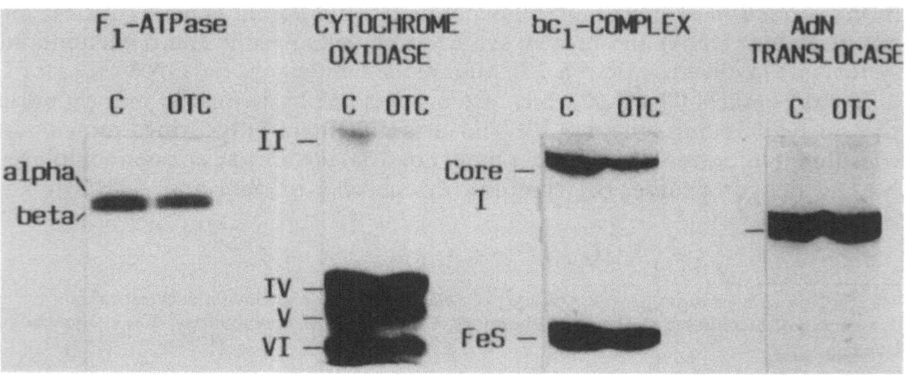

Fig. 1. Lowered contents of several rat liver mitochondrial proteins upon in vivo inhibition of mitochondrial protein synthesis. Oxytetracycline was administered by continuous i.v. infusion (constant serum level of the drug of 15 μg/ml) for 15 days. Equal amounts of liver mitochondria from control (*C*) and treated (*OTC*) animals were electrophoresed, transferred onto nitrocellulose and processed with antisera. The immune complexes were decorated with radioiodinated protein A and radiographed. The positions of the subunits against which the antisera were raised are indicated

1). The same tendency is found in resting and regenerating liver and in rapidly growing Zajdela hepatoma (Table 1).

In vivo inhibition of mitochondrial protein synthesis in rats leads also to a decrease in the amount of NADH dehydrogenase [13], the proteins of which are also of dual origin [9,10]. Relative decrease in the contents of its intra- and extramitochondrially synthesized peptides has not been studied so far.

From the above results it may appear that not only the synthesis of intra- but also of the extramitochondrially synthesized partner proteins is blocked by the in vivo inhibition of mitochondrial protein synthesis. This possibility was checked using a rat liver-derived, in vitro protein-synthesizing system translating endogenous mRNA [14]. The system consisting essentially of rat liver postmitochondrial supernatant was found to synthesize mitochondrial proteins in a stable precursoric form. Quantitative immunoadsorption of several representative proteins newly synthesized in the translation system revealed that the synthesis of extramitochondrially formed partner proteins is not preferentially affected by prolonged inhibition of mitochondrial protein synthesis (Fig. 2).

Pulse-chase labeling of the partner proteins in the liver-derived translation system revealed that the in vivo inhibition of mitochondrial protein synthesis does not lead to a preferential degradation of the newly synthesized partner proteins outside mitochondria.

The possibility existed that the reduced mitochondrial content of extramitochondrially synthesized partner proteins upon the inhibition of mitochondrial protein synthesis results from an altered transport of newly synthesized partner proteins into mitochondria. This was examined by pulse and pulse-chase labeling of whole Zajdela hepatoma cells followed by rapid digitonin fractionation into cytosolic and particulate fractions and quantitative immunoadsorption of the proteins from the fractions. It was found that the ability of mitochondria to transport the precursors of partner proteins and to convert them into the mature

Table 1. Decrease in contents of cytochrome oxidase, bc_1 complex and ATP synthase upon in vivo inhibition of mitochondrial protein synthesis[a]

Tissue/treatment	Cytochrome oxidase (%)	bc_1 Complex (%)	ATP synthase (%)	AdN translocase
Regenerating liver / 3 days	45	21	8	None
Resting liver / 3 weeks	72	47	19	None
Zajdela hepatoma / 4 days	52	25	10	None

[a]Average values from 3 experiments are presented. Oxytetracycline was administered as given in Fig. 1 for the indicated time. Mitochondrial contents of the enzyme complexes were determined by quantitative immunoblotting using antisera directed against subunits II and IV-VI of cytochrome oxidase, Core I, FeS protein and cytochrome c_1 of the bc_1 complex and to the three largest subunits of ATP synthase. The regions of the blots containing radiolabeled protein A bound to immune complexes were directly counted and the sum of radioactivity for respective enzymes was evaluated. Cytochrome b content was determined spectrophotometrically; its decrease correlated with that of the immunochemically estimated constituents of the bc_1 complex.

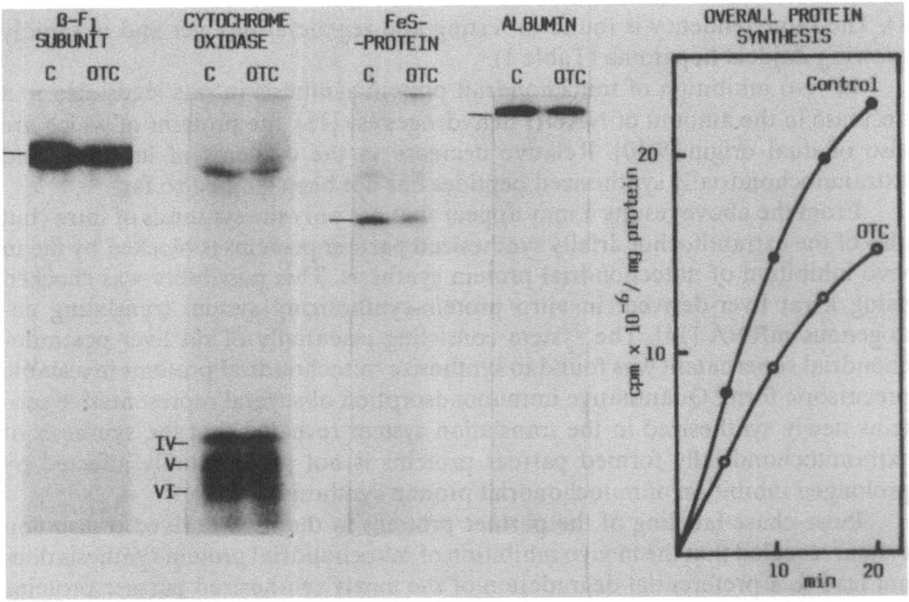

Fig. 2. In vivo inhibition of mitochondrial protein synthesis does not preferentially affect the synthesis of extramitochondrially synthesized mitochondrial partner proteins. Proteins labeled with ^{35}S-methionine in the translation system from regenerating livers of control (*C*) and oxytetracycline-treated (*OTC*) rats were immunoadsorbed with an excess of appropriate antisera, electrophoresed and fluorographed. Antisera against the β-subunit of ATP synthase, whole cytochrome oxidase, FeS protein of the bc_1 complex and albumin were used. The graph on the *right* illustrates the rate of overall protein synthesis in the translation system from control and oxytetracycline-treated animals

forms is not diminished upon the prolonged inhibition of mitochondrial protein synthesis (Fig. 3).

Sonically disrupted liver mitochondria from control and oxytetracycline-treated rats degraded the in vitro synthesized precursor of the FeS protein and β-F_1 subunit in an identical way. This indicates that a mitochondrial protease activity degrading extramitochondrially synthesized partner proteins is not induced by the in vivo inhibition of mitochondrial protein synthesis.

These results strongly indicate that the mitochondrial content of extramitochondrially synthesized partner proteins is regulated at the organelle level, probably by the availability of their intramitochondrially synthesized partner proteins. This implies the lack of a tight regulatory link in the formation of intra- and extramitochondrially synthesized partner proteins. It appears that under inhibited mitochondrial protein synthesis the partner proteins of extramitochondrial origin are normally synthesized and transported into mitochondria. Here, lacking their intramitochondrially synthesized partners they cannot be assembled into functional complexes and are eventually degraded.

The data also show and for the first time analyze and quantitate at the peptide constituent level the markedly decreased contents of several mitochondrial enzyme complexes upon the inhibition of mitochondrial protein synthesis in mammalian cells.

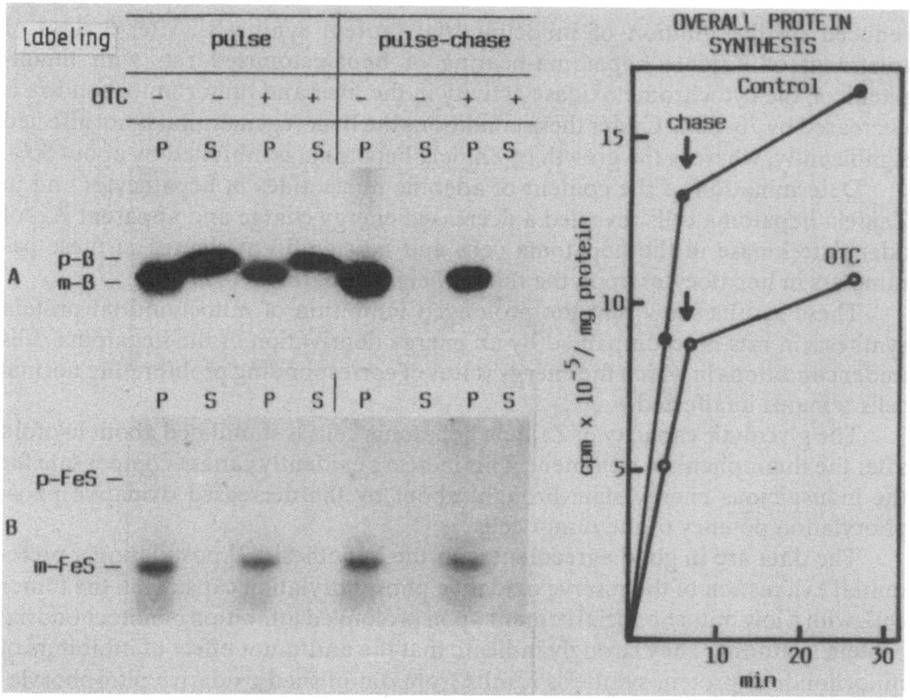

Fig. 3. In vivo inhibition of mitochondrial protein synthesis does not affect mitochondrial transport of extramitochondrially synthesized partner proteins. Control and in vivo oxytetracycline-treated Zajdela hepatoma cells were pulse (5 min) or pulse-chase (5 min pulse followed by 20 min chase) labeled with ^{35}S-methionine. The cells were permeabilized with digitonin and fractionated into soluble (*S*) and particulate (*P*) fractions. The β-F_1 subunit and FeS protein of the bc_1 complex were immunoadsorbed from the fractions, electrophoresed and fluorographed. *p* precursoric; *m* mature form. Graph on the *right* shows the rate of overall cellular protein synthesis

Energy Deprivation of Tumor Cells upon in Vivo Inhibition of Mitochondrial Protein Synthesis. It has been shown that prolonged inhibition of mitochondrial protein synthesis in vivo blocks the growth of several experimental tumors in rats without affecting the multiplication of normal cells [3–5,15]. To explain this effect the following hypothesis [3] has been put forward. Due to a low mitochondrial content, the rapidly growing tumors possess a low reserve of oxidative phosphorylation capacity. Consequently, under inhibited mitochondrial protein synthesis, when the formation of functional mitochondria is blocked, the tumor cells reach the state of energy limitation earlier than normal cells which generally have a high oxidative phosphorylation capacity. We checked the validity of this hypothesis directly by determining the contents of some of the oxidative phosphorylation constituents and by estimating the energy state of rapidly growing Zajdela hepatoma cells exposed to thiamphenicol, an efficient inhibitor of mitochondrial protein synthesis.

As detected by immunochemical and enzymatic determination of representative mitochondrial enzymes, the content of mitochondria in Zajdela hepatoma cells is about 20% of that in hepatocytes of regenerating rat liver. As indicated above, this low oxidative phosphorylation capacity of the tumor cells can be further reduced by the inbition of mitochondrial protein synthesis. After 3 days of treatment of Zajdela hepatoma-bearing or hepatectomized rats with thiamphenicol, the cytochrome oxidase activity in the liver and tumor mitochondria is decreased by 70–80%. Under these conditions the liver regeneration is not affected significantly, whereas the growth of Zajdela hepatoma is inhibited by about 50%.

Determination of the content of adenine nucleotides in hepatocytes and in Zajdela hepatoma cells revealed a decreased energy charge and apparent K_{eq} of adenylate kinase in the hepatoma cells and no significant change in these parameters in hepatocytes upon the thiamphenicol treatment (Table 2).

These results show that the prolonged inhibition of mitochondrial protein synthesis in rats is accompanied by an energy deprivation of the hepatoma cells under conditions in which the energy status of corresponding proliferating normal cells remains unaffected.

The glycolytic capacity of Zajdela hepatoma cells is stimulated about twofold after the thiamphenicol treatment. This increase evidently cannot compensate for the inauspicious energy state brought about by the decreased oxidative phosphorylation potency of the tumor cells.

The data are in good agreement with the hypothesis [3] postulating a preferential exhaustion of the reserve oxidative phosphorylation capacity in the tumor cells with a low mitochondrial content upon prolonged inhibition of mitochondrial protein synthesis. They strongly indicate that the antitumor effect of inhibitors of mitochondrial protein synthesis results from diminished oxidative phosphorylation capacity and consequent energy deprivation of the tumor cells.

2.2 *Inhibition of Mitochondrial Protein Transport*

One of the important steps in mitochondrial biogenesis is the transport of extramitochondrially synthesized mitochondrial proteins from the site of their synthesis into the organelles. The vast majority of mitochondrial proteins have to cross one or two mitochondrial membranes before they reach their final destination. Most of them are synthesized in the form of larger precursors with an extension of about 20 amino acids at their N-terminus [review in 7,16]. This

Table 2. Effect of in vivo inhibition of mitochondrial protein synthesis on growth and energy status of rat hepatocytes and Zajdela hepatoma cells[a]

Cells	Thiamphenicol treatment	Cytochrome oxidase activity (%)	Energy charge	Approx. K_{eq} of adenylate kinase	Inhibition of growth (%)
Hepatocytes	–	100	0.85	2.29	–
	+	22.7	0.84	2.18	5.1 ±4.0
Zajdela hepatoma	–	100	0.84	2.16	
	+	22.0	0.71	0.73	53.3 ±6.1

[a] Thiamphenicol was administered i.p. in 12 h intervals (800 mg/kg body weight). The treatment started 3 h after 70% hepatectomy or 66 h after the tumor inoculation and lasted for 72 h.

extension (signal sequence) plays an important role in the recognition of the organelles and in the transport of precursors through mitochondrial membrane(s). After fulfilling its role in the transport the signal sequence is cleaved within mitochondria by a specific protease (maturase) giving rise to the mature form of the protein [review in 17–20].

There are only a few inhibitors of the mitochondrial protein transport. Among these are Rhodamine 123 and Rhodamine 6G [21,22], dyes which selectively accumulate inside mitochondria [23,24]. Rhodamines 123 and 6G exert a strong anticarcinoma activity both in vitro and in vivo [25–28]. The antitumor effect of Rhodamine 123 correlates with the ability of the cells to retain the accumulated drug within mitochondria [25–27]. However, the basis of the antitumor effect of mitochondriotropic Rhodamines remains unclear.

We found [22] that in Zajdela hepatoma cells exposed to Rhodamine 6G the newly synthesized mitochondrial proteins are not processed to the mature forms and accumulate in the cytosol (Fig. 4).

Also, in a reconstituted in vitro system [29] Rhodamine 6G inhibits transport and maturation of newly synthesized mitochondrial proteins (Fig. 5). The inhibitory effect of Rhodamine 6G on mitochondrial protein transport is not caused by collapsing the mitochondrial membrane potential by the drug.

By testing the effect of Rhodamine 6G on proteolytic processing of newly synthesized mitochondrial proteins by the mitochondrial matrix fraction, it was found that the drug inhibits the mitochondrial maturase activity [29] (Fig. 6).

The data obtained show that the anticarcinoma drug Rhodamine 6G blocks mitochondrial protein transport and that this effect results from the inhibition of mitochondrial maturase activity by the drug.

It is very likely that similarly to the inhibition of mitochondrial protein synthesis, the inhibition of mitochondrial biogenesis at the mitochondrial protein transport level affects preferentially the cells with a low capacity of oxidative phosphorylation by modifying adversely their energy state.

Taken together, the data strongly indicate that different steps in mitochondrial biogenesis could serve as a suitable target in cancer chemotherapy.

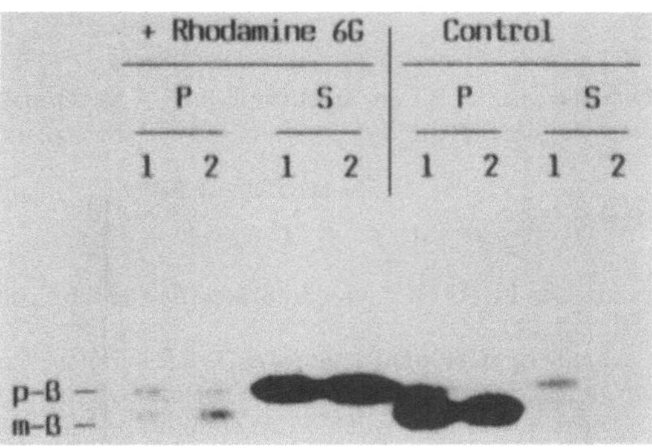

Fig. 4. Precursor of the β-F_1 subunit accumulates in the cytosol of Rhodamine 6G-treated Zajdela hepatoma cells. The hepatoma cells were pulse (10 min; *lanes 1*) or pulse-chase (10 min pulse followed by 10 min chase; *lanes 2*) labeled with ^{35}S-methionine in the absence or presence of 40 μM Rhodamine 6G. The cells were permeabilized with digitonin and fractionated into particulate (*P*) and soluble (*S*) fractions. The β-F_1 subunit was immunoadsorbed from the fractions, electrophoresed and fluorographed. *p* precursoric; *m* mature form

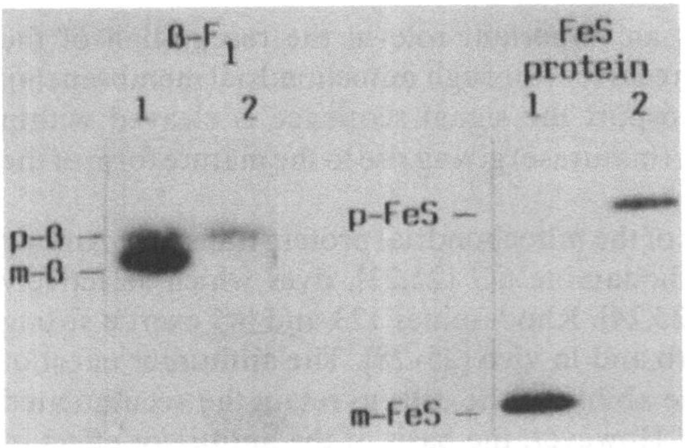

Fig. 5. Rhodamine 6G inhibits protein transport into isolated mitochondria. Proteins synthesized in rabbit reticulocyte lysate programmed with free polysomes from rat liver were labeled with ^{35}S-methionine. Isolated rat liver mitochondria were incubated with the labeled lysate in the absence (*lanes 1*) or presence (*lanes 2*) of 5 μM Rhodamine 6G. Mitochondria were reisolated and used for immunoadsorption of the β-F_1 subunit and FeS protein of the bc_1 complex with monospecific antisera. The immunoadsorbed material was electrophoresed and fluorographed. *p* precursoric; *m* mature form

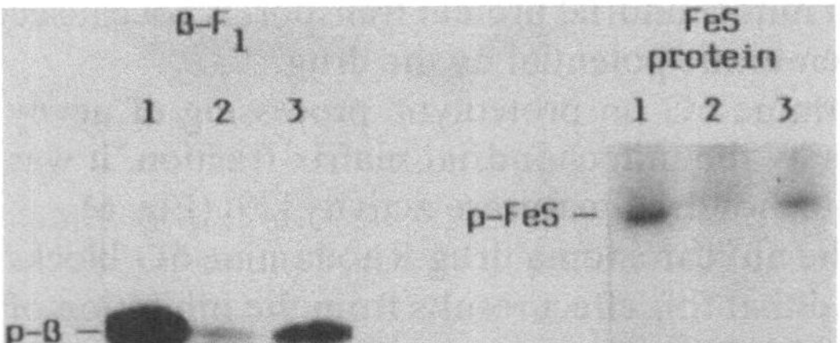

Fig. 6. Rhodamine 6G inhibits proteolytic processing of precursory forms of mitochondrial proteins by the mitochondrial matrix fraction. Rabbit reticulocyte lysate programmed with rat liver-free polysomes was labeled with ^{35}S-methionine. The labeled lysate was kept on ice (*1*) or incubated 20 min at 30° C with rat liver mitochondrial matrix in the absence (*2*) or presence (*3*) of 1 mM Rhodamine 6G. The β-F_1 subunit and FeS protein of the bc_1 complex were immunoadsorbed from the samples, electrophoresed and fluorographed. *P* precursoric; *m* mature form

Acknowledgements. We are indebted to Prof. A.M. Kroon (University of Groningen) for his helpful interest and for enabling us to perform some of the experiments in his laboratory.

References

1. Pedersen PL (1978) Tumor mitochondria and the bioenergetics of cancer cells. Progr Exp Tumor Res 22:190–274
2. Aisenberg AC (1961) The glycolysis and respiration of tumors. Academic Press, New York London
3. Van den Bogert C, Dontje BHJ, Wybegna JJ, Kroon AM (1981) Arrest of in vivo proliferation of Zajdela tumor cells by inhibition of mitochondrial protein synthesis. Cancer Res 41:1943–1947
4. Van den Bogert C, Dontje BHJ, Kroon AM (1985) The antitumour effect of doxycycline on a T-cell leukemia in the rat. Leukemia Res 9:617–623
5. Van den Bogert C, Dontje BHJ, Kroon AM (1983) Arrest of in vivo growth of a solid Leydig cell tumor by prolonged inhibition of mitochondrial protein synthesis. Cancer Res 43:2247–2251

6. Schatz G, Mason TL (1974) The biosynthesis of mitochondrial proteins. Annu Rev Biochem 54:51–81
7. Nelson BD (1987) Biogenesis of mammalian mitochondria. Curr Top Bioenerg 15:221–272
8. Anderson S, Bankier AT, Barrell BG, de Bruijn MMHL, Coulson AR, Drouin J, Eperon IC, Nierlich DP, Roe BA, Sanger F, Schrier PH, Smith AJM, Staden R, Young IG (1981) Sequence and organization of human mitochondrial genome. Nature (London) 290:457–465
9. Chomyn A, Mariottini P, Cleeter MWJ, Ragan CI, Matsuno-Yagi Y, Hatefi Y, Doolitle RF, Attardi G (1985) Six unidentified reading frames of human mtDNA encode components of the respiratory-chain NADH dehydrogenase. Nature (London) 314:592–597
10. Chomyn A, Cleeter MJW, Ragan CI, Riley M, Doolitle RF, Attardi G (1986) URF6, last unidentified reading frame of human mtDNA, codes for an NADH-dehydrogenase subunit. Science 234:614–618
11. de Vries H, Kroon AM (1970) On the effect of chloramphenicol and oxytetracycline on the biogenesis of mammalian mitochondria. Biochim Biophys Acta 204: 531–541
12. Kolarov J, Kužela Š, Wielburski A, Nelson BD (1981) Characterization of mitochondrial translation products in rat liver and rat hepatoma. FEBS Lett 126:61–65
13. Van den Bogert C, Holtrop M, de Vries H, Kroon AM (1985) Specific inhibition of mitochondrial protein synthesis influences the amount of complex I in mitochondria of rat liver and *Neurospora crassa* directly. FEBS Lett 192:225–229
14. Eisenstein RS, Harper AE (1984) Characterization of a protein synthesizing system from rat liver. Translation of endogenous and exogenous messenger RNA. J Biol Chem 259:9922–9928
15. Kužela Š, Luciaková K (1983) Energy deprivation of Zajdela hepatoma cells upon in vivo treatment with thiamphenicol. Neoplasma 30:651–657
16. Roise D, Schatz G (1988) Mitochondrial presequences. J Biol Chem 263:4509–4511
17. Schatz G (1987) Sir Hans Krebs lecture: signals guiding proteins to specific intramitochondrial locations. Eur J Biochem 165:1–6
18. Douglas MG, McCaramon MT, Vassarotti A (1986) Targeting proteins into mitochondria. Bacteriol Rev 50:166–178
19. Hurt EC, van Loon APGM (1986) How proteins find mitochondria and intramitochondrial compartments. Trends Biochem Sci 11:204–207
20. Morita T, Mori M, Ikeda F, Tatibana M (1982) Transport of carbamyl phosphate synthetase I and ornithine carbamylase into mitochondria. J Biol Chem 257:10547–10550
21. Kolarov J, Nelson BD (1984) Import and processing of cytochrome bc_1 subunits in isolated hepatoma ascites cells. Eur J Biochem 144:387–392
22. Gear ARL (1974) Rhodamine 6G a potent inhibitor of mitochondrial oxidative phosphorylation. J Biol Chem 249:3628–3637
23. Johnson LV, Walsh ML, Chen LB (1980) Localization of mitochondria in living cells with rhodamine 123. Proc Natl Acad Sci USA 77:990–994
24. Bernal SD, Lampidis TJ, McIsaac RM, Chen LB (1983) Anticarcinoma activity in vivo of rhodamine 123, a mitochondria-specific dye. Science 222:169–171
25. Lampidis TJ, Bernal SD, Summerhayes IC, Chen LB (1982) Selective toxicity of rhodamine 123 in carcinoma cells in vitro. Cancer Res 43:716–720
26. Herr HW, Huffman JL, Huryk R, Heston WDW, Melamed MR, Whitmore JrWF (1988) Anticarcinoma activity of rhodamine 123 against a murine renal adenocarcinoma. Cancer Res 48:2061–2063
27. Wilkie D, Fearon K (1985) Mitochondria and cancer. In: quagliariello E, Slater EC, Palmieri F, Saccone C, Kroon AM (eds) Achievements and perspectives of mitochondrial research, vol 2. Biogenesis. Elsevier, Amsterdam, pp 437–444
28. Kužela Š, Joste V, Nelson BD (1986) Rhodamine 6G inhibits matrix-catalyzed processing of precursors of rat liver mitochondrial proteins. Eur J Biochem 154:553–557

6 Lysosomes and Peroxisomes in Health and Disease

The Complex of β-Galactosidase, Neuraminidase and "Protective Protein" in Lysosomes: Molecular Characterization of the "Protective Protein"

A. D'AZZO, N. GILLEMANS and N. GALJART

Human lysosomal β-D-galactosidase and N-acetyl-α-neuraminidase (sialidase) are hydrolytic enzymes responsible for the degradation of a variety of natural and synthetic substrates (for review see [1,2]). They are associated in their functional state with a third glycoprotein, referred to as "protective protein" [3,4]. These three glycoproteins copurify as a complex, the molecular basis of which is so far poorly understood.

Mutations at the structural genes coding for β-galactosidase and neuraminidase cause deficient or severely altered enzyme activities, with consequent accumulation of undegraded substrate(s) in the lysosomes. The resulting metabolic storage disorders associated with a single β-galactosidase deficiency are G_{M1}-gangliosidosis and Morquio B [2,5], whereas sialidosis is due to an isolated neuraminidase deficiency [6]. These disorders have a recessive mode of inheritance and are characterized by a wide spectrum of clinical manifestations [2,7]. A third autosomal recessive disorder, currently designated galactosialidosis [8], is due to a primary defect of the "protective protein" which leads to a combined β-galactosidase and neuraminidase deficiency [9]. Different clinical phenotypes have been observed, ranging from very severe early infantile forms, fatal within childhood, to late infantile and juvenile/adult variants with milder pathological features and later onset of the disease (Table 1; for reviews see [8,10,11]).

Evidence for the existence of the protective protein came from biochemical studies on human lysosomal β-galactosidase and neuraminidase. In human liver and cultured fibroblasts active β-galactosidase is present as a 64 kD monomeric polypeptide and as a high molecular weight aggregate of 600–700 kD [12,13]. Table 2 summarizes the properties of β-galactosidase in galactosialidosis cells. The residual enzyme activity is about 10% of the normal level and only the monomeric form of β-galactosidase is present. It was originally postulated that intralysosomal accumulation of sialylated compounds, due to a primary defect of neuraminidase, caused inhibition of β-galactosidase activity [9]. Subsequent studies demonstrated, however, that increased proteolysis of a normally synthesized β-galactosidase polypeptide accounts for the low residual activity of the enzyme [14]. The primary (molecular) defect in galactosialidosis was eventually identified studying the biosynthesis and intracellular processing of β-galactosidase.

In normal human fibroblasts the first immunoprecipitable component of β-galactosidase is a precursor of 85 kD. This precursor form is trimmed to the size

Department of Cell Biology and Genetics, Erasmus University, P.O. Box 1738, 3000 DR Rotterdam, The Netherlands

A. Azzi, Z. Drahota and S. Papa
Molecular Basis of Membrane-Associated Diseases

Table 1. Clinical features of galactosialidosis patients

Early infantile type
Kidney involvement, hepatosplenomegaly, heart failure, and other abnormalities result in early death or fetal hydrops. 10% residual β-galactosidase activity; neuraminidase activity not detectable.
Late infantile type
Age of onset 1–3 years, coarse facial features, skeletal abnormalities, hepatosplenomegaly, visual problems and slight mental retardation during childhood. 10–15% residual β-galactosidase activity; neuraminidase activity 1–4% of normal level.
Juvenile/adult type
Nearly normal development until adulthood, slowly progressing mental retardation, ataxia, myoclonus, seizures, skeletal abnormalities, visual problems, heart involvement. Oligosacchariduria, 10–15% residual β-galactosidase activity; neuraminidase activity not detectable.

Table 2. Properties of β-galactosidase in galactosialidosis cells

– Residual activity	:	10% of control value
– Molecular weight	:	66 kD monomer no aggregation
– Km and pH optimum	:	as normal enzyme
– Turnover time	:	less than 1 day
– Rate of synthesis	:	0.5 pmol/day/mg cell protein^{-1}
– Enhanced proteolytic degradation in lysosomes		

of the mature enzyme through a number of posttranslational steps [3]. These immunoprecipitation experiments also revealed the presence of 32 and 20 kD polypeptides consistently coprecipitated with anti-human β-galactosidase antiserum. The significance of the 32 kD protein became apparent when it was found to be deficient in fibroblasts from an early infantile galactosialidosis patient [3]. Subsequently, it was demonstrated that the majority of active β-galactosidase is present together with the 32 and 20 kD polypeptides in the aforementioned high molecular weight aggregate [4].

The 32 kD component is synthesized as a 54 kD precursor which is also recovered extracellularly [3]. This secreted precursor, taken up by galactosialidosis fibroblasts, is processed intracellularly and promotes multimerization of monomeric β-galactosidase [4,15]. It became clear that at least the 32 kD protein, thereafter called "protective protein", is essential for the aggregation of mature β-galactosidase molecules, rendering the enzyme more stable [4]. The importance of the 20 kD polypeptide has been unclear until now. Table 3 summarizes the biochemical heterogeneity for the protective protein deficiency that has been described in galactosialidosis patients with different clinical phenotypes [16].

Concomitant studies on lysosomal neuraminidase clarified its relation with the protective protein and β-galactosidase. In bovine testis and human placenta neuraminidase copurifies with β-galactosidase and the protective protein.

Table 3. Biochemical heterogeneity of the protective protein in galactosialidosis patients

Early infantile :	absence of cross-reactive material for the 'protective protein'
Late infantile :	normal synthesis of 'protective protein' precursor which is glycosylated, phosphorylated but very poorly cleaved to the mature size
Juvenile/adult :	reduced amount of 'protective protein' precursor, not cleaved to the mature size
Mode of inheritance: autosomal recessive	

Furthermore, neuraminidase activity depends on the presence of the protective protein [17,18]. These findings together strongly supported the idea that the three glycoproteins exist in a functional complex.

We have isolated the cDNA encoding human protective protein in order to study the primary structure and expression of this protein as well as to investigate the mechanism of interaction between the components of the complex and the molecular nature of the mutations in galactosialidosis patients. Polyclonal antibodies raised against the purified 32 kD polypeptide identified a cDNA clone (Hu54) of 1.85 kb (Fig. 1) that hybridized with two independent oligonucleotide probes. Colinearity of the DNA sequence with amino terminal and tryptic peptide sequences confirmed the identity of the clone. The cDNA recognizes an mRNA of 2 kb in normal fibroblasts (Fig. 2). This RNA species is absent in fibroblasts of an early infantile (E.I.) galactosialidosis patient but present in fibroblasts of a patient with the late infantile (L.I.) form of the disease. These results demonstrate different mutations that are present in the two clinical forms of galactosialidosis tested. Figure 2 also shows that the parents of the E.I. galactosialidosis patient have reduced amounts of the 2 kb mRNA. They are consanguineous [19] and therefore likely to carry the same allelic mutation. These data demonstrate heterozygosity for the protective protein deficiency, for which no direct enzymatic assay is available.

The sequence of the protective protein cDNA comprises 1825 nucleotides. An ATG translational start codon is found 6 nucleotide from the 5′ end. To verify the potential usage of this start codon an in vitro transcription and translation assay was performed, after which radiolabeled proteins were immunoprecipitated using anti-32 kD antibodies and resolved by SDS-PAGE. In Fig. 3 the results are shown of translation assays that were done using different RNA sources. A major product of 54 kD is immunoprecipitated when the coding strand (sense) of the Hu54 cDNA

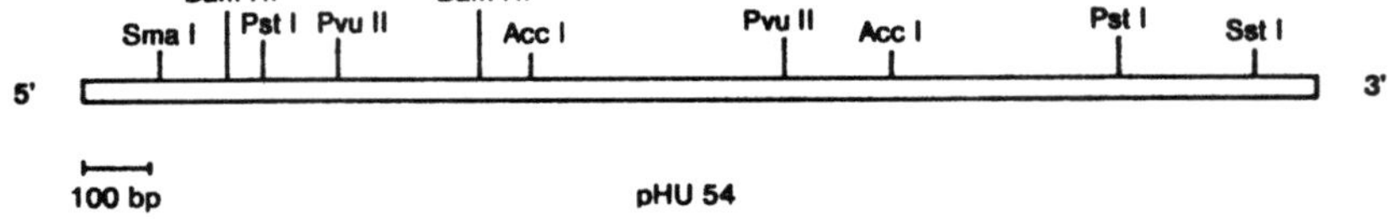

Fig. 1. Partial restriction map of the cDNA encoding the 54 kD precursor of the protective protein

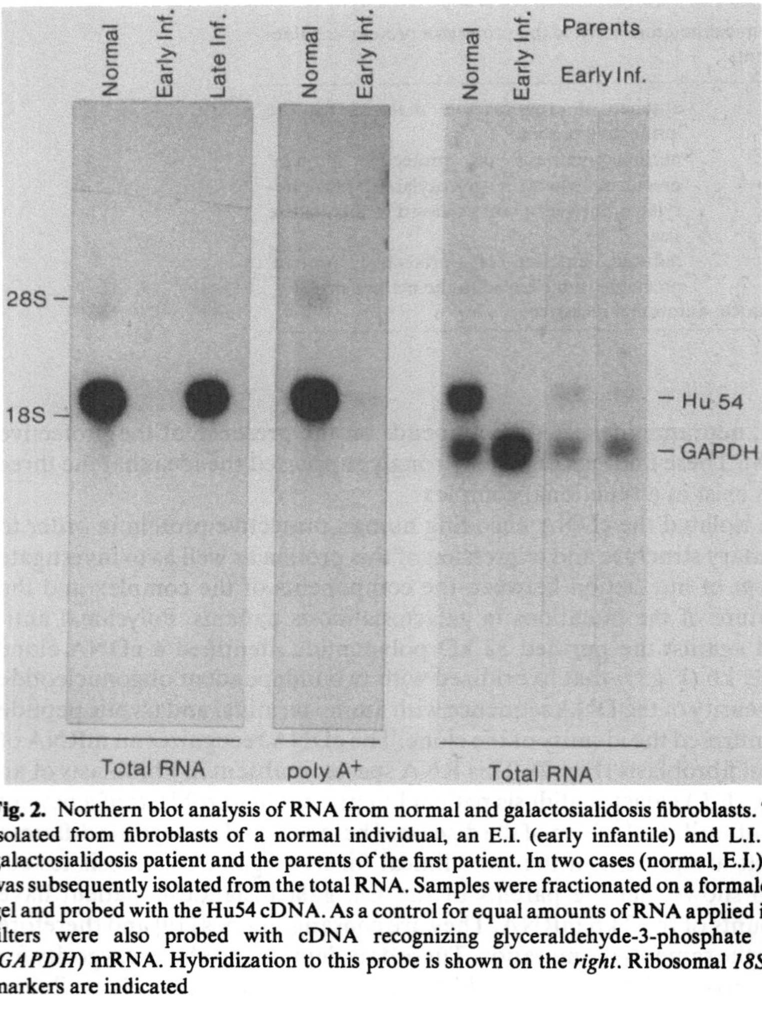

Fig. 2. Northern blot analysis of RNA from normal and galactosialidosis fibroblasts. Total RNA was isolated from fibroblasts of a normal individual, an E.I. (early infantile) and L.I. (late infantile) galactosialidosis patient and the parents of the first patient. In two cases (normal, E.I.) poly(A)$^+$ RNA was subsequently isolated from the total RNA. Samples were fractionated on a formaldehyde-agarose gel and probed with the Hu54 cDNA. As a control for equal amounts of RNA applied in each lane, the filters were also probed with cDNA recognizing glyceraldehyde-3-phosphate dehydrogenase (*GAPDH*) mRNA. Hybridization to this probe is shown on the *right*. Ribosomal *18S* and *28S* RNA markers are indicated

is transcribed and translated. The size of this molecule correlates with that of the protein immunoprecipitated after translation of total or poly(A)$^+$ RNA from normal fibroblasts. It appears, therefore, that the isolated cDNA contains the entire coding region for the protective protein.

In Fig. 4 some features of the Hu54 cDNA and the predicted amino acid sequence of the protective protein are depicted. The ATG start codon is the beginning of an open reading frame (ORF) of 1440 nucleotides which is flanked by a 3′ end untranslated region of 363 nucleotides, followed by a poly(A) tail of 10 nucleotides. A potential polyadenylation signal is found shortly upstream of the poly(A) tail. The ORF encodes a precursor form of the protein, that includes a signal sequence of 28 residues and the 32 kD as well as the 20 kD components of 298 and 154 amino acids respectively. Both polypeptides carry one N-linked glycosylation site.

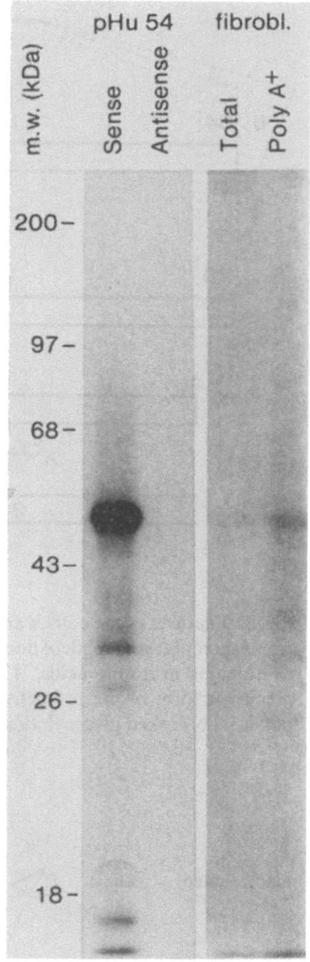

Fig. 3. Cell-free translation of in vitro transcribed RNA and fibroblast RNA. Protective protein mRNA, transcribed in vitro from the Hu54 cDNA coding (*sense*) strand and non-coding (*antisense*) strand as well as total and poly(A)$^+$ RNA from fibroblasts, were translated in vitro in a rabbit reticulocyte lysate system. Radiolabeled proteins were immunoprecipitated using anti-32 kD antibodies and resolved on a 12.5% SDS polyacrylamide gel. Proteins were visualized by fluorography. Molecular size markers are indicated on the *left*

These findings identify the 20 kD polypeptide as an integral part of mature protective protein. After removal of the signal sequence, the precursor molecule undergoes a further proteolytic cleavage which converts the protein into its mature and active form: a heterodimer of 32 and 20 kD polypeptides, held together by disulfide bridges.

In order to show that the Hu54 cDNA directs the synthesis of a functional protective protein, transient expression of the cDNA in COS1 cells [20] was performed. After transfection, a 54 kD precursor protein is produced in these cells, which is partially secreted into the medium. Galactosialidosis fibroblasts endocytose this COS1-derived precursor and are able to process it correctly into 32 and 20 kD components. Moreover, β-galactosidase and neuraminidase activities are increased severalfold in cells that contain this COS-derived protective protein as compared to non-treated cells. We therefore conclude that the Hu54 cDNA encoded protein is biologically functional.

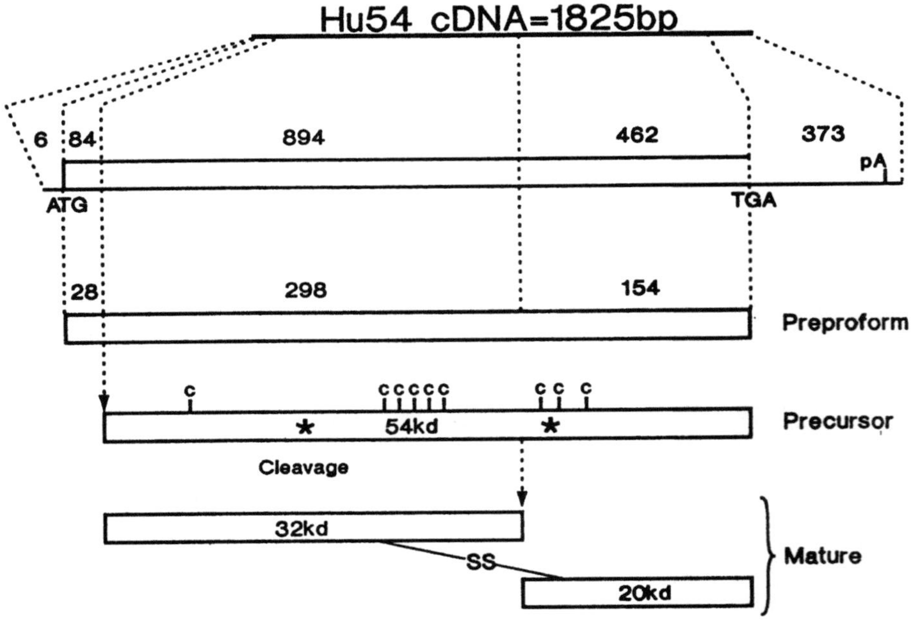

Fig. 4. Features of the cDNA and primary structure and processing of the protective protein. *Top bar: Numbers* represent length of domains in nucleotides. *Middle* and *lower bars*: Numbers represent length of domains in amino acids. *ATG* start codon; TGA stop codon; *pA* polyadenylation signal. The precursor form of the protective protein is depicted with the position of cysteine residues (*c*) and potential N-linked glycosylation sites (*asterisks*)

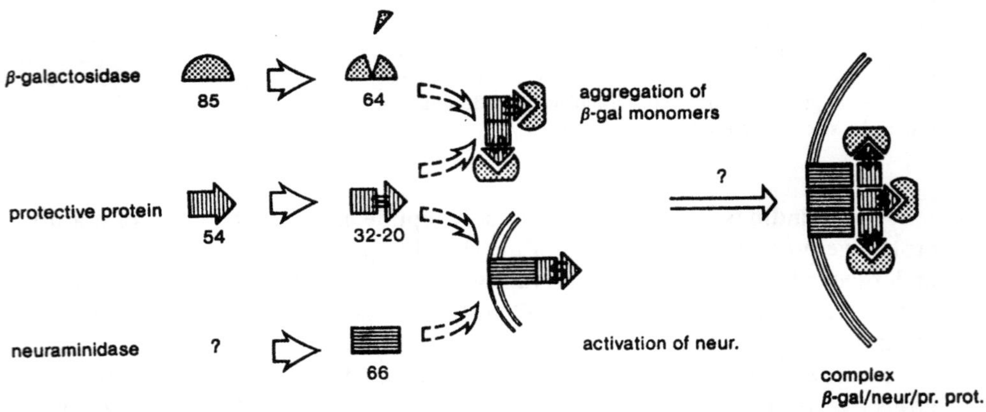

Fig. 5. Model for formation of β-galactosidase/neuraminidase/protective protein complex in normal cells. *Numbers* indicate the molecular sizes of precursor and mature forms of the different components of the complex

The predicted amino acid sequence of the protective protein has revealed a striking homology to two yeast serine proteases. Both of these enzymes are carboxypeptidases [21,22] but function in different subcellular compartments and are involved in different proteolytic pathways. The homology suggests a putative protease activity for the protective protein. We can envisage that this activity is responsible for specific proteolytic modifications of β-galactosidase and neuraminidase at their respective C-termini, resulting in the stabilization of the first and activation of the second enzyme. Complex formation of the three proteins, as delineated in Fig. 5, could represent the final functional state, after which slow proteolytic degradation commences. Studies of the interaction of the three proteins will become feasible once the genes coding for neuraminidase and β-galactosidase have been cloned.

Acknowledgements. We would like to thank Mirko Kuit and Cor van Dijk for excellent photographic work and Diana Heinsius for typing of this manuscript.

References

1. Corfield AP, Michalski JC, Schauer R (1981) The substrate specificity of sialidases from microorganisms and mammals. In: Tettamanti G, Durand P, DiDonato S (eds) Sialidases and sialidoses. Ermes, Milan, pp 1–70
2. O'Brien JS (1983) The gangliosidoses. In: Stanbury JB, Wijngaarden JB, Fredrickson DS, Goldstein JL, Brown MS (eds) The metabolic basis of inherited disease. McGraw-Hill, New York, pp 945–969
3. d'Azzo A, Hoogeveen AT, Reuser AJJ, Robinson D, Galjaard H (1982) Molecular defect in combined β-galactosidase and neuraminidase deficiency in man. Proc Natl Acad Sci USA 79:4535–4539
4. Hoogeveen AT, Verheijen FW, Galjaard H (1983) The relation between human lysosomal β-galactosidase and its protective protein. J Biol Chem 258:12143–12146
5. Groebe H, Krins M, Schmidberger H, von Figura K, Harzer K, Kresse H, Paschke E, Sewell A, Ullrich K (1980) Morquio syndrome (mucopolysaccharidosis IV-B) associated with β-galactosidase deficiency. Report of two cases. Am J Human Genet 32:258–272
6. Cantz M, Gelher J, Spranger J (1977) Mucolipidosis I: increased sialic acid content and deficiency of an α-N-acetylneuraminidase in cultured fibroblasts. Biochem Biophys Res Commun 74:732–738
7. Cantz M, Mendla K, Baumkötter J (1984) Sialidosis and related disorders: the nature of the sialidase defect. In: Vanier MT (ed) Recent progress in neurolipidoses and allied disorders. Mérieux, Lyon, pp 137–145
8. Andria G, Strisciuglio P, Pontarelli G, Sly WS, Dodson WE (1981) Infantile neuraminidase and β-galactosidase deficiencies (galactosialidosis) with mild clinical courses. In: Tettamanti G, Durand P, DiDonato S (eds) Sialidases and sialidoses. Ermes, Milan, pp 379–395
9. Wenger DA, Tarby TJ, Wharton C (1978) Macular cherry-red spots and myoclonus with dementia: coexistent neuraminidase and β-galactosidase deficiencies. Biochem Biophys Res Commun 82:589–595
10. Galjaard H, d'Azzo A, Hoogeveen AT, Verheijen FW (1984) Combined β-galactosidase-sialidase deficiency in man: genetic defect of a "protective protein". In: Barranger JA, Brady RO (eds) Molecular basis of lysosomal storage disorders. Academic Press, New York London, pp 113–132
11. Suzuki Y, Sakuraba H, Yamanaka T, Ko YM, Iimori Y, Okamura Y, Hoogeveen AT (1984) Galactosialidosis: a comparative study of clinical and biochemical data on 22 patients. In: Arina M, Suzuki Y, Yabuuchi H (eds) The developing brain and its disorders. Univ Tokyo Press, pp 161–175

12. Norden AGW, Tennant L, O'Brien JS (1974) Ganglioside G_{M1} β-galactosidase A: purification and studies of the enzyme from human liver. J Biol Chem 249:7969–7976
13. Hoeksema HL, van Diggelen OP, Galjaard H (1979) Intergenic complementation after fusion of fibroblasts from different patients with β-galactosidase deficiency. Biochim Biophys Acta 566:72–79
14. van Diggeln OP, Hoogeveen AT, Smith PJ, Reuser AJJ, Galjaard H (1982) Enhanced proteolytic degradation of normal β-galactosidase in the lysosomal storage disease with combined β-galactosidase and neuraminidase deficiency. Biochim Biophys Acta 703:69–76
15. Hoogeveen AT, d'Azzo A, Brossmer R, Galjaard H (1981) Correction of combined β-galactosidase/neuraminidase deficiency in human fibroblasts. Biochem Biophys Res Commun 103:292–300
16. Palmeri S, Hoogeveen AT, Verheijen FW, Galjaard H (1986) Galactosialidosis: molecular heterogeneity among distinct clinical phenotypes. Am J Human Genet 38:137–148
17. Verheijen FW, Brossmer R, Galjaard H (1982) Purification of acid β-galactosidase and acid neuraminidase from bovine testis: evidence for an enzyme complex. Biochem Biophys Res Commun 108:868–875
18. Verheijen FW, Palmeri S, Hoogeveen AT, Galjaard H (1985) Human placental neuraminidase. Activation, stabilization and association with β-galactosidase and its protective protein. Eur J Biochem 149:315–321
19. Kleijer WJ, Hoogeveen AT, Verheijen FW, Niermeijer MF, Galjaard H, O'Brien JS, Warner TG (1979) Prenatal diagnosis of sialidosis with combined β-galactosidase and neuraminidase deficiency. Clin Genet 16:60–61
20. Gluzman Y (1981) SV40-transformed simian cells support the replication of early SV40 mutants. Cell 23:175–1182
21. Dmochowska A, Dignard D, Henning D, Thomas DY, Bussey H (1987) Yeast *KEX1* gene encodes a putative protease with a carboxypeptidase B-like function involved in killer toxin and α-factor precursor processing. Cell 50:573–584
22. Valls LA, Hunter CP, Rothman JH, Stevens TH (1987) Protein sorting in yeast: the localization determinant of yeast vacuolar carboxypeptidase Y resides in the propeptide. Cell 48:887–897

The Role of Activator Proteins in Glycolipid Degradation and Storage Diseases

E. CONZELMANN

1 Introduction

1.1Activator Proteins

The lysosomal catabolism of glycosphingolipids is accomplished by the sequential action of several hydrolases. Some of these enzymes are water-soluble and it has been shown that they cannot attack their membrane-bound substrates directly but require the assistance of small non-enzymic proteins, so-called activator proteins. These activator proteins bind to the glycolipid substrate, extract it from the membrane and form a water-soluble activator/lipid complex which is believed to be the true substrate of the enzymic reaction [1–4].

The first such activator protein was discovered by Mehl and Jatzkewitz [5] who found that, after partial purification, porcine arylsulfatase A had lost its ability to degrade its physiological substrate, sulfatide, but remained fully active towards the artificial water-soluble substrates. Addition of another, enzymically inactive, fraction to the assay restored the activity against the glycolipid. The protein cofactor that mediated the interaction between the enzyme and the lipid substrate was termed "the activator protein for cerebroside sulfate sulfatase". An analogous cofactor for sulfatide catabolism was also found in human liver [6] and was later purified to homogeneity by Fischer and Jatzkewitz [7] and identified as an acidic glycoprotein (isoelectric point at pH 4.3) with a molecular weight of 22 000 daltons.

A similar requirement for small non-enzymic activator proteins was demonstrated by Li and co-workers for the degradation of ganglioside G_{M2} by β-hexosaminidase A [8], of ganglioside G_{M1} by β-galactosidase and of globotriaosylceramide by α-galactosidase [9].

The physiological significance of these activator proteins became evident when it was recognized that an unusual enzymic variant of infantile G_{M2} gangliosidosis, without enzyme defect (variant AB) is caused by the deficiency of the activator protein for lysosomal ganglioside G_{M2} hydrolysis [10] (see below).

A few years later, Stevens et al. [11] found that, similarly, the deficiency of the activator protein needed for sulfatide degradation was the underlying cause for an unusual variant of metachromatic leukodystrophy. With immunochemical methods, Inui et al. [12] demonstrated, in fibroblasts of the same patient, a deficiency of the activator protein for hydrolysis of ganglioside G_{M1} by β-

Institut für Organische Chemie und Biochemie der Universität Bonn, Gerhard-Domagk-Str. 1, D-5300 Bonn 1, FRG

A. Azzi, Z. Drahota and S. Papa
Molecular Basis of Membrane-Associated Diseases

galactosidase. Both activators as well as the one for degradation of globotriaosylceramide by α-galactosidase were also found to be deficient in the urine of these patients, with immunochemical as well as with enzymic methods [13], indicating that the protein cofactors for enzymic degradation of sulfatide, ganglioside G_{M1} and globotriaosylceramide may be identical.

In a recent study that was prompted by these observations [14], the identity of these three activators was demonstrated directly, by immunochemical methods and by copurification of the three activities. For example, ion-exchange chromatography of a partially purified activator preparation from urine, on an HPLC column, resolved the activators into two major peaks and one shoulder (Fig. 1). All three activator activities determined, that is G_{M1}, sulfatide and GTC activator coincided precisely. In contrast, the activator for degradation of ganglioside G_{M2} by β-hexosaminidase A was clearly separated from the others. Similar studies conducted in parallel by others arrived at the same results [15].

It thus seems that two activator proteins of this type are required for lysosomal glycolipid catabolism. One of them, the "G_{M2} activator" promotes specifically the

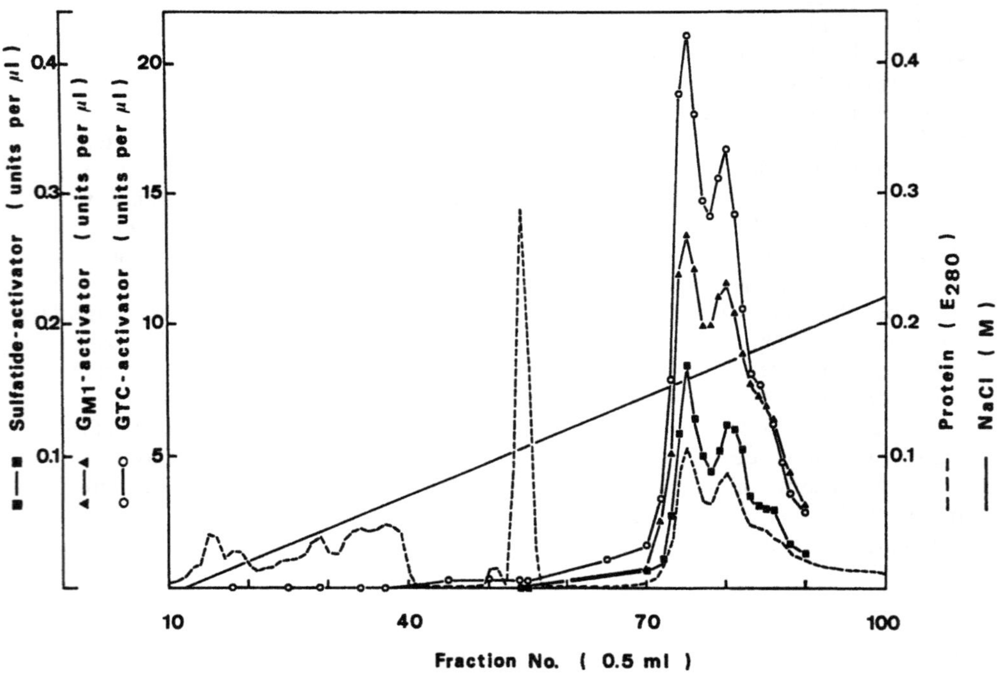

Fig. 1. Ion-exchange HPLC of purified sulfatide activator from urine. A preparation of sulfatide activator that had been purified from human urine by gel filtration over Sephadex G-50 and hydrophobic chromatography over octyl-Sepharose was lyophilized, dissolved in 50 µl of 10 mM phosphate buffer, pH 6.0, and loaded on a TSK 545 DEAE column. The column was eluted with a linear NaCl gradient (0–0.3 M). Fractions of 0.5 ml were collected and assayed for protein content and for their capability to stimulate the enzymatic degradation of ganglioside G_{M1}, sulfatide and globotriaosylceramide by the respective hydrolases [14]

degradation of ganglioside G_{M2} and a few related glycolipids by β-hexosaminidase A. The other one is much less specific with respect to lipid substrates as well as enzymes. For historic reasons (and also pathobiochemical ones, see below) it is usually named "the sulfatide activator". Other names proposed are "SAP-1" (for "sphingolipid activating protein" [16]) or "dispersin" [17].

An entirely different type of protein cofactor was described to enhance the activity of membrane-associated hydrolases on their lipid substrates, e.g., glucosylceramide or galactosylceramide [18–22]. These factors activate the enzyme itself, they promote the degradation also of the water-soluble substrates. Their physiologic role is not yet really understood. Sarmientos et al. [23] showed that, under appropriate conditions, even highly purified glucosylceramidase is fully active on membrane-bound glucosylceramide, without any need for "activating factors". On the other hand, Christomanou and co-workers found that one of the peptide cofactors for glucosylceramidase was absent from the fibroblasts of a patient who had died from Gaucher disease in spite of normal glucosylceramidase activity [24].

1.2 Lysosomal Storage Diseases

As outlined above, the lysosomal catabolism of glycoconjugates proceeds sequentially, by the stepwise removal of monosaccharide units from the non-reducing end by a number of glycosidases, each of which is specific for one sugar in a certain anomeric linkage (α or β). If, due to a genetic defect, one of these enzymes is deficient, the degradation of the affected compounds comes to a halt at this stage. The only partially degraded substances can hardly (or not at all) be discharged from the lysosomes but accumulate there. In the case of glycolipid substrates, this problem is compounded by the insolubility of these membrane components which are practically unable to traverse the lysosomal membrane. As a consequence, the undergradable material precipitates in the lysosomes, causing them to swell (and probably also to increase in number). Although the storage compounds are normal, non-toxic components of the cell, their excessive accumulation may interfere with intracellular transport and metabolism and may thus lead to a storage disease (for reviews, see, e.g., [25–27]).

The lysosomal hydrolases (and their inherited defects) occur in all organs and cell types (except erythrocytes) but the glycolipids accumulate predominantly in those organs where they are mainly synthesized. (A notable exception are again blood cells, which are degraded in the reticuloendothelial system.) These amphiphiles precipitate together with other membrane components within the cells in which they were synthesized. Only small amounts of the accumulating lipids leave the cells; they may be detected in cerebrospinal fluid and in urine. Thus, defects in ganglioside catabolism will mainly lead to storage in neurons, whereas a defect in sulfatide degradation affects mainly the white matter, resulting in leukodystrophy (for review, see, e.g., [26–29]). Besides the severe forms of these neurolipidoses with onset of symptoms in early or late infancy, rapid progression and death in early childhood, a wide range of milder cases exists, with onset of symptoms at any time between childhood and adult age.

2 The G_{M2} Activator Protein

2.1 Properties of the G_{M2} Activator

The G_{M2} activator is a small acidic protein (pI = 4.8) with a molecular weight of approx. 22 kD [3]. The similar molecular weights found with gel filtration and SDS electrophoresis indicate a monomeric structure of the protein. After metabolic labeling of cultured skin fibroblasts with [^{3}H]leucine, a precursor of 24 kD could be precipitated from culture medium, whereas only the mature 22 kD form could be detected intracellularly [30]. The glycoprotein nature of G_{M2} activator was demonstrated by metabolic labeling with [^{3}H]mannose [30]. In subcellular fractionation studies with human fibroblasts, the G_{M2} activator was found to co-distribute precisely with lysosomal marker enzymes [31]. The gene of G_{M2} activator was mapped to chromosome 5 in the human genome [32].

The activator protein is water-soluble and can be extracted from tissues with water or buffer solutions, without detergents. However, it binds strongly to hydrophobic chromatography media such as octyl-Sepharose [3], indicating the presence of a hydrophobic groove or pocket that is accessible from the outside. The formation of water-soluble 1:1 complexes of the G_{M2} activator with gangliosides G_{D1a}, G_{M1}, G_{M2}, G_{M3} and glycolipid G_{A2} was demonstrated with isoelectric focusing, gel filtration, gel electrophoresis and ultracentrifugation techniques [4]. The strength of the interaction between the activator and the glycolipids decreased in the order $G_{M2} > G_{M1} \approx G_{D1a} \gg G_{M3} \approx G_{A2}$. This indicates some specificity of the carbohydrate recognition site for the glycolipid substrate, ganglioside G_{M2}. The negative charge of sialic acid contributes to the binding of this ganglioside to the G_{M2} activator. Li et al. [33] showed that the reduction of its carboxylate group to the alcohol or conversion to the methyl ester completely blocked the degradation of the ganglioside in the presence of the activator.

The hydrolysis of water-soluble substrates such as 4-methylumbelliferyl-N-acetyl-β-D-glucosaminides by hexosaminidases is not stimulated by the G_{M2} activator [3]. The G_{M2} activator acts primarily as a co-substrate with G_{M2} ganglioside and does not really "activate" the enzyme.

2.2 Role of the G_{M2} Activator in Glycolipid Catabolism

Several glycolipids contain in their oligosaccharide chain β-glycosidically linked N-acetyl-D-galactosamine residues. In the course of lysosomal catabolism, these sugars are hydrolyzed off by an enzyme that accepts also N-acetyl-β-D-glucosaminides as substrates and is therefore named "N-acetyl-β-D-hexosaminidase" or briefly "hexosaminidase". In human tissues, mainly two isoenzymes of lysosomal hexosaminidase are found which, on the basis of their different isoelectric points, are referred to as hexosaminidases A (for acidic) and B (for basic). Hexosaminidase A is a dimer of two subunits, α and β, which are encoded on different chromosomes (15 and 5, respectively) but are structurally very similar. Hexosaminidase B is a homodimer of two β subunits. A dimer of α subunits (hexosaminidase S) is usually also present in low amounts but appears to be of little physiologic relevance (for reviews, see [29,34,35]).

Both subunits, α and β, carry an active site but each has different substrate specificities [36]. Neutral, water-soluble substrates such as neutral oligosaccharides or the artificial p-nitrophenyl and 4-methylumbelliferyl glycosides are hydrolyzed much faster by the β than by the α subunit. On the other hand, the α subunit recognizes also negatively charged substrates such as oligosaccharides with a penultimate charged (uronic acid) residue, as might be derived from dermatan sulfate or chondroitin sulfates, or with a negative charge on the terminal N-acetylglucosamine residue itself as in keratan sulfate oligosaccharides or in the artificial 6-sulfo-N-acetylglucosaminides [37].

The major lipid substrates of hexosaminidases, ganglioside G_{M2} and its asialo residue, G_{A2} are in vivo only hydrolyzed by hexosaminidase A but not by the B isoenzyme as evidenced by their accumulation in Tay-Sachs disease (see below). When assayed in vitro, in the presence of detergents, hexosaminidase B cleaves G_{A2} twice as fast as hexosaminidase A [38] and has also some activity on G_{M2}, depending on the detergent used. In contrast, in assays with the G_{M2} activator the B isoenzyme is practically inactive with any of the glycolipid substrates [3].

From our present knowledge, a model can be suggested for the mechanism by which the G_{M2} activator stimulates glycolipid breakdown in the lysosomes [28]. This model is schematically summarized in Fig. 2. The hydrophilic moiety of ganglioside G_{M2} on the membrane surface cannot be attacked directly by hexosaminidase A. It can, however, be recognized and bound by a specific region on the G_{M2} activator protein. In a second step, the ceramide residue of this lipid is pulled out of the membrane and folds into a hydrophobic groove of the activator. The resulting activator/lipid complex is fully water-soluble as is the free activator. This extraction process is, of course, reversible. The activator binds to a specific recognition site of hexosaminidase A, most probably in such a way that the terminal residue of the glycolipid is correctly positioned in the active center of the α-subunit of the enzyme. After hydrolysis is completed, the activator/product complex is released, diffuses away and the lipid product may be reinserted into the membrane or perhaps "handed over" to the next activator on the degradative pathway.

The degradation of glycolipids with longer oligosaccharide chains seems to depend less strongly on the presence of the activator, presumably because the carbohydrate moiety protrudes far enough into the aqueous phase for the terminal sugar residue to be accessible to the enzyme. Kidney globoside (GalNacβ1 $\rightarrow$ 3Galα1 $\rightarrow$ 4Galβ1 $\rightarrow$ 4Glcβ1 $\rightarrow$ 1′Cer) was found to accumulate in Sandhoff disease (G_{M2} gangliosidosis variant 0) [39] where both hexosaminidases, A and B, are deficient, but not in the activator deficiency variant (see below), indicating that it has to be hydrolyzed by one of the hexosaminidases but the G_{M2} activator is not essential. This contrasts with the in vitro results: in the absence of detergents, globoside is only poorly attacked by the hexosaminidases. Addition of the G_{M2} activator strongly stimulates its hydrolysis by hexosaminidase A but not by the B isoenzyme [3]. The hydrolysis of ganglioside G_{D1a}-GalNac, a minor storage compound in G_{M2} gangliosidoses [40], by hexosaminidase is not accelerated by the G_{M2} activator. G_{D1a}-GalNac, when presented as a component of liposomes, is directly attacked by hexosaminidase A, while the B enzyme is almost inactive (E. Meier and K. Sandhoff, in preparation). The accumulation of this compound in

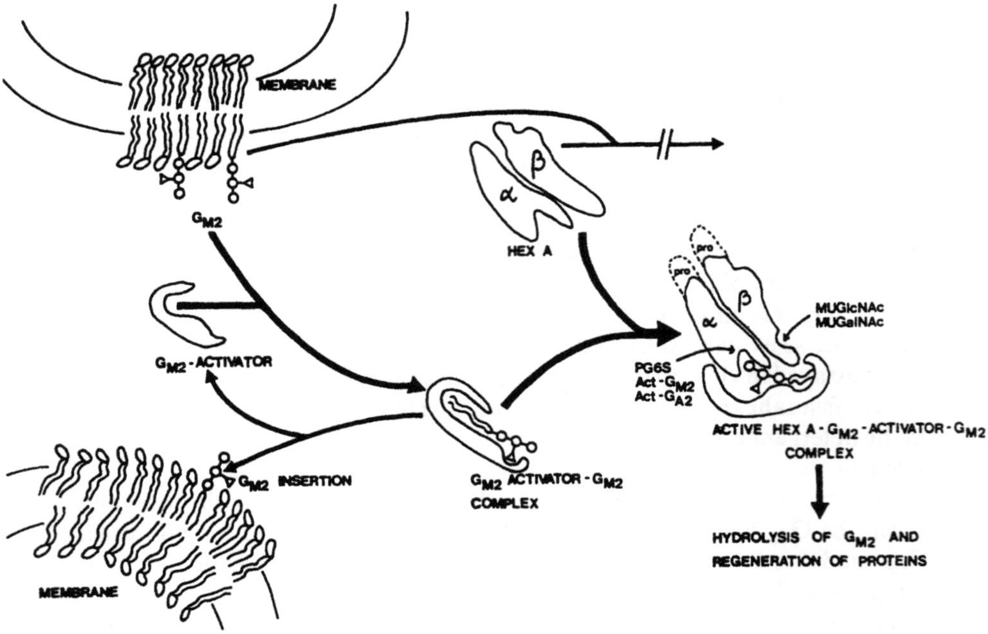

Fig. 2. Model for the lysosomal catabolism of ganglioside G_{M2}. Hexosaminidase A does not directly interact with the membrane-bound ganglioside. Instead, the activator protein extracts the glycolipid and the resulting (water-soluble) activator/lipid complex is the substrate for the enzymatic reaction. After the reaction, the product, ganglioside G_{M3}, is reinserted into the membrane and the activator is available for another round of catalysis [28]

Tay-Sachs disease also indicates that it is a poor substrate for hexosaminidase B, owing presumably to the negative charge of the sialic acid in close proximity to the terminal N-acetylgalactosamine.

A ganglioside G_{M2} derivative selectively deacylated in the ceramide moiety ("lyso-G_{M2}", II^3-sialylgangliotriaosylsphingosine) is found in small amounts in the brain tissue of G_{M2} gangliosidosis patients but not in normal tissues [41,42] and is probably formed by the unspecific action of some amidase on ganglioside G_{M2} when the latter is present in unphysiologically high concentrations. Lyso-G_{M2} is not bound by the G_{M2} activator, probably due to electrostatic repulsion between the free sphingosine amino group and another positively charged group at the corresponding site in the protein [43]. N-Acetylation of lyso-G_{M2} abolishes its positive charge and restores binding by the activator protein.

2.3 Role of the G_{M2} Activator in Storage Diseases

For the degradation of ganglioside G_{M2}, three polypeptides are needed, i.e., the α and β subunits of hexosaminidases and the G_{M2} activator protein. Inherited defects in any one of them are known to impair lysosomal ganglioside catabolism. Accordingly, three different biochemical variants can be distinguished:

α-Subunit Deficiency. A deficiency or defect of the α-subunit of hexosaminidase A blocks the formation of intact hexosaminidase A (and the minor isoenzyme hexosaminidase S, which is an α_2 dimer) while the unaffected β-subunits can still associate to form hexosaminidase B. Biochemically, this disorder is characterized by the presence of normal or even elevated levels of hexosaminidase B and has therefore been referred to as "variant B of G_{M2} gangliosidosis" [39]. The eponym "Tay-Sachs disease" usually refers to the infantile form of this enzymic variant. The major storage compounds are ganglioside G_{M2} (10–12% dry brain weight) and its asialo derivative, G_{A2} (0.7–1.4% dry weight) [39]. It is evident that in accordance with the in vitro findings, hexosaminidase B is practically unable to interact with the G_{M2} activator and to degrade ganglioside G_{M2}. The accumulation of glycolipid G_{A2} is interesting for two reasons: On the one hand, hexosaminidase B degrades this neutral glycolipid twice as fast as hexosaminidase A when assayed in the presence of detergents [38] but is almost inactive when measured with the activator instead of detergents [3], demonstrating that, at least in the case of hexosaminidases, glycolipid catabolism is governed by the specificity of the activator-enzyme interaction. On the other hand, the small activity still observed for the action of hexosaminidase B on glycolipid G_{A2} in the presence of the activator (some 3% of that of hexosaminidase A [3]) correlates nicely with the observation that in this variant glycolipid G_{A2} storage is considerably less than in the β-subunit deficiency, where both hexosaminidases are missing (see below).

β-Subunit Deficiency. Patients with a deficiency of the β-subunit, which is common to both major hexosaminidase isoenzymes, A and B, lack both activities [39,44]. This variant has therefore been designated as variant O (for "zero" hexosaminidase activity). The eponym "Sandhoff disease" relates to the fact that this enzymic variant was first distinguished from classical Tay-Sachs disease (variant B) by Sandhoff et al. [44]. The minor residual hexosaminidase activity found in the tissues of these patients is due to α_2 dimers ("hexosaminidase S"). The question whether the G_{M2} activator interacts with hexosaminidase S to any significant extent is probably obsolete since it was shown that the α-subunits are transported to the lysosomes only after association with the β-subunits [45]. Presumably, hexosaminidase S is mainly located outside the lysosomes. As a consequence of the complete hexosaminidase deficiency all substrates, including glycoprotein oligosaccharides and glycosaminoglycan fragments, are stored. Therefore, the patients suffer from symptoms of a severe mucopolysaccharidosis in addition to the G_{M2} gangliosidosis. In brain tissue, levels of ganglioside G_{M2} are almost as high as in the B variant, the storage of glycolipid G_{A2} is even more prominent (2–4% dry weight) [39]. Characteristic of variant 0 is the storage of kidney globoside in visceral tissues [39,44].

G_{M2} Activator Deficiency (Variant AB). A third variant, so far rarely diagnosed, is caused by the deficiency of the G_{M2} activator protein [10,30,31,46–48]. The hexosaminidase isoenzymes are not affected in these patients [39,49] and are both present in normal or elevated amounts, hence the name "variant AB". Historically, the demonstration of the deficiency of a non-enzymic protein co-factor in this variant attracted attention to the role of activator proteins in lysosomal lipid catabolism [10]. It not only proved *that* activator proteins do have some physiologic

importance but the analysis of the storage compounds revealed, in addition, for the degradation of *which* compounds the G_{M2} activator is necessary: In brain tissue, ganglioside G_{M2} (15% dry weight) and glycolipid G_{A2} (2.8%) were found to accumulate [39]. No other storage compound was hitherto reported. G_{D1a}-GalNAc seems to be normal (E. Meier, personal communication), lyso-G_{M2} was not yet looked for. The level of globoside in visceral tissues was found to be normal [39].

All this information, taken together, indicates that the physiologic importance of the G_{M2} activator is to promote the degradation of glycosphingolipids with oligosaccharide chains of up to three sugars and with N-acetylgalactosamine at the non-reducing end.

The specific interaction between activator protein and enzyme is also of considerable importance for the enzymic diagnosis of variant forms of G_{M2} gangliosidosis, particularly late-onset forms, in which the enzyme is not totally absent but has a small but detectable residual activity. Assays with synthetic substrates, or with ganglioside G_{M2} in the presence of detergents, may in such cases yield misleading results. Residual activities towards ganglioside G_{M2}, determined in skin fibroblast extracts in the presence of the G_{M2} activator, were found to correlate well with the clinical status of the patient [50]. The discrepancy between the assay methods is evident, of course, where the same probands were assayed with different methods as in the case of the adult-type G_{M2} gangliosidosis patients described by Navon et al. [51]. Before the onset of symptoms, these patients were regarded as asymptomatic adults with low hexosaminidase A [52], the finding of heterozygote values for ganglioside G_{M2} hydrolysis by skin fibroblast extracts [53] seemed to explain the discrepancy. After the patients had developed neurologic symptoms in their early thirties, these cells were re-examined with an activator-based assay and were found to have residual activities of 2–4% of normal controls [50], typical of adult-type G_{M2} gangliosidosis, variant B.

3 The Sulfatide Activator (SAP-1)

3.1 Properties of the Sulfatide Activator

The activator protein known as the "sulfatide activator" was first purified by Fischer and Jatzkewitz [7] and identified as a small acidic glycoprotein (pI = 4.3). On gel filtration, it appeared to have a molecular mass of approx. 22,000 daltons [7] while on SDS-PAGE the molecular masses found were much smaller, between 6000–8000 daltons [54] and 10000 daltons [16], indicating an oligomeric structure of the protein. Whether the oligomer consists of identical or different peptides is still an open question. The fact that in the purified protein only one N-terminus was found [55] would argue for only one type of subunit but other explanations are still possible. In biosynthetic labeling studies with cultured skin fibroblasts, Fujibayashi and Wenger [56] demonstrated that the sulfatide activator is synthesized as a 53 kD precursor which is co-translationally glycosylated to a 65 kD product. Further glycosylation yields a 70 kD form which is within a few hours proteolytically processed via intermediates of 52 and 35 kD to peptides with 8–13 kD. The cDNA of the precursor protein was cloned and its sequence reported [55]. Chemical sequencing of the entire mature protein [57] revealed a number of errors in the published cDNA sequence. A corrected sequence was meanwhile obtained by O'Brien et al. [58]. This cDNA codes, in addition to the sulfatide activator, for two other known proteins [57]. One is the activator protein for glucosylceramidase which Christomanou and co-workers [24] reported to be deficient in an unusual variant of Gaucher disease (activator A1) and which turned out to be identical with the co-glucosidase (SAP-2) of Wenger and co-workers [58]. The other one (component C) is a protein that is persistently found to contaminate preparations of the G_{M2} activator. It shows considerable homology with both the sulfatide activator and the A1 activator. Its physiological role is still unknown. Assuming that the primary transcript, the precursor protein, is cleaved into the various proteins in the lysosome, Fürst et al. [57] could account for all processing intermediates observed by Fujibayashi and Wenger [56]. The still open question is whether this is another example of one gene coding for several proteins with different functions or whether the functional sulfatide activator consists of several polypeptides that are derived from one precursor by proteolytic processing (and may be held together by non-covalent forces or by disulfide bridges).

The lysosomal localization of the sulfatide activator was shown in rat liver [59]. (In human tissues this was not yet examined but there can be little doubt that it is also lysosomal). The gene for the sulfatide activator was mapped to chromosome 10 [60].

The mechanism by which the sulfatide activator promotes the degradation of sulfatide and other glycolipids by the respective hydrolases seems to be essentially the same as for the G_{M2} activator [1,2]. The activator protein is water-soluble but binds strongly to hydrophobic chromatography media [61]. It was found to form stoichiometric complexes with a variety of lipids, glycerophospholipids as well as sphingolipids [2,62,63]. Experimental studies [33,64] as well as theoretical calculations [65] suggest that unlike the G_{M2} activator, the sulfatide activator does not require a negative charge on the lipid substrate. A three-domain binding model was proposed for the interaction of this protein with glycolipids [65]. Apparently, this activator protein acts mainly as a physiologic solubilizer, with little specificity towards lipid substrate as well as enzyme.

It cannot, however, be excluded that the activator binds to some of the enzymes in a more specific way (arylsulfatase A might be a likely candidate). This question could be answered if, for instance, MLD variants were identified whose arylsulfatase A has a high residual activity with artificial substrates but is poorly active with sulfatide in the presence of the activator. The methodology to assay the latter activity in cultured skin fibroblasts was meanwhile developed [66] but no such cases were yet detected.

3.2 Role of the Sulfatide Activator in Glycolipid Catabolism

The fact that an activator promotes a certain reaction in vitro does not necessarily mean that it is also needed for the same reaction in vivo (cf. the degradation of globoside discussed above). This raises the question of the physiologic role of the sulfatide activator protein: For which reactions is it indispensable, which ones are accelerated to some extent (and to what extent) but would also proceed sufficiently well without it, and on which ones does it have no influence at all?

The inherited deficiency of this protein leads to a variant of metachromatic leukodystrophy (MLD) with onset at juvenile age [11]. It can thus be safely assumed that the sulfatide activator is also in vivo indispensable for the hydrolysis of sulfatide by arylsulfatase A. Much less information is available on the importance of this activator for the degradation of the other glycolipids, notably ganglioside G_{M1} and globotriaosylceramide. Of the few cases of sulfatide activator deficiency diagnosed since 1979 [67], no postmortem examination could as yet be performed and data on the accumulation of lipids in the tissues are not yet available. In their urine, the patients excrete besides sulfatide several neutral glycolipids including digalactosylceramide and globotriaosylceramide [13]. The lack of ganglioside excretion does not necessarily indicate a normal ganglioside catabolism since these compounds are synthesized and hence stored mainly in the brain. Nor does the lack of typical gangliosidosis symptoms exclude a moderate accumulation as these symptoms may manifest later and may be masked by the more severe MLD symptoms. (The same is, of course, true for the symptomology of Fabry disease that might be caused by the accumulation of globotriaosylceramide.)

The availability of skin fibroblasts from these patients enabled studies on the degradation of exogenously added radiolabeled glycolipids in cell culture. A first series of experiments was conducted with ganglioside G_{D1a} [68]. In a typical experiment, the ganglioside, which was tritium-labeled in the sphingoid base, was dissolved in the cell culture medium containing 0.3% fetal calf serum. After feeding the cells for various times they were harvested, the lipids were extracted, separated into neutral and acidic species by ion-exchange chromatography and the individual fractions were analyzed by thin-layer chromatography/fluorography.

It had previously been shown that almost all the sphingoid base liberated from gangliosides is re-used for the synthesis of sphingomyelin or is incorporated into glycerophospholipids (as fatty acid), so the amount of neutral lipids formed can be regarded to reflect the overall degradation rate [48]. From the data listed in Table 1 it is evident that in the sulfatide activator-deficient cells the degradation of ganglioside G_{D1a} is significantly retarded but not completely blocked. The end

Table 1. Incorporation and metabolism of ganglioside G_{D1a} in cultured skin fibroblasts from normal control subjects and activator-deficient MLD patients[a]

Time (days)	Incorporation (nmol/mg protein)		Degradation (% of radioactivity in neutral lipids)	
	normal	act.-def.	normal	act.-def.
1	5.4	6.5	35.2	26.2
3	10.4	11.1	41.3	29.7
5	11.1	11.4	48.6	36.8

[a] Fibroblast cultures were incubated with 50 μM of [sphingosine-^{3}H]ganglioside G_{D1a} dissolved in the culture medium. After the times indicated, the cells were harvested and trypsinized, to remove adhering ganglioside micelles. The lipids were extracted with organic solvents and separated into neutral and acidic components by ion-exchange chromatography [48].

products formed by biosynthetic re-utilization of liberated [^{3}H]sphingoid, i.e., sphingomyelin and the phosphoglycerolipids, were at all time points found to be reduced in the activator-deficient cells, while lactosylceramide (the first uncharged lipid intermediate in the catabolic pathway of the ganglioside), which is hardly detectable in normal cells, attained a rather high steady-state level (10–20% of the neutral lipids).

In the acidic fraction, the radioactivity was mainly (> 95%) in gangliosides. Figure 3 shows the levels of the intermediates of ganglioside G_{D1a} catabolism, the monosialogangliosides G_{M1}, G_{M2} and G_{M3}. Their level was generally higher in activator-deficient than in control cells. However, contrary to the expectations raised by the in vitro results, G_{M1} and G_{M2} were only moderately elevated (up to three-fold) and reached a steady-state plateau after 3 days, whereas G_{M3} kept accumulating even after 5 days of incubation. Control experiments with cells from a late-infantile MLD patient (with arylsulfatase A deficiency) showed that the accumulation of G_{M3} and LacCer as well as the reduced catabolism of gangliosides in general were not secondary effects of sulfatide accumulation but were direct consequences of the sulfatide activator deficiency. Also, feeding of the purified sulfatide activator to the deficient cells reduced the concentrations of ganglioside G_{M3} and LacCer considerably.

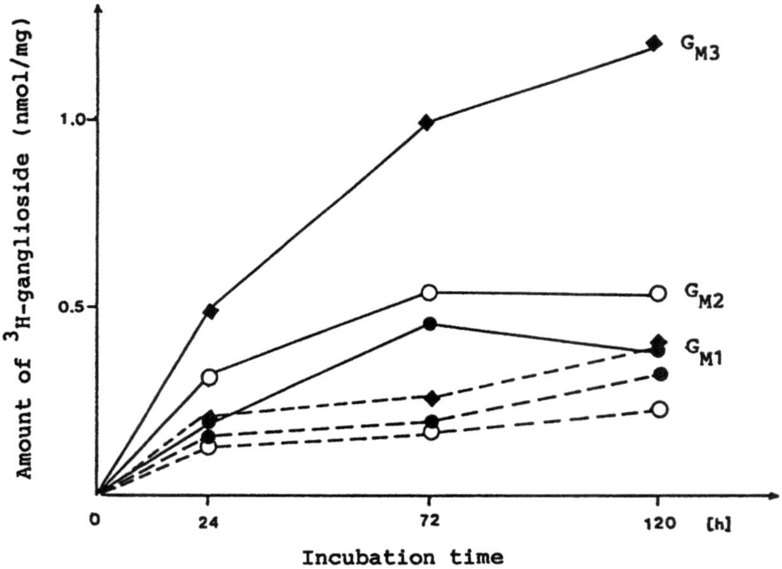

Fig. 3. Concentrations of individual gangliosides after feeding [sphingosine-^{3}H]ganglioside G_{D1a} for various times. Cells were incubated with 50 μM [sphingosine-^{3}H]ganglioside G_{D1a} in the culture medium for the times indicated. Lipids were extracted, separated and quantified essentially as described by Sonderfeld et al. [48]. *Squares* ganglioside G_{M3}; *open circles* ganglioside G_{M2}; *filled circles* ganglioside G_{M1}. *Solid lines* activator-deficient cells; *broken lines* normal controls [68]

It appears that at least one of the properties of this protein observed in vitro, i.e., the promotion of the hydrolysis of ganglioside G_{M1} by β-galactosidase, is of little importance in vivo since the degradation of ganglioside G_{M1} seems to proceed fairly rapidly in the absence of the activator. The accumulation of ganglioside G_{M1} in activator-deficient cells is less than that of ganglioside G_{M2}, the hydrolysis of which is known to be mediated by another protein, the G_{M2} activator. However, some cell types or tissues may have a higher load of glycolipids to digest than the fibroblasts in the present experiments and may not be able to cope with the influx rate in the absence of the activator protein.

Conversely, the activator may ensure a normal ganglioside catabolism with reduced β-galactosidase levels, as demonstrated by Paschke and Kresse [69] for Morquio disease type B. Patients with this disease have β-galactosidase levels below 5% of controls when assayed with water-soluble substrates such as oligosaccharides or the artificial chromogenic and fluorogenic substrates (or ganglioside G_{M1} in the presence of detergents). When ganglioside G_{M1} was employed as substrate in the presence of the sulfatide activator, without detergents, residual activities of 20–50% of normal were found, explaining why these patients suffer from a mild mucolipidosis variant but not from G_{M1} gangliosidosis.

The accumulation of ganglioside G_{M3} is most interesting, particularly because the nature of the sialidase that degrades gangliosides in the lysosome is still controversial. So far, only one sialidase has been found in lysosomes but this enzyme was believed to act specifically on water-soluble oligosaccharides [35,70]. If this enzyme also degrades gangliosides but only in the presence of the sulfatide activator, the failure to demonstrate any lysosomal ganglioside sialidase activity, e.g., on subcellular fractionation [71,72], may simply be attributed to erroneous assay conditions. It could indeed be shown now that in the presence of purified sulfatide activator lysosomal sialidase cleaves ganglioside substrates in vitro (R. Fingerhut, M. Lieser, F. Verheijen, M. Cantz, E. Conzelmann and K. Sandhoff, in preparation). Preliminary observations on ganglioside metabolism in sialidosis cells (S. Sonderfeld et al., in preparation) tend to support this notion.

For the degradation of lactosylceramide the sulfatide activator is probably not indispensable but still very helpful. In the absence of the activator, the lipid must be present at a much higher concentration for a normal degradation rate to be achieved. This observation also re-opens the question as to which β-galactosidase, the water-soluble lysosomal β-galactosidase or the membrane-associated galactosylceramide β-galactosidase (galactocerebrosidase) degrades this compound in vivo.

The observation discussed above that the primary precursor contains at least two more polypeptides, which are closely related to the protein regarded as the sulfatide activator proper [57,58], raises a number of additional questions. What is the role of the other proteins? Are they all deficient in the "sulfatide activator deficiency" patients? If so, why do these patients not accumulate glucosylceramide in their liver? Conversely, which of the proteins are deficient in the Gaucher patient that lacks the "A1" activator? Do the cells of this patient still possess the "sulfatide activator"? (If not, why did she not have MLD symptoms?) These questions are presently under intensive study.

Computer search of cDNA sequence data libraries afforded further surprising information: The cDNA for the human sulfatide activator precursor shares some 78% homology with the sulfated glycoprotein of rat Sertoli cells [73]. It cannot be excluded that the primary product of the human gene (which is now merely regarded as the precursor of the sulfatide activator) serves some other important function and that the slightly hydrophobic fragments which its lysosomal degradation yields just happen to facilitate glycolipid catabolism in a rather unspecific way.

References

1. Fischer G, Jatzkewitz H (1977) The activator of cerebroside sulphatase: binding studies with enzyme and substrate demonstrating the detergent function of the activator protein. Biochim Biophys Acta 481:561–572
2. Fischer G, Jatzkewitz H (1978) The activator of cerebroside sulphatase: a model of the activation. Biochim Biophys Acta 528:69–76
3. Conzelmann E, Sandhoff K (1979) Purification and characterization of an activator protein for the degradation of glycolipids G_{M2} and G_{A2} by hexosaminidase A. Hoppe-Seylers Z Physiol Chem 360:1837–1849
4. Conzelmann E, Burg J, Stephan G, Sandhoff K (1982) Complexing of glycolipids and their transfer between membranes by the activator protein for degradation of lysosomal ganglioside G_{M2}. Eur J Biochem 123:455–464
5. Mehl E, Jatzkewitz H (1964) Eine Cerebrosidsulfatase aus Schweineniere. Hoppe-Seylers Z Physiol Chem 339:260–276
6. Jatzkewitz H, Stinshoff K (1973) An activator of cerebroside sulfatase in human normal liver and in cases of congenital metachromatic leukodystrophy. FEBS Lett 32:129–131
7. Fischer G, Jatzkewitz H (1975) The activator of cerebroside sulphatase: purification from human liver and identification as a protein. Hoppe-Seylers Z Physiol Chem 356:605–613
8. Li Y-T, Mazzotta MY, Wan C-C, Orth R, Li S-C (1973) Hydrolysis of Tay-Sachs ganglioside by β-hexosaminidase A of human liver and urine. J Biol Chem 248:7512–7515
9. Li S-C, Wan C-C, Mazzotta MY, Li Y-T (1974) Requirement of an activator for the hydrolysis of sphingoglycolipids by glycosidases from human liver. Carbohydr Res 34:189–193
10. Conzelmann E, Sandhoff K (1978) AB variant of infantile G_{M2}-gangliosidosis: deficiency of a factor necessary for stimulation of hexosaminidase A-catalyzed degradation of ganglioside G_{M2} and glycolipid G_{A2}. Proc Natl Acad Sci USA 75:3979–3983
11. Stevens RL, Fluharty AL, Kihara H, Kaback MM, Shapiro LJ, Marsh B, Sandhoff K, Fischer G (1981) Cerebroside sulfatase activator deficiency induced metachromatic leukodystrophy. Am J Human Genet 33:900–906
12. Inui K, Emmett M, Wenger DA (1983) Immunological evidence for deficiency of an activator protein for sulfatide sulfatase in a variant form of metachromatic leukodystrophy. Proc Natl Acad Sci USA 80:3074–3077
13. Li S-C, Kihara H, Serizawa S, Li Y-T, Fluharty AL, Mayes JS, Shapiro LJ (1985) Activator protein required for the enzymatic hydrolysis of cerebroside sulfate. J Biol Chem 260:1867–1871
14. Vogel A, Fürst W, Abo-Hashish MA, Lee-Vaupel M, Conzelmann E, Sandhoff K (1987) Identity of the activator proteins for the enzymic hydrolysis of sulfatide, ganglioside G_{M1} and globotriaosylceramide. Arch Biochem Biophys 259:627–638
15. Banks MN, Herbert SJ, Wynn CH (1987) Activator proteins for sulphatide hydrolysis and G_{M1}-ganglioside hydrolysis. Probable identity on the basis of their co-purification, properties, ligand-binding and immunochemical interactions. Glycoconjugate J 4:157–170
16. Inui K, Wenger DA (1984) Biochemical, immunological and structural studies on a sphingolipid activator protein. Arch Biochem Biophys 233:556–564
17. Li S-C, Li Y-T (1986) Activator proteins (protein cofactors) for the catabolism of glycosphingolipids. In: Freysz L, Dreyfus H, Massarelli R, Gatt S (eds) Enzymes of lipid metabolism II, NATO ASI Ser, vol A 116. Plenum, New York, pp 307–314

18. Ho MW, O'Brien JS (1971) Gaucher's disease: deficiency of 'acid' β-glucosidase and reconstruction of enzyme activity in vitro. Proc Natl Acad Sci USA 68:2810–2813
19. Peters SP, Coyle P, Coffee CJ, Glew RH, Kuhlenschmidt MS, Rosenfeld L, Lee YC (1977) Purification and properties of a heat-stable glucocerebrosidase-activating factor from control and Gaucher spleen. J Biol Chem 252:563–573
20. Wenger DA, Sattler M, Roth S (1982) A protein activator of galactosylceramide α-galactosidase. Biochim Biophys Acta 712:639–649
21. Iyer SS, Berent SL, Radin NS (1983) The cohydrolases in human spleen that stimulate glucosylceramide β-glucosidase. Biochim Biophys Acta 748:1–7
22. Basu A, Glew RH, Daniels LB, Clark LS (1984) Activators of spleen glucocerebrosidase from controls and patients with various forms of Gaucher's disease. J Biol Chem 259:1714–1719
23. Sarmientos F, Schwarzmann G, Sandhoff K (1986) Specificity of human glucosylceramide β-glucosidase towards synthetic glucosylsphingolipids inserted into liposomes. Kinetic studies in a detergent-free system. Eur J Biochem 160:527–535
24. Christomanou H, Aignesberger A, Linke RP (1986) Immunochemical characterization of two activator proteins stimulating enzymic sphingomyelin degradation in vitro. Absence of one of them in a human Gaucher disease variant. Biol Chem Hoppe-Seyler 367:879–890
25. Neufeld EF, Lim TW, Shapiro LJ (1975) Inherited disorders of lysosomal metabolism. Annu Rev Biochem 44:357–376
26. Sandhoff K (1977) The biochemistry of sphingolipid storage diseases. Angew Chem (Int Edn) 16:273–285
27. Callahan JW, Lowden JA (eds) (1981) Lysosomes and lysosomal storage diseases. Raven, New York
28. Sandhoff K, Conzelmann E (1984) The biochemical basis of gangliosidoses. Neuropediatrics 15 (Suppl):85–92
29. Sandhoff K, Conzelmann E, Neufeld EF, Kaback MM, Suzuki K (1989) The G_{M2} gangliosidoses. In: Scriver CR, Beaudet AL, Sly WS, Valle D (eds) The metabolic basis of inherited disease, 6th edn. McGraw-Hill, New York (in press)
30. Burg J, Banerjee A, Sandhoff K (1985) Molecular forms of G_{M2} activator protein. A study on its biosynthesis in human skin fibroblasts. Biol Chem Hoppe-Seyler 366:887–891
31. Banerjee A, Burg J, Conzelmann E, Carroll M, Sandhoff K (1984) Enzyme-linked immunosorbent assay for the ganglioside G_{M2}-activator protein. Screening of normal human tissues and body fluids, of tissues of G_{M2} gangliosidosis, and for its subcellular localization. Hoppe-Seylers Z Physiol Chem 365:347–356
32. Burg J, Conzelmann E, Sandhoff K, Solomon E, Swallow DM (1985) Mapping of the gene coding for the human G_{M2} activator protein to chromosome 5. Ann Human Genet 49:41–45
33. Li S-C, Serizawa S, Li Y-T, Nakamura K, Handa S (1984) Effect of modification of sialic acid on enzymic hydrolysis of gangliosides G_{M1} and G_{M2}. J Biol Chem 259:5409–5410
34. Mahuran DA, Novak A, Lowden JA (1985) The lysosomal hexosaminidase isozymes. In: Rattazzi MC, Scandalios JG, Whitt GS (eds) Isozymes: current topics in biological and medical research, vol 12. Liss, New York, pp 229–288
35. Conzelmann E, Sandhoff K (1987) Enzymes of lysosomal glycoprotein and glycolipid degradation. Adv Enzymol 60:89–216
36. Kytzia H-J, Sandhoff K (1985) Evidence for two different active sites on human β-hexosaminidase A. Interaction of G_{M2} activator protein with β-hexosaminidase A. J Biol Chem 260:7568–7572
37. Kresse H, Fuchs W, Glössl J, Holtfrerich D, Gilberg W (1981) Liberation of N-acetylglucosamine-6-sulfate by human β-N-acetylhexosaminidase A. J Biol Chem 256:12926–12932
38. Sandhoff K, Conzelmann E, Nehrkorn H (1977) Specificity of human liver hexosaminidases A and B against glycosphingolipids G_{M2} and G_{A2}. Purification of the enzymes by affinity chromatography employing specific elution. Hoppe-Seylers Z Physiol Chem 358:779–787
39. Sandhoff K, Harzer K, Wässle W, Jatzkewitz H (1971) Enzyme alterations and lipid storage in three variants of Tay-Sachs disease. J Neurochem 18:2469–2489
40. Iwamori M, Nagai Y (1979) Ganglioside composition of brain in Tay-Sachs disease: increased amounts of G_{D2} and N-acetyl-β-D-galactosaminyl G_{D1a} ganglioside. J Neurochem 32:767–777
41. Neuenhofer S, Conzelmann E, Schwarzmann G, Egge H, Sandhoff K (1986) Occurrence of lysoganglioside lyso-G_{M2} (II3-Neu-5-Ac-gangliotriaosylsphingosine) in G_{M2} gangliosidosis brain. Biol Chem Hoppe-Seyler 367:241–244

42. Rosengren B, Månsson J-E, Svennerholm L (1987) Composition of gangliosides and neutral glycosphingolipids in classical Tay-Sachs and Sandhoff disease: more lyso-G_{M2} in Sandhoff disease? J Neurochem 49:834–840
43. Neuenhofer S, Schwarzmann G, Egge H, Sandhoff K (1985) Synthesis of lysogangliosides. Biochemistry 24:525–532
44. Sandhoff K, Andreae U, Jatzkewitz H (1968) Deficient hexosaminidase activity in an exceptional case of Tay-Sachs disease with additional storage of kidney globoside in visceral organs. Pathol Eur 3:278–285
45. Proia RL, d'Azzo A, Neufeld EF (1984) Association of α- and β-subunits during the biosynthesis of β-hexosaminidase in cultured human fibroblasts. J Biol Chem 259:3350–3354
46. Hechtman P, Gordon BA, Ng Ying Kin NMK (1982) Deficiency of the hexosaminidase A activator protein in a case of G_{M2} gangliosidosis; variant AB. Pediatr Res 16:217–222
47. Hirabayashi Y, Li Y-T, Li S-C (1983) The protein activator specific for the enzymic hydrolysis of G_{M2} ganglioside in normal human brain and in brains of three types of G_{M2} gangliosidosis. J Neurochem 40:168–175
48. Sonderfeld S, Conzelmann E, Schwarzmann G, Burg J, Hinrichs U, Sandhoff K (1985) Incorporation and metabolism of ganglioside G_{M2} in skin fibroblasts from normal and G_{M2} gangliosidosis subjects. Eur J Biochem 149:247–255
49. Conzelmann E, Sandhoff K, Nehrkorn H, Geiger B, Arnon R (1978) Purification, biochemical and immunological characterization of hexosaminidase A from variant AB of infantile G_{M2} gangliosidosis. Eur J Biochem 84:27–33
50. Conzelmann E, Kytzia H-J, Navon R, Sandhoff K (1983) Ganglioside G_{M2} N-acetyl-β-D-galactosaminidase activity in cultured fibroblasts of late-infantile and adult G_{M2} gangliosidosis patients and of healthy probands with low hexosaminidase levels. Am J Human Genet 35:900–913
51. Navon R, Argov Z, Brandt N, Sandbank U (1981) Adult G_{M2} gangliosidosis in association with Tay-Sachs disease, a new phenotype. Neurology 31:1397–1401
52. Navon R, Geiger B, Ben-Yoseph Y, Rattazzi MC (1976) Low levels of β-hexosaminidase A in healthy individuals with apparent deficiency of this enzyme. Am J Human Genet 28:339–349
53. Tallman JF, Brady RO, Navon R, Padeh B (1974) Ganglioside catabolism in hexosaminidase A deficient adults. Nature (London) 252:254–255
54. Fischer G (1977) Anreicherung und Charakterisierung des Aktivators der Sulfatase A. Thesis, Univ Stuttgart
55. Dewji NN, Wenger DA, O'Brien JS (1987) Nucleotide sequence of cloned cDNA for human sphingolipid activator protein 1 precursor. Proc Natl Acad Sci USA 84:8652–8656
56. Fujibayashi S, Wenger DA (1986) Biosynthesis of the sulfatide/G_{M1} activator protein (SAP-1) in control and mutant cultured skin fibroblasts. Biochim Biophys Acta 875:554–562
57. Fürst W, Machleidt W, Sandhoff K (1988) The precursor of sulfatide activator protein is processed to three different proteins. Biol Chem Hoppe-Seyler 369:317–328
58. O'Brien JS, Kretz KA, Dewji NN, Wenger DA, Esch F, Fluharty AL (1988) Coding of two sphingolipid activator proteins (SAP-1 and SAP-2) by same genetic locus. Science 241:1098–1101
59. Mraz W, Fischer G, Jatzkewitz H (1976) The activator of cerebroside sulfatase. Lysosomal localization. Hoppe-Seylers Z Physiol Chem 357:1181–1191
60. Inui K, Kao F-T, Fujibayashi S, Jones C, Morse HG, Law ML, Wenger DA (1985) The gene coding for a sphingolipid activator protein, SAP-1, is on human chromosome 10. Hum Genet 69:197–200
61. Conzelmann E, Sandhoff K (1987) Activator proteins for lysosomal glycolipid hydrolysis. Meth Enzymol 138:792–815
62. Wenger DA, Inui K (1984) Studies on the sphingolipid activator protein for enzymatic hydrolysis of G_{M1} ganglioside and sulfatide. In: Barranger A, Brady RO (eds) Molecular basis of lysosomal storage disorders. Academic Press, New York London, Orlando, pp 61–78
63. Dubois G, Masson M, Rosenthal C (1986) The specificity of cerebroside sulfatase activator. In: Freysz L, Dreyfus H, Massarelli R, Gatt S (eds) Enzymes of lipid metabolism II, NATO ASI Ser, vol A 116. Plenum, New York, pp 349–352
64. Mitsuyama T, Gasa S, Nojima T, Taniguchi N, Makita A (1985) Purification and properties of galactosylceramide sulfatase activator from human liver. J Biochem 98:605–613
65. Wynn CH (1986) A triple-binding-domain model explains the specificity of the interaction of a sphingolipid activator protein (SAP-1) with sulfatide, G_{M1}-ganglioside and globotriaosylceramide. Biochem J 240:921–924

66. Lee-Vaupel M, Conzelmann E (1987) Assay for cerebroside sulfate (sulfatide) sulfatase in cultured skin fibroblasts with the natural activator protein. Clin Chim Acta 168:55–68
67. Shapiro LJ, Aleck KA, Kaback MM, Itabashi H, Desnick RJ, Brand N, Stevens RL, Fluharty AL, Kihara H (1979) Metachromatic leukodystrophy without arylsulfatase A deficiency. Pediatr Res 13:1179–1181
68. Conzelmann E, Lee-Vaupel M, Sandhoff K (1988) The physiological roles of activator proteins for lysosomal glycolipid degradation. In: Salvayre R, Douste-Blazy L, Gatt S (eds) Lipid storage disorders: biological, medical aspects. Plenum, New York pp 323–332
69. Paschke E, Kresse H (1982) Morquio disease, type B: activation of G_{M1} β-galactosidase by G_{M1} activator protein. Biochem Biophys Res Commun 109:568–575
70. Cantz M (1982) Sialidoses. In: Schauer R (ed) Sialic acids: chemistry, metabolism and function. Springer, Berlin Heidelberg New York, pp 307–320
71. Zeigler M, Bach G (1981) Cellular localization of neuraminidases in cultured human fibroblasts. Biochem J 198:505–508
72. Van Dessel G, De Wolf M, Lagrou A, Hilderson H, Dierick W (1984) Characterization, purification and subcellular localization of bovine thyroid sialidases. J Biochem 96:937–947
73. Collard MW, Sylvester SR, Tsuruta JK, Griswold MD (1988) Biosynthesis and molecular cloning of sulfated glycoprotein 1 secreted by rat Sertoli cells: sequence similarity with the 70-kilodalton precursor to sulfatide/G_{M1} activator. Biochemistry 27:4557–4564

Solute Translocation Across the Lysosome Membrane: Physiology, Pathology and Pharmacology

J.B. LLOYD, T. KOOISTRA, S. FORSTER, S.J. BIRD and G.P. IVESON

1 Introduction

Lysosomes were discovered, almost 40 years ago, because their limiting membrane is impermeable to certain low molecular weight solutes. 2-Phosphoglycerol is such a substance and, as it is also a substrate for lysosomal acid phosphatase, lysosomes incubated with 2-phosphoglycerol hydrolyse it only if their membrane is ruptured. De Duve has described in a number of publications how this initial observation led on to the characterization of lysosomes as a distinct class of subcellular organelles (see, for example, De Duve 1969 and 1979).

Interest in lysosomes burgeoned in the 1960s and 1970s. As the metabolic pathways occurring within them were delineated, their role in the catabolism of biopolymers was increasingly understood. A surprisingly large number of human genetic diseases were traced to absent or defective lysosomal hydrolases, drawing attention away from the lysosome membrane and towards metabolic and pathologic considerations. More recently attention has focused on the biogenesis of the lysosomal enzymes and the subtleties of vesicular trafficking in both the prelysosomal and the endocytic pathways.

The permeability characteristics of the lysosome membrane are, however, of central importance to the normal and abnormal functioning of the organelle. The membrane needs to be, and indeed is, impermeable to biopolymers, but permeable to their digestion products; otherwise, substrates would escape the attention of the hydrolases, or products would accumulate in the lysosome. Conversely, inborn errors of lysosomal enzymes lead to lysosomal storage only because the membrane is impermeable to the defective enzyme's usual substrate(s).

The mechanisms by which metabolites and inorganic ions cross the lysosome membrane have received little attention until very recently. Ten years ago two major reviews (Goldman 1976; Reijngoud and Tager 1977) found that there was insufficient evidence for any firm conclusions to be drawn, and that there was no positive reason for postulating the existence of membrane porters. This is no longer the case, and Table 1 lists the lysosome-membrane porters for which there is strong or at least good evidence. It is already clear that some (at least) of these porters are essential for normal lysosome function, as human diseases exist that can be traced to a defective porter.

Cellular Pharmacology Research Group, Department of Biological Sciences, University of Keele, Staffordshire, ST5 5BG, UK

A. Azzi, Z. Drahota and S. Papa
Molecular Basis of Membrane-Associated Diseases

Table 1. Metabolite carriers in the lysosome membrane

Species carried	Remarks
H^+	Vectorial, ATP-driven
Cystine	Km(cystine) 0.5 mM
Lysine, Arginine	Km(Arg) 0.32 mM
Proline	Km(Pro) 70 μM
Alanine, serine, proline	Km(Pro) 10 μM
Alanine, serine, threonine	
Tyrosine, phenylalanine, tryptophan, leucine, histidine	Km(Leu,Phe) 5–15 μM
Aspartate, glutamate	Km(Asp,Glu) 10 μM
Neutral sugars	Km(Glc) 50 mM
Sialic acid	
Nucleosides	Km(adenosine) 8 mM
Vitamin B_{12}	

In this chapter we describe our own recent experiments in this field of study, setting our findings in the context of other relevant work. The reader is referred to a recent review (Forster and Lloyd 1988) for a more extended account of this background information.

2 Physiology

It is not axiomatic that every metabolite released from lysosomes requires a porter for its translocation. This is particularly the case for non-electrolytes, where passive diffusion is a plausible alternative mechanism. We have therefore sought to determine the range of compounds that may cross by passive diffusion and, by implication, identify those metabolites for which a porter must be envisaged.

The rates at which different solutes diffuse unassisted across a biological membrane depend solely on their physicochemical characteristics. Extensive studies on the plasma membranes of erythrocytes have shown correlations between permeability and oil-water partition coefficients, which in turn correlate (inversely) with a solute's ability to form hydrogen bonds (Stein 1967; Diamond and Wright 1969). We have chosen to study the permeability of the rat liver lysosome membrane to a range of uncharged compounds containing only carbon, hydrogen and oxygen. The majority of the compounds studied are non-physiological, and we argue that a good correlation between permeability and purely physicochemical features is a strong argument for a passive-diffusional mechanism. If the correlation is poor, solute-specific porters would need to be invoked.

For this study we needed an experimental method that could be applied to a large number of solutes that are unavailable in radiolabelled form. We chose the osmotic-protection method, which has been used by several previous investigators and which has the practical advantage of not needing a highly purified preparation of lysosomes. In this technique lysosomes are suspended and incubated in a 0.25 M solution of the test solute. At intervals the intactness of the lysosomes is assessed by measuring the availability of some suitable lysosomal enzyme to a non-permeant substrate, in our case a fluorogenic substrate for N-acetylhexosaminidase. Solutes typically show one of three basic patterns (Fig. 1). A non-penetrant solute (pattern A) provides long-term osmotic protection for the lysosomes: the non-penetrant solute in the medium is iso-osmotic with resident non-penetrant substances within the lysosomes. A rapidly penetrant solute (pattern C) enters the lysosomes so quickly that the lysosomes swell and burst immediately on resuspension. More slowly penetrant solutes (patterns B) provide initial osmotic protection, but this is lost more or less rapidly as the solute flows into the lysosome interior.

These three patterns are well illustrated by our results with poly(ethyleneglycol) oligomers. These molecules have the structure shown, and are

$$HO-[-CH_2-CH_2-O-]_n-H$$

commercially available for values of *n* from 1 to 6 and also as polydisperse mixtures of given molecular weight average. Lysosomes suspended at 25°C in a 0.25 M solution of a poly(ethyleneglycol) whose molecular weight average is above 600 displayed pattern A, as described above: they were as stable as when suspended in 0.25 M sucrose or mannitol, two known non-penetrant substances. When the molecular weight of the poly(ethyleneglycol) was between 400 and 600, pattern B was seen and, when the molecular weight was below 300 ($n = 1$ to 6), pattern C.

These results indicate that the rat liver lysosome is rapidly penetrated by low oligomers of poly(ethyleneglycol). An alternative explanation that these oligomers break lysosomes by a detergent-like effect was excluded by showing that lysosomes were stable if incubated in a solution containing both 0.25 M sucrose and 0.25 M poly(ethyleneglycol). The mechanism of penetration is not conclusively shown by these experiments, but passive diffusion is the most plausible, since poly-(ethyleneglycol)s are wholly non-physiological substances with little resemblance to any product of lysosomal metabolism, and also because of the correlation between increasing molecular weight and decreasing ease of penetration.

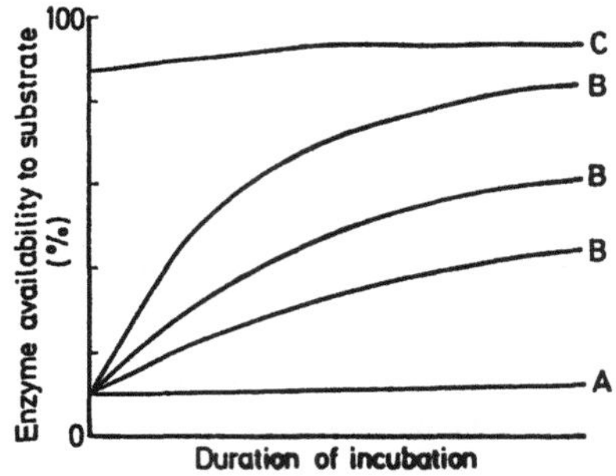

Fig. 1. In osmotic-protection experiments, a lysosome preparation is resuspended in an aqueous 0.25 M solution of the solute under investigation. At appropriate time intervals after resuspension the activity of a chosen lysosomal enzyme is measured under conditions that preserve or immediately disrupt the integrity of any lysosomes that had remained intact. The ratio of these activities, expressed as a percentage, is graphed against the duration of exposure to the test solute. Results in osmotic-protection experiments typically fall into one of three basic patterns, *A*, *B* and *C* (see text for details)

The poly(ethyleneglycol)s differ from one another in only one respect, the number of ethyleneoxide units. Every member of the series has two terminal hydroxy-groups. We reasoned that if the oligomers were penetrating the lysosome membrane by passive diffusion, other molecules of similar size containing only aliphatic alcohol and ether groups should also penetrate, and at similar rates. Furthermore, it should be possible to probe the reason for the decreasing rate of penetration with increasing molecular size: the relevant parameter could be either size per se or the accompanying increase in hydrophilicity contributed by each additional ether linkage. The latter explanation receives a degree of plausibility from the observation that poly(ethyleneglycol) is rapidly penetrant up to and well beyond the molecular weight (182) of mannitol, a non-penetrant species. Figure 2 shows the results we have obtained in our experiments to date. The compounds tested are placed within a matrix that indicates their molecular weight and their hydrogen-bonding capacity. It is very striking that, without exception, substances whose hydrogen-bonding capacity is below 7.5 are rapidly penetrant, those between 7.5 and 11.5 are more slowly penetrant, and those above 11.5 non-penetrant. At least within the range of substances so far tested, molecular weight is not a direct determinant of the capacity to penetrate.

It must be concluded from these results that passive diffusion across the lysosome membrane is a viable mechanism for the translocation of some non-electrolytes. Relatively few physiological substances would be predicted to diffuse, but those that would include the neutral sugars, and it is noteworthy that the

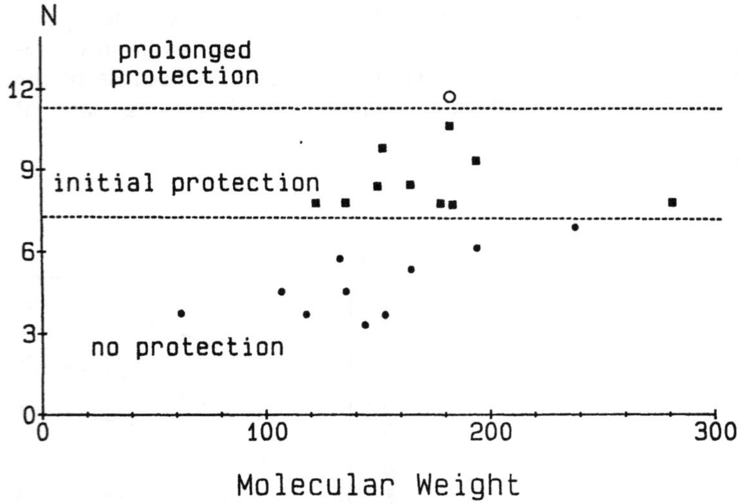

Fig. 2. Osmotic-protection data on 20 aliphatic non-electrolytes whose only functional groups are alcohol and ether moieties. The ordinate axis shows the substances' hydrogen-bonding capacities, calculated by adding together the contributions of the oxygen atoms. For this purpose each hydroxy-group was assigned a value of 2.0 and each C-O-C group 0.8; the first of these values is derived directly from the work of Stein (1967) and Diamond and Wright (1969), while the second was chosen to reflect Diamond and Wright's (1969) assertion that two ether oxygens are "slightly less potent than one hydroxyl". The osmotic-protection data we obtained are classified into one or other of three typical patterns: A(○), B(■) and C(●)

pentoses and the hexoses would be expected from their hydrogen-bonding capacities (8.8 and 10.8 respectively) to display pattern-B behaviour in osmotic-protection experiments, and this was found to be so experimentally. It would be incorrect to deduce from this result, however, that pentose and/or hexose release from lysosomes in vivo is entirely by passive diffusion. Evidence exists for a facilitated diffusion system carrying neutral sugars across the rat liver lysosome membrane (Maguire et al. 1983) but its efficacy in promoting net sugar flow would be diminished under the conditions in osmotic-protection experiments. In the cell, however, carrier-mediated transport of sugars may be quantitatively as important, or more important, than the diffusional component.

As will be explained more fully below, the osmotic-protection method may not reveal the presence of a metabolite-porter in the lysosome membrane. It does, however, provide valuable information on the likely limits of passive diffusion, indicating that solutes with a hydrogen-bonding capacity greater than about 11.5 are unlikely to diffuse at a significant rate. If such a compound can be shown to cross the lysosome membrane, a porter must be postulated. If no porter exists, such compounds will be unable to cross; di- and oligo-saccharides appear to be in this category.

Turning now to electrolytes, passive diffusion seems much less likely. Ginsburg and Stein (1987) calculate that the zwitterionic form of a neutral amino acid such as glycine or L-alanine or L-valine has a hydrogen-bonding capacity of about 11. Our early (Lloyd 1971) and recent osmotic-protection experiments with these neutral amino acids indicate a slow rate of entry, consistent with this value. Even slower, but still significant, uptake rates are seen with β-alanine, α-aminoisobutyric acid and D-valine. The rather minor differences found in these experiments between physiological and non-physiological amino acids at first appears inconsistent with the known presence of carriers for neutral amino acids in the lysosome membrane (Pisoni et al. 1987). The explanation probably lies in the very low Km values (below 100 μM) reported for these carriers: assuming that they are bidirectional and symmetrical, they will rapidly become saturated with substrate in osmotic-protection experiments and thus contribute little to net inflow of amino acid into the lysosome. This is the reason why osmotic-protection experiments may give a good indication of the rate of passive diffusion even when a solute-porter is present.

One of us reported many years ago (Lloyd 1971) that dipeptides afford very much less prolonged osmotic protection to lysosomes. Goldman (1973) subsequently confirmed and extended this finding, showing that the observed lysosome rupture was due to rapid influx of dipeptides into lysosomes, combined, in the cases of some (but not all) dipeptides, with intralysosomal hydrolysis to relatively poorly permeant amino acids. The rapid entry of dipeptides into lysosomes is a surprising finding since, like amino acids, dipeptides are doubly charged species. In contrast to amino acids, however, dipeptides such as glycylglycine can adopt a partially charge-neutralized conformation, and we conjectured that this might account for their rapid entry. Support for this explanation was provided by our finding that cyclic amides such as δ-valerolactam displayed pattern-C behaviour in osmotic-protection experiments, indicating very rapid passage across the lysosome membrane. However, when we tested 5-aminopentanoic and 6-aminohexanoic acids, compounds as capable as glycylglycine of adopting charge-neutralized conformations, the results indicated only slow uptake (Bird et al. 1988). We are therefore led to conclude that the passage of glycylglycine across the lysosome membrane is probably porter-assisted. This conclusion is greatly strengthened by our further observation that, while many dipeptides display pattern-B behaviour in osmotic-protection experiments, their D-enantiomers display pattern-A behaviour, indicating that they are essentially non-penetrant. Figure 3 illustrates this phenomenon. It is not yet clear whether the L-dipeptide carrier is specific for dipeptides or is an amino acid carrier of broader specificity. However, it is probable that its Km for dipeptides is well into the millimolar range: this follows from our comments above concerning the ability of carriers to effect net translocation of solute in osmotic-protection experiments.

What can we conclude from these experiments? First, that non-electrolytes can diffuse into and out of lysosomes, provided that their hydrogen-bonding capacity does not exceed about 11. Diffusible compounds will include glycerol and (probably) monoacylglycerols, as well as a wide range of xenobiotics, which may be delivered to lysosomes by targeted drug-delivery systems. Secondly, that charged species, particularly those bearing two or more charges, will be impermeant unless specific porters are present in the membrane. It is noteworthy that the porters thus far identified in the lysosome membrane are mostly carriers for doubly or multiply charged molecules. Thirdly, there is an intermediate class of compounds, some bearing single or partial charges, where diffusion may be a slow but adequate mechanism, supplemented in some cases by porters. Hexoses, hexosamines, hexonic and hexuronic acids, and nucleosides come to mind, although the sialic acid storage that occurs in Salla disease indicates that diffusion of this substance is negligible.

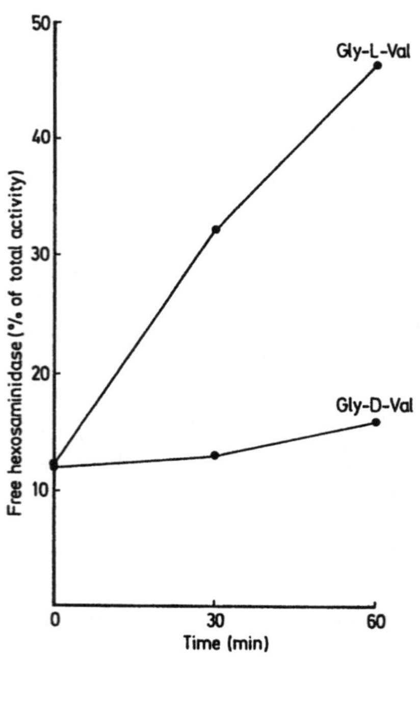

Fig. 3. Progressive rupture of rat liver lysosomes when incubated at 25°C in 0.25 M glycyl-L-valine or glycyl-D-valine, pH 7.0, measured by the loss of latency of N-acetylhexosaminidase

3 Pathology

Twenty years ago, the elucidation of metabolic pathways in the lysosome was greatly assisted by investigations of genetic defects of lysosomal enzymes. Similarly, several of the metabolite porters in the lysosome membrane have come to light because of clinical conditions in which these porters are defective. Indeed, the only firm evidence for the porters for sialic acid and vitamin B_{12} comes from the existence of these conditions.

The first metabolite porter to be recognized was that for cystine. Absence of this porter from human lysosome membranes is responsible for the Mendelian recessive autosomal disorder cystinosis, in which cystine accumulates, and in some cells crystallizes, in the lysosomes. Many questions still remain to be answered about cystinosis, the most fundamental perhaps being the nature of the causal link between lysosomal cystine overload and the serious tissue pathology that results. Work in our laboratory failed to show any major impairment of either pinocytosis or intralysosomal proteolysis in cystinotic cells (Kooistra and Lloyd 1986).

A question that has been only partially answered concerns the origin of the cystine that accumulates in the lysosomes of cystinotic cells. Studies by Thoene and Lemons (1982) indicated that such cells cultured in vitro and depleted of their cystine stores by cysteamine (see below) could re-accumulate cystine by incubation in the presence of either cystine-containing proteins or free cystine. Cysteine, whether protein-bound or free, was ineffective. The reaccumulation from disulphide-containing proteins presumably follows their pinocytosis and intralysosomal digestion, but this seems to imply that proteolysis in lysosomes proceeds without concomitant reduction of cystine residues. There is good evidence, however, that disulphide residues are reduced during proteolysis (Kooistra et al. 1982; Mego 1984). A possible solution to this dilemma (Lloyd 1986) sees cytoplasmically-derived cysteine as the physiological reducing agent for intralysosomal reduction of disulphide linkages in proteins. If this proposal is correct, the cystine that accumulates in cystinotic lysosomes is not derived directly from cystine residues in pinocytosed protein but from the cysteine molecules that serve to reduce these residues.

The reaccumulation of cystine from exogenous free cystine is also puzzling, as recognized by Thoene and Lemons (1982). This is because the lysosome membrane of cystinotic cells is not permeable to cystine and also because any cystine entering the cytoplasm will be immediately reduced to cysteine by the high concentration of reduced glutathione. Pinocytosis of cystine constitutes an alternative route for exogenous cystine to reach the lysosomes, avoiding both the glutathione and the necessity to cross the lysosome membrane, but Thoene and Lemons (1982) showed that the rate of fluid-phase pinocytosis in cystinotic fibroblasts was inadequate to account for the observed rate of cystine reaccumulation. There are thus apparently cogent arguments against each of the two possible mechanisms of reaccumulation from extracellular free cystine. We have recently examined this question by studying the effects, on cystine reaccumulation, of colchicine, which inhibits pinocytosis in human fibroblasts (Kooistra and Lloyd 1985a), and of glutamate, which shares with cystine a porter in the fibroblast plasma membrane (Bannai and Kitamura 1980). The results were very clear: glutamate inhibited reaccumulation but colchicine did not, indicating that the first phase of uptake of exogenous cystine is the passage across the plasma membrane. Since, as explained above, this cystine will be rapidly reduced to cysteine, and in any case could not cross the lysosome membrane of cystinotic cells, it seems inescapable that the cystine that reaccumulates arrives in the lysosomes of cystinotic cells in the reduced form. Its conversion into cystine is explained by its participation in protein degradation (as discussed above), although the possibility of some autoxidation of cysteine in lysosomes cannot be discounted. Indeed, the necessity for a cystine porter in lysosomes may in part be due to the autoxidation of cysteine molecules that find their way into lysosomes. This account (illustrated in Fig. 4) seems complete except in one respect: why is exogenous cysteine less effective than exogenous cystine in causing reaccumulation of cystine, if the cystine is reduced to cysteine *en route* to the lysosomes? Presumably because the plasma membrane porter for cystine is more effective than that for cysteine.

4 Pharmacology

A major difference between the inborn errors affecting lysosomal enzymes and those affecting lysosomal membrane porters is the size of the stored product. In the former class of diseases it is chiefly macromolecules that accumulate, and rational approaches to therapy must seek to replace the missing enzyme or to stimulate

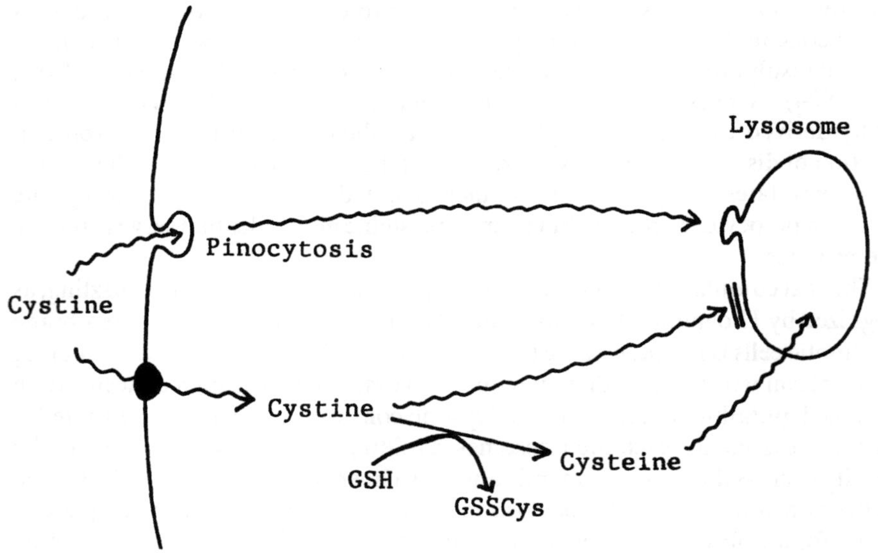

Fig. 4. The proposed route of cystine reaccumulation from exogenous cystine by cystine-depleted cystinotic fibroblasts in culture. Cystine crosses the plasma membrane on the cystine-glutamate porter and in the cytoplasm is reduced to cysteine by glutathione (*GSH*). Cysteine enters the lysosome, where it is reoxidized by participation in the reduction of cystine residues in proteins and/or by autoxidation. Cytoplasmic cystine cannot itself enter the lysosome, owing to the absence of the cystine porter in cystinotic cells. Direct transfer of exogenous cystine to the lysosome, by pinocytosis, was found experimentally not to be a quantitatively significant route

exocytosis. Membrane porters, by contrast, serve to translocate the products of intralysosomal digestion, which are all quite small molecules, such as monosaccharides, amino acids, dipeptides and nuclosides. As discussed above, many of these substances are probably too hydrophilic to diffuse passively across the lysosome membrane; hence the necessity for porters. However, these metabolites, unlike macromolecules, are not much beyond the limits for passive diffusion, and we consider now the possibility of pharmacologic intervention aimed at a moderate increase in the permeability properties of the lysosome membrane.

Cystinosis is a disease ripe for this approach. Cystine, the stored product, has a molecular weight of only 240, and its failure to cross the lysosome membrane by passive diffusion presumably reflects it possession of four charged groups. However, before considering whether the lysosome membrane could be rendered permeable to cystine, we first discuss the medical therapy currently used for cystinosis. This involves the chronic administration of cysteamine, a substance that depletes the lysosomal cystine stores of cystinotic cells both in vitro and in vivo. Cysteamine reduces cystine, non-enzymically it is assumed, to cysteamine-cysteine mixed disulphide and cysteine (Fig. 5). In order for this reaction to be effective, cysteamine must be able to enter the lysosome and the two reaction products must be able to leave. This appears to be so: osmotic-protection experiments showed that cysteamine and cysteine are permeant (Kooistra et al. 1982) and the mixed disulphide is carried on the lysosomal membrane porter for lysine (Pisoni et al. 1985) with which it has a close structural resemblance (see Fig. 5).

Cysteamine therapy of cystinosis appears to be very successful clinically, and has recently been refined by substituting S-phosphocysteamine, a more palatable derivative. We have, however, noted a potentially harmful effect of cysteamine: in cultured human fibroblasts it massively depletes lysosomal β-galactosidase, apparently because this enzyme is unstable if exposed to thiol at a raised pH (Kooistra

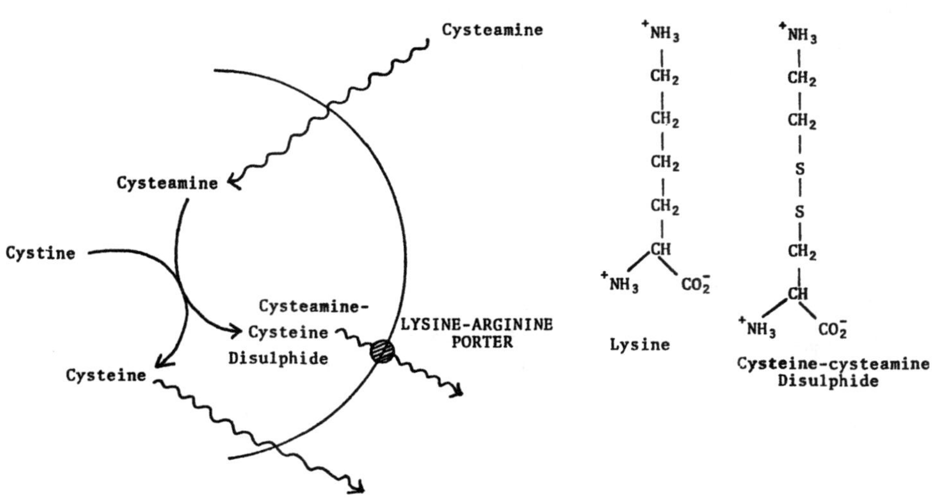

Fig. 5. The mechanism of cysteamine-induced depletion of cystine from the lysosomes of cystinotic cells. Cysteine-cysteamine disulphide is a close structural analogue of lysine and is carried out of the lysosome on the lysine-arginine porter

and Lloyd 1985b). If cysteamine has this effect on cystinotic cells in vivo, it could lead to a secondary lysosomal storage phenomenon resembling GM_1 gangliosidosis.

Our first attempt to modify the lysosome membrane of cystinotic cells was based on the observations of Firestone et al. (1979). These authors designed a range of 'lysosomotropic detergents', long-chain alkyl derivatives of weak bases, which should be protonated at the pH of the lysosomes but uncharged at pH 7.4. These compounds were predicted to be cytotoxic, since they should accumulate in lysosomes and effect lysosome membrane dissolution by a detergent effect. Several members of the series, such as N-dodecylmorpholine and N-dodecylimidazole, were indeed cytotoxic, and good evidence was adduced indicating that their mechanism of action was that on which their design was based (Miller et al. 1983). Since detergents such as Triton X-100, in subcytolytic concentrations, induce pores in plasma membranes (Bashford et al. 1986), we hoped to identify concentrations of lysosomotropic detergents that would permeabilize lysosomes to cystine without allowing the escape of macromolecules. We showed that, although the application of N-dodecylimidazole (20 μg/ml) to cystinotic cells in culture led to a rapid depletion of the stored cystine, this effect was accompanied by a greatly increased release of lactate dehydrogenase from the cells (Forster et al. 1987, 1988). Lower concentrations of N-dodecylimidazole were less cytotoxic, but these had little effect on cystine levels. Parenthetically, our results offered no clear support for the underlying concept that lysosomal damage precedes and causes cell death induced by the lysosomotropic detergents, and subsequently others have also cast doubt on this hypothesis [Cabantchik et al. (1989)].

More recently we have examined the effect of induced changes in membrane fluidity on the cystine levels of cystinotic cells. It is reported that cells cultured at below-normal temperatures increase the proportion of unsaturated fatty acids in their membranes, thus maintaining membrane fluidity at the lower temperature. We reasoned that cystinotic cells grown at decreased temperature might rapidly lose their cystine when returned to normal growth conditions. Figure 6 shows the results of our initial experiments. It may be seen that growth at 28°C increases

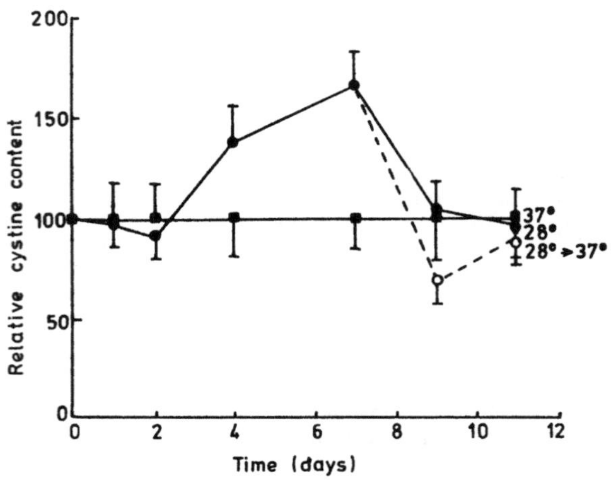

Fig. 6. Cystine content of cystinotic cells maintained at 28°C for 11 days, or at 28°C for 7 days followed by 4 days at 37°C. Values are shown as percentages of the cystine content of cells maintained throughout at 37°C. Each point on the graph represents the mean of 6 determinations

cystine stores initially, but that these return to more typical values after a period of 9 days. Return to 37°C accelerates the loss of the extra cystine stored. These data show that the cystine concentration of cystinotic cells responds to growth temperature, but do not in themselves prove that this results from alterations in membrane fluidity. We are pursuing this matter further by examining the effect of supplementing the cells' incubation medium with various fatty acids. Meanwhile the results, showing an *increase* in cystine stored when the growth temperature is decreased, bring into sharp focus a question that has been largely ignored, that of how the cystine level in cystinotic cells is maintained. Is the supra-normal but stable level of cystine achieved by cystine diffusing out of the lysosomes or by the presence of an alternative porter? Lemons et al. (1986) concluded from some experiments on cystine loss from cells maintained at elevated temperatures that another porter may exist. These questions are at present unresolved.

5 Conclusions

The discovery of metabolite-porters in the lysosome membrane and of human diseases resulting from their congenital deficiency has opened a new chapter in research on lysosomes. Although the claim (Anonymous 1986) that lysosomal transport defects may prove as important as deficiencies of lysosomal hydrolases is perhaps exaggerated, there is without doubt much yet to be discovered.

Acknowledgements. We thank the National Fund for Research into Crippling Diseases, the National Kidney Research Fund, the West Midlands Regional Health Authority, and the Science and Engineering Research Council for their support of the work reported above.

References

Anonymous (1986) Lysosomal storage diseases. Lancet ii:898–899

Bannai S, Kitamura E (1980) Transport interaction of L-cystine and L-glutamate in human diploid fibroblasts in culture. J Biol Chem 255:2372–2376

Bashford CL, Alder GM, Menestrina G, Micklem KJ, Murphy JJ, Pasternak CA (1986) Membrane damage by hemolytic viruses, toxins, complements, and other cytotoxic agents. J Biol Chem 261:9300–9308

Bird SJ, Forster S, Lloyd JB (1988) Mechanism of dipeptide translocation across the lysosome membrane. Biochem Soc Trans 16:42–43

Cabantchik ZI, Silfen J, Firestone RA, Krugliak M, Nissani E, Ginsburg H (1989) Effects of lysosomotropic detergents on the human malarial parasite *Plasmodium falciparium* in in vitro culture. Biochem Pharmacol (in press)

De Duve C (1969) The lysosome in retrospect. In: Dingle JT, Fell HB (eds) Lysosomes in biology and pathology, vol 1. Elsevier North Holland Biomedical Press, Amsterdam, pp 3–40

De Duve C (1979) The mitochondrial fraction. Biochem Soc Trans 7:823–835

Diamond JM, Wright EM (1969) Biological membranes: the physical basis of ion and nonelectrolyte selectivity. Annu Rev Physiol 31:581–646

Firestone RA, Pisano JM, Bonney RJ (1979) Lysosomotropic agents. 1. Synthesis and cytotoxic action of lysosomotropic detergents. J Med Chem 22:1130–1133

Forster S, Lloyd JB (1988) Solute translocation across the mammalian lysosome membrane. Biochim Biophys Acta (in press)

Forster S, Scarlett L, Lloyd JB (1987) The effect of lysosomotropic detergents on the permeability properties of the lysosome membrane. Biochim Biophys Acta 924:452–457

Forster S, Scarlett L, Lloyd JB (1988) Effects of a lysosomotropic detergent on cystinotic fibroblasts. Biochem Soc Trans 16:43–44

Ginsburg H, Stein WD (1987) Biophysical analysis of novel transport pathways induced in red blood cell membranes. J Membrane Biol 96:1–10

Goldman R (1973) Dipeptide hydrolysis within intact lysosomes in vitro. FEBS Lett 33:208–212

Goldman R (1976) Ion distribution and membrane permeability in lysosomal suspensions. In: Dingle JT, Dean RT (eds) Lysosomes in biology and pathology, vol 5. Elsevier North Holland Biomedical Press, Amsterdam, pp 309–336

Kooistra T, Lloyd JB (1985a) Degradation of insulin by human fibroblasts: effects of inhibitors of pinocytosis and lysosomal activity. Int J Biochem 17:805–811

Kooistra T, Lloyd JB (1985b) A study of the low β-galactosidase activity in cystinotic fibroblasts: effects of cysteamine. Clin Chim Acta 145:9–16

Kooistra T, Lloyd JB (1986) Pinocytosis and degradation of exogenous proteins by cystinotic fibroblasts. Biochim Biophys Acta 887:182–188

Kooistra T, Millard PC, Lloyd JB (1982) Role of thiols in degradation of proteins by cathepsins. Biochem J 204:471–477

Lemons RM, Pisoni RL, Christensen HN, Thoene JG (1986) Elevated temperature produces cystine depletion in cystinotic fibroblasts. Biochim Biophys Acta 884:429–434

Lloyd JB (1971) A study of permeability of lysosomes to amino acids and peptides. Biochem J 121:245–248

Lloyd JB (1986) Disulphide reduction in lysosomes. Biochem J 237:271–272

Maguire GA, Docherty K, Hales CN (1983) Sugar transport in rat liver lysosomes. Biochem J 212:211–218

Mego JL (1984) Role of thiols, pH and cathepsin D in the lysosomal catabolism of serum albumin. Biochem J 218:775–783

Miller DK, Griffiths E, Lenard J, Firestone RA (1983) Cell killing by lysosomotropic detergents. J Cell Biol 97:1841–1851

Pisoni RL, Thoene JG, Christensen HN (1985) Detection and characterization of carrier-mediated cationic amino acid transport in lysosomes of normal and cystinotic human fibroblasts. J Biol Chem 260:4791–4798

Pisoni RL, Flickinger KS, Thoene JG, Christensen HN (1987) Characterization of carrier-mediated transport systems for small neutral amino acids in human fibroblast lysosomes. J Biol Chem 262:6010–6017

Reijngoud D-J, Tager JM (1977) The permeability properties of the lysosomal membrane. Biochim Biophys Acta 472:419–449

Stein WD (1967) The movement of molecules across cell membranes. Academic Press, New York London

Thoene JG, Lemons RM (1982) Cystine accumulation in cystinotic fibroblasts from free and protein-linked cystine but not cysteine. Biochem J 208:823–830

The Peroxisomal β-Oxidation Systems: Characteristics and (Dys) Functions in Man

R.J.A. WANDERS[1,4], C.W.T. VAN ROERMUND[1], M.J.A. VAN WIJLAND[1], R.B.H. SCHUTGENS[1], H. VAN DEN BOSCH[2] and J.M. TAGER[3]

1 Introduction

Peroxisomes are now known to be present in virtually every mammalian cell except the mature erythrocyte. They were first described by Rhodin in 1954 as "spheric or oval bodies" present in the cytoplasm of mouse proximal kidney tubules [1]. Evidence that these "microbodies" were, indeed, different from the other sub-cellular organelles known at that time, came from cell fractionation experiments by de Duve and co-workers (see [2] for review). The identification of catalase and several H_2O_2-producing oxidases (D-aminoacid oxidase, urate oxidase, glycollate oxidase) within this organelle prompted de Duve and co-workers to introduce the name "peroxisome".

For a long time the functional significance of peroxisomes in mammalian metabolism remained unclear. In 1976, however, Lazarow and de Duve [3] described the presence of a fatty acid β-oxidation system in rat liver peroxisomes. Their study was inspired by earlier work of Cooper and Beevers [4] who found that glyoxysomes, organelles closely related to peroxisomes (they contain catalase and several H_2O_2-producing oxidases) are capable of oxidizing fatty acids.

Before that time fatty acid β-oxidation was generally assumed to occur only in mitochondria. Present knowledge indicates that non-mitochondrial β-oxidation is, in fact, much more widely distributed in nature than β-oxidation in mitochondria, which is mainly restricted to the animal kingdom. Non-mitochondrial β-oxidation has been detected in animals, plants and eukaryotic micro-organisms (see Kunau et al. [5] for discussion).

In recent years much has been learned about the physiological importance of the peroxisomal β-oxidation system. In particular, the recognition of a group of inherited diseases in man caused by a dysfunction in one or more steps involved in peroxisomal β-oxidation has contributed greatly to our current understanding of the role of peroxisomes in the oxidation of fatty acids and other compounds and has elucidated several aspects of the functional organization of the peroxisomal β-oxidation system.

[1]Department of Pediatrics, University Hospital Amsterdam, Meibergdreef 9, 1105 AZ Amsterdam, The Netherlands
[2]Department of Biochemistry, State University Utrecht, Padualaan 9, 3584 CH Utrecht, The Netherlands
[3]Department of Biochemistry, University of Amsterdam, Meibergdreef 15, 1105 AZ Amsterdam, The Netherlands
[4]To whom correspondence should be addressed

A. Azzi, Z. Drahota and S. Papa
Molecular Basis of Membrane-Associated Diseases

2 Enzymic Organization of the Mitochondrial and Peroxisomal β-Oxidation Systems

Fatty acids are transported between organs in the form of unesterified fatty acids complexed to albumin or in the form of triacylglycerols which are subsequently hydrolyzed. Following their uptake into cells via a mechanism which is not well understood, the long-chain fatty acids are activated to the corresponding CoA-esters. This occurs via a long-chain acyl-CoA synthetase which is present in the outer mitochondrial membrane, the membrane of peroxisomes and the membrane of the endoplasmic reticulum [6–9]; these three enzymes are indistinguishable as judged by several physicochemical, catalytic and immunological properties [6,10].

The acyl-CoA thioesters are carried across the mitochondrial inner membrane by a concerted mechanism involving L-carnitine, carnitine palmitoyltransferase I (CPT I), the carnitine: acylcarnitine translocator and carnitine palmitoyltransferase II (CPT II). Short-chain and medium-chain fatty acids with ten carbon atoms or less are taken up by mitochondria as free fatty acids independently of carnitine; this is followed by their activation in the mitochondrial matrix (see [11] for further information).

Once inside the mitochondrion the acyl-CoA esters are oxidized via sequential steps of dehydrogenation, hydration, dehydrogenation and thiolytic cleavage. Three different acyl-CoA dehydrogenases catalyze the conversion of short-chain, medium-chain or long-chain thioesters, respectively, to the corresponding 2-*trans*-enoyl-CoA esters. In addition to these three enzymes, there are two acyl-CoA dehydrogenases involved in dehydrogenation of isovaleryl-CoA and 2-methyl-branched-chain acyl-CoA esters, respectively. In all these dehydrogenases tightly bound FAD is the prosthetic group. Enzyme-bound $FADH_2$ is re-oxidized by transfer of electrons to a second flavoprotein named electron-transferring flavoprotein (ETF) which, in turn, donates its electrons to an iron-sulphur flavoprotein called ETF: ubiquinone oxidoreductase. Finally, the latter enzyme feeds its electrons into the mitochondrial respiratory chain at the level of ubiquinone (see [12]).

The second step in β-oxidation involves hydration of 2-*trans*-enoyl-CoA to L-3-hydroxyacyl-CoA, a reaction catalyzed by enoyl-CoA hydratase which also catalyzes hydration of 2-*cis*-enoyl-CoA to D-3-hydroxyacyl-CoA. Two distinct hydratases have been identified, of which the short-chain enoyl-CoA hydratase or crotonase is best characterized (for discussion see [12]). In the third step the L-3-hydroxyacyl-CoA ester is dehydrogenated to the 3-ketoacyl-CoA compound, a reaction catalyzed by one of two L-3-hydroxyacyl-CoA dehydrogenases, which show overlapping chain-length specificities (see [11,12] for discussion). Finally, in the fourth step of β-oxidation the 3-ketoacyl-CoA ester is cleaved to yield acetyl-CoA and an acyl-CoA ester shortened by two carbon atoms. This reaction is catalyzed by 3-ketoacyl-CoA thiolases, commonly referred to as thiolases. In mitochondria two thiolases are present, one specific for acetoacetyl-CoA and the other displaying a broad specificity ([13,14]; see [12] for discussion).

Peroxisomal β-oxidation, like mitochondrial β-oxidation, proceeds via successive steps of dehydrogenation, hydration, dehydrogenation and thiolytic cleavage. The enzymes involved, however, are quite distinct from their mitochondrial counterparts, except for the long-chain fatty acyl-CoA synthetase. This enzyme, present on the cytosolic site of the peroxisomal membrane [9], is identical to the enzyme present on the mitochondrial outer membrane and the membrane of the endoplasmic reticulum [6,10]. Recently evidence has been brought forward for the existence of a separate acyl-CoA synthetase involved in activation of very long-chain fatty acids [15–18]. This enzyme activity appears to be present in peroxisomes and the endoplasmic reticulum but not in mitochondria [17,18]. As discussed below there is evidence suggesting a tight coupling between activation of very long-chain fatty acids on the peroxisomal membrane and their subsequent β-oxidation within the peroxisome [19].

It is generally assumed that fatty acyl-CoA esters synthesized on the outer site of the peroxisomal membrane enter the peroxisome unhindered. Indeed, peroxisomal β-oxidation appears to show no requirement for carnitine and peroxisomes are devoid of carnitine palmitoyltransferase activity. The first step in peroxisomal β-oxidation is catalyzed by acyl-CoA oxidase, which, like the mitochondrial acyl-CoA dehydrogenases, contains FAD as prosthetic group [3,20–23]. Reoxidation of enzyme-bound $FADH_2$, however, occurs via direct interaction with molecular oxygen yielding H_2O_2. Acyl-CoA oxidase is not reactive with short-chain acyl-CoA esters such as butyryl-CoA and hexanoyl-CoA. As a consequence, fatty acyl-CoA esters are not oxidized to completion but only chain-shortened.

The next two reactions in peroxisomal β-oxidation are catalyzed by a bifunctional polypeptide harbouring both enoyl-CoA hydratase and L-3-hydroxyacyl-CoA dehydrogenase activities [24–26]. It is referred to as the bifunctional protein [24–26]. Finally, the last reaction in peroxisomal β-oxidation is catalyzed by peroxisomal 3-ketoacyl-CoA thiolase with a native MW of 89 kD (Table 1). This thiolase exhibits virtually no activity with acetoacetyl-CoA but is highly active with longer-chain substrates [27,28].

Although peroxisomes lack carnitine palmitoyl-CoA transferase activity, they do possess medium-chain carnitine acyltransferase activity [29,30] and carnitine acetyltransferase activity [29,31]. Farrell and Bieber [32] suggested that these enzymes are involved in the transport of the products of peroxisomal β-oxidation (acetyl-CoA plus medium-chain acyl-CoA esters) to the mitochondria for further oxidation.

Both mitochondria and peroxisomes contain the enzymic equipment to catalyze the β-oxidation of unsaturated fatty acids. The double bonds present in naturally occurring unsaturated fatty acids are usually of the *cis*-configuration. According to the classic concepts of Stoffel and co-workers [33,34], these double bonds are handled by the two enzymes, Δ^3-*cis*-Δ^2-*trans*-enoyl-CoA isomerase and 3-hydroxyacyl-CoA epimerase. It is now clear, however, that this scheme is no longer tenable. Instead, there is overwhelming evidence that degradation of unsaturated fatty acids requires the participation of 2,4-dienoyl-CoA reductase and Δ^3-*cis*-Δ^2-*trans*-enoyl-CoA isomerase (see [12,35] for a detailed discussion). The isomerase is required for the degradation of all fatty acids with double bonds extending from odd-numbered carbon atoms, whereas the reductase is necessary for the removal of all double bonds extending from even-numbered carbon atoms. Recent studies by Kunau and Schulz [35] have clearly shown that degradation of 2-*trans*-4-*cis*-decadienoyl-CoA occurs predominantly (more than 98%) via the

Table 1. Comparative aspects of the enzymes involved in mitochondrial and peroxisomal β-oxidation

	MW (kD)	Subunit structure	Ref.
Peroxisomes			
Acyl-CoA oxidase	139	A_2, ABC, B_2C_2	[22]
Bifunctional protein	78	Monomer	[25]
3-Ketoacyl-CoA thiolase	89	Dimer	[28]
Long-chain acyl-CoA synthetase	76	Monomer	[10]
2,4-Dienoyl-CoA reductase	120	Tetramer	[37]
2,3-Enoyl-CoA isomerase	60	Dimer	[39]
Carnitine octanoyl transferase	66	Monomer	[30]
Mitochondria			
Short-chain acyl-CoA dehydrogenase	169	Tetramer	[40]
Medium-chain acyl-CoA dehydrogenase	182	Tetramer	[40]
Long-chain acyl-CoA dehydrogenase	168	Tetramer	[40]
Enoyl-CoA hydratase	161	Hexamer	[25]
L-3-Hydroxyacyl-CoA dehydrogenase	65	Tetramer	[41]
3-ketoacyl-CoA thiolase	154	Tetramer	[27]
Long-chain acyl-CoA synthetase	76	Monomer	[10]
2,4-dienoyl-CoA reductase	124	Tetramer	[36]
2,3-enoyl-CoA isomerase	90	Dimer	[12]
Carnitine palmitoyltransferase	300	Tetramer	[30]
Carnitine acetyltransferase	67	Monomer	[42]

revised (reductase-dependent) pathway. This is true for mitochondria as well as peroxisomes [35]. The mitochondrial and peroxisomal 2,4-dienoyl-CoA reductases are different enzyme proteins [36,37] as are the mitochondrial and peroxisomal Δ^3-*cis*-Δ^2-*trans*-enoyl-CoA isomerases [38,39]. In Table 1 information about the mitochondrial and peroxisomal enzyme proteins involved in β-oxidation is given.

3 Induction of Peroxisomal β-Oxidation

One of the most striking features of the peroxisomal β-oxidation system is that its capacity can be increased several-fold by a multitude of substances. These include clofibrate and other hypolipidemic drugs, plasticizers such as di-(2-ethylhexyl)phthalate (DEHP), perfluorinated fatty acids [43], high fat diets [44], valproic acid [45], phytol [46] and other compounds. Furthermore, hyperthyroidism [47], vitamin E deficiency [48], cold adaptation [49], and adrenocorticotropin [50] have also been reported to induce peroxisomal β-oxidation. Mannaerts et al. [51] have shown that clofibrate stimulates not only peroxisomal β-oxidation but also mitochondrial β-oxidation, although to a much lower extent.

4 Energetic Aspects of Mitochondrial and Peroxisomal β-Oxidation

Mitochondrial β-oxidation is energetically much more favourable than peroxisomal β-oxidation. This is due to the fact that the energy released in the first and third step of mitochondrial β-oxidation is conserved yielding two and three molecules of ATP, respectively, whereas the energy released in the first step of peroxisomal β-oxidation is released as heat. The NADH generated at the third step of peroxisomal β-oxidation, however, can be oxidized and coupled to the synthesis of ATP by shuttling of the reducing equivalents into the mitochondria. It would thus be energetically favourable if oxidation of the major fatty acids palmitate and oleate would proceed in mitochondria rather than in peroxisomes. Available evidence suggests that this is, indeed, the case [51–53].

5 The Inborn Errors of Peroxisomal β-Oxidation

In recent years several inherited diseases in man have been recognized in which there is an impairment in peroxisomal β-oxidation. In the Zellweger syndrome and the other related peroxisome deficiency disorders, morphologically distinguishable peroxisomes are virtually absent. In these patients the peroxisomal β-oxidation enzyme proteins are deficient, as shown by immunoblotting [54–56]. In three other diseases, i.e. X-linked adrenoleukodystrophy, acyl-CoA oxidase deficiency and peroxisomal thiolase deficiency peroxisomes are present in normal amounts but there is a specific deficiency of the peroxisomal very long-chain fatty acyl-CoA synthetase [19], acyl-CoA oxidase [58] and peroxisomal thiolase [59], respectively (Table 2; see [60] for review). Interestingly, in X-linked adrenoleukodystrophy the very long-chain fatty acids activating activity present in microsomes was found to be normal [19]; this activity is apparently not able to support peroxisomal β-

Table 2. Biochemical abnormalities in the inborn errors of peroxisomal β-oxidation[a]

Metabolites in body fluids	Zellweger syndrome	X-linked adreno-dystrophy	Acyl-CoA oxidase deficiency	Peroxisomal thiolase deficiency
Very long-chain fatty acids	Elevated	Elevated	Elevated	Elevated
Di- and trihydroxycholestanoate	Elevated	Normal	Normal	Elevated
Pipecolic acid	Elevated	Normal	Normal	
Phytanic acid	Elevated	Normal	Normal	Normal
Peroxisomes				
In liver	(Virtually) absent	Present	Present	Present
Plasmalogen biosynthesis				
DHAPAT	Deficient	Normal	Normal	Normal
Alkyl DHAP synthase	Deficient	Normal	Normal	Normal
De novo plasmalogen biosynthesis	Impaired	Normal	Normal	Normal
Peroxisomal β-oxidation				
Activity (C26:0 as substrate)	Deficient	Deficient	Deficient	Deficient
Enzyme proteins				
Acyl-CoA oxidase	Deficient	Normal	Deficient	Normal
Bifunctional protein	Deficient	Normal	Normal	Normal
Peroxisomal thiolase	Deficient	Normal	Normal	Deficient

[a]For details the reader is referred to the information in [54–59]. Abbreviations: DHAP, dihydroxyacetone phosphate, DHAPAT, acyl-CoA: dihydroxyacetone phosphate acyltransferase.

oxidation. This suggests that the metabolism of very long-chain fatty acids is functionally compartmentalized within the cell and that the two synthetases have distinct functions, the peroxisomal enzyme providing acyl-CoA esters for β-oxidation in the peroxisome and the microsomal enzyme acyl-CoA esters for integration into cellular lipids. A similar conclusion was reached by Numa [61] in studies with the yeast *Candida lipolytica*.

6 Physiological Function of the Peroxisomal β-Oxidation System

In the last few years it has become clear that the peroxisomal β-oxidation system is not just an auxiliary system assisting mitochondrial β-oxidation in times of increased fatty acid supply, but is instead involved in the oxidative chain-shortening of various compounds. At present it is known that the following compounds are oxidised in peroxisomes.

6.1 Saturated Very Long-Chain Fatty Acids

Mainly through work from the group of Moser [62,63] it has become clear that the initial steps involved in the oxidation of tetracosanoic acid [lignoceric acid (C24:0)] and hexacosanoic acid [cerotic acid (C26:0)] proceed in peroxisomes as first envisaged by Bremer [64]. In a recent study by Singh et al. [17] C26:0 β-oxidation was, indeed, found to proceed almost exclusively in peroxisomes. Surprisingly, mitochondria from rat liver [17] and human fibroblasts [65] were found to oxidize the CoA-ester of hexacosanoic acid at an appreciable rate. These results imply that the compartmentation of the very long-chain fatty acyl-CoA esters, generated in the endoplasmic reticulum, prevents their utilization not only by peroxisomes ([19] and discussion therein) but also by mitochondria (see Numa [61]).

Since most of these studies were carried out in different subcellular fractions isolated from rat liver or cultured skin fibroblasts, we decided to try and obtain information on the relative contribution of peroxisomes and mitochondria to very long-chain fatty acid β-oxidation by carrying out experiments in intact cells. In various studies inhibitors of the mitochondrial electron transport chain such as KCN, antimycin and rotenone have been used for this purpose. However, inhibition of mitochondrial electron flow not only eliminates mitochondrial β-oxidation by blocking re-oxidation of enzyme-bound $FADH_2$ and NADH, generated at the first and third steps of β-oxidation, respectively, but also leads to lowered ATP levels [51]. Furthermore, re-oxidation of NADH generated during peroxisomal β-oxidation is blocked. In order to circumvent these difficulties we decided to use 2-[5-(4-chlorophenyl)-pentyl]-oxirane-2-carboxylate (POCA), a powerful inhibitor of carnitine palmitoyltransferase. This compound inhibits mitochondrial β-oxidation by blocking uptake of fatty acyl-CoA esters into mitochondria. The results of Table 3 show that oxidation of hexacosanoic acid is practically insensitive towards POCA, whereas oxidation of palmitate is inhibited progressively by increasing concentrations of POCA. These results indicate that at least the first cycle of hexacosanoic acid β-oxidation takes place in peroxisomes. Apparently, under in vivo conditions mitochondria contribute only little to C26:0

Table 3. Effect of increasing concentrations of POCA on the oxidation of palmitic acid and hexacosanoic acid in intact human skin fibroblasts[a]

Concentration of POCA added (mM)	Rate of fatty acid β-oxidation (pmol min^{-1} mg^{-1})	
	Palmitate	Hexacosanoate
0	87.2	2.54
0.01	60.4	2.41
0.05	30.3	2.48
0.10	25.5	2.61
0.50	9.7	2.43

[a] Cultured skin fibroblasts from a control subject were incubated in a standard reaction medium containing 2 mM carnitine plus 0.01 mM [1-^{14}C]palmitic acid or [1-^{14}C]-hexacosanoic acid for 60 min at 37°C and the production of radiolabelled CO_2 plus water-soluble products measured as described before [18].

β-oxidation despite the fact that in homogenates mitochondria are capable of oxidizing hexacosanoyl-CoA [17,65].

Further evidence in favour of a predominant role of peroxisomes in chain-shortening of very long-chain fatty acids (notably C24:0 and C26:0), is obtained from studies with fibroblasts from patients with a deficiency of one or more of the peroxisomal β-oxidation enzyme proteins (see Table 2). Indeed, in fibroblasts from patients with a virtually complete deficiency of acyl-CoA oxidase, oxidation of hexacosanoic acid is strongly impaired [58].

6.2 Mono- and Polyunsaturated Fatty Acids

Inspired by Bremer's suggestion that the peroxisomal β-oxidation system might be involved in the chain-shortening of certain substrates not easily oxidized by mitochondria [64], Christiansen [66] provided evidence that peroxisomes contribute significantly to the oxidation of erucic acid (C22:1), a finding subsequently confirmed (see [67] for discussion).

In 1981 Dommes et al. [68] reported the presence of 2,4-dienoyl-CoA reductase activity in peroxisomes, suggesting that peroxisomes have the capacity to β-oxidize polyunsaturated fatty acids. Studies with isolated peroxisomes and intact hepatocytes have clearly shown that polyunsaturated fatty acids can, indeed, be chain-shortened in peroxisomes (see e.g. [69]). Recent studies by Hovik and Osmundsen [70] on the β-oxidation of long-chain fatty acids with chain lengths of 20 and 22 carbon atoms and possessing from zero to six double bonds have shown that in general the peroxisomal β-oxidation system shows a preference for the more polyunsaturated fatty acids. The quantitative role of peroxisomes and mitochondria to the oxidation of these fatty acids remains to be established. At present we are studying the contribution of peroxisomes and mitochondria to the oxidation of various polyunsaturated fatty acids in rat liver and human skin fibroblasts. Use will be made of POCA, on the one hand, and mutant cell lines (e.g. acyl-CoA oxidase deficient fibroblasts), on the other hand, to address this question.

6.3 *Dicarboxylic Acids*

Available evidence suggests that the oxidation of long- and medium-chain dicarboxyl-CoA esters can occur both in mitochondria and peroxisomes [71,72], although the quantitative contribution of the two organelles has not been established. Activation of long-chain dicarboxylic acids occurs at the endoplasmic reticulum [73].

6.4 *Di- and Trihydroxycholestanoic Acid*

According to current concepts, the final steps in the biosynthesis of the bile acids, cholic acid and chenodeoxycholic acid, involves oxidative cleavage of the side chain of tri- and dihydroxycholestanoic acid, respectively [74]. Studies in the early 1960s already indicated that chain-shortening of trihydroxycholestanoic acid (THCA) is brought about via a β-oxidation process yielding propionyl-CoA and choloyl-CoA [75]. Although it was long believed that mitochondria are the site of THCA oxidation, it is now clear that this is primarily brought about in peroxisomes, as first shown by Pedersen and Gustafsson [76]. Further studies by Kase et al. [77,78], both in rat liver and human liver, have provided convincing evidence for the primary role of peroxisomes in the oxidative chain-shortening of THCA and DHCA. Additional in vivo and in vitro studies on bile acid formation in Zellweger patients are in line with this concept [79]. Indeed, we recently found that β-oxidation of trihydroxycholestanoyl-CoA is strongly deficient in livers from Zellweger patients (controls: 2.1–3.7 nmol min^{-1} g^{-1} (n = 3); Zellweger patients: $<$ 0.1 nmol min^{-1} g^{-1} (n = 3); Casteels and van Roermund, unpublished results).

In contrast to the situation observed in peroxisomal very long-chain fatty acid β-oxidation, where activation occurs at the peroxisomal membrane, activation of THCA and DHCA takes place in the endoplasmic reticulum [80], apparently followed by transport of the CoA-esters to the peroxisomes.

6.5 *Prostaglandins E2 and F2α*

When prostaglandins are injected into animals or man, the urinary metabolites consist mainly of the dinor and tetranor compounds, suggesting that β-oxidation has occurred. Experiments with crude mitochondrial fractions have, indeed, demonstrated that prostaglandins were subject to one or two β-oxidation cycles [81]. Since the mitochondrial fractions used by Hamberg [81] were strongly contaminated, Diczfalusy et al. [82] and Schepers et al. [83] reinvestigated the subcellular site of prostaglandin F2α and E2 β-oxidation, respectively. Both prostaglandins were found to be oxidized efficiently by isolated peroxisomes [82,83]. Prostaglandin E2 was also found to be oxidized by mitochondria [83]. Surprisingly, mitochondrial oxidation of prostaglandin E2 required no carnitine, suggesting that the CoA-ester is able to penetrate through the mitochondrial membrane. The relative contribution of peroxisomes and mitochondria to overall prostaglandin oxidation remains to be established.

6.6 *Xenobiotics with an Acyl Side Chain*

Recent evidence suggests that peroxisomes play an important role in the β-oxidation of the acyl side chain of xenobiotics, thus allowing them to be excreted as shortened, more polar compounds [84,85].

6.7 *Reflections on the Role of Peroxisomal β-Oxidation*

On the basis of the data described above, the picture emerges that the major naturally occurring fatty acids are primarily oxidized in mitochondria, thus giving rise to an optimal yield of ATP for diverse cellular processes. This is in line with the intricate control mechanisms in mitochondrial β-oxidation. Peroxisomal β-oxidation, on the other hand, which is not obligatorily coupled to the energy demands of the cell, is ideally suited to catalyze the chain-shortening of a variety of compounds due to its relatively simple organization, absence of an intricate transport mechanism, and rapid and extensive induction by a variety of agents. If this hypothesis is, indeed, true it can be expected that many compounds known to be β-oxidized are in fact β-oxidized in peroxisomes. To test this hypothesis we will study the β-oxidation of diverse compounds such as vitamin E [86] and leukotrienes [87].

Finally, peroxisomes may turn out to be very important in detoxification reactions, serving to catalyze the β-oxidative chain-shortening of various compounds; this can be followed by conjugation of the chain-shortened compounds, e.g. to glutathione, and removal of the products from the body via the bile. This hypothesis is currently under investigation in our laboratory. It is interesting to note in this respect that peroxisomes do contain glutathione S-transferase activity [88].

7 The Peroxisomal β-Oxidation System: One Set of β-Oxidation Enzyme Proteins?

In mitochondria several acyl-CoA dehydrogenases catalyze the dehydrogenation of acyl-CoA esters depending upon whether the CoA-ester involved is a long-chain, medium-chain or short-chain acyl-CoA. The question arises whether the same set of peroxisomal β-oxidation enzyme proteins is responsible for the β-oxidation of the diversity of compounds descrived above. There is some evidence to suggest that this is not true. Recent data indicate that different oxidases are involved in the chain-shortening of very long-chain fatty acids, on the one hand, and trihydroxycholestanoic acid, on the other hand. This is first concluded from the finding that THCA β-oxidation is normal in the patient characterized by an acyl-CoA oxidase deficiency, in contrast to the deficient oxidation of very long-chain fatty acids [58]. Recent studies by Casteels et al. [80] agree with this view. These authors carried out competition experiments in rat liver peroxisomes using THCA-CoA and palmitoyl-CoA and observed virtually no inhibition of THCA-CoA β-oxidation by palmitoyl-CoA [80]. Taken together, these results clearly show the involvement of two different oxidases. On the other hand, the finding of elevated levels of di- and trihydroxycholestanoic acid and very long-chain fatty acids in serum from a patient with a peroxisomal 3-oxoacyl-CoA thiolase deficiency [59], suggests the involvement of one thiolase at least in the degradation of these compounds.

Acknowledgements. The research described in this work was supported by grants from the Netherlands Organization for Scientific Research, under the auspices of the Netherlands Foundation for Medical and Health Research (MEDIGON) and the Princess Beatrix Fund (The Hague, The Netherlands). The help of Simone Majoor in the expert preparation of the manuscript is gratefully acknowledged.

References

1. Rhodin J (1954) Correlation of ultrastructural organisation and function in normal and experimentally changed proximal tubule cells of the mouse kidney. PhD Thesis, Aktiebolaget Godvil, Stockholm
2. De Duve C, Baudhuin P (1966) Peroxisomes (microbodies) and related particles. Physiol Rev 46:323–357
3. Lazarow P, de Duve C (1976) A fatty acyl-CoA oxidizing system in rat liver peroxisomes: enhancement by clofibrate, a hypolipidemic drug. Proc Natl Acad Sci USA 73:2043–2046
4. Cooper TG, Beevers H (1969) β-oxidation in glyoxysomes from castor bean endosperm. J Biol Chem 244:3514–3520
5. Kunau WH, Kionka C, Ledebur A, Mateblowski M, Moreno de la Garza M, Schultz-Borchard U, Thieringer R, Veenhuis M (1987) β-oxidation systems in eukaryotic microorganisms. In: Fahimi HD, Sies H (eds) Peroxisomes in biology and medicine. Springer, Berlin Heidelberg New York Tokyo, pp 128–140
6. Tanaka T, Hosaka T, Hoshimaru M, Numa S (1979) Purification and properties of long-chain acyl coenzyme A synthetase from rat liver. Eur J Biochem 98:165–172
7. Shindo Y, Hashimoto T (1978) Acyl-coenzyme A synthetase and fatty acid oxidation in rat liver peroxisomes. J Biochem 84:1177–1181
8. Krisans SK, Mortensen RM, Lazarow PB (1980) Acyl-CoA synthetase in rat liver peroxisomes. J Biol Chem 255:9599–9607
9. Mannaerts GP, Van Veldhoven P, Van Broekhoven A, Van de Broek G, De Beer LJ (1982) Evidence that peroxisomal acyl-CoA synthetase is located at the cytoplasmic site of the peroxisomal membrane. Biochem J 204:17–23
10. Miyazawa S, Hashimoto T, Yokota S (1985) Identity of long-chain acyl coenzyme A synthetase of microsomes, mitochondria, and peroxisomes in rat liver. J Biochem 98:723–733
11. Bremer J, Osmundsen H (1984) Fatty acid oxidation and its regulation. In: Numa S (ed) Fatty acid metabolism and its regulation. Elsevier, Amsterdam, pp 113–154
12. Schulz H (1985) Oxidation of fatty acids. In: Vance DE, Vance FE (eds) Biochemistry of lipids and membranes. Benjamin/Cummings, Menlo Park, CA, pp 116–142
13. Middleton B (1973) The oxoacyl CoA thiolases of animal tissues. Biochem J 132:717–730
14. Staack H, Binstock JF, Schulz H (1978) Purification and properties of a pig heart thiolase with broad chain length specificity and comparison of thiolases from heart and *Escherichia coli*. J Biol Chem 253:1827–1831
15. Bhusnan A, Singh RP, Singh I (1986) Characterization of rat brain microsomal acyl-coenzyme A ligase: different enzymes for the synthesis of palmitoyl-CoA and lignoceroyl-CoA. Arch Biochem Biophys 246:374–380
16. Wanders RJA, van Roermund CWT, van Wijland MJA, Schutgens RBH, Schram AW, van den Bosch H, Tager JM (1987) Studies on the peroxisomal oxidation of palmitate and lignocerate in rat liver. Biochim Biophys Acta 919:21–25
17. Singh H, Derwas N, Poulos A (1987) Very long chain fatty acid β-oxidation by rat liver mitochondria and peroxisomes. Arch Biochem Biophys 359:382–390
18. Wanders RJA, van Roermund CWT, van Wijland MJA, Schutgens RBH, Heikoop J, van den Bosch H, Schram AW, Tager JM (1987) Peroxisomal fatty acid β-oxidation in relation to the accumulation of very long chain fatty acids in peroxisomal disorders. J Clin Invest 80:1778–1783

19. Wanders RJA, van Roermund CWT, van Wijland MJA, Schutgens RBH, van den Bosch H, Schram AW, Tager JM (1988) Direct demonstration that the deficient oxidation of very long chain fatty acids in X-linked adrenoleukodystrophy is due to an impaired ability of peroxisomes to activate very long chain fatty acids. Biochem Biophys Res Commun 153:618–624
20. Osumi T, Hashimoto T (1978) Acyl-CoA oxidase of rat liver: a new enzyme for fatty acid oxidation. Biochem Biophys Res Commun 83:479–485
21. Inestrosa NC, Bronfman M, Leighton F (1979) Detection of peroxisomal fatty acyl-coenzyme A oxidase activity. Biochem J 182:779–788
22. Osumi T, Hashimoto T, Ui N (1980) Acyl-CoA oxidase of rat liver: a new enzyme for fatty acid oxidation. J Biochem 87;1735–1746
23. Inestrosa NC, Bronfman M, Leighton M (1980) Purification of the peroxisomal fatty acyl-Coa oxidase from rat liver. Biochem Biophys Res Commun 95:7–12
24. Osumi T, Hashimoto T (1979) Peroxisomal β-oxidation system of rat liver. Copurification of enoyl-CoA hydratase and 3-hydroxyacyl-CoA dehydrogenase. Biochem Biophys Res Commun 89:580–584
25. Furuta S, Miyazawa S, Osumi T, Hashimoto T, Ui N (1980) Properties of mitochondrial and peroxisomal enoyl-CoA hydratase from rat liver. J Biochem 88;1059–1070
26. Osumi T, Hashimoto T (1980) Purification and properties of mitochondrial and peroxisomal 3-hydroxyacyl-CoA dehydrogenase from rat liver. Arch Biochem Biophys 203:372–383
27. Miyazawa S, Osumi T, Hashimoto T (1980) The presence of a new 3-oxoacyl-CoA thiolase in rat liver peroxisomes. Eur J Biochem 103:589–596
28. Miyazawa S, Furuta S, Osumi T, Hashimoto T, Ui N (1981) Properties of peroxisomal 3-ketoacyl-Coa thiolase from rat liver. J Biochem 90:511–519
29. Markwell MAK, Tolbert NE, Bieber LL (1976) Comparison of the carnitine acyltransferase activities from rat liver peroxisomes and microsomes. Arch Biochem Biophys 176:479–488
30. Miyazawa S, Ozasu H, Osumi T, Hashimoto T (1983) Purification and properties of carnitine octanoyltransferase and carnitine palmitoyltransferase from rat liver. J Biochem 94:529–542
31. Markwell MAK, McGroarty EJ, Bieber LL, Tolbert NE (1973) The subcellular distribution of carnitine acyltransferases in mammalian liver and kidney. A new peroxisomal enzyme. J Biol Chem 248:3426–3452
32. Farrell SO, Bieber LL (1983) Carnitine octanoyltransferase of mouse liver peroxisomes: properties and effect of hypolipidemic drugs. Arch Biochem Biophys 222:123–132
33. Stoffel W, Ditzer R, Caesar H (1964) Der Stoffwechsel der ungesättigten Fettsäuren III. Zur β-Oxidation der Mono- und Polyenfettsäuren. Der Mechanismus der enzymatischen Reaktionen an Δ^3-cis-enoyl-CoA-Verbindungen. Hoppe-Seylers Z Physiol Chem 339:167–181
34. Stoffel W, Caesar H (1965) Der Stoffwechsel der ungesättigten Fettsäuren V. Zur β-Oxidation der Mono- und Polyenfettsäuren. Der mechanismus der enzymatischen Reaktionen an Δ^2-cis-enoyl-CoA-Verbindungen. Hoppe-Seylers Z Physiol Chem 341:76–83
35. Kunau WH, Schulz H (1987) β-oxidation of unsaturated fatty acids: a revised pathway. Trends Biochem Sci 12:403–406
36. Dommes V, Kunau WH (1984) Purification and properties of acyl coenzyme A dehydrogenases from bovine liver. J Biol Chem 259:1789–1797
37. Kimura C, Kondo A, Koeda N, Yamanaka H, Mizugaki M (1984) Studies on the metabolism of unsaturated fatty acids. XV. Purification and properties of 2,4-dienoyl-CoA reductase from rat liver peroxisomes. J Biochem 96:1463–1469
38. Kärki T, Hakkola E, Hassinen IE, Hiltunen JK (1987) β-oxidation of polyunsaturated fatty acids in peroxisomes. Subcellular distribution of Δ^3, Δ^2-enoyl-CoA isomerase activity in rat liver. FEBS Lett 215:228–232
39. Palosaari PM, Autio-Harmainen H, Sormunen R, Hassinen IE, Hiltunen JK (1988) β-oxidation of polyunsaturated fatty acids. Peroxisomal Δ^3-Δ^2-enoyl-CoA isomerase in rat liver. In: IUB-Congr, Prague (Abstr)
40. Furuta S, Miyazawa S, Hashimoto T (1981) Purification and properties of rat liver acyl-CoA dehydrogenases and electron transfer flavoprotein. J Biochem 90:1739–1750
41. Osumi T, Hashimoto T (1980) Purification and properties of mitochondrial and peroxisomal 3-hydroxyacyl-CoA dehydrogenase from rat liver. Arch Biochem Biophys 203:372–383
42. Miyazawa S, Ozasa H, Furuta S, Osumi T, Hashimoto T (1983) Purification and properties of carnitine acetyltransferase from rat liver. J Biochem 93:439–451

43. Ikeda Y, Dabrowski C, Tanaka K (1983) Separation and properties of five distinct acyl-CoA dehydrogenases from rat liver mitochondria. Identification of a new 2-methyl-branched chain acyl-CoA dehydrogenase. J Biol Chem 258:1066–1076
44. Neat CE, Thomassen MS, Osmundsen H (1981) Effects of high-fat diets on hepatic fatty acid oxidation in the rat. Biochem J 196:149–159
45. Horie S, Suga T (1985) Enhancement of peroxisomal β-oxidation in the liver of rats and mice treated with valproic acid. Biochem Pharmacol 34:1357–1362
46. Van den Branden C, Vamecq J, Wijbo I, Roels F (1986) Phytol and peroxisome proliferation. Pediatr Res 20:411–415
47. Just WW, Hartl FU (1983) Stimulation of peroxisomal fatty acid β-oxidation by thyroid hormones. Hoppe-Seylers Z Physiol Chem 264:1541–1547
48. Reddy JK, Lalwani ND, Dabholkar AS, Reddy MK, Qureshi SA (1981) Proliferation of peroxisomes and vitamin E deficiency. Biochem Int 3:41–49
49. Nedergaard J, Alexson S, Cannon B (1980) Cold adaptation in the rat: increased brown fat peroxisomal β-oxidation relative to maximal mitochondrial oxidative capacity. Am J Physiol 239:C208–C216
50. Black VA, Russo JJ (1980) Hormone-dependent changes in peroxisomal enzyme activity in guinea pig adrenal. Am J Anat 159:85–120
51. Mannaerts GP, DeBeer LJ, Thomas J, De Schepper PJ (1979) Mitochondrial and peroxisomal fatty acid oxidation in liver homogenates from control and clofibrate treated rats. J Biol Chem 254:4585–4595
52. Ide T, Ontko JA (1981) Increased secretion of very low density lipoprotein triglyceride following inhibition of long chain fatty acid oxidation in isolated rat liver. J Biol Chem 256:10247–10255
53. Foerster EC, Führenkemper T, Rabe U, Graf P, Sies H (1981) Peroxisomal fatty acid oxidation as detected by H_2O_2 production in intact perfused liver. Biochem J 196:705–712
54. Tager JM, Ten Harmsen van de Beek WA, Wanders RJA, Hashimoto T, Heymans HSA, van den Bosch H, Schutgens RBH, Schram AW (1985) Peroxisomal β-oxidation enzyme proteins in the Zellweger syndrome. Biochem Biophys Res Commun 126:1269–1275
55. Lazarow PB, Black V, Shio H, Fujiki Y, Hajra AK, Datta NS, Bangaru BS, Dancis J (1985) Zellweger syndrome: biochemical and morphological studies on two patients treated with clofibrate. Pediatr Res 19:1356–1364
56. Suzuki Y, Orii T, Mori M, Tatibana M, Hashimoto T (1986) Deficient activities and proteins of peroxisomal β-oxidation enzymes in infants with Zellweger syndrome. Clin Chim Acta 156:191–196
57. Chen WW, Watkins PP, Osumi T, Hashimoto T, Moser HW (1987) Peroxisomal β-oxidation enzyme proteins in adrenoleukodystrophy: distinction between X-linked and neonatal adrenoleukodystrophy. Proc Natl Acad Sci USA 84:1425–1428
58. Poll-Thé BT, Roels F, Ogier H, Scotto J, Vamecq J, Schutgens RBH, Wanders RJA, van Roermund CWT, van Wijland MJA, Schram AW, Tager JM, Saudubray JM (1988) A new peroxisomal disorder with enlarged peroxisomes and a specific deficiency of acyl-CoA oxidase (pseudoneonatal adrenoleukodystrophy). Am J Human Genet 42:422–434
59. Schram AW, Goldfischer S, van Roermund CWT, Brouwer-Kelder EM, Collins J, Hashimoto T, Heymans HSA, van den Bosch H, Schutgens RBH, Tager JM, Wanders RJA (1987) Human peroxisomal 3-oxoacyl-coenzyme A thiolase deficiency. Proc Natl Acad Sci USA 84:2494–2497
60. Wanders RJA, Heymans HSA, Schutgens RBH, Barth PG, van den Bosch H, Tager JM (1988) Peroxisomal disorders in neurology. J Neurol Sci 88:1–39
61. Numa S (1981) Two long-chain acyl coenzyme A synthestases: their different roles in fatty acid metabolism and its regulation. Trends Biochem Sci 6:113–115
62. Kawamura N, Moser HW, Kishimoto Y (1981) Very long chain fatty acid oxidation in rat liver. Biochem Biophys Res Commun 99:1216–1225
63. Singh I, Moser AB, Goldfischer S, Moser HW (1984) Lignoceric acid is oxidized in the peroxisome: implications for the Zellweger cerebrohepato-renal syndrome and adrenoleukodystrophy. Proc Natl Acad Sci USA 81:4203–4207
64. Bremer J (1977) Carnitine and its role in fatty acid metabolism. Trends Biochem Sci 2:207–209
65. Singh H, Derwas N, Poulos a (1987) β-oxidation of very long chain fatty acids and their coenzyme A derivatives by human skin fibroblasts. Arch Biochem Biophys 254:526–533
66. Christiansen RZ (1978) The effect of clofibrate feeding on hepatic fatty acid metabolism. Biochim Biophys Acta 530:314–324

67. Bremer J, Norum KR (1982) Metabolism of very long chain monounsaturated fatty acids (C22:1) and the adaptation to their presence in the diet. J Lipid Res 23:243–256
68. Dommes V, Baumgart C, Kunau WH (1981) Degradation of unsaturated fatty acids in peroxisomes. J Biol Chem 256:8259–8262
69. Hiltunen JK, Kärki T, Hassinen IE, Osmundsen H (1986) β-oxidation of polyunsaturated fatty acids by rat liver peroxisomes. A role for 2,4-dienoyl-coenzyme A reductase in peroxisomal β-oxidation. J Biol Chem 261:16484–16493
70. Hovik R, Osmundsen H (1987) Peroxisomal β-oxidation of long-chain fatty acids possessing different extents of unsaturation. Biochem J 247:531–535
71. Kolvraa S, Gregersen N (1986) In vitro studies on the oxidation of medium chain dicarboxylic acids in rat liver. Biochim Biophys Acta 876:515–525
72. Vamecq J, Draye JP (1987) Interaction between the and β-oxidation of fatty acids. J Biochem 102:225–234
73. Vamecq J, de Hoffmann E, Van Hoof F (1985) The microsomal dicarboxylyl-CoA synthetase. Biochem J 230:683–693
74. Danielsson H, Sjövall J (1975) Bile acid metabolism. Annu Rev Biochem 44:233–253
75. Masui T, Staple E (1966) The formation of bile acids from cholesterol. J Biol Chem 241:3889–3893
76. Pedersen JI, Gustafsson J (1980) Conversion of 3α, 7α, 12α-trihydroxy-5β-cholestanoic acid into cholic acid by rat liver peroxisomes. FEBS Lett 121:345–348
77. Kase BF, Björkhem I, Pedersen JI (1983) Formation of cholic acid from 3α, 7α, 12α-trihydroxy-5β-cholestanoic acid by rat liver peroxisomes. J Lipid Res 24:1560–1567
78. Kase BF, Prydz K, Björkhem I, Pedersen JI (1986) In vitro formation of bile acids from di- and trihydroxy-5β-cholestanoic acid in human liver peroxisomes. Biochim Biophys Acta 877:37–42
79. Kase BF, Pedersen JI, Strandvik B, Björkhem I (1985) In vivo and in vitro studies on the formation of bile acids in patients with the Zellweger syndrome. Evidence that peroxisomes are of importance in the normal biosynthesis of both cholic acid and chenodeoxycholic acid. J Clin Invest 76:2393–2402
80. Casteels M, Schepers L, Van Eldere J, Eyssen H, Mannaerts GP (1988) Inhibition of 3α, 7α, 12α-trihydroxy-5β-cholestanoic acid oxidation and of bile acid secretion in rat liver by fatty acids. J Biol Chem 263:4654–4661
81. Hamberg M (1968) Metabolism of prostaglandins in rat liver mitochondria. Eur J Biochem 6:135–146
82. Diczfalusy U, Alexson SEH, Pedersen JI (1987) Chain-shortening of prostaglandin F2α by rat liver peroxisomes. Biochem Biophys Res Commun 144:1206–1215
83. Schepers L, Casteels M, Vamecq J, Parmentier G, Van Veldhoven PP, Mannaerts GP (1988) β-oxidation of the carboxyl side chain of prostaglandin E2 in rat liver peroxisomes and mitochondria. J Biol Chem 263:2724–2731
84. Yamada J, Horie S, Watanabe T, Suga T (1984) Participation of peroxisomal β-oxidation system in the chain-shortening of a xenobiotic acyl compound. Biochem Biophys Res Commun 125:123–128
85. Yamada J, Itoh S, Horie S, Watanabe T, Suga T (1986) Chain-shortening of a xenobiotic acyl compound by the peroxisomal β-oxidation system in rat liver. Biochem Pharmacol 35:4363–4368
86. Simon EJ, Gross CS, Milhorat AT (1985) The metabolism of vitamin E. J Biol Chem 221:797–805
87. Stene DO, Murphy RC (1988) Metabolism of leukotriene E4 in isolated rat hepatocytes. J Biol Chem 263:2773–2778
88. Tsukamoto T, Ohno K, Yokota S, Fujiki Y (1987) Biogenesis of 15 kD integral membrane protein of rat liver peroxisomes, a common endomembrane protein. In: Tada K, Orii T, Hashimoto T, Fujiki Y (eds) 4th Int Congr Inborn errors of metabolism Abstr, p 20

Genetic Heterogeneity in Inherited Disorders with a Generalized Impairment of Peroxisomal Functions: Visualization by Immunofluorescence Microscopy of Peroxisome Assembly After Somatic Cell Fusion of Complementary Cell Lines

S. BRUL[1], E.A.C. WIEMER[1], M. OOSTHUIZEN[1], A. WESTERVELD[2], R. VAN DRIEL[1], A. STRIJLAND[1], R.B.H. SCHUTGENS[3], H. VAN DEN BOSCH[4], R.J.A. WANDERS[3], A.W. SCHRAM[1] and J.M. TAGER[1]

1 Introduction

The cerebro-hepato-renal (Zellweger) syndrome is a rare familial disease that is characterized by a variety of biochemical and clinical aberrations [1–6]. The most striking biochemical abnormality in this disorder is the absence of morphologically distinguishable peroxisomes [7], and the deficiency of several peroxisomal enzymes [2–6] and enzyme proteins [8]. In recent years other inherited diseases in man have been recognized in which there is a generalized impairment of peroxisomal functions [2–6,9,10]. These diseases include the infantile form of Refsum disease (IRD), the neonatal form of adrenoleukodystrophy (NALD) and hyperpipecolic acidaemia [2–6,9,10]. The existence of such diseases, which usually result in early death of the patients, stresses the importance of peroxisomes in cellular metabolism [9,10].

The role of peroxisomes in mammalian metabolism has long remained an enigma. Only recently has it been shown that essential steps in the β-oxidation of very long-chain fatty acids [11–13] and prostaglandins [14], in the biosynthesis of ether phospholipids [15] and bile acids [16,17], in the catabolism of pipecolic acid [18,19] and, in man, in the catabolism of glyoxylate [20,21], take place in these organelles.

We have shown by complementation analysis after somatic cell fusion that fibroblasts from patients deficient in peroxisomes can be divided into at least four complementation groups [22], indicating that at least four gene products are required for peroxisome biogenesis. Complementation was assessed in the heterokaryons by measuring the increase in activity of acyl-CoA: dihydroxyacetone-phosphate acyltransferase, which is a peroxisomal enzyme [15] and is deficient in the parental cell lines [22]. A second biochemical method of assessing complementation is to measure the intracellular localization of catalase in the hete-

[1]Laboratory of Biochemistry, University of Amsterdam, P.O. Box 20151, 1000 HD Amsterdam, The Netherlands
[2]Division of Cell Biology and Genetics, Erasmus University Rotterdam, P.O. Box 1738, 3000 DR Rotterdam, The Netherlands
[3]Department of Pediatrics, University Hospital University of Amsterdam, Meibergdreef 9, 1105 AZ Amsterdam, The Netherlands
[4]Laboratory of Biochemistry, University of Utrecht, Padualaan 8, 3584 CH Utrecht, The Netherlands

A. Azzi, Z. Drahota and S. Papa
Molecular Basis of Membrane-Associated Diseases

rokaryons. After fusion of complementary cell lines deficient in peroxisomes, catalase, which is initially present in the cytosol, is incorporated into particles, presumably peroxisomes [22,23].

In this chapter we report on the visualization of the complementation process by immunofluorescence microscopy.

2 Materials and Methods

2.1 Cell Lines

Cultured skin fibroblasts were obtained from patients with the cerebro-hepato-renal syndrome (cell line GOM85AD), the neonatal form of adrenoleukodystrophy (cell line AAL85AD) and the infantile form of Refsum disease (cell line BOV84AD) (see [22]).

2.2 Materials

Ham F-10 medium and Dulbecco's modified Eagle's medium (DMEM) were obtained from Gibco (Glasgow, UK). Polyethylene glycol (mol. wt. 1000) was purchased from Merck (Darmstadt, FRG), biotinylated donkey anti-(rabbit IgG) (biotin-DAR) and streptavidin-fluorescein isothiocyanate (streptavidin-FITC) from Amersham (UK). In order to visualize catalase molecules in human fibroblasts, an anti-(bovine liver catalase) antiserum was used. As shown previously [8], the antibodies to the bovine liver enzyme cross-react with the homologous human protein.

2.3 Methods

Cells were cultured in a 1:1 mixture of Ham F-10 and DMEM containing 10% (v/v) foetal calf serum. The gas phase was 5% CO_2-95% air. Cells grown either on cover slips or in 25 cm^2 culture flasks were fused with polyethyleneglycol, exactly as described in [22]. Particle-bound catalase was assayed biochemically as described in [24].

The procedure for staining fibroblasts with the anti-(catalase) antiserum was as follows. Cells grown on cover slips were washed three times with phosphate-buffered saline (PBS) and incubated for 20 min with 2% (w/v) paraformaldehyde in PBS containing 0.2% (v/v) Triton X-100. Subsequently, the cover slips were washed three times with PBS-Triton X-100, and incubated for 10 min with PBS containing 0.2% (v/v) Triton X-100 and 0.1 M NH_4Cl in order to block free aldehyde groups. After this the cover slips were extensively washed with PBS and either stored in PBS plus sodium azide and protease inhibitors or stained immediately with the anti-(catalase) antiserum according to the following protocol. Fixed cells on cover slips were incubated for 30 min in a 30 μl drop of PBS containing 10 mg/ml bovine serum albumin and a 1:250 dilution of the primary anti-(catalase) antiserum. Next, the cover slip was washed three times with PBS and incubated for 30 min in PBS-bovine serum albumin containing a 1:100 dilution of biotin-DAR. After washing the cover slip three times with PBS, it was finally incubated for 30 min in a 1:100 dilution of streptavidin-FITC in PBS-bovine serum albumin. The cover slip was washed extensively with PBS and mounted on a microscope slide in a mounting medium consisting of 9 vol of glycerol, 1 vol of PBS and 1 mg/ml paraphenylenediamine. The cover slip was sealed with nail varnish and examined in a Leitz fluorescence microscope with a No. 4 blue filter (excitation maximum 466 nm, emission maximum 530 nm).

3 Results and Discussion

3.1 Biochemical Assessment of Complementation

Table 1 shows measurements of the percentage particle-bound catalase at different time intervals after fusion of the Zellweger cell line GOM85AD with the neonatal adrenoleukodystrophy cell line AAL85AD, and of GOM85AD with the infantile Refsum disease cell line BOV84AD. Cell line BOV84AD belongs to complementation group 2, GOM85AD to complementation group 3 and AAL85AD to complementation group 4 [22]. These biochemical observations show that complementation between cell lines GOM85AD and AAL85AD is rapid and cycloheximide-insensitive, in accordance with earlier results obtained with another combination of cell lines [23]. Thus, after fusion of these cell lines the assembly of peroxisomes is not dependent on protein synthesis, indicating that the components required for the assembly must be present in a stable form in the parental cell lines. In contrast, in fusions between cell lines GOM85AD and BOV84AD formation of catalase-positive particles is cycloheximide-sensitive, indicating that it is dependent on protein synthesis.

Table 1. Kinetics of the increase in particle-bound catalase after fusion of complementary peroxisome-deficient cell lines[a]

Cell lines fused	Time after fusion (h)	Particle-bound catalase (% of total)	
		No cycloheximide	+ Cycloheximide
ZS2 × NALD	1.5	28	22
	3	35	36
	6	42	46
ZS2 × IRD	24	46	48
	1.5	10	< 5
	3	25	< 5
	6	60	< 5
	24	48	< 5
NALD × NALD	6	< 5	
ZS2 × ZS2	6	< 5	
IRD × IRD	6	< 5	
Control × control	6	65	

[a] Human fibroblasts were fused exactly as described in [22]. The multinucleate cells were harvested at intervals after fusion and the percentage of particle-bound catalase was assessed as described in [24]. Cell lines: NALD: AAL85AD, complementation group 4; ZS2: GOM85AD, complementation group 3; IRD: BOV84AD, complementation group 2 [22].

3.2 Visualization of Peroxisomes by Indirect Immunofluorescence

When control fibroblasts were immunolabelled with anti-(catalase) as described in Section 2, a punctate fluorescence was observed (Fig. 1A), representing catalase present predominantly in peroxisomes (cf. [25–27]). A similar pattern of punctate fluorescence was obtained in multinucleate cells arising after treatment of control cells with polyethyleneglycol (Fig. 1B). In cells from a patient with the Zellweger syndrome (cell line GOM85AD) a diffuse fluorescence was seen (Fig. 1C), in accordance with the fact that catalase is present in the cytosol and is not concentrated in particles in such cell lines; a diffuse fluorescence was also observed in multinucleate cells obtained in a self-fusion control experiment (Fig. 1D). Similar results were obtained with all the other peroxisome-deficient cell lines tested (not shown).

3.3 Morphological Assessment of Complementation

In the experiment of Fig. 2 heterokaryons formed by fusion of cell lines GOM85AD and AAL85AD were cultured either in the absence (Fig. 2A) or in the presence (Fig. 2B) of cycloheximide and stained with the anti-(catalase) antiserum at different time intervals after fusion. Each panel in the figure shows a typical example of about 100 heterokaryons examined. The fusion efficiency, defined as the percentage of total nuclei in multinucleate cells, was about 80%. There was an increase with time up to 3–6 h in the extent of punctate fluorescence in the fused cells. The emergence of the punctate fluorescence labelling pattern correlated with the increase in particle-bound catalase as assessed biochemically (Table 1), and therefore clearly represents peroxisome assembly.

Culture in the presence of cycloheximide of heterokaryons formed by fusion of the cell lines GOM85AD and AAL85AD had no effect on the emergence of punctate fluorescence (Fig. 2B) in agreement with the biochemical observations (Table 1). This indicates that the components required for complementation to occur must have been present in the parental cell lines, at least one of which must have contained peroxisomal ghost-like structures (see [24]). The presence of such structures has been inferred from the results of immunoblotting experiments, indicating that peroxisomal integral membrane proteins are not deficient in the Zellweger syndrome ([28]; and A.W. Schram and W. Just, unpublished observations). Santos et al. [27] have recently provided immunocytochemical evidence for the presence of peroxisomal ghosts in fibroblasts from patients with the Zellweger syndrome. Recently we also obtained immunocytochemical evidence for the existence of such ghosts in fibroblasts from patients lacking functional peroxisomes.

In heterokaryons formed after fusion of GOM85AD and BOV85AD fibroblasts a clear increase in particle-bound catalase was observed only in the absence of cycloheximide (Table 1; see also [24]). Similarly, punctate fluorescence was only observed in the heterokaryons if cycloheximide was omitted from the culture medium (not shown). Thus, with this combination of cell lines protein synthesis is required for complementation to occur.

Application of the immunofluorescence technique in complementation studies has two important advantages as compared with the biochemical meas-

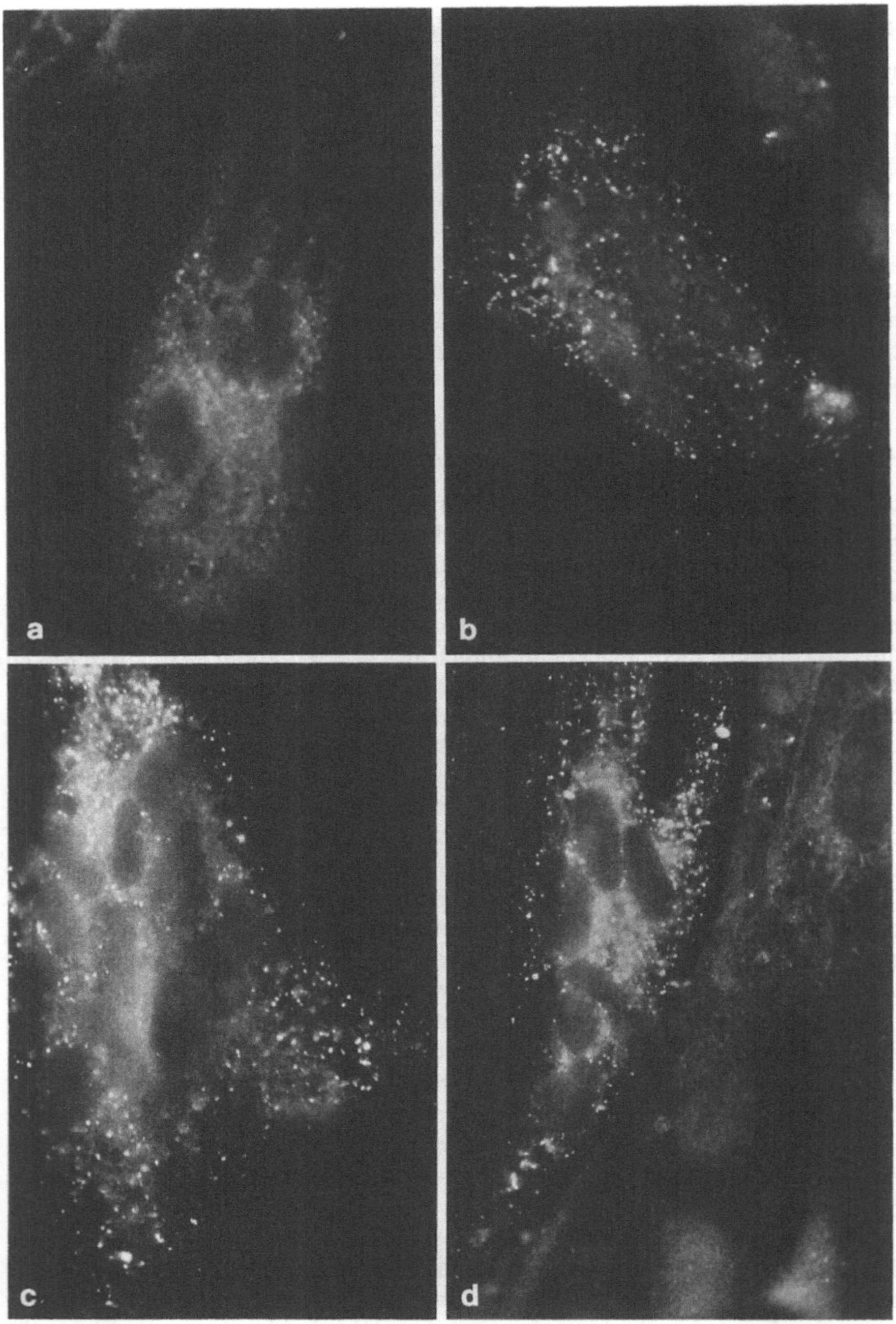

Fig. 1A–D. Visualization of peroxisomes using indirect immunofluorescence. Fibroblasts (**A,C**) or multinucleate cells obtained 6 h after treatment of the fibroblasts with polyethylene glycol (**B,D**) were processed for immunofluorescence microscopy with an anti-(catalase) antiserum, as described in section 2. **A, B** Fibroblasts from a control subject. **C, D** Fibroblasts from a Zellweger patient (GOM85AD) from complementation group 3 (see [22])

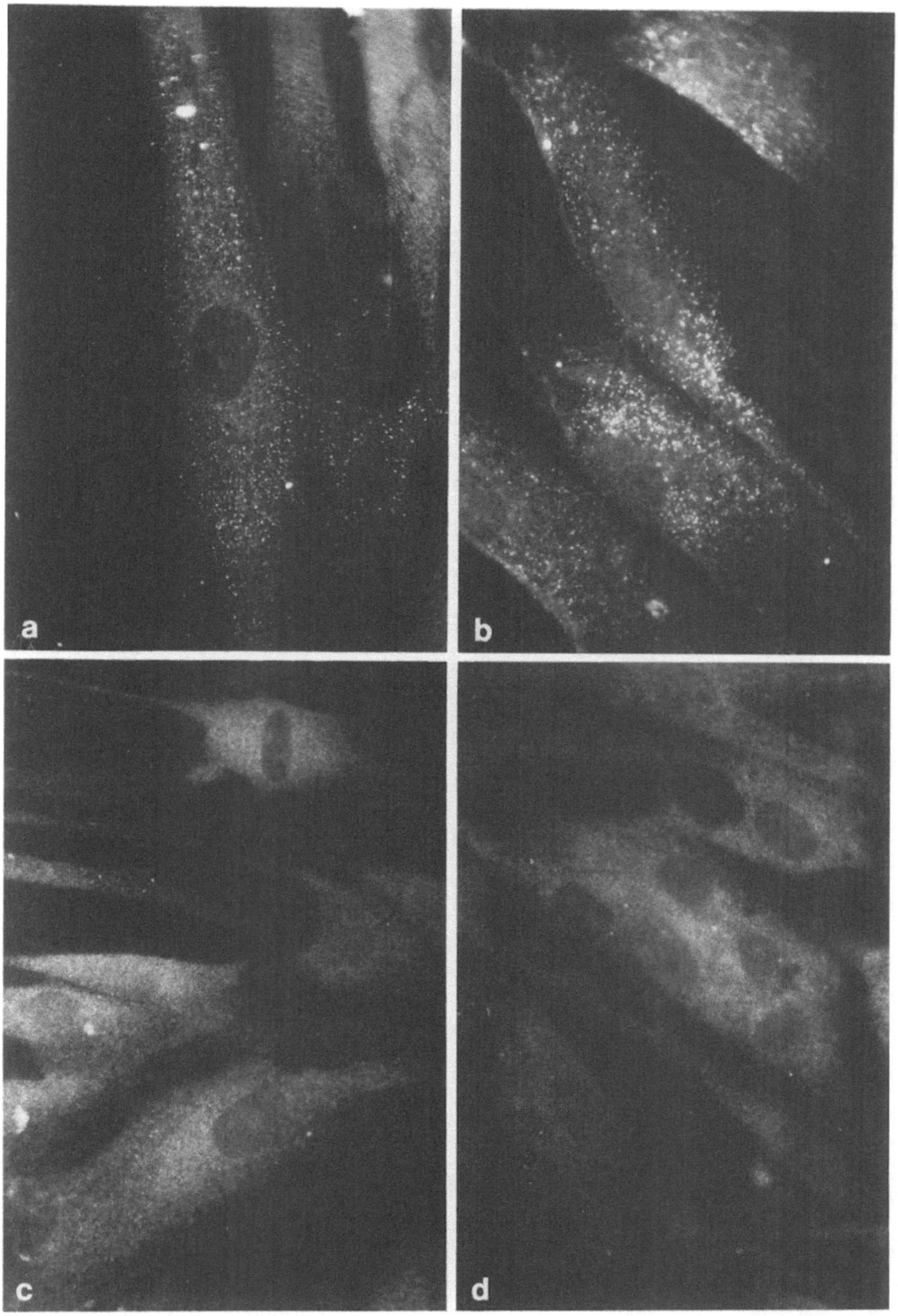

Fig. 2(pp. 425 and 426). Visualization by immunofluorescence of the assembly of peroxisomes after fusion of complementary cell lines deficient in peroxisomes. Fibroblast cell lines NALD (AAL85AD) and ZS2 (GOM85AD) were fused with polyethyleneglycol as described in [22]. After fusion, the heterokaryons were fixed at the time intervals indicated with 2% (w/v) paraformaldehyde and 0.2% (v/v) Triton X-100. The fixed multinucleate cells were processed further for immunofluorescence microscopy using an anti-(catalase) antiserum as described in Section 2. The fused cells were cultured either in the absence (p. 425) or in the presence (p. 426) of 2.8 mg/ml cycloheximide, and fixed after 1.5 h (**A**), 3 h (**B**), 6 h **C**) or 24 h (**D**)

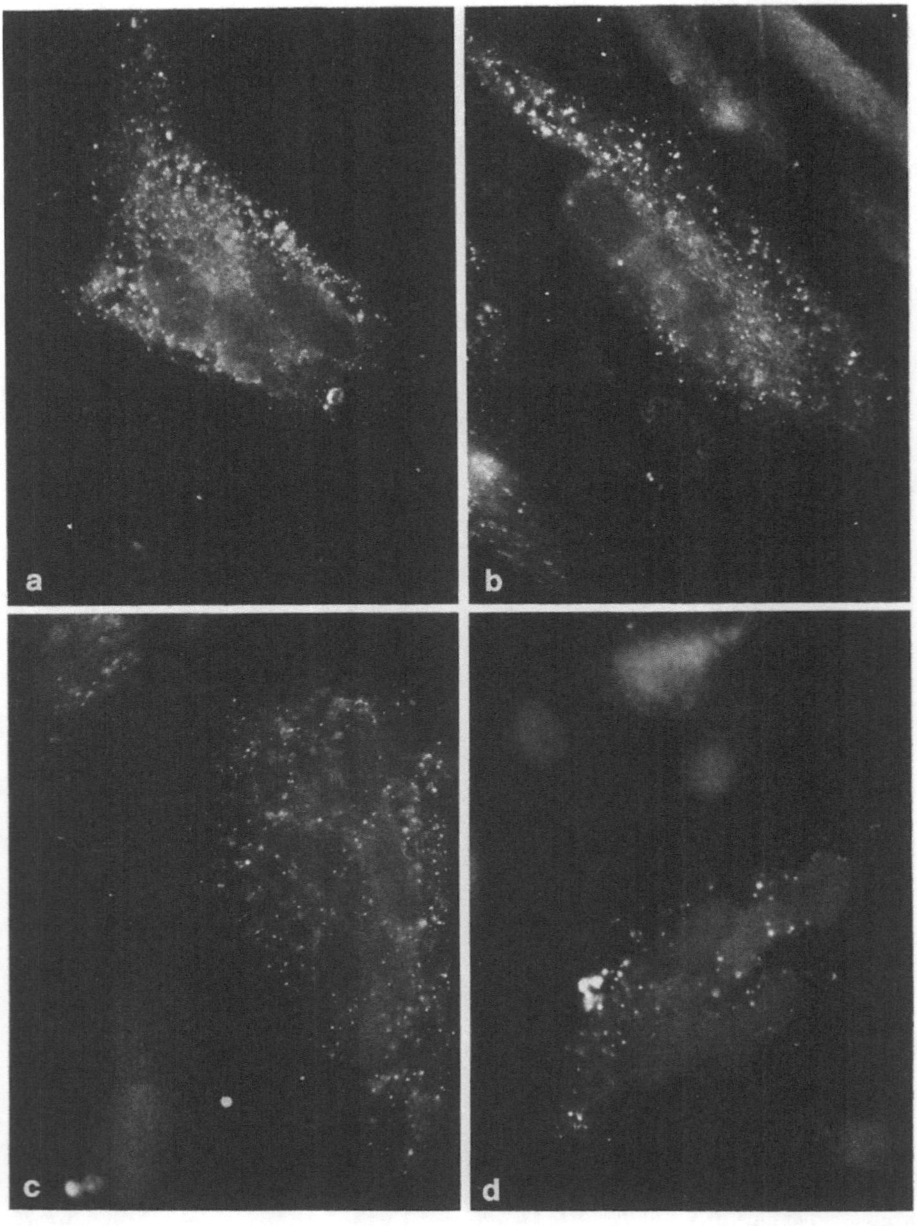

Fig. 2 (continued). Visualization by immunofluorescence of the assembly of peroxisomes after fusion of complementary cell lines deficient in peroxisomes. Fibroblast cell lines NALD (AAL85AD) and ZS2 (GOM85AD) were fused with polyethyleneglycol as described in [22]. After fusion, the heterokaryons were fixed at the time intervals indicated with 2% (w/v) paraformaldehyde and 0.2% (v/v) Triton X-100. The fixed multinucleate cells were processed further for immunofluorescence microscopy using an anti-(catalase) antiserum as described in Section 2. The fused cells were cultured either in the absence (p. 425) or in the presence (p. 426) of 2.8 mg/ml cycloheximide, and fixed after 1.5 h (**A**), 3 h (**B**), 6 h **C**) or 24 h (**D**)

urements we have employed in previous studies [22,23]. Firstly, it should enable one to visualize the assembly of peroxisomes, in particular the conversion of peroxisomal ghosts to functional peroxisomes able to incorporate catalase into the particles and thus able to confer structure-linked latency on the enzyme (see [22]). Secondly, the immunofluorescence technique provides a means of assaying single cells after microinjection of putative complementation factors. This should facilitate identification and purification of the gene products deficient in the Zellweger syndrome and similar disorders. Studies along these lines are in progress.

Acknowledgements. The authors wish to thank Hans Geuze, Piet Borst and Gilbert Keller for helpful discussions and Wendy van Noppen, Marcel Dutrieux and Elsbeth Vlugt-van Daalen for their help in the preparation of this manuscript. Supported by a grant from the Netherlands Organization for Scientific Research (NWO) under the auspices of the Netherlands Foundation for Medical and Health Research (MEDIGON).

References

1. Kelley RI (1983) Review: the cerebro-hepato-renal syndrome of Zellweger, morphologic and metabolic aspects. Am J Med Genet 16:503–517
2. Goldfischer S, Reddy JK (1984) Peroxisomes (microbodies) in cell pathology. Int Rev Exp Pathol 26:45–84
3. Moser HW (1987) New approaches in peroxisomal disorders. Dev Neurosci 9:1–18
4. Zellweger H (1987) The cerebro-hepato-renal (Zellweger) syndrome and other peroxisomal disorders. Dev Med Child Neurol 29:821–829
5. Schutgens RBH, Heymans HSA, Wanders RJA, Van Den Bosch H, Tager JM (1986) Review. Peroxisomal disorders: a newly recognized group of genetic diseases. Eur J Pediatr 144:430–440
6. Schutgens RBH, Wanders RJA, Schram AW, Van Den Bosch H, Heymans HSA, Schrakamp G, Tager JM (1987) Peroxisomal disorders: biochemical characteristics and genetic relationship between the different diseases. In: Vogel F, Sperling K (eds) Human genetics. Springer, Berlin Heidelberg New York Tokyo, pp 360–368
7. Goldfischer S, Moore CL, Johnson AB, Spiro AJ, Valsamis MP, Wisnewski HK, Ritch RH, Norton WT, Rapin I, Gartner LM (1973) Peroxisomal and mitochondrial defects in the cerebro-hepato-renal syndrome. Science 182:62–64
8. Tager JM, Ten Harmsen Van Der Beek WA, Wanders RJA, Hashimoto T, Heymans HSA, Van Den Bosch H, Schutgens RBH, Schram AW (1985) Peroxisomal β-oxidation enzyme proteins in the Zellweger syndrome. Biochem Biophys Res Commun 126:1269–1275
9. Goldfischer S, Collins J, Rapin I, Coltoff-Schiller B, Chang C-H, Nigro M, Black VH, Javitt NB, Moser HW, Lazarow PB (1985) Peroxisomal defects in neonatal-onset and X-linked adrenoleukodystrophies. Science 227:67–70
10. Scotto JM, Hadchouel M, Odievre M, Landat MH, Saudubray JM, Dulac O, Beucler I, Bean P (1982) Infantile phytanic acid storage disease, a possible variant of Refsum's disease: three cases including ultrastructural studies of the liver. J Inherit Metab Dis 5:83–90
11. Lazarow PB, De Duve C (1976) A fatty acyl-CoA oxidizing system in rat liver peroxisomes; enhancement by clofibrate, a hypolipidemic drug. Proc Natl Acad Sci USA 73:2043–2046
12. Bremer J (1977) Carnitine and its role in fatty acid metabolism. Trends Biochem Sci 2:207–209
13. Singh I, Moser AE, Goldfischer S, Moser HW (1984) Lignoceric acid is oxidized in the peroxisome: implications for the Zellweger cerebro-hepato-renal syndrome and adrenoleukodystrophy. Proc Natl Acad Sci USA 81:4203–4207
14. Diczfalusy U, Alexson SEH, Pedersen JI (1987) Chain-shortning of prostaglandin $F_2\alpha$ by rat liver peroxisomes. Biochem Biophys Res Commun 144:1206–1213
15. Hajra AK, Bishop JE (1982) Glycerolipid biosynthesis in peroxisomes via the acyl dihydroxyacetonephosphate pathway. Ann NY Acad Sci 386:170–182

16. Pedersen JI, Gustafsson J (1980) Conversion of 3α, 7α, 12α-trihydroxy-5β-cholestanoic acid into cholic acid by rat liver peroxisomes. FEBS Lett 121:345–348
17. Hagey LR, Krisans SK (1982) Degradation of cholesterol to propionic acid by rat liver peroxisomes. Biochem Biophys Res Commun 107:834–841
18. Wanders RJA, Romeyn GJ, van Roermund CWT, Schutgens RBH, Van Den Bosch H, Tager JM (1988) Identification of L-pipecolic oxidase in human liver and its deficiency in the Zellweger syndrome. Biochem Biophys Res Commun (in press)
19. Mihalik SJ, Rhead WJ (1988) Pipecolic acid is oxidized to alpha aminoadipic acid in the peroxisome of man and monkey. Trans Am Soc Neurochem 19:72 (Abstr)
20. Noguchi T, Takada Y (1979) Peroxisomal localization of alanine: glyoxylate aminotransferase in human liver peroxisomes. Biochem Biophys Res Commun 144:1206–1213
21. Danpure CJ, Jennings PR (1986) Peroxisomal alanine: glyoxylate aminotransferase deficiency in primary hyperoxaluria type 1. FEBS Lett 201:20–24
22. Brul S, Westerveld A, Strijland A, Wanders RJA, Schram AW, Heymans HSA, Schutgens RBH, Van Den Bosch H, Tager JM (1988) Genetic heterogeneity in the cerebro-hepato-renal (Zellweger) syndrome and other inherited disorders with a generalized impairment of peroxisomal functions: a study using complementation analysis. J Clin Invest 81:1710–1715
23. Brul S, Wiemer EAC, Westerveld A, Strijland A, Wanders RJA, Schram AW, Heymans HSA, Schutgens RBH, Van Den Bosch H, Tager JM (1988) Kinetics of the assembly of peroxisomes after fusion of complementary cell lines from patients with the cerebro-hepato-renal (Zellweger) syndrome and related disorders. Biochem Biophys Res Commun 152:1083–1089
24. Wanders RJA, Kos M, Roest B, Meijer AJ, Schrakamp G, Heymans HSA, Tegelaers WHH, Van Den Bosch H, Schutgens RBH, Tager JM (1984) Activity of peroxisomal enzymes and intracellular distribution of catalase in the Zellweger syndrome. Biochem Biophys Res Commun 123:1054–1061
25. Keller G-A, Gould S, Deluca M, Subramani S (1987) Firefly luciferase is targeted to peroxisomes in mammalian cells. Proc Natl Acad Sci USA 84:3264–3268
26. Gould SJ, Keller GA, Subramani S (1987) Identification of a peroxisomal targeting signal at the carboxy terminus of firefly luciferase. J Cell Biol 105:2923–2931
27. Santos MJ, Imanaka T, Shio H, Small GM, Lazarow PB (1988) Peroxisomal membrane ghosts in Zellweger syndrome – aberrant organelle assembly. Science 239:1536–1538
28. Lazarow PB, Fujiki Y, Small GM, Watkins P, Moser H (1986) Presence of the peroxisomal 22-kDa integral membrane protein in the liver of a person lacking recognizable peroxisomes (Zellweger syndrome). Proc Natl Acad Sci USA 83:9193–9196

Subject Index

MIX
Papier aus verantwortungsvollen Quellen
Paper from responsible sources
FSC® C105338
FSC
www.fsc.org

If you have any concerns about our products,
you can contact us on
ProductSafety@springernature.com

In case Publisher is established outside the EU,
the EU authorized representative is:
Springer Nature Customer Service Center GmbH
Europaplatz 3, 69115 Heidelberg, Germany

Printed by Libri Plureos GmbH
in Hamburg, Germany